The School Mathematics Project

When the SMP was founded in 1961, its main objective was to devise radically new secondary school mathematics courses to reflect, more adequately than did the traditional syllabuses, the up-to-date nature and usages of mathematics. The first texts produced embodied new courses for O-level and A-level, and SMP GCE examinations were set up, available to schools through any of the GCE examining boards.

Since its beginning the SMP has continued to develop new materials and approaches to the teaching of mathematics. Further series of texts have been produced to meet new needs, and the original books are revised or replaced in the light of changing circumstances and experience in the classroom.

The SMP A-level course is now covered by *Revised Advanced Mathematics Books 1, 2* and *3*. Five shorter texts cover the material of the various sections of the A-level examination SMP Further Mathematics. The SMP Additional Mathematics syllabus has been revised and a new text replaces the original two books at this level.

The six Units of *SMP 7–13*, designed for pupils in that age-range, provide a course which is widely used in primary schools, middle schools and the first two years of secondary schools. A useful preliminary to Unit 1 of *SMP 7–13* is *Pointers*, a booklet for teachers which offers suggestions for mathematical activities with young children.

There is now a range of SMP materials for the eleven to sixteen age-range. The SMP O-level course is covered by *Books 1, 2* and *New Books 3, 4, 5*. *Books A–G* and *X, Y, Z*, together with the booklets of the *SMP Calculator Series*, also cover the O-level course, while *Books A–H* provide a CSE course for which most CSE boards offer a suitable examination.

SMP 11–16, designed to cater for about the top 85% of the ability range, is the newest SMP secondary school course, providing varied materials which facilitate the provision of a differentiated curriculum to match the varying abilities of pupils. Publication of this course began in 1983 and will be complete in 1988.

Teacher's Guides accompany all these series.

The SMP has produced many other texts, and teachers are encouraged to obtain each year from Cambridge University Press, The Edinburgh Building, Shaftesbury Road, Cambridge CB2 2RU, the full list of SMP publications currently available. In the same way, help and advice may always be sought by teachers from the Executive Director at the SMP Office, Westfield College, Kidderpore Avenue, London NW3 7ST. SMP syllabuses and other information may be obtained from the same address.

The SMP is continually evaluating old work and preparing for new. The effectiveness of the SMP's work depends, as it always has done, on the comments and reactions received from a wide variety of teachers – and also from pupils – using SMP materials. Readers of the texts can, therefore, send their comments to the SMP in the knowledge that they will be valued and carefully studied.

THE SCHOOL MATHEMATICS
PROJECT

REVISED
ADDITIONAL
MATHEMATICS

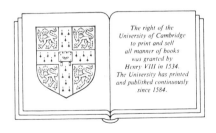

The right of the
University of Cambridge
to print and sell
all manner of books
was granted by
Henry VIII in 1534.
The University has printed
and published continuously
since 1584.

CAMBRIDGE UNIVERSITY PRESS

Cambridge
London   New York   New Rochelle
Melbourne   Sydney

Published by the Press Syndicate of the University of Cambridge
The Pitt Building, Trumpington Street, Cambridge CB2 1RP
32 East 57th Street, New York, NY 10022, USA
10 Stamford Road, Oakleigh, Melbourne 3166, Australia

SMP *Additional Mathematics Book Part 1* and *Part 2* first published 1966, 1968 respectively
Metricated editions 1970
Part 1 reprinted 1973, 1974, 1977, 1978
Part 2 reprinted 1972, 1974, 1977, 1979
Revised and combined as SMP *Revised Additional Mathematics* 1985

Printed in Great Britain by the University Press, Cambridge

ISBN 0 521 27843 0

(ISBN 0 521 07876 8 *Additional Mathematics Book Part 1*)
(ISBN 0 521 07877 6 *Additional Mathematics Book Part 2*)

Contents

Preface

The preface to the original SMP *Additional Mathematics Book* pointed out that there are three main groups of pupils for whom the course is relevant: 'fifth-form' students who have mastered the main school course and wish to study more mathematics before embarking on an A-level course; sixth-form students who enjoy mathematics and have the time to explore some of its developments; sixth-form scientists who are not taking an A-level course in mathematics but who wish to study some more mathematics to support their A-level courses. In writing this new book the authors have borne in mind these same groups of pupils, and have aimed to present the mathematics in a style which will make the course accessible to these different constituencies.

Additional Mathematics provides a good opportunity for pupils to study mathematics on their own. It is hoped that they will be able to work from this new text with guidance, but with a minimum of teaching.

As with other SMP texts, the mathematical content of this book provides a body of mathematical knowledge which can be used by pupils preparing for a variety of examinations. However, it is inevitable that the content and structure of the book should reflect the current SMP Additional Mathematics (AO) syllabus and examination. Although schools use the course (and examination) in a variety of ways in sixth forms, it is apparent that for many schools the major role of the course is to extend the mathematical horizons of the more able fifth-form pupils. Some of the topics in the course introduce ideas which are studied in more depth and detail in many A-level courses, and in this respect the course can make a useful 'bridge' to A-level mathematics. However, there are other significant mathematical ideas which are worthy of study by these pupils which are rarely taken up in A-level courses. Taken together, these ideas can form the basis of an enriching study for fifth-form pupils whether or not they intend to proceed to an A-level mathematics course.

Like its predecessor, the present text is in two parts, but the structure is different. Part One now brings together many of the significant new ideas to form a 'core' of knowledge; in Part Two a number of topics are explored. The mathematics in the core contains a number of immediate extensions of the SMP O-level course, especially in algebra and trigonometry, together with an introduction to calculus and applications of these ideas, particularly to kinematics. The topics in Part Two are independent of each other, but in many cases they rely on a knowledge of the core and extend some of the ideas in it. In constructing a course teachers can select topics which take up the particular interests of their pupils.

The emphasis is on the understanding of the basic ideas rather than acquiring elaborate techniques, although, of course, proficiency in straightforward techniques is valuable and essential as a basis for progress. It is chiefly for this reason that the Leibniz notation for differentiation is not used in Part One. The authors feel that, besides introducing even more new notation to be mastered at the beginning of a new development, the early use of this notation can obscure, rather than clarify, the basic concepts.

Some exercises contain a number of questions marked with an asterisk. These should be worked through by the student as they carry forward the argument of the text, either developing what has gone before or introducing new ideas to be taken up in the next section of the chapter.

Where appropriate, exercises have been constructed with 'parallel' sets of questions. A single ruled line separates groups of parallel questions; a double line indicates that questions of a different type, or on a different topic, follow. The second set of parallel questions can be used for revision or for further practice of a technique when a student has had little success with the first group.

Two sets of revision exercises are provided in Part One: the first based on Chapters 1–5, the second covering the whole of the 'core'. A revision exercise is provided at the end of each of the five topics.

Throughout the book it is assumed that a scientific (not necessarily programmable) calculator will be used by the student. There are many places where a computer can be used to advantage to illuminate the mathematics.

Revised Additional Mathematics has been written by

Bill Bailey	David Holland
Neil Bibby	Sue Robinson
Maureen Bownas	Timothy Lewis
David Cundy	Tony Thomas
John Davis	

and edited by David Cundy and John Hersee.

The authors wish to express their thanks to those who have helped by their advice and criticism, particularly the teachers and pupils who have tested the material in draft form, and to the Oxford and Cambridge Schools Examination Board for permission to reprint questions from examination papers set for the SMP and MEI syllabuses.

PART ONE

1

Kinematics and differentiation

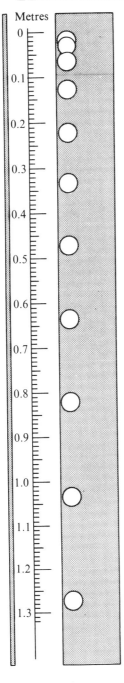

Metres

1. THE MOTION OF A FALLING BODY

Kinematics is the study of motion, in particular the relationship between time and position in space. It is this study which gives rise to the notions of displacement, velocity and acceleration. We shall find that we need a new mathematical tool if we are to be able to consider anything other than the very simplest kinds of motion: this tool is differentiation. We start with a new look at a familiar event.

Figure 1 is drawn from a photograph of a golf-ball falling freely under gravity.

The exposures were made at regular intervals of $\frac{1}{20}$ of a second. It is clear that in each successive interval the distance that the golf-ball falls increases; that is, the speed of the golf-ball is increasing during its fall.

Suppose that we wish to find the speed of the golf-ball at some time after its release. We shall start by trying to answer the question 'What is the speed of the golf-ball three seconds after release?'

Before we can do this we shall need to know *how far* it has fallen at any instant of time after it is released.

Let s be the function which tells us the number of metres fallen t seconds after release, so that after t seconds the golf-ball has fallen $s(t)$ metres. It is clear from the photograph that the graph of s is not a straight line, since the distance fallen in successive intervals increases. In Exercise A we investigate, and identify, the function s.

Exercise A

*1   By using the scale next to the photograph, complete Table 1, giving the distance fallen correct to the nearest 0.01 m.

Table 1

Time of fall (seconds)	Distance fallen (metres)
0	0
0.05	0.01
0.10	0.05
0.15	0.11
0.20	
0.25	
0.30	
0.35	
0.40	
0.45	
0.50	

*2   Here is an extract from the complete table:

Table 2

t	$s(t)$
.	.
.	.
0.15	0.11
.	.
×3 .	. ×?
.	.
.	.
0.45	1.01
.	.
.	.

(a) The multiplier from $t = 0.15$ to $t = 0.45$ is 3. What is the corresponding multiplier for $s(t)$ (to the nearest integer)?
(b) Choose four other pairs of values of t which exceed 0.1. Work out the multiplier for t, and the corresponding multiplier for $s(t)$ (to the nearest integer). Present the results as in Table 3.

Table 3

Values of t	t-multiplier	$s(t)$-multiplier
0.15, 0.45	3	
0.15, 0.3	2	

*3 Calculate the values of t^2, and, using a scale of 0.01 units to 1 cm for t^2 and 0.05 units to 1 cm for $s(t)$, draw a graph of the function $t^2 \to s(t)$. What conclusions may reasonably be drawn from this graph?

2. DISTANCE FALLEN

It should be clear from your results in question 2 that the $s(t)$-multiplier is always the square of the t-multiplier, which indicates that

$$s(t) \propto t^2.$$

This is borne out by the form of the graph of $s(t)$ against t^2, which is linear. So we deduce that

$$s(t) = kt^2,$$

where k is a constant, equal to the gradient of the graph in question 3. The value of k is roughly 5, so we may write

$$s(t) \approx 5t^2.$$

In order to investigate the speed of the golf-ball, we shall work with an idealised model of the situation and assume that the function $s : t \to 5t^2$ gives the exact distance fallen from rest at any instant. We shall then no longer be restricted to the finite set of positions and times given by the multi-flash photograph.

The graph of the function is a continuous curve which associates each instant of time with the corresponding distance fallen (Figure 2).

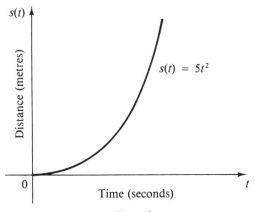

Figure 2

We shall often be working with an idealised model in this way. When we give a formula for the distance covered by a train in t seconds, for example, it should be remembered that this is a model and that predictions based on it should be interpreted with care – particularly with regard to the number of significant figures quoted and the set of values of t for which the model may be regarded as valid.

Exercise B

1 (*a*) Use the distance function $s : t \rightarrow 5t^2$ to complete Table 4, comparing the experimental results, found in Exercise A question 1, with those calculated using the function s.

Table 4

t (seconds)	Distance fallen (metres)	$5t^2$
0		
0.05		
0.10		
0.15		
0.20		
0.25		
0.30		
0.35		
0.40		
0.45		
0.50		

2 Use the function $s : t \rightarrow 5t^2$ to work out how far the golf-ball falls during each fifth of the first second (Table 5).

Table 5

Time interval (seconds)	Distance fallen (metres)
$t = 0$ to $t = 0.2$	
$t = 0.2$ to $t = 0.4$	
$t = 0.4$ to $t = 0.6$	
$t = 0.6$ to $t = 0.8$	
$t = 0.8$ to $t = 1$	

3. AVERAGE VELOCITIES

Now that we can calculate the distance fallen at any time after the golf-ball is released, we can return to the original problem given in Section 1: 'What is the speed of the golf-ball three seconds after release?'

We could start by calculating the average speed from $t = 3$ to $t = 5$. During this time the distance gone is

$$5 \times 5^2 - 5 \times 3^2 \text{ m} = 80 \text{ m}.$$

So that the average speed during this two-second interval is

$$\tfrac{80}{2} \text{ m s}^{-1} \quad \text{or} \quad 40 \text{ m s}^{-1}.$$

However, we can get a much better idea of the speed three seconds after release if we consider how far it falls in a shorter interval of time than that from $t = 3$ to $t = 5$. Suppose we now consider an interval of only 1 second, from $t = 3$ to $t = 4$. During this time the golf-ball falls

$$5 \times 4^2 - 5 \times 3^2 \text{ m} = 35 \text{ m}.$$

So the average speed is

$$\tfrac{35}{1} \text{ m s}^{-1} \quad \text{or} \quad 35 \text{ m s}^{-1}.$$

In this way we can proceed to consider smaller and smaller intervals of time, and so get an increasingly accurate idea of the speed after three seconds.

Exercise C

*1 Use the distance function $s(t) = 5t^2$ to calculate the average speed over time intervals from $t = 3$ to $t = 3 + h$, taking h successively as $1, 0.1, 0.01$, etc. (The first calculation was done above, and the second is done for you below.) Set out your working as in Table 6, and write down all the significant digits from your calculator display. Extend the table as far as $h = 10^{-4}$.

Table 6

Time interval	h, length of time interval (seconds)	Distance fallen (metres)	Average speed (metres per second)
$t = 3$ to $t = 4$	1	$5 \times 4^2 - 5 \times 3^2 = 35$	$\tfrac{35}{1} = 35$
$t = 3$ to $t = 3.1$	0.1	$5 \times 3.1^2 - 5 \times 3^2 = 3.05$	$\tfrac{3.05}{0.1} = 30.5$
$t = 3$ to $t = 3.01$	0.01	$5 \times 3.01^2 - 5 \times 3^2$	

*2 In question 1 we have considered intervals of time from $t = 3$ to $t = 3 + h$, with h positive. The intervals could equally well start before $t = 3$ and end at $t = 3$. In this question we calculate average speeds over time intervals from $t = 3 - h$ to $t = 3$, again taking h successively as 1, 0.1, 0.01, etc., so the time intervals considered are from $t = 2$ to $t = 3$, then from $t = 2.9$ to $t = 3$, then from $t = 2.99$ to $t = 3$, and so on. Copy and complete Table 7, proceeding as in question 1.

Table 7

Time interval	h, length of time interval (seconds)	Distance fallen (metres)	Average speed (metres per second)
$t = 2$ to $t = 3$	1	$5 \times 3^2 - 5 \times 2^2 \quad = 25$	$\frac{25}{1} = 25$
$t = 2.9$ to $t = 3$	0.1	$5 \times 3^2 - 5 \times 2.9^2 =$	
$t = 2.99$ to $t = 3$	0.01	$5 \times 3^2 -$	

3   On planet Omega the distance in metres fallen by an object in t seconds is modelled by $s(t) = 3t^2$. Use this function to calculate the average speed over time intervals from $t = 5$ to $t = 5 + h$, taking h successively as 1, 0.1, 0.01, etc. Set out your working as in Table 8, writing down all the significant digits from your calculator display. Extend the table as far as $h = 10^{-4}$.

Table 8

Time interval	h, length of time interval (seconds)	Distance fallen (metres)	Average speed (metres per second)
$t = 5$ to $t = 6$	1	$3 \times 6^2 - 3 \times 5^2 = 33$	$\frac{33}{1} = 33$

4   Calculate the average speed for the distance function $s : t \to 3t^2$ over time intervals from $t = 1 - h$ to $t = 1$, taking h successively as 1, 0.1, 0.01, etc.

4. VELOCITY AT AN INSTANT

Is it possible to deduce anything from the results for the average speeds which you found in Exercise C? Consider question 1 first. Your results should have been like Table 9.

We can see from this table that there is a connection between the time interval and the average speed. If we take a time interval of h seconds, the average speed is $(30 + 5h)$ metres per second. So as we make h, the length of the time interval, smaller and smaller, the average speed after 3 seconds becomes closer and closer to 30 m s^{-1}. This is because reducing the value of h makes $5h$ smaller, and thus makes $(30 + 5h)$ closer to 30. Moreover, providing we make h sufficiently small, we can make $(30 + 5h)$ as close to 30 as we please.

Table 9

Time interval	h, length of time interval (seconds)	Average speed (metres per second)
$t = 3$ to $t = 4$	1	35
$t = 3$ to $t = 3.1$	0.1	30.5
$t = 3$ to $t = 3.01$	0.01	30.05
$t = 3$ to $t = 3.001$	0.001	30.005
.	.	.
.	.	.
.	.	.
.	.	.

Let us now consider the results of Exercise C question 2. You should have obtained the results of Table 10.

Table 10

Time interval	h, length of time interval (seconds)	Average speed (metres per second)
$t = 2$ to $t = 3$	1	25
$t = 2.9$ to $t = 3$	0.1	29.5
$t = 2.99$ to $t = 3$	0.01	29.95
$t = 2.999$ to $t = 3$	0.001	29.995
.	.	.
.	.	.
.	.	.

In this case we can see that the average speed from $t = 3 - h$ to $t = 3$ is $(30 - 5h)$ metres per second.

In summary, then, for time intervals from $t = 3$ to $t = 3 + h$ the average speed is $(30 + 5h)$ m s^{-1}, and for time intervals from $t = 3 - h$ to $t = 3$ the average speed is $(30 - 5h)$ m s^{-1}.

Now, for all positive h,

$$30 - 5h < 30 < 30 + 5h.$$

It thus seems reasonable to state that the *instantaneous* speed when $t = 3$ is 30 m s^{-1}. This speed is the unique value which lies between the two sets of *average* speeds, and the value to which they both become increasingly close as h approaches zero. In this way we define the instantaneous speed as the value approached by the average speed, as we consider ever-decreasing time intervals. We speak of 30 as 'the *limit* of $30 + 5h$ as h approaches zero'. It is also the limit of $30 - 5h$ as h approaches zero.

Is it possible to extend the method we have developed in order to deduce the instantaneous speed at times other than $t = 3$? In the next exercise we shall try to do this, and then try to generalise the results we have found.

Exercise D

*1 (a) Using the distance function $s : t \to 5t^2$ determine the average speed from $t = 2$ to $t = 2 + h$, taking h successively as $1, 0.1, 0.01$ etc. Set out your working in a table as in Exercise C. Try to give a general form in terms of h for the average speed over this interval of time. Does this average speed approach a limit as h approaches zero?

   (b) Repeat part (a), but for the time interval from $t = 2 - h$ to $t = 2$, again setting out your working as a table. Again give a general form for the average speed. What is the limit of this average speed as h approaches zero?

   (c) From your answers to parts (a) and (b), deduce the instantaneous speed of a freely-falling golf-ball, when $t = 2$.

*2 Use a similar method to that of question 1 to find the instantaneous speed when $t = 1$.

5. THE DERIVED FUNCTION

When developing a piece of mathematics we often investigate particular cases and then try to observe a pattern behind the various results. We have already done this: in calculating the average speeds from $t = 3$ to $t = 3 + h$ in Section 4, we saw that the results were all of the form $30 + 5h$. Now, in Exercise D, we have deduced average speeds over other intervals of time, and we can look for a pattern here too.

Before we look at these results, it will be convenient to introduce some new notation. We shall use $[2, 2 + h]$ to denote 'the interval from $t = 2$ to $t = 2 + h$', and speak of 'the average speed over $[2, 2 + h]$' rather than 'the average speed from $t = 2$ to $t = 2 + h$'. Thus $[1, 2]$ denotes the interval from $t = 1$ to $t = 2$, and so on.

Table 11 summarises the results of Exercise C and Exercise D.

Table 11

Time t for which speed is required	Average speed over $[t - h, t]$	Average speed over $[t, t + h]$	Limit of average speeds, as h approaches zero
1	$10 - 5h$	$10 + 5h$	10
2	$20 - 5h$	$20 + 5h$	20
3	$30 - 5h$	$30 + 5h$	30

If we look at the first and fourth columns it becomes clear that there is an obvious pattern.

In effect, we have deduced a new function, the function

$$t \to 10t,$$

which tells us the instantaneous speed at time t of a freely-falling object. We have deduced this speed-function from a distance-function s, which was given by

$$s : t \to 5t^2.$$

We call the speed-function the *derived function* of s, and denote it by s'. So the distance-function

$$s : t \to 5t^2$$

gives rise to the speed-function

$$s' : t \to 10t,$$

and $10t$ is called the *derivative* of $5t^2$.

The process of finding the derived function is called *differentiation*. It will now be useful to make clear the stages involved in the process of differentiation by considering another function.

Example 1
Differentiate the function $s : t \to t^3$.

We require the speed of a particle which moves in such a way that after t seconds it has covered a distance of t^3 metres.

We start by considering $t = 2$, say (Table 12).

Table 12

h, length of time interval, (seconds)	Average speed over the time interval $[2, 2+h]$ (metres per second)			
1.0	$\dfrac{s(3) - s(2)}{1}$	$= \dfrac{3^3 - 2^3}{1}$	$= \dfrac{27 - 8}{1}$	$= 19$
0.1	$\dfrac{s(2.1) - s(2)}{0.1}$	$= \dfrac{2.1^3 - 2^3}{0.1}$	$= \dfrac{9.261 - 8}{0.1}$	$= 12.61$
0.01	$\dfrac{s(2.01) - s(2)}{0.01}$	$= \dfrac{2.01^3 - 2^3}{0.01}$	$= \dfrac{8.120601 - 8}{0.01}$	$= 12.0601$
0.001	$\dfrac{s(2.001) - s(2)}{0.001}$	$= \dfrac{2.001^3 - 2^3}{0.001}$	$= \dfrac{8.012006001 - 8}{0.001}$	$= 12.006001$
.	.	.	.	.
.	.	.	.	.
.	.	.	.	.

In this way we get a good idea of the limit of the average speeds from the instant when $t = 2$ to some later instant; we also need to investigate the average speeds from some earlier instants to the instant when $t = 2$. In doing this, it is useful to consider *negative* values of h. So, with $t = 2$, if $h = {}^-0.1$, the interval is $[1.9, 2]$, and if $h = {}^-0.01$, the interval is $[1.99, 2]$, and so on. For these intervals we obtain Table 13.

Table 13

h (seconds)	Average speed over $[2+h, 2]$ (metres per second)
-0.1	$\dfrac{s(2)-s(1.9)}{0.1} = \dfrac{2^3-1.9^3}{0.1} = \dfrac{8-6.859}{0.1} = 11.41$
-0.01	$\dfrac{s(2)-s(1.99)}{0.01} = \dfrac{2^3-1.99^3}{0.01} = \dfrac{8-7.880599}{0.01} = 11.9401$
-0.001	$\dfrac{s(2)-s(1.999)}{0.001} = \dfrac{2^3-1.999^3}{0.001} = \dfrac{8-7.98800...}{0.001} = 11.994001$
.	
.	
.	

It is clear that the values of the average speed, for both positive and negative h, are converging to 12 m s^{-1}. From this and from the earlier examples you will have noticed that as we make h smaller it becomes easier to deduce the limit of the average speeds. We could thus cut down our work considerably by immediately proceeding to $h = 0.01$ or 0.001. In many cases the required limit will be obvious; if it is not, we have to use still smaller values of h until we are certain of the limit. In Exercise E you are recommended to cut down your working in this way.

Exercise E

*1   Continue the differentiation of $s : t \to t^3$. Consider the average speeds over $[1, 1.001]$, $[3, 3.001]$ and $[4, 4.001]$ and present your results as in Table 14.

   Guess the limits of the average speeds as h approaches zero, and hence deduce the derived function, s'.

Table 14

t (seconds)	Average speed over $[t, t+0.001]$	Limit of average speeds (metres per second)
1	12.006...	12
2		
3		
4		

*2   Use the technique of question 1 to find the derived function of $s : t \to t^4$.

*3   Use the technique of question 1 to find the derived function of $s : t \to t^2$.

*4   (a) Complete Table 15, summarising the results so far.

Table 15.

s	s'
$t \to t^2$	
$t \to t^3$	
$t \to t^4$	

(b) By studying the table, try to guess the derived function of $s : t \to t^n$.

*5   (a) For the distance function $s : t \to t^5$, calculate the average speeds over [1, 1.001], [2, 2.001], [3, 3.001] and [4, 4.001].

(b) Use the result of question 4(b) to investigate whether these average speeds correspond to the values of your guess at the derived function of $s : t \to t^5$, for $t = 1, 2, 3$ and 4.

6   Repeat question 5 for the function $s : t \to t^6$.

6. GENERALISATIONS

The results of Exercise E questions 4 and 5 suggest that

$$s : t \to t^n \quad \Rightarrow \quad s' : t \to n t^{n-1}.$$

A general proof of this result is given (for positive integers n) in Chapter 5. Notice in particular that it is true for $n = 1$, that is

$$s : t \to t \quad \Rightarrow \quad s' : t \to 1 \times t^0 = 1$$

since for this distance-function the speed is always 1 m s^{-1}.

We have already seen that for the function

$$s : t \to t^2$$

we obtained

$$s' : t \to 2t,$$

while for

$$s : t \to 5t^2$$

we obtained

$$s' : t \to 10t = 5 \times 2t.$$

This strongly suggests that if

$$f : t \to kg(t),$$

then

$$f' : t \to kg'(t).$$

A general proof of this rule is also given in Chapter 5. An important special case is that of constant speed, for example

$$s : t \to 6t \quad \Rightarrow \quad s' : t \to 6.$$

Notice also that

$$s : t \to 6 \quad \Rightarrow \quad s' : t \to 0,$$

corresponding to a stationary object.

Example 2
Differentiate $f: t \to 12t^7$.

Since we know that the derivative of t^7 is $7t^6$, we obtain
$$f': t \to 12 \times 7t^6$$
or
$$f': t \to 84t^6.$$

Example 3
Differentiate $f: t \to t^2 + t^3$.

We work out $\dfrac{f(t+0.001) - f(t)}{0.001}$ for $t = 1, 2, 3$ (Table 16).

Table 16

t	$\dfrac{f(t+0.001) - f(t)}{0.001}$			
1	$\dfrac{1.001^2 - 1^2 + 1.001^3 - 1^3}{0.001}$	$= \dfrac{1.001^2 - 1^2}{0.001}$	$+ \dfrac{1.001^3 - 1^3}{0.001}$	$= 2.001 + 3.003\ldots$
2	$\dfrac{2.001^2 - 2^2 + 2.001^3 - 2^3}{0.001}$	$= \dfrac{2.001^2 - 2^2}{0.001}$	$+ \dfrac{2.001^3 - 2^3}{0.001}$	$= 4.001 + 12.006\ldots$
3	$\dfrac{3.001^2 - 3^2 + 3.001^3 - 3^3}{0.001}$	$= \dfrac{3.001^2 - 3^2}{0.001}$	$+ \dfrac{3.001^3 - 3^3}{0.001}$	$= 6.001 + 27.009\ldots$

We can see that differentiation from scratch is unnecessary. For sufficiently small h, the values of f' correspond to $2t + 3t^2$, as Table 17 makes clear.

Table 17

t	$\dfrac{f(t+0.001) - f(t)}{0.001}$	$2t + 3t^2$
1	$2.001 + 3.003$	$2 + 3$
2	$4.001 + 12.006$	$4 + 12$
3	$6.001 + 27.009$	$6 + 27$

So when we differentiate the sum of two functions, we just differentiate each function separately and add the results together.

Example 4
Differentiate $g: t \to t^5 + t^{11}$.

The derivative of t^5 is $5t^4$ and that of t^{11} is $11t^{10}$.
So
$$g': t \to 5t^4 + 11t^{10}.$$

These two rules may be used in combination.

Example 5
Differentiate $h:t\to 10t^7-4t^{13}+9$.

This function may be written as
$$h:t\to 10t^7+{}^-4t^{13}+9.$$
So
$$h':t\to 10\times 7t^6+{}^-4\times 13t^{12}+0$$
or
$$h':t\to 70t^6-52t^{12}.$$

Exercise F

1 Differentiate the following functions:
   (a) $t\to t^{12}$; (b) $t\to 8t^9$; (c) $t\to 20t^4$;
   (d) $t\to 14t^6$; (e) $t\to 7$.

2 Differentiate the following functions:
   (a) $x\to x^2+x^5$; (b) $x\to x^8+x^7$;
   (c) $x\to x^{70}+x^{100}$; (d) $x\to x^{46}+17$.

3 Differentiate the following functions:
   (a) $t\to 5t^{11}+3t^6$; (b) $t\to 9t^4-3t^8$;
   (c) $t\to 8t^{15}+\tfrac{1}{2}t^{12}$; (d) $t\to 40t^3+101$.

4 A stone is thrown vertically upwards so that its height, $h(t)$ metres, t seconds after it leaves the thrower's hand is modelled by the function
   $$h:t\to 20t-5t^2.$$
   (a) Find the derived function, h'.
   (b) Use h' to find the speed of the stone when $t=0,1,2,3,4$.

---

5 Differentiate the following functions:
   (a) $t\to t^{13}$; (b) $t\to{}^-15t^5$; (c) $t\to 9t^8$;
   (d) $t\to 0.8t^{10}$; (e) $t\to 100$.

6 Differentiate the following functions:
   (a) $x\to x^4+x^3$; (b) $x\to x^{10}+x^6$;
   (c) $x\to x^{65}+x^{40}$; (d) $x\to x^{30}+41$.

7 Differentiate the following functions:
   (a) $t\to 4t^{15}+2t^3$; (b) $t\to 7t^5-4t^{11}$;
   (c) $t\to\tfrac{1}{3}t^{18}+4t^7$; (d) $t\to 53+25t^4$.

8 A car moves away from a set of traffic lights so that the distance, $s(t)$ metres, covered after t seconds is modelled by
   $$s(t)=0.2t^3.$$
   (a) Find its speed after t seconds.
   (b) What will its speed be when $t=1,2,3$?

===

9 After a period of t seconds, a train has covered a distance of $5t+0.1t^2$ metres from the terminus. How long does it take to reach a speed of 35 m s^{-1}?

10 A rock climber loses his grip. For two seconds he falls freely, and then his safety-rope arrests his fall. The distance he falls, in metres, is given by
   $$s(t)=5t^2,\qquad\qquad 0\leqslant t\leqslant 2,$$
   and
   $$s(t)={}^-10t^2+60t-60,\quad 2<t\leqslant 3.1.$$

Find his speed after two seconds, and find when he is stationary. When $t = 3.1$, he alights on a ledge. What is his speed at this time?

11 A Saturn rocket is launched from Cape Canaveral, so that after t seconds its height, $h(t)$ metres, is given by

$$h(t) = 20t^2, \quad 0 \leqslant t \leqslant 150.$$

After $2\frac{1}{2}$ minutes the first stage is jettisoned.
   (a) Find the height and speed of the rocket when the first stage is jettisoned.
   (b) After how long is the speed 1000 m s^{-1}?

7. SUMMARY

Suppose that a particle is moving in a straight line, so that after t seconds it is $s(t)$ metres away from a fixed point.

For the time interval $[t, t+h]$, the average speed is

$$\frac{s(t+h) - s(t)}{h} \text{ m s}^{-1}.$$

The limit, as h approaches zero, of $\dfrac{s(t+h) - s(t)}{h}$ is denoted by $s'(t)$, and this corresponds to the instantaneous speed at time t. The function s' which gives the speed is called the derived function of s.

The process of finding the derived function is called differentiation.

Techniques: (1) $\quad f : t \to t^n$

$\Rightarrow \quad f' : t \to nt^{n-1}.$

(2) $\quad f : t \to kg(t)$

$\Rightarrow \quad f' : t \to kg'(t).$

(3) $\quad f : t \to g(t) + h(t)$

$\Rightarrow \quad f' : t \to g'(t) + h'(t).$

2

The quadratic function

1. TRANSFORMATIONS

We shall call the function $f: x \to x^2$ the *standard* quadratic function. Figure 1 shows the graph of $y = x^2$. The shape formed by this graph is called a *parabola*. The 'nose' of the curve (which in this graph is at the origin) is called the *vertex*.

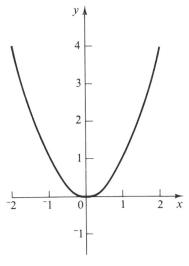

Figure 1

Functions such as $f: x \to 2x^2$, $f: x \to x^2 - 3$, $f: x \to x^2 + 3x$, $f: x \to 2x^2 - 3x + 2$ are also quadratic functions. We shall see that the graphs of all quadratic functions can be regarded as transformations of the graph of the standard quadratic function shown in Figure 1. All the graphs will be parabolas, but the vertices will not necessarily be at the origin.

(a) Translations parallel to the y-axis

The graphs of $y = x^2 + 2$ and $y = x^2 - 3$ are images of the standard graph under the translations $\begin{bmatrix} 0 \\ 2 \end{bmatrix}$ and $\begin{bmatrix} 0 \\ -3 \end{bmatrix}$ respectively (Figure 2).

This can easily be seen by looking at the tables of values for these two functions and comparing them with that for the standard quadratic function (Table 1).

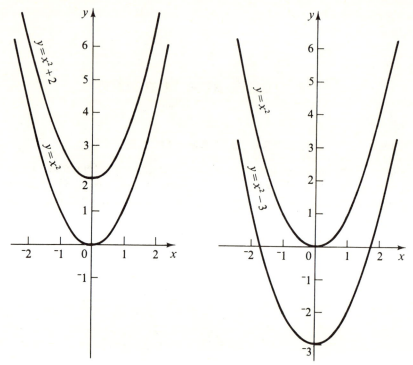

Figure 2

Table 1

x	-3	-2	-1	0	1	2	3
x^2	9	4	1	0	1	4	9
x^2+2	11	6	3	2	3	6	11
x^2-3	6	1	-2	-3	-2	1	6

Clearly the values for x^2+2 are each 2 greater than the corresponding values for x^2, and the values for x^2-3 are each 3 less than the corresponding values for x^2.

(b) Translations parallel to the x-axis

Figure 3 shows the standard graph ($y = x^2$) and the graph of $y = (x-3)^2$. Notice that the latter is the image of the standard graph under a translation $\begin{bmatrix} 3 \\ 0 \end{bmatrix}$. It is important to remember that although the sign in the expression $(x-3)^2$ is *negative*, the resulting translation is in the direction of the *positive* x-axis. In the same way, the graph of $y = (x+2)^2$ is the image of the standard graph under a translation of 2 units in the direction of the *negative* x-axis.

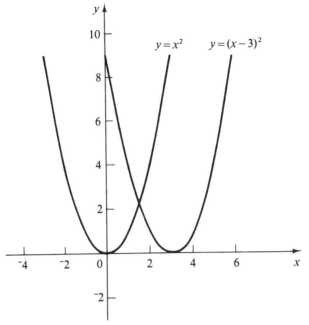

Figure 3

Table 2

x		$^-3$	$^-2$	$^-1$	0	1	2	3	4	5	6
$x-3$		$^-6$	$^-5$	$^-4$	$^-3$	$^-2$	$^-1$	0	1	2	3
x^2		9	4	1	0	1	4	9	16	25	36
$(x-3)^2$		36	25	16	9	4	1	0	1	4	9

We can check this by comparing the tables of values for the function $f:x \to x^2$ and $g:x \to (x-3)^2$ (Table 2).

It is easy to see that the values for $(x-3)^2$ are found by shifting the values for x^2 three places to the *right*.

It may seem that there is some contradiction between the ways in which the translations are obtained, but if we write

$$y = x^2 - 3 \quad \text{as} \quad y + 3 = x^2$$

it is clear that replacing $y+3$ by y would give us the standard parabola, and that the graph of $y+3 = x^2$ is therefore the image of the standard parabola under the translation $\begin{bmatrix} 0 \\ -3 \end{bmatrix}$.

A similar argument demonstrates that $y = (x+3)^2$ is the image of the standard parabola under the translation $\begin{bmatrix} -3 \\ 0 \end{bmatrix}$.

Combining these two ideas, we see that, in general, the graph of

$$y - b = (x - a)^2$$

is the image of the standard parabola under the translation $\begin{bmatrix} a \\ b \end{bmatrix}$.

(c) Reflection in the x-axis

The graph of $y = {}^-x^2$ is the reflection of the standard graph in the x-axis (Figure 4).

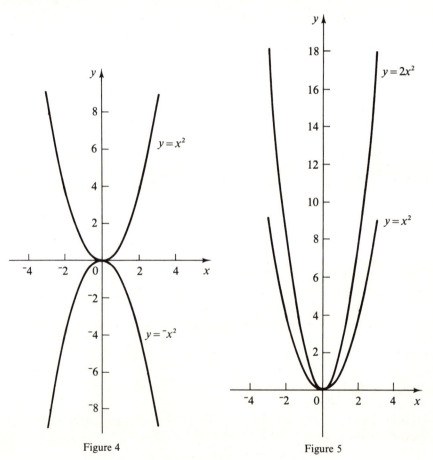

Figure 4   Figure 5

(d) Stretches parallel to the y-axis

The graph of $y = 2x^2$ is the image of the standard graph under a one-way stretch of factor 2 parallel to the y-axis (Figure 5).

2. COMBINING TRANSFORMATIONS

The equation $y = 2x^2 + 1$ represents two transformations applied to the standard graph: first, a stetch of factor 2 parallel to the y-axis, and then a translation of $\begin{bmatrix} 0 \\ 1 \end{bmatrix}$. The two stages are illustrated in Figure 6.

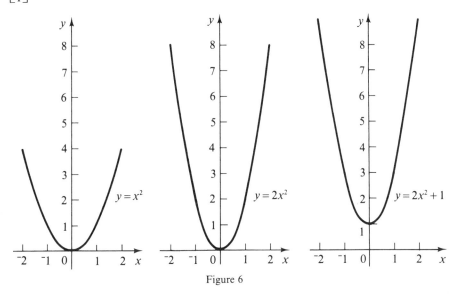

Figure 6

It is, of course, important to get the order of the transformations correct.

Exercise A

In questions 1–10 and 11–20, sketch the standard graph $y = x^2$, together with the graphs indicated.

1 $y = x^2 + 3$;
  $y = x^2 - 1$.

2 $y = {}^-x^2$;
  $y = {}^-x^2 + 2$;
  $y = {}^-x^2 - 3$.

3 $y = 3x^2$;
  $y = \frac{1}{2}x^2$.

4 $y = \frac{1}{4}x^2$;
  $y = {}^-\frac{1}{4}x^2$.

5 $y = (x + 1)^2$;
  $y = (x - 2)^2$;
  $y = {}^-(x - 2)^2$.

6 $y = (x - 1)^2$;
  $y = 2(x - 1)^2$.

7 $y = 2x^2$;
  $y = 2x^2 - 3$.

8 $y = (x + 2)^2$;
  $y = (x + 2)^2 + 4$.

9 $y = (x - 3)^2$;
  $y = \frac{1}{2}(x - 3)^2$;
  $y = \frac{1}{2}(x - 3)^2 + 5$.

10 $y = (x + 3)^2$; $y = 2(x + 3)^2$; $y = {}^-2(x + 3)^2$; $y = {}^-2(x + 3)^2 - 5$.

11 $y = x^2 - 2$;
  $y = x^2 + 1$.

12 $y = {}^-x^2$;
  $y = {}^-x^2 - 1$;
  $y = {}^-x^2 + 4$.

13 $y = \frac{1}{3}x^2$;
  $y = 2x^2$.

14 $y = 2x^2$;
  $y = {}^-2x^2$.

15 $y = (x - 4)^2$;
  $y = (x + \frac{1}{2})^2$;
  $y = {}^-(x - 4)^2$.

16 $y = (x + 2)^2$;
  $y = \frac{1}{2}(x + 2)^2$.

17 $y = \frac{1}{2}x^2$;   **18** $y = (x-3)^2$;   **19** $y = (x+1)^2$;
      $y = \frac{1}{2}x^2 + 1$. $y = (x-3)^2 - 2$. $y = 3(x+1)^2$;
                                                                                           $y = {}^{-}3(x+1)^2$.

20 $y = (x-1)^2$; $y = \frac{1}{2}(x-1)^2$; $y = {}^{-}\frac{1}{2}(x-1)^2$; $y = {}^{-}\frac{1}{2}(x-1)^2 + 2$.

In questions 21 and 22 (Figures 7 and 8), write down the equations of the quadratic graphs shown.

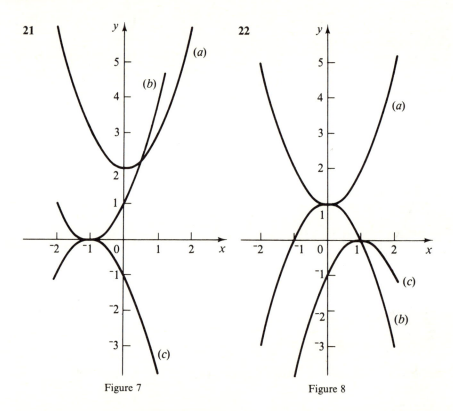

Figure 7   Figure 8

The following questions illustrate that the idea of applying transformations to the standard quadratic graph can also be used with straight-line graphs.

23 On the same diagram draw the graphs of $y = 2x$ and $y = 2x + 4$. Verify that the latter is the image of $y = 2x$ under the translation $\begin{bmatrix} 0 \\ 4 \end{bmatrix}$.

24 On the same diagram draw the graphs of $y = x$ and $y = 2x$. Verify that the latter is the image of $y = x$ under a one-way stretch with factor 2 parallel to the y-axis.

25 On the same diagram draw the graphs of $y = \frac{1}{2}x$ and $y = {}^{-}\frac{1}{2}x$. Verify that the latter is the image of $y = \frac{1}{2}x$ under reflection in the x-axis.

26 On the same diagram draw the graphs of $y = x$ and $y = 2(x+2)$. Verify that the latter is the image of $y = x$ under

    (a) first, a translation $\begin{bmatrix} -2 \\ 0 \end{bmatrix}$, (b) next, a one-way stretch with factor 2 parallel to

    the y-axis.

27 On the same diagram draw the graphs of $y = 2x-1$ and $y = {}^-2x+1$ (which can be written $y = {}^-(2x-1)$). Verify that each graph is the reflection of the other in the x-axis.

In each of questions 28–31, *sketch* the graphs on the same diagrams.

28 $y = \frac{1}{2}x$ and $y = \frac{1}{2}x-2$.

29 $y = x-5$ and $y = 2(x-5)$.

30 $y = x$, $y = x-1$ and $y = {}^-(x-1)$.

31 $y = x$, $y = x+1$, $y = 2(x+1)$ and $y = {}^-2(x+1)$.

3. PERFECT SQUARES

By using algebra, we can write the equation $y = (x+3)^2$ in another form:

$$y = (x+3)^2$$
$$\Rightarrow \quad y = (x+3)(x+3)$$
$$\Rightarrow \quad y = x^2+3x+3x+9$$
$$\Rightarrow \quad y = x^2+6x+9.$$

From the *original* form of the equation we could tell that the graph of $y = (x+3)^2$ is the image of the standard quadratic graph under the translation $\begin{bmatrix} -3 \\ 0 \end{bmatrix}$, but if we had been *given* the equation in the form $y = x^2+6x+9$, it would not have been so easy to spot the transformation concerned.

Similarly, we can rewrite the equation $y = (x-2)^2-5$ as

$$y = (x-2)(x-2)-5$$
$$= x^2-2x-2x+4-5$$
$$= x^2-4x-1,$$

but, given the form $y = x^2-4x-1$, it is not easy to see that the graph is the standard graph translated by both $\begin{bmatrix} 2 \\ 0 \end{bmatrix}$ and $\begin{bmatrix} 0 \\ -5 \end{bmatrix}$. In order to find these translations it is necessary to reverse the algebraic argument above and rewrite $y = x^2-4x-1$ as $y = (x-2)^2-5$. This process, which we now discuss, is known as *completing the square*.

First, we need to remember the formula

$$(x+a)^2 = x^2+2ax+a^2. \tag{1}$$

This formula applies for any number a; for example, when $a = 3$, then $2a = 6$ and $a^2 = 9$, so that $(x+3)^2 = x^2+6x+9$, as we saw above. When we reverse the process, for example by examining the expression $x^2-10x+25$, we see that it is a version of the formula above, with $2a = {}^-10$ and $a^2 = 25$, so that $x^2-10x+25 = (x-5)^2$.

Exercise B

In questions 1–6 and 7–12 rewrite the expressions without brackets.

1 $(x+1)^2$.

2 $(x-3)^2$.

3 $(y+4)^2$.

4 $(z-6)^2$.

5 $(t+1\frac{1}{2})^2$.

6 $(k-2\frac{1}{2})^2$.

7 $(x-1)^2$.	**8** $(x+2)^2$.	**9** $(y-4)^2$.
10 $(z+5)^2$.	**11** $(t-\frac{1}{2})^2$.	**12** $(x-3\frac{1}{2})^2$.

In questions 13–18 and 19–24, rewrite the expressions as squares.

13 $x^2+14x+49$.	**14** y^2-4y+4.	**15** z^2+2z+1.
16 $x^2+5x+6\frac{1}{4}$.	**17** $t^2-11t+30\frac{1}{4}$.	**18** $k^2+k+\frac{1}{4}$.
19 $x^2-8x+16$.	**20** $y^2+18y+81$.	**21** $z^2+12z+36$.
22 $x^2+3x+2\frac{1}{4}$.	**23** $t^2-5t+6\frac{1}{4}$.	**24** $k^2+7k+12\frac{1}{4}$.

4. COMPLETING THE SQUARE

Not every quadratic expression is an example of formula (1) above. Consider the expression $x^2+6x+16$: if it is to fit formula (1), then $2a = 6$ and $a^2 = 16$; but it is not possible to have $a = 3$ and $a = 4$ simultaneously. In such a case it is best to proceed as follows:

(i) take $2a = 6$ rather than $a^2 = 16$;

(ii) write out formula (1) with $a = 3$; this gives $(x+3)^2 = x^2+6x+9$;

(iii) convert x^2+6x+9 into the required $x^2+6x+16$ by adding $(+7)$.

Thus we finally arrive at the equation

$$x^2+6x+16 = (x+3)^2+7.$$

Example 1

Rewrite the expression $x^2-10x+6$ in the form $(x\pm a)^2\pm b$.

We observe that the middle term is $(-10x)$, so that the first part of the required expression is $(x-5)^2$.

$$(x-5)^2 = x^2-10x+25.$$

We require $\qquad\qquad x^2-10x+6$.

Hence $\qquad\qquad x^2-10x+6 = (x-5)^2-19$.

Exercise C

In questions 1–8 and 9–16, rewrite the expressions in the form $(p\pm a)^2\pm b$.

1 $x^2+14x+40$.	**2** y^2-8y+2.	**3** z^2+2z-3.
4 t^2+6t-2.	**5** k^2-6k-2.	**6** $x^2-4x+10$.
7 x^2+5x+6.	**8** $y^2-9y+16$.	
9 $x^2+10x+10$.	**10** y^2-6y+5.	**11** z^2-4z+1.
12 t^2+2t+2.	**13** $k^2-12k-5$.	**14** x^2+8x-1.
15 x^2-3x+2.	**16** y^2+y-1.	

In questions 17–19 and 20–22, by completing the square, sketch the graphs as translations of the graph of $y = x^2$.

17 $y = x^2-12x+32$.	**18** $y = x^2+4x-5$.	**19** $y = x^2-3x+3$.
20 $y = x^2+6x+8$.	**21** $y = x^2-2x-3$.	**22** $y = x^2+5x+1$.

5. SYMMETRY

The graph of $y = x^2$ has a vertical axis of symmetry (the y-axis), and so have its images under reflection in the x-axis, stretches parallel to the y-axis, and translations. Thus *all* quadratic graphs have a vertical axis of symmetry (which need not, of course, be the y-axis itself).

Consider the graph of $y = x^2 + 2x - 3$. Since

$$x^2 + 2x - 3 = (x+1)^2 - 4,$$

the graph of $y = x^2 + 2x - 3$ is the image of the standard graph under the translation $\begin{bmatrix} -1 \\ -4 \end{bmatrix}$. The image of the origin is therefore $(^-1, ^-4)$, and so the vertex is at this point and the equation of the axis of symmetry is $x = ^-1$. The graph is sketched in Figure 9.

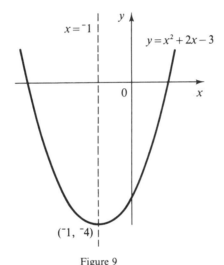

Figure 9

When completing the square to find the vertex and axis of symmetry of a graph such as $y = ^-x^2 + 4x - 5$, it is best to proceed as follows:

$$y = ^-x^2 + 4x - 5$$
$$\Rightarrow \quad ^-y = x^2 - 4x + 5$$
$$= x^2 - 4x + 4 + 1$$
$$= (x - 2)^2 + 1$$
$$\Rightarrow \quad y = ^-(x - 2)^2 - 1.$$

This form of the equation tells us that the graph is the image of the standard graph under

(i) the translation $\begin{bmatrix} 2 \\ 0 \end{bmatrix}$, (ii) reflection in the x-axis, (iii) the translation $\begin{bmatrix} 0 \\ -1 \end{bmatrix}$,

and this can be seen in the sketch in Figure 10.

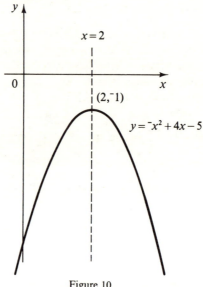

Figure 10

Since the standard graph had been reflected, it is now 'upside down', and has a *maximum* point instead of a *minimum* point. The form of the equation still tells us the axis of symmetry ($x = 2$) and the coordinates of the *turning-point*, (2, ⁻1). (A maximum point or a minimum point of a graph is called a turning-point. In this case, the ⁻x^2 term tells us that the turning-point is a maximum.)

Exercise D

For each of the following graphs $y = f(x)$, (*a*) find the equation of the axis of symmetry, (*b*) find the coordinates of the turning-point, stating whether it is a maximum or a minimum, (*c*) *sketch* the graph.

1 $y = x^2 - 2x - 1$.	**2** $y = ⁻x^2 - 4x - 3$.	**3** $y = x^2 + 6x + 12$.
4 $y = ⁻x^2 + 4x - 7$.	**5** $y = x^2 + 8x + 15$.	**6** $y = ⁻x^2 + 4x - 4$.

7 $y = x^2 - 2x + 2$.	**8** $y = x^2 + 6x + 7$.	**9** $y = ⁻x^2 + 5x - 8$.
10 $y = x^2 - 3x + 1$.	**11** $y = ⁻x^2 + 7x - 26$.	**12** $y = x^2 - 9x + 12\frac{1}{4}$.

6. QUADRATIC EQUATIONS

An equation such as $x^2 = 8x - 11$ is called a quadratic equation. We can solve such an equation by means of a graph. One method is as follows:

(i) Rewrite the equation as $x^2 - 8x + 11 = 0$. (When an equation is written like this, with one side zero, we shall say it is *written in zero form*.)

(ii) Draw the graph of $y = x^2 - 8x + 11$. (See Figure 11.)

(iii) Read off the values of x where the graph cuts the x-axis. These values are $x = 1.8$ and $x = 6.2$ approximately, so that 1.8 and 6.2 are both (approximate) solutions to the given equation.

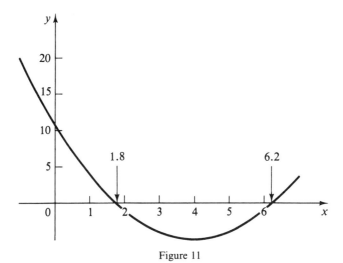

Figure 11

The method depends on the fact that for any value of x we choose, the corresponding y-coordinate is the value of $x^2 - 8x + 11$. At the points where the graph crosses the x-axis, $y = 0$, so the x-coordinates of these points are the values of x which make $x^2 - 8x + 11 = 0$; that is, they are the solutions of the equation.

Check:

$$\text{for } x = 1.8, \quad x^2 = 3.24 \quad \text{and} \quad 8x - 11 = 3.4;$$
$$\text{for } x = 6.2, \quad x^2 = 38.44 \quad \text{and} \quad 8x - 11 = 38.6.$$

The check suggests that we have found the solutions of the equation with reasonable accuracy.

The solutions of an equation are often called its *roots*. Once we have written the equation in zero form, i.e. as $f(x) = 0$, its roots obviously depend on the position of the graph of $f(x)$, since we can read off the roots from the points where the graph cuts the x-axis.

There are three possible ways in which a quadratic graph may be related to the x-axis, as illustrated in Figure 12.

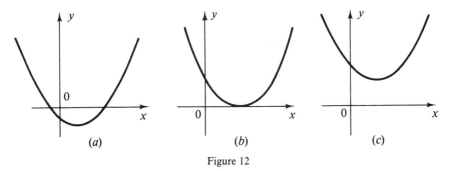

Figure 12

In Figure 12(a) the graph cuts the x-axis in two places, so the corresponding equation has two roots, as in the previous example; in (b) the graph 'just touches' the x-axis, and the corresponding equation has just one root; in (c) the graph does not intersect the x-axis at all, and the corresponding equation will have no roots.

Exercise E

Rewrite the following equations in zero form, i.e. as $f(x) = 0$. In each case use the process of completing the square to sketch the graph of $y = f(x)$, and from your sketch either (a) *estimate* the roots of the equation, or (b) deduce that the equation has no roots.

1 $x^2 + 8 = 6x.$ **2** $x^2 + 26 = 10x.$ **3** $x^2 = 8x - 16.$

4 $x^2 - 2x = 3.$

5 $x^2 + 2x = 8.$ **6** $x^2 - 4x = 5.$ **7** $x^2 + 5 = 4x.$

8 $x^2 + 1 = 2x.$

7. ALGEBRAIC SOLUTION

One disadvantage of the graphical solution of equations is that it takes quite a long time to draw the necessary graphs. Fortunately, for quadratic equations there is an alternative method using the process of completing the square. We use the equation $x^2 = 8x - 11$ to illustrate the method.

First, rewrite the equation in zero form as

$$x^2 - 8x + 11 = 0.$$

Next, rewrite the left-hand side of the equation in 'completing the square' form:

$$(x - 4)^2 - 5 = 0.$$

Now, rewrite it as $\qquad (x - 4)^2 = 5.$

From tables or a calculator we find that $\sqrt{5} \approx 2.24$, but we must remember that $(^-2.24)^2 \approx 5$ also.

So we deduce that $\qquad$ either $\quad x - 4 \approx 2.24,$

$$\text{or} \quad x - 4 \approx {}^-2.24.$$

These two simple equations give us the two solutions to the original equation:

$$x \approx 6.24 \quad \text{or} \quad x \approx 1.76.$$

(Compare these with the graphical solutions in Section 6.)

Example 2

Find the roots of the equation $6x - x^2 = 4$.

$$6x - x^2 = 4 \quad \Leftrightarrow \quad x^2 - 6x + 4 = 0 \text{ (notice that we make the } x^2 \text{ term } positive)$$
$$\Leftrightarrow \quad (x - 3)^2 - 5 = 0$$
$$\Leftrightarrow \quad (x - 3)^2 = 5$$
$$\Leftrightarrow \quad x - 3 \approx \pm 2.24$$
$$\Leftrightarrow \quad x \approx 3 \pm 2.24$$
$$\Leftrightarrow \quad x \approx 5.24 \text{ or } 0.76.$$

The roots of the equation are 5.24 and 0.76 to 2 decimal places.

Exercise F

Solve the following equations.

1 $x^2 = 9$.	2 $(x-1)^2 = 9$.	3 $(x-3)^2 = 7$.
4 $(x-5)^2 = 2$.	5 $x^2 - 2x - 8 = 0$.	6 $y^2 + 4y + 3 = 0$.
7 $z^2 = 8z + 20$.	8 $2 = 3x - x^2$.	9 $t^2 - 4t = 2$.
10 $x^2 + 3x = 4.75$.	11 $y^2 - 9y = 6$.	12 $5z - z^2 = 1$.

13 $(x+2)^2 = 16$.	14 $(y+1)^2 = 4$.	15 $(z+2)^2 = 3$.
16 $(t+1)^2 = 13$.	17 $x^2 + 4x - 5 = 0$.	18 $y^2 + 4y + 4 = 0$.
19 $z^2 = 6z + 5$.	20 $2 = x^2 - x$.	21 $x^2 = 14 - 5x$.
22 $x^2 = 11 - 5x$.	23 $y^2 + 7 + 6y = 0$.	24 $3z - z^2 - 1 = 0$.

8. THE QUADRATIC FORMULA

In all the above examples and exercises, the equations have contained a term x^2 but never $2x^2$ or $5x^2$. To deal with such equations, we can divide through by 2 or 5, say, to reduce the equation to a type we are familiar with. Similarly expressions containing $2x^2$ can be handled by taking out a factor of 2.

Example 3
Write $2x^2 + 6x + 5$ in the form $a(x+p)^2 + q$.

$$y = 2x^2 + 6x + 5 \quad \Rightarrow \quad \tfrac{1}{2}y = x^2 + 3x + \tfrac{5}{2}$$

so we consider $(x + \tfrac{3}{2})^2$.

Since $(x + \tfrac{3}{2})^2 = x^2 + 3x + \tfrac{9}{4}$, we deduce that

$$\tfrac{1}{2}y = (x + \tfrac{3}{2})^2 - \tfrac{9}{4} + \tfrac{5}{2}$$
$$= (x + \tfrac{3}{2})^2 + \tfrac{1}{4}$$

and hence $\qquad\qquad y = 2(x + \tfrac{3}{2})^2 + \tfrac{1}{2}.$

Example 4
Solve the equation $3x^2 + 4x - 7 = 0$.

First, divide by 3:

$$x^2 + \tfrac{4}{3}x - \tfrac{7}{3} = 0.$$

Next, complete the square:

$$(x + \tfrac{2}{3})^2 - \tfrac{25}{9} = 0.$$

Hence $\qquad\qquad x = \tfrac{-2}{3} \pm \tfrac{5}{3}$

and the roots are 1 and $\tfrac{-7}{3}$.

The working in such a problem, however, gets very heavy, and it is more usual to use a formula which will deal with all equations of the type $ax^2 + bx + c = 0$, whatever the values of a, b and c. The formula is

$$x = \frac{-b \pm \sqrt{(b^2 - 4ac)}}{2a}.$$

A proof is given on p. 34.

Example 5

Solve the equation $3x^2 + 4x - 7 = 0$.

$$a = 3, \quad b = 4, \quad c = {}^-7.$$
$$b^2 - 4ac = 16 - 4(3)({}^-7)$$
$$= 16 + 84$$
$$= 100.$$

From the formula, $$x = \frac{{}^-4 \pm \sqrt{100}}{2(3)}$$

$$= \frac{{}^-4 + 10}{6} \quad \text{or} \quad \frac{{}^-4 - 10}{6}$$

$$= 1 \quad \text{or} \quad {}^-\tfrac{7}{3}.$$

Example 6

Solve the equation $5t - t^2 = 1$.

Rewrite the equation in zero form with positive t^2:

$$t^2 - 5t + 1 = 0.$$
$$a = 1, \quad b = {}^-5, \quad c = 1$$
$$b^2 - 4ac = ({}^-5)^2 - 4(1)(1)$$
$$= 25 - 4$$
$$= 21.$$

From the formula $$t = \frac{{}^-({}^-5) \pm \sqrt{21}}{2(1)}$$

$$\approx \frac{5 \pm 4.58}{2}$$

$$= \frac{9.58}{2} \quad \text{or} \quad \frac{0.42}{2}$$

$$= 4.79 \quad \text{or} \quad 0.21.$$

Exercise G

Solve the following equations by using the formula.

1 $2x^2 + 5x + 2 = 0$.	**2** $3y^2 + 5y + 2 = 0$.	**3** $4z^2 - 9z + 2 = 0$.
4 $2t^2 + 3t = 1$.	**5** $3s - 5s^2 + 7 = 0$.	**6** $x^2 = 1 - x$.
7 $2x^2 + 0.4x - 1.8 = 0$.	**8** $3x^2 - 6.2x + 2.1 = 0$.	**9** $4.6t = 3.2 - 0.8t^2$.
10 $6x^2 - x - 2 = 0$.	**11** $4y^2 + 5y - 6 = 0$.	**12** $2z^2 - 5z - 3 = 0$.
13 $2t^2 = 5t + 3$.	**14** $3s = 2 - 2s^2$.	**15** $4x^2 - 3x - 5 = 0$.
16 $y(y + 2) = 9$.	**17** $6x^2 + 7.2x + 1.6 = 0$.	**18** $2.5 = 8.4t - 0.3t^2$.

9. FACTORS

Although it is normally necessary to 'complete the square' or use the formula to solve a quadratic equation by algebraic means, it is sometimes possible to use a much quicker method. You will already know that if the product of two numbers is zero, then at least one of the numbers themselves must be zero, and vice versa; in symbols,

$$PQ = 0 \quad \Leftrightarrow \quad \text{either } P = 0 \quad \text{or} \quad Q = 0 \quad \text{(or both).}$$

So, for example, $\qquad x^2 - 5x = 0 \quad \Leftrightarrow \quad x(x - 5) = 0$

$$\Leftrightarrow \quad x = 0 \quad \text{or} \quad x - 5 = 0$$
$$\Leftrightarrow \quad x = 0 \quad \text{or} \quad x = 5.$$

If we happen to have an equation written in the form such as

$$(x - 2)(2x + 3) = 0$$

we can immediately deduce that either $\quad x - 2 = 0$

$$\text{or} \qquad 2x + 3 = 0$$

so that either $x = 2$ or $x = {}^-1.5$.

If we multiply together the expressions in brackets in the example above, using the following layout:

and then add together the two terms enclosed in the loop (the two terms in x), we get

$$2x^2 - x - 6 = 0.$$

Given the equation in this form, we could solve it using the formula, but if we could somehow reverse the process of multiplication and return the equation to the form $(x - 2)(2x + 3) = 0$ we could solve the equation by using the process just described. This 'reversal' of the algebraic multiplication is called 'factorisation', and the expressions in brackets, $(x - 2)$ and $(2x + 3)$, are called *factors* of $2x^2 - x - 6$. It is not always practically possible to find factors, but it is usually worthwhile looking for them, since if we do find them we have a very quick solution of the equation concerned.

Factorisation is a matter of intelligent guesswork. We expect the factors (if any) to be of the form $(px \pm q)$, where p and q are whole numbers.

Example 7
Factorise the expression $x^2 - 5x + 4$.

When we set up the multiplication table, we require it to look like this:

Since the only simple way to produce x^2 is $x \cdot x$, and the only simple ways using integers to produce $(^+4)$ are $(^+2)(^+2)$, $(^-2)(^-2)$, $(^+4)(^+1)$, and $(^-4)(^-1)$, there are only four likely possibilities for the table:

x	$^+2$		x	$^-2$		x	$^+4$		x	$^-4$
x	x^2	^+2x	x	x^2	^-2x	x	x^2	^+4x	x	x^2	^-4x
$^+2$	^+2x	$^+4$	$^-2$	^-2x	$^+4$	$^+1$	^+x	$^+4$	$^-1$	^-x	$^+4$

Checking the term in x, we see that of the four possibilities, only the last one 'works'; hence

$$x^2 - 5x + 4 = (x-4)(x-1).$$

Example 8

Use the factor method to solve the equation $3x^2 = 2 - 5x$.

First write the equation in zero form:

$$3x^2 + 5x - 2 = 0.$$

Next, find the factors of $3x^2 + 5x - 2$:
to produce $3x^2$, we shall need $3x$ and x;
to produce $^-2$, we shall need either $(^+1)$ and $(^-2)$ or $(^-1)$ and $(^+2)$, and *any one of these four numbers* might be 'attached' to the term $3x$ so the possibilities are:

$$(3x+1),\ (x-2);$$
$$(3x-2),\ (x+1);$$
$$(3x-1),\ (x+2);$$
$$(3x+2),\ (x-1).$$

By checking each of these pairs in turn, we find that the third pair gives $5x$ and so is the one required, so that the equation can be written as

$$(3x-1)(x+2) = 0$$
$$\Leftrightarrow \quad \text{either} \quad 3x-1=0 \quad \text{or} \quad x+2=0$$
$$\Leftrightarrow \qquad\qquad\qquad x = \tfrac{1}{3} \quad \text{or} \quad x = ^-2.$$

So the roots of the equation are $\tfrac{1}{3}$ and $^-2$.

A word of warning. If you try to factorise the expression $x^2 - 2x - 1$ by this method, it is likely to take you a long time; the factors are (approximately) $(x+0.41)(x-2.41)$. If you cannot find factors easily, it is best to fall back on one of the other methods you know for solving a quadratic equation. In the following exercise, however, the equations have been deliberately chosen so that the factors involve only integers.

Exercise H

1   Multiply:
    (a) $(x+2)(x-5)$;   (b) $(x-2)(x+5)$;   (c) $(x-2)(x-5)$;
    (d) $(3x+1)(x+4)$; (e) $(3x+1)(x-4)$; (f) $(3x-1)(x+4)$;
    (g) $(3x-1)(x-4)$.

2 Factorise:
(a) x^2+5x+6; (b) $x^2+2x-24$; (c) $x^2-6x-16$;
(d) x^2-8x; (e) x^2-3x-4; (f) $2x^2+7x+3$;
(g) $2x^2+5x+3$; (h) $2x^2+5x-3$; (i) $3x^2-x$;
(j) $2x^2-7x-4$; (k) $3x^2-14x+8$.

3 Solve the following equations by the factor method:
(a) $x^2-5x=0$; (b) $y^2+2y-3=0$; (c) $z^2+3z=10$;
(d) $2t^2+5t+2=0$; (e) $2x^2=5x+3$; (f) $2y^2=2-3y$;
(g) $x^2-7x=18$; (h) $x^2=x+6$; (i) $2z^2=z+1$;
(j) $3t^2+3=10t$; (k) $x(x-1)=2$; (l) $y(y+2)=8$;
(m) $x^2=6x-9$.

4 Multiply:
(a) $(y-3)(y+2)$; (b) $(y+3)(y-2)$; (c) $(y-3)(y-2)$;
(d) $(2z+3)(z-1)$; (e) $(2z-3)(z+1)$; (f) $(2z+1)(z-3)$;
(g) $(2z-1)(z+3)$.

5 Factorise:
(a) t^2-7t+6; (b) t^2-5t-6; (c) t^2-t-6;
(d) $x^2-2x-15$; (e) $y^2+4y-32$; (f) $2z^2+11z+5$;
(g) $2z^2+7z+5$; (h) $3x^2+x-2$; (i) $5y^2-3y-2$;
(j) $3t^2+t-4$; (k) $7x^2+2x-5$.

6 Solve the following equations by the factor method:
(a) $x^2-3x-4=0$; (b) $y^2-10y+21=0$; (c) $z^2+7z=0$;
(d) $5t^2=14t+3$; (e) $4+3t^2=7t$; (f) $3x^2=7x+6$;
(g) $21y=2y^2+10$; (h) $z(z+3)=10$; (i) $7x^2=19x+6$;
(j) $2y^2+9y+10=0$.

7 Factorise:
(a) $6x^2-5x-6$; (b) $8y^2+18y+9$; (c) $12z^2+z-6$.

8 Solve by the factor method:
(a) $(2x+1)(x-4)+10=0$; (b) $x+2=\dfrac{8}{x}$; (c) $3x=\dfrac{5x+2}{x}$.

10. SUMMARY

Transformations

Compared with the standard quadratic graph, $y=x^2$,

(a) $y=x^2+b$ is the result of a translation $\begin{bmatrix} 0 \\ b \end{bmatrix}$,

(b) $y=(x-a)^2$ is the result of a translation $\begin{bmatrix} a \\ 0 \end{bmatrix}$,

(c) $y-b=(x-a)^2$ is the result of a translation $\begin{bmatrix} a \\ b \end{bmatrix}$,

(d) $y=\,^-x^2$ is the result of a reflection in the x-axis,
(e) $y=k.x^2$ is the result of a one-way stretch of factor k parallel to the y-axis.

Completing the square

The expression $x^2 + bx + c$ can be rewritten as

$$(x+p)^2 + q$$

where $p = \frac{1}{2}b$ and q can be found by comparing the two versions of the expression.

Quadratic equations

(i) The roots of a quadratic equation $x^2 + bx + c = 0$ can be found by drawing the graph of $y = x^2 + bx + c$ and reading off the x-values of the points where the graph cuts the x-axis.

(ii) The graph of $y = (x+p)^2 + q$ will cut the x-axis
  (a) at two points, if $q < 0$,
  (b) at one point if $q = 0$,
  (c) not at all if $q > 0$;

thus the equation $(x+p)^2 + q = 0$ will have two roots, one root or no roots according to the value of q.

(iii) The equation $x^2 + bx + c = 0$ can be solved by:
  (a) rewriting as $(x+p)^2 + q = 0$ and then as $(x+p)^2 = {}^-q$;
  (b) finding the square roots of each side of this equation (provided that the right-hand side is not negative).

(iv) The equation $ax^2 + bx + c = 0$ has solutions

$$x = \frac{{}^-b \pm \sqrt{(b^2 - 4ac)}}{2a}$$

provided that the expression $b^2 - 4ac$ is not negative.

Factors

(i) A quadratic expression $ax^2 + bx + c$ can sometimes be written in factorised form $(px+q)(rx+s)$, where p, q, r, s are integers.

(ii) When a quadratic equation $ax^2 + bx + c = 0$ has been written in factorised form $(px+q)(rx+s) = 0$, the roots of the equation are

$$\frac{{}^-q}{p}, \quad \frac{{}^-s}{r}.$$

11. PROOF OF THE QUADRATIC EQUATION FORMULA

$$ax^2 + bx + c = 0 \implies x^2 + \frac{b}{a}x = \frac{{}^-c}{a}$$

$$\implies x^2 + \frac{b}{a}x + \frac{b^2}{4a^2} = \frac{{}^-c}{a} + \frac{b^2}{4a^2}$$

$$\implies \left(x + \frac{b}{2a}\right)^2 = \frac{{}^-4ac}{4a^2} + \frac{b^2}{4a^2}$$

$$\Rightarrow \qquad x + \frac{b}{2a} = \pm\frac{\sqrt{(b^2 - 4ac)}}{2a}$$

$$\Rightarrow \qquad x = \frac{-b \pm \sqrt{(b^2 - 4ac)}}{2a}.$$

Miscellaneous exercise

1   Solve the following equations by whatever seems the most suitable method:
    (a) $x^2 + 4x = 21$;   (b) $y^2 + 4y = 19$;   (c) $2x^2 + 9x = 20$;
    (d) $2y^2 + 9y = 18$;   (e) $3z^2 + 14z + 15 = 0$;   (f) $z^2 - z - 9 = 0$;
    (g) $t - t^2 + 20 = 0$;   (h) $4x^2 + 3 = 8x$;   (i) $4x^2 + 3 = 10x$;
    (j) $t^2 = 18t - 74$;   (k) $y^2 + 1.2y = 0.28$.

2   State the coordinates of the points where the graph of $y = (x-2)(3-x)$ cuts
    (a) the x-axis, (b) the y-axis.
        Sketch the graph, and indicate on your diagram how it can be used to estimate
    the roots of the equation $(x-2)(3-x) = {}^-3$.
        Solve the equation algebraically, giving your answers corrected to 1 decimal
    place.   (SMP)

3   The function f is given by $f : x \to 5 - 3x - x^2$. Use the formula for solving quadratic
    equations to calculate, correct to 2 significant figures, the roots of $f(x) = 0$.
        $f(x)$ may be written in the form $(a+x)(b-x)$. Write down the values of a and
    b, correct to 2 significant figures, and hence state the solution set for $f(x) > 0$.
                                                                          (SMP)

4   Find the coordinates of the turning-point of the graph $y = x^2 - 4x + 3$. Sketch the
    graph for values of x in the interval ${}^-1 \leqslant x \leqslant 5$.
        Find the greatest value of k for which the quadratic equation

$$x^2 - 4x + k = 0$$

    has a solution.

5   Sketch the graph of $y = x^2 + 3x + 0.5$ and state the coordinates of its turning-point.
        Solve the equation $x^2 + 3x + 0.5 = 0$.
        State the set of values of x for which $x^2 + 3x + 0.5 \geqslant 4.5$.

6   Express $x^2 - 2x - 3$ in the form $(x-p)^2 - q$. Hence find (a) the turning-point and
    (b) the axis of symmetry of the curve $y = x^2 - 2x - 3$. Sketch this curve and indicate
    clearly on your sketch your results for parts (a) and (b).
        Solve algebraically $x^2 - 2x = 4$ and label points on your sketch graph which
    correspond to your answers.   (SMP)

7   State the coordinates of the points where the graph of $y = (x+1)(x-2)$ cuts (a) the
    x-axis, (b) the y-axis. Sketch the graph and indicate on your diagram how it can
    be used to estimate the roots of the equation

$$(x+1)(x-2) = x.$$

    Solve this equation algebraically, giving your answer to 2 decimal places.

8   Solve the simultaneous equations:

$$y = x + 4,$$
$$2y = x^2.$$

Illustrate on a graph the region satisfying all of the inequalities

$$y \leqslant x+4,$$
$$y \geqslant \tfrac{1}{2}x^2,$$
$$x \leqslant 3,$$

by shading the regions that are excluded.

9 Sketch the graph of the equation

$$y = x^2 + x,$$

marking the coordinates of the points where the curve cuts the x-axis. On the same diagram sketch the lines

(a) $y = 3x$,
(b) $y = 3x - 1$,
(c) $y = 3x - 2$.

By solving three pairs of simultaneous equations, find the coordinates of the points (if any) where each of the given lines intersects the parabola. What is the relationship between the line (b) and the parabola?

10 If n is an integer (which is not necessarily positive), then $n-1$, n, $n+1$ are three consecutive integers. In each of the following cases form an equation for n, and solve it.

(a) The three consecutive integers add up to 96.
(b) The sum of the first integer with twice the second and three times the third is 86.
(c) If the first and third integers are multiplied together, and the product added to the second, the answer is 41.
(d) The sum of the squares of the three integers is 365.   (MEI)

11

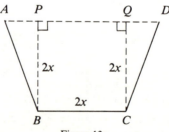

Figure 13

A long thin strip of metal is bent to form a length of guttering. Figure 13 shows a cross-section $ABCD$ of the gutter, which has a vertical line of symmetry. The gutter is $2x$ cm wide at the base and $2x$ cm deep.

(i) The original strip of metal was 38 cm wide. Write down in terms of x an expression for the length AB.
(ii) The top of the gutter is 22 cm wide. Write down in terms of x an expression for the length AP.
(iii) By using the theorem of Pythagoras on the triangle APB, write down an equation involving x, and solve it.   (MEI)

12 Figure 14 shows an isosceles triangle ABC inscribed in a circle, centre O. The line AO produced meets BC at right-angles at N. The radius of the circle is r cm; $AB = t$ cm; $BN = s$ cm; $ON = h$ cm.

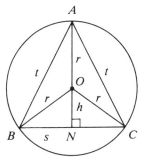

Figure 14

Use the theorem of Pythagoras to write down (a) an expression for s^2 in terms of r and h; (b) an expression for t^2 in terms of r, h and s. Hence show that

$$t^2 = 2r(r+h).$$

Solve this equation for r when $t = 12$ and $h = 1$.   (MEI)

13 An *open* rectangular cardboard box is 7 cm high, and its length is 5 cm greater than its breadth. If the breadth is x cm, write down expressions for:
   (a) its length;
   (b) the area of each of the two smaller sides;
   (c) the area of each of the two larger sides;
   (d) the area of the base.
From your answers to (b), (c) and (d) write down and simplify as far as possible an expression for the total surface area of the box.
   The total surface area is 500 cm². Write down an equation for x and solve it.

14

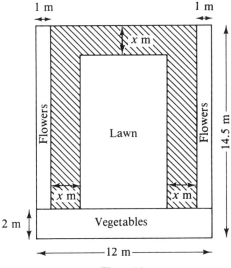

Figure 15

A house has a rectangular garden 14.5 m long and 12 m wide. The owner decides to grow vegetables at the bottom of his garden and flowers along each side. The rest of the garden will consist of a rectangular lawn surrounded on three sides by a concrete path x m wide, as shown shaded in Figure 15, which also gives all necessary dimensions.

- (a) Calculate the area of garden *not* devoted to vegetables or flowers.
- (b) Write down the dimensions of the lawn in terms of x.
- (c) Write down an expression involving x for the area of concrete to be laid.
- (d) The owner acquires cheaply a load of concrete sufficient to lay a path of total area 48 cm². Write down an equation for x and solve it.   (MEI)

3

Vectors and relative motion

One important practical use of vectors is in solving problems of navigation. In order to arrive at his destination, the navigator of a ship has to allow for winds and currents; if a space shuttle is to make a successful rendezvous with a space station already in orbit, the launch has to allow for the motion of the station (and, indeed, for the motion of the Earth). In this chapter we shall develop some of the principles underlying the solution of such problems; first, however, we shall revise and consolidate relevant work on vectors.

1. REPRESENTATION OF VECTORS

A vector can be defined algebraically or geometrically. Algebraically, a column of two numbers such as $\begin{bmatrix} 4 \\ -3 \end{bmatrix}$ is a two-dimensional vector; we can add two such column-matrices according to a very simple rule which you will already have met – for example,

$$\begin{bmatrix} 4 \\ -3 \end{bmatrix} + \begin{bmatrix} 2 \\ 5 \end{bmatrix} = \begin{bmatrix} 6 \\ 2 \end{bmatrix}.$$

The most familiar example of a geometrical vector is a translation. Two translations can be added (combined) to give another translation. We shall discuss the *addition* of geometrical vectors in the next section – for the moment we shall simply recall that a translation can be represented by directed line-segment, in which the direction of the line-segment represents the direction of the translation and the length of the line-segment represents the distance of the translation.

Figure 1 shows a number of such line-segments. $\underset{\sim}{QP}$, for example, represents the translation $\begin{bmatrix} 4 \\ -3 \end{bmatrix}$, and $\underset{\sim}{QQ}$ represents the translation $\begin{bmatrix} 1 \\ 2 \end{bmatrix}$. Notice, however, that $\underset{\sim}{AB}$, $\underset{\sim}{CD}$ and $\underset{\sim}{EF}$ *also* represent the translation $\begin{bmatrix} 4 \\ -3 \end{bmatrix}$, so that $\underset{\sim}{AB}$, $\underset{\sim}{CD}$, $\underset{\sim}{EF}$ and $\underset{\sim}{QP}$ are all different names for the *same* vector, which could equally well be described as '5 units at 37° below the x-axis'. In the same way, $\underset{\sim}{QQ}$, $\underset{\sim}{PR}$, $\underset{\sim}{FS}$, and $\underset{\sim}{TU}$ are all different names for the vector $\begin{bmatrix} 1 \\ 2 \end{bmatrix}$. It is very important to remember that a vector can be represented by different line-segments; so long as we ensure that it has the correct length and direction, we can represent a vector on a diagram in any position which we find convenient.

Rather than choosing a particular directed line-segment on a diagram to name

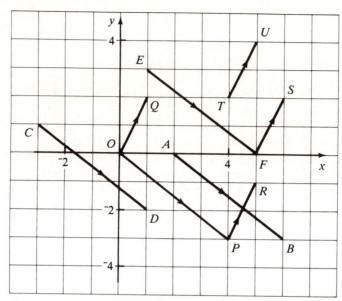

Figure 1

a vector, we often denote the vector by a single letter; so we could use the letter
p to denote the vector represented by AB or EF, and then we would write

$$\mathbf{p} = AB \quad \text{(or, of course, } \mathbf{p} = EF, \text{ etc.).}$$

We then say that OP, CD, AB, ... are different *representations* of the vector **p**.

In Figure 1 we can easily construct a triangle of which CD is the hypotenuse
and hence calculate by using Pythagoras' formula that the length of the line CD
is $\sqrt{(3^2+4^2)} = 5$ units. If we think of CD as a representation of a vector, we
say that the *magnitude* of the vector is 5 units. In the same way, the directed
line-segment OQ, which represents the vector $\begin{bmatrix} 1 \\ 2 \end{bmatrix}$, has a magnitude of $\sqrt{(1^2+2^2)}$
units, i.e. approximately 2.24 units. In general, the magnitude of a vector is the
length of the line-segment representing it (to some suitable scale), and if the
vector is written as a column-matrix $\begin{bmatrix} a \\ b \end{bmatrix}$ then its magnitude is $\sqrt{(a^2+b^2)}$.

Exercise A

Questions 1–5 refer to Figure 1. Use the points labelled in that diagram to write down
your answers.

1 Write down another representation of each of the following vectors:
   (a) CQ, (b) SR, (c) SE.

2 Write down each of the vectors in question 1 in column-matrix form.

3 Calculate the magnitudes of each of the vectors in question 1.

4 $OT = \mathbf{p}$. Express **p** in column-matrix form and write down another representation
   of **p**.

5 $PB = \mathbf{q}$. Write down another representation of **q**.

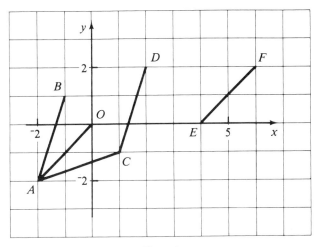

Figure 2

Questions 6–11 refer to Figure 2.

6 Write down another representation of each of the following vectors:
   (*a*) CD, (*b*) CA, (*c*) FE.

7 Write down each of the vectors in question 6 in column-matrix form.

8 Calculate the magnitudes of the vectors in question 6.

9 $AC = \mathbf{x}$. Write $\mathbf{x}$ in column-matrix form and write down *two* other representations of $\mathbf{x}$.

10 $OC = \mathbf{y}$. Write down another representation of $\mathbf{y}$.

11 $DE = \mathbf{z}$. Write down another representation of $\mathbf{z}$.

2. ADDITION AND SUBTRACTION

The sum of two vectors is obtained by choosing two segments which represent the vectors and which are placed 'head-to-tail', that is, one segment starts from where the other leaves off. For example, in order to find $OP + OQ$, we must find representations of OP and OQ which lie 'head-to-tail'. Since $OQ = PR$, and PR starts from where OP leaves off, it will be convenient to use OP and PR as our segments. We then *define*

$$OP + PR = OR.$$

This is illustrated in Figure 3, which is extracted from Figure 1.

In Figure 3, the sum of OP and PR is shown as OR. (Notice the arrow used for the sum vector.) Since $PR = OQ$, we have

$$OP + OQ = OR.$$

Notice also that since $OP = EF$ and $OQ = FS$, we could equally well have represented $OP + OQ$ by $EF + FS$, in which case the sum vector is represented by ES. It is clear that $ES = QR$, so that either way of representing the sum gives the same result.

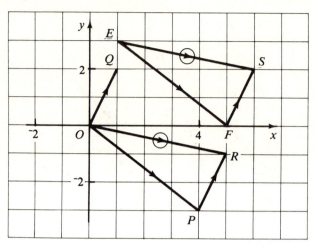

Figure 3

We can check the reasonableness of the argument by using column-matrices. $QP = \begin{bmatrix} 4 \\ -3 \end{bmatrix}$ and $OQ = \begin{bmatrix} 1 \\ 2 \end{bmatrix}$, so that $QP + OQ = \begin{bmatrix} 5 \\ -1 \end{bmatrix}$, and we see in the diagram that the sum vector is indeed $\begin{bmatrix} 5 \\ -1 \end{bmatrix}$, whether we represent it by QR or by ES.

Exercise B

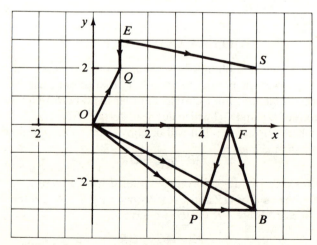

Figure 4

In Figure 4, denote vectors by single letters as follows:
  $\mathbf{p} = FP$, $\mathbf{q} = PB$, $\mathbf{r} = OP$, $\mathbf{s} = OQ$, $\mathbf{t} = OF$,
  $\mathbf{u} = EQ$, $\mathbf{v} = FB$, $\mathbf{w} = OB$, $\mathbf{x} = ES$.

1 Write the following vector sums as single letters:

  (*a*) **p**+**q**; (*b*) **q**+**r**; (*c*) **p**+**s**;

  (*d*) **p**+**t**; (*e*) **r**+**s**; (*f*) **t**+**u**.

2 Write each of the vector additions of question 1 in column-matrix form.

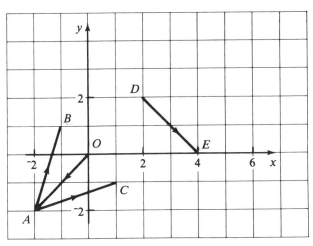

Figure 5

Questions 3 and 4 refer to Figure 5. Denote vectors by single letters as follows:

$\widetilde{AB} = \mathbf{a}, \quad \widetilde{OA} = \mathbf{b}, \quad \widetilde{AC} = \mathbf{c}, \quad \widetilde{DE} = \mathbf{d},$

$\widetilde{OB} = \mathbf{e}, \quad \widetilde{AD} = \mathbf{f}, \quad \widetilde{OC} = \mathbf{g}, \quad \widetilde{OD} = \mathbf{h}.$

3 Write the following vector sums as single letters:

  (*a*) **a**+**b**; (*b*) **a**+**c**; (*c*) **a**+**d**;

  (*d*) **b**+**c**; (*e*) **b**+**f**; (*f*) **c**+**e**.

4 Write each of the vector additions in question 3 in column-matrix form.

3. VECTOR TRIANGLES

So far in our discussion we have defined the vectors with reference to pairs of points placed on a square grid with coordinate axes, but this is of course not necessary. So long as we keep in mind some reference direction, such as 'north' or 'straight upwards', we can represent any vector by drawing a line-segment with an arrow pointing in the correct direction and ensuring that it has the correct length (on some suitable scale). For example, if **a** is the vector '3 km north-east' and **b** is the vector '2 km west' we can represent **a** and **b** as in Figure 6.

If we now wish to find the sum **a**+**b** we simply 'move' one of the line-segments until it is 'head-to-tail' with the other and complete the triangle, as in Figure 7. We can now measure the bearing and length of the sum vector, which is about 2.1 km on a bearing of 3°.

This method is known as the *triangle law of addition*. With a little practice it is quite easy to draw the triangle straight away instead of first drawing the

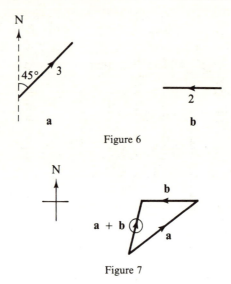

Figure 6

Figure 7

vector representations separately and then moving one of them 'head-to-tail' with the other.

Before we can deal with the subtraction of vectors we must define the *negative* of a vector. This is very simple: if **a** is a vector, then its negative is a vector of the same magnitude but in the opposite direction. The negative of **a** is written as ⁻**a**. (See Figure 8.)

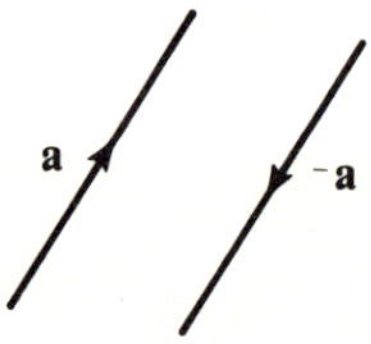

Figure 8

Having defined the negative of a vector, we can define subtraction of vectors in the following way:

To subtract a vector, add its negative.

This definition can be written in symbols:

$$\mathbf{a} - \mathbf{b} = \mathbf{a} + (^-\mathbf{b}).$$

For example, suppose **a** and **b** are represented as in Figure 9.

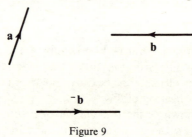

Figure 9

Then ⁻**b** is as shown and **a**−**b** is **a**+(⁻**b**), which is shown in Figure 10.

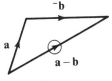

Figure 10

There is a particularly simple way of finding the difference between two vectors (i.e. the result of subtracting one from the other) if we represent the given vectors

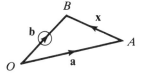

Figure 11

by line-segments *starting from the same point*. In Figure 11, vectors **a** and **b** are represented by directed line-segments $O\underset{\sim}{A}$ and $O\underset{\sim}{B}$ both starting at O. We denote the vector $A\underset{\sim}{B}$ by **x**. From the triangle law of addition, we know that

$$\mathbf{a}+\mathbf{x} = \mathbf{b}$$

so that
$$\mathbf{x} = \mathbf{b}-\mathbf{a}.$$

Notice carefully the *order* of the letters. The vector **b**−**a** is represented by the directed line-segment *from A to B*. To find **a**−**b** we must go from B to A, i.e. we use $B\underset{\sim}{A}$ to represent **a**−**b**.

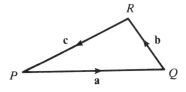

Figure 12

Suppose that three vectors, **a**, **b**, **c**, can be represented by the three sides of a triangle, as shown in Figure 12. Clearly the three successive displacements, P to Q, Q to R, R to P will take a point which started at P back to P again. When three (or more) vectors are related like this we say that their sum is the *zero vector*, and we can write

$$\mathbf{a}+\mathbf{b}+\mathbf{c} = \mathbf{0}.$$

Notice that **0** is a *vector*, not a number.

The sum of a vector and its negative is also the zero vector:

$$\mathbf{a}+(⁻\mathbf{a}) = \mathbf{0}$$

or
$$\mathbf{a}-\mathbf{a} = \mathbf{0}$$

as we might expect.

What about $\mathbf{a}+\mathbf{a}$? This clearly results in a vector which has the same direction as $\mathbf{a}$ but has twice the magnitude (or length). It is natural to call this vector $2\mathbf{a}$.

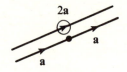

Figure 13

(See Figure 13.) This suggests a simple and sensible definition for multiplication of a vector by a *number*: the vector $k\mathbf{a}$ (where k is a positive number) is a vector which has the same direction as $\mathbf{a}$ but has k times the magnitude of $\mathbf{a}$. If k is negative then the vector $k\mathbf{a}$ has ^-k times the magnitude of $\mathbf{a}$ but is in the opposite direction.

Exercise C

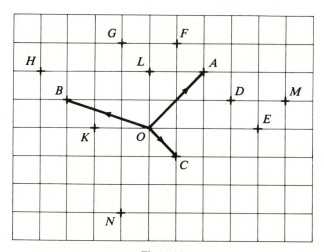

Figure 14

In Figure 14, $OA = \mathbf{a}$. $OB = \mathbf{b}$, $OC = \mathbf{c}$.

1   Write down in the form XY vectors representing:
   (a) $2\mathbf{c}$;   (b) $3\mathbf{c}$;   (c) $^-3\mathbf{c}$.

2   Write down in the form OX the following vectors:
   (a) $\mathbf{a}+\mathbf{b}$;   (b) $(\mathbf{a}+\mathbf{b})+\mathbf{c}$;   (c) $\mathbf{b}+\mathbf{c}$;
   (d) $\mathbf{a}+(\mathbf{b}+\mathbf{c})$;   (e) $\mathbf{a}+\mathbf{c}$;   (f) $(\mathbf{a}+\mathbf{c})+\mathbf{b}$;
   (g) $\mathbf{a}-\mathbf{b}$;   (h) $\mathbf{a}-\mathbf{c}$;   (i) $\mathbf{c}-\mathbf{a}$;
   (j) $\mathbf{b}-\mathbf{c}$;   (k) $\mathbf{a}+2\mathbf{c}$.

3   Express the following vectors in the form $\mathbf{p}-\mathbf{q}$:
   (a) AC;   (b) AB;   (c) BC.

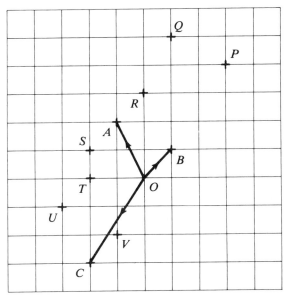

Figure 15

In Figure 15, $OA = \mathbf{x}$, $OB = \mathbf{y}$, $OC = \mathbf{z}$.

4 Write down in the form OX the following vectors:
(a) $\mathbf{x}+\mathbf{y}$; (b) $\mathbf{x}-\mathbf{y}$; (c) $\mathbf{x}+\mathbf{z}$;
(d) $\mathbf{x}-\mathbf{z}$; (e) $\mathbf{y}+\mathbf{z}$; (f) $\mathbf{y}-\mathbf{z}$;
(g) $\mathbf{x}+\mathbf{y}+\mathbf{z}$.

5 Express the following vectors in the form $\mathbf{p}-\mathbf{q}$:
(a) BA; (b) BC; (c) AC.

4. RELATIVE VELOCITY

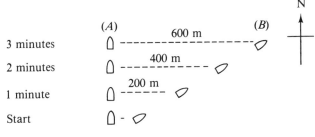

Figure 16

Figure 16 represents a sequence of 'aerial snapshots' of two boats. Boat A is travelling north and boat B is travelling north-east. At the start the two boats are together, but as time goes on they get farther and farther apart; however, it so happens (in this example!) that *B* is travelling at just the right speed to keep it always due east of *A*.

A passenger in boat *A*, idly enjoying the trip, may be totally unaware that he is travelling north. As far as he is concerned, he is watching boat *B* moving away from *him* in an easterly direction, and their distance apart increasing at a rate of 200 metres every minute. In these circumstances we say that *the velocity of B relative to A is* 200 *metres per minute east.*

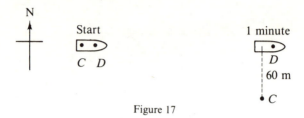

Figure 17

Two people are sitting in a boat on a river which is flowing due east. One of them, *C*, dives into the water and swims directly towards the south bank. The other, *D*, sits and watches her, allowing the boat to drift with the stream.

If *C* has no landmark to aim for but simply 'points' herself due south as she swims, she will be carried downstream. However, the boat will also be carried downstream at the same rate so that *D* will continue to see *C* due south of him. If after one minute *D* sees *C* 60 m due south of him, he will reckon that *C*'s velocity *relative to the boat* is 1 m s^{-1} due south. (See Figure 17.)

Since the boat is drifting, that is, not moving relative to the water, we can deduce that the velocity of *C* relative to the *water* is 1 m s^{-1} due south. This would also be the case if, for example, *C* dived in from the north bank and started swimming south.

In the first example above we saw that the velocity of boat *B* relative to boat *A* was 200 m min^{-1} due east. We can represent this velocity by a directed line-segment:

$$\xrightarrow{\hspace{2cm}\text{200 m min}^{-1}\hspace{2cm}}$$

Similarly, we can represent the velocity of the swimmer in the second example relative to the water by a directed line-segment:

These suggest that we can treat velocities as *vectors*. To justify this, we examine the first example a little more closely, using coordinate axes.

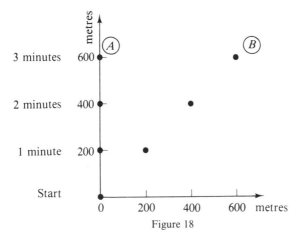

Figure 18

We know that the velocity of boat A is 200 m min^{-1} due north, which we can represent either by a directed line-segment

200 m min^{-1}

or by the column-matrix $\begin{bmatrix} 0 \\ 200 \end{bmatrix}$. Now the *position* vectors of boat A after 1, 2, 3 minutes are

$$\begin{bmatrix} 0 \\ 200 \end{bmatrix}, \begin{bmatrix} 0 \\ 400 \end{bmatrix}, \begin{bmatrix} 0 \\ 600 \end{bmatrix}$$

which can be written as

$$1 \times \begin{bmatrix} 0 \\ 200 \end{bmatrix}, \quad 2 \times \begin{bmatrix} 0 \\ 200 \end{bmatrix}, \quad 3 \times \begin{bmatrix} 0 \\ 200 \end{bmatrix}$$

so that we can get the displacement of boat A from the origin if we multiply the velocity vector by the time elapsed, which is reasonable. In the same way, the velocity of boat B can be written as a column-matrix $\begin{bmatrix} 200 \\ 200 \end{bmatrix}$ and we can find its displacement from the origin after, say, 3 minutes if we calculate

$$3 \times \begin{bmatrix} 200 \\ 200 \end{bmatrix} = \begin{bmatrix} 600 \\ 600 \end{bmatrix}.$$

To summarise: if an object is moving with constant velocity (i.e. its speed and direction of motion remain unchanged) then

(displacement vector) = (velocity vector) × (time elapsed).

There is another very important idea which can be seen in this example. In our earlier discussion we saw that the velocity of B relative to A is 200 m min^{-1} due east, which we have already represented by a directed line-segment

$$\xrightarrow{\text{200 m min}^{-1}}$$

but which can also be represented by the column-matrix $\begin{bmatrix} 200 \\ 0 \end{bmatrix}$. Now the velocity vectors of A and B are $\begin{bmatrix} 0 \\ 200 \end{bmatrix}$ and $\begin{bmatrix} 200 \\ 200 \end{bmatrix}$ respectively, and $\begin{bmatrix} 200 \\ 200 \end{bmatrix} - \begin{bmatrix} 0 \\ 200 \end{bmatrix} = \begin{bmatrix} 200 \\ 0 \end{bmatrix}$; that is, the vector representing the velocity of B relative to A can be found by subtracting the velocity vector of A from the velocity vector of B:

(velocity vector of B relative to A)

$$= \text{(velocity vector of } B) - \text{(velocity vector of } A).$$

Exercise D

1 The velocity of aircraft P is $\begin{bmatrix} 450 \\ 300 \end{bmatrix}$ and the velocity of aircraft Q is $\begin{bmatrix} 500 \\ -200 \end{bmatrix}$. What is the velocity of (*a*) Q relative to P, (*b*) P relative to Q? (The units are km h^{-1}.)

2 A train is travelling due north at 120 km h^{-1} when a helicopter passes overhead travelling due east at 90 km h^{-1}. What is the velocity of the helicopter relative to the train? Give your answer as a speed on a certain bearing.

3 A liner is moving north-west at 20 knots when a powerboat passes close by the stern travelling north-east at 35 knots. What is the velocity of the liner relative to the powerboat?

4 A snooker ball has velocity $\begin{bmatrix} 20 \\ 0 \end{bmatrix}$ cm s^{-1} and a second ball has a velocity of $\begin{bmatrix} 20 \\ 6 \end{bmatrix}$ cm s^{-1}. What is the velocity of the second ball relative to the first? If the first ball started at the point with coordinates $(0, 60)$ and the second started simultaneously at the origin, what time will elapse before they collide? (Assume that neither ball is interrupted.)

5 A dodgem car is travelling west at 3 m s^{-1} when it is struck by another one travelling south at 4 m s^{-1}. What is the velocity of the second one relative to the first just before impact?

6 An Inter-City train is travelling due west at 100 km h^{-1} as it passes under a bridge. On the bridge a London Transport train is moving south-west at 40 km h^{-1}. What is the velocity of the London Transport train relative to the Inter-City?

5. ADDING VELOCITIES

In the previous section we saw that we can find a *relative* velocity vector by subtracting one velocity vector from another:

(velocity of A relative to B) = (velocity of A) − (velocity of B).

In many practical applications we *know* the velocity of B and the relative velocity, and we wish to *find* the velocity of A.

Example 1

A man wishes to row from the north to the south bank of a straight stretch of river in which the current is flowing at 3 m s⁻¹ due east. The river is 100 m wide and the man can row at 2 m s⁻¹. If he rows steadily towards the opposite bank, in what direction and with what speed will he actually travel, and how far downstream from his starting-point will he arrive?

We know the velocity of the water: 3 m s⁻¹ due east, which we may write as $\begin{bmatrix} 3 \\ 0 \end{bmatrix}$.

We *want* to know the velocity of the man in the boat: let us call it $\begin{bmatrix} x \\ y \end{bmatrix}$. We also know the velocity of the boat relative to the water: 2 m s⁻¹ due south, which we can write as $\begin{bmatrix} 0 \\ -2 \end{bmatrix}$. Now we use the formula for relative velocity:

(velocity of boat relative to water) = (velocity of boat) − (velocity of water);

$$\begin{bmatrix} 0 \\ -2 \end{bmatrix} = \begin{bmatrix} x \\ y \end{bmatrix} - \begin{bmatrix} 3 \\ 0 \end{bmatrix}$$

from which we deduce that

$$\begin{bmatrix} x \\ y \end{bmatrix} = \begin{bmatrix} 0 \\ -2 \end{bmatrix} + \begin{bmatrix} 3 \\ 0 \end{bmatrix} = \begin{bmatrix} 3 \\ -2 \end{bmatrix}.$$

We can illustrate this with a vector triangle (Figure 19). Careful examination

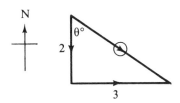

Figure 19

of this triangle shows that it 'fits with common sense'; the velocity of 2 m s⁻¹ southwards is due to the rower's efforts, the velocity of 3 m s⁻¹ eastwards is due to the current and the sum vector shows the final result – the man travels diagonally across the river.

The magnitude (length) of the sum vector gives us his actual speed: by Pythagoras' theorem this is $\sqrt{(2^2 + 3^2)} = 3.6$ m s⁻¹ approximately. By trigonometry, $\theta° = \tan^{-1}\tfrac{3}{2} = 56°$ to the nearest degree, so that he actually travels on a bearing of about $(180° - 56°) = 124°$.

To find how far downstream he lands it is simplest to argue as follows: If there were no current, the boat would go straight across at 2 m s⁻¹, so that it would travel the 100 m width of the river in 50 seconds. Now the current makes no difference to his crossing time, since its only effect is to sweep the boat downstream, so that it still takes him 50 seconds to cross even when the current is flowing. Since the current is flowing at 3 m s⁻¹, then during these 50 seconds it will sweep him downstream 150 metres, so that he will land 150 m east of his starting-point.

Notice how in the above example we rearranged the relative velocity formula to read:

(velocity of boat) = (velocity of boat relative to water)+(velocity of water).

When we want to find the 'true' velocity of the boat it is usually easiest to think of it as a combination of the rower's effort (the velocity relative to the water) and the effect of the current (the velocity of the water).

Example 2

A light aircraft has an airspeed of 250 km h^{-1} (i.e. its speed relative to the air is 250 km h^{-1}). The pilot sets a course due north (i.e. her compass shows that she is 'pointing' due north as she flies). The wind is blowing from the north-west at 80 km h^{-1}. In what direction and at what speed does the aircraft actually travel?

(Velocity of aircraft) = (velocity of aircraft relative to air)+(velocity of air).

The velocity of the aircraft relative to the air is 250 km h^{-1} north; the velocity of the air (i.e. the wind) is 80 km h^{-1} south-east. We draw a diagram showing the sum of these two vectors (Figure 20). $\underset{\sim}{AB}$ represents the relative velocity and

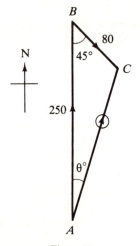

Figure 20

$\underset{\sim}{BC}$ represents the velocity of the air; the sum vector $\underset{\sim}{AC}$ represents the actual velocity of the aircraft. We need to find the magnitude of $\underset{\sim}{AC}$, which will give us the actual speed of the aircraft, and the angle $\theta°$, which will give us the bearing on which it actually travels.

From an accurate scale drawing we find that $AC \approx 200$ and $\theta \approx 16$.

So the aircraft travels at about 200 km h^{-1} on a bearing of about 16°.

Exercise E

1   A helicopter is flying on a course due east with an airspeed of 120 km h^{-1}. The wind is blowing from the north at 50 km h^{-1}. In what direction and at what speed does the helicopter actually travel?

2   A boat is heading west at 12 knots. The current is running at 8 knots in a north-westerly direction. In what direction and at what speed does the boat actually travel?

3   A train is travelling at 80 km h^{-1} when a child throws an empty soft-drink can out of the window at right-angles to the train at 15 km h^{-1}. If the can strikes an unfortunate passer-by, at what speed does it strike him? (Neglect the effect of wind resistance.)

4   A river is flowing due north at 5 km h^{-1}. A swimmer who can swim at 2 km h^{-1} dives in from the west bank and swims towards the opposite bank. In what direction and with what speed does she actually travel? If the river is 100 m wide, how long will it take her to cross, and how far north of her starting-point will she land?

5   Icarus is flying on a bearing of 060° with an airspeed of 20 km h^{-1}. The wind is blowing from the west at 30 km h^{-1}. On what bearing and at what speed does he actually travel?

6   An interceptor jet plane is travelling at 1200 km h^{-1} when it fires a missile at a velocity relative to the plane of 500 km h^{-1} at 30° to the line of flight. At what speed does the missile actually travel?

6. POINTS OF VIEW

We have seen that the phrase 'the velocity of *A* relative to *B*' refers to the velocity of *A* from the point of view of a (sometimes imaginary) person who is moving with *B*; for example, the velocity of a swimmer relative to the water is her speed and direction of motion as seen by someone in a boat which is drifting with the current, and the velocity of a missile relative to an aircraft is its speed and direction of motion as seen by someone in the aircraft.

In the last exercise the phrase 'at what speed does it actually travel' carried the unwritten words 'from the point of view of someone on the ground' or 'from the point of view of someone on dry land.' We have made the unstated assumption that the ground, i.e. the Earth, is in some way fixed and that the 'actual' velocity of an object is its velocity relative to the Earth. This is all very well for motion taking place on or near the surface of the Earth, but we know that in fact the Earth *is* moving and that when scientists calculate the trajectory of a space probe to Jupiter they have to take into account the velocity of the Earth relative to Jupiter. We cannot really talk about the 'actual' velocity of an object at all; *every* velocity is relative to *something*, and it is often helpful to make this clear when we write down the solutions of problems. For example:

(velocity of boat relative to bank) = (velocity of boat relative to water)
$+$(velocity of water relative to bank);

(velocity of missile relative to ground) = (velocity of missile relative to aircraft)
$+$(velocity of aircraft relative to ground);

(velocity of probe relative to Jupiter) = (velocity of probe relative to Earth)
+ (velocity of Earth relative to Jupiter).

In order to start solving relative velocity problems we need first to construct sentences such as these, at least in our minds. The 'shape' of such sentences is easier to see if we write them in a shortened form:

boat/bank   = boat/*water*   + *water*/bank;
missile/ground = missile/*aircraft* + *aircraft*/ground;
probe/Jupiter = probe/*Earth*   + *Earth*/Jupiter.

7. 'AIMING OFF'

Example 3

A river is flowing east at 4 m s⁻¹. A boat which travels through the water at 8 m s⁻¹ leaves the south bank and heads due north. In what direction will it travel relative to the bank?

This is a type of problem we have already met:

(boat relative to bank) = (boat relative to water) + (water relative to bank).

The velocities are shown in Figure 21.

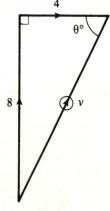

Figure 21

By trigonometry, $\theta° = \tan^{-1} \frac{8}{4} = 63°$ to the nearest degree, so that the boat travels in a direction making an angle of about 63° with the bank. We could also calculate its speed relative to the bank:

$$v = \sqrt{(8^2 + 4^2)} = 8.9 \text{ to 2 s.f.}$$

so that someone watching from the bank would see the boat travelling at about 8.9 m s⁻¹.

Clearly the boat is being carried downstream. Suppose that the skipper wanted to land at the point on the north bank immediately opposite her starting-point;

she knows that if she heads the boat due north the current will carry her east of her objective, so she 'aims off', that is, she heads the boat slightly upstream. In mathematical terms, she wants to make her vector triangle such a shape that the *sum vector* points north.

How can she do this? The vector representing the velocity of the water relative to the bank is something over which she has no control – it must remain a vector of magnitude 4 pointing east. Assuming that she keeps her *speed* relative to the water at 8 m s^{-1}, all she can do is change the *direction* of the vector representing her velocity relative to the water by pointing it west of north: she is planning to get a vector triangle looking like Figure 22.

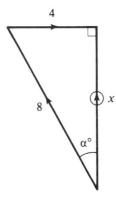

Figure 22

The velocity of the water relative to the bank is 4 m s^{-1} east, as it must be; the sum vector, giving the boat's velocity relative to the bank, is pointing due north, as required; the *magnitude* of the boat's velocity relative to the water is 8 m s^{-1}, as given – all that remains is to find the *direction* of this vector:

$$\alpha° = \sin^{-1} \tfrac{4}{8} = 30°$$

so that if the skipper 'aims off' by 30° upstream she will actually travel due north relative to the bank.

Notice that the magnitude of the sum vector is given by

$$x = \sqrt{(8^2 - 4^2)} = 6.9 \text{ approximately,}$$

so that the watcher on the bank will see the boat travelling more slowly than in the previous case.

Example 4
An aircraft whose airspeed (i.e. its speed relative to the air) is 500 km h^{-1} wishes to travel from point P to point Q which is 1000 km due west of P. If the wind is blowing from the south-west at 100 km h^{-1}, in what direction must the pilot head, and how long will the journey take?

(Aircraft relative to ground)

$$= \text{(aircraft relative to air)} + \text{(air relative to ground)}.$$

If the pilot heads west, i.e. he sets the velocity of the aircraft relative to the *air* due west, the vector triangle will look like Figure 23. The wind is carrying

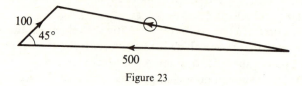

Figure 23

him north of the direction he wants to go, so he must 'aim off' somewhat south. The correct vector triangle is as in Figure 24. (Notice that the speed and direction

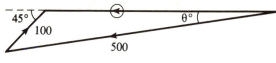

Figure 24

of the wind are as given, the speed of the aircraft relative to the air is as given, and the direction of the sum vector is west, as required.) The rest of the solution is a matter of calculation or accurate drawing. Here is one way of calculating a solution:

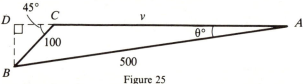

Figure 25

Referring to Figure 25,

$$BD = 100 \sin 45° \approx 70.71; \quad \theta° = \sin^{-1}\frac{70.71}{500} \approx 8.1°,$$

so that the pilot must head the aircraft on a bearing of about 262°.

$$AD = 500 \cos \theta° \approx 494.97, \quad DC = 100 \cos 45° \approx 70.71,$$

so $$AC \approx 424.26,$$

that is, the speed of the aircraft relative to the ground is about 424 km h⁻¹. Since the length of the journey is 1000 km, it will take

$$\frac{1000}{424.26} = 2.4 \text{ hours to 2 s.f.}$$

Exercise F

1 A helicopter which can fly at 150 km h^{-1} in still air is to fly from airport P to airport Q which lies 100 km east of P. The wind is blowing from the south at 40 km h^{-1}. In what direction should the helicopter head in order to fly straight from P to Q? What will be its speed relative to the ground? How long will the journey take?

2 A boy who can swim at 4 km h^{-1} is beside a river flowing at 3 km h^{-1}. At what angle to the bank must he swim
   (*a*) to travel straight across,
   (*b*) to intercept an object floating past, just level with him as he dives in?
   (SMP)

3 An aircraft with an airspeed of 200 km h^{-1} has to fly to an airport south-west of its present position. If the wind is blowing at 70 km h^{-1} from the east, find the course which the pilot must set, and the speed of the aircraft relative to the ground.

4 A trainee pilot leaves Amwick aerodrome to fly to Burford, which lies 250 km away on a bearing of 060°. He flies at an airspeed of 250 km h^{-1} on a course of 060°, having failed to allow for a steady wind blowing at 60 km h^{-1} from the north.
   (*a*) Calculate where he will be in relation to Burford after one hour.
   (*b*) If he now correctly alters course and flies directly to Burford, calculate how long it will take him, assuming the same wind velocity and airspeed.
   (*c*) Find the course he should set to return to Amwick from Burford, still assuming the same wind velocity and airspeed. (MEI)

5 A canoeist who can paddle at 2 m s^{-1} in still water is on a river which is flowing at 1.5 m s^{-1}. He wishes to make a 'ferry-glide', that is, to go straight across the river on a track perpendicular to banks and current. Calculate the direction in which he must paddle. How long will he take to cross if the river is 60 m wide? (SMP)

6 A ship is to travel 30 nautical miles due north. If her speed in still water is 8 knots and there is a current of 3 knots flowing in a direction 300°, find the direction in which the ship must head and the time the journey will take.

7 A helicopter flies at 150 km h^{-1} in still air. On a certain day the wind is blowing at 30 km h^{-1} from the east.
   (*a*) How long altogether would the helicopter take to fly directly to a point 90 km east and then back?
   (*b*) Find on what bearing the helicopter should head in order to fly directly to a point 90 km due north, and how long this journey will take.
   (*c*) On what bearing should the helicopter head in order to fly due south? (MEI)

8 A river which is 80 m wide is flowing at 2.5 m s^{-1}, and a boy sets out from a point O rowing at 1.5 m s^{-1} relative to the water to cross the river. On an accurate diagram mark the set of all possible positions the boy may be in after 20 seconds.
   From your diagram find:
   (*a*) the actual distance travelled by the boy in these 20 seconds if he heads directly across the river;
   (*b*) the possible directions in which he could head the boat so that his actual track makes an angle of 20° with the bank;
   (*c*) the direction that he should head so as to be carried downstream as little as possible. (MEI)

8. SUMMARY

A vector can be represented to some suitable scale by a directed line-segment. Equal line-segments in the same direction all represent the same vector.

Vectors can be added by the triangle law. If PQ and QR (Figure 26) are 'head-to-tail' line-segments representing vectors, then

$$PQ + QR = PR.$$

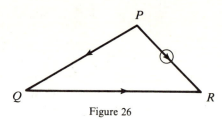

Figure 26

The negative of a vector **a** is a vector of the same magnitude as **a** but in the opposite direction.

Subtraction of vectors is defined by

$$\mathbf{a} - \mathbf{b} = \mathbf{a} + (^-\mathbf{b}).$$

If vectors **a** and **b** are represented by QA and QB, then the vector $\mathbf{a} - \mathbf{b}$ is represented by BA.

If the velocities of P and Q relative to the same base (such as the ground) are **p** and **q**, then the velocity of Q relative to P is $\mathbf{q} - \mathbf{p}$.

Problems on relative velocity can be solved by constructing a triangle of vectors using the principle:

(velocity of P relative to Q) + (velocity of Q relative to the Earth)
$$= \text{(velocity of } P \text{ relative to the Earth)}$$

or suitable variations on this formula.

Miscellaneous exercise

1   $\mathbf{a} = \begin{bmatrix} 3 \\ 1 \end{bmatrix}, \mathbf{b} = \begin{bmatrix} 1 \\ -2 \end{bmatrix}.$

Represent the vectors **a** and **b** on a diagram by drawing line-segments QA and QB respectively. Complete the parallelogram $OAPB$ and draw the diagonals OP and AB intersecting at X.

(i) Write down in column-matrix form the vectors represented by the displacements: (*a*) QP; (*b*) QX; (*c*) AX.

(ii) Express in terms of the vectors **a** and **b** each of the three displacements given in (i).   (MEI)

2   *ABCDEF* is a regular hexagon, centre O. The vectors QD and QC are denoted by **p** and **q** respectively.

(i) Write down the following vectors in terms of **p** and **q**:
(*a*) ED; (*b*) FC; (*c*) EC.

(ii) The vector **p** can be written in the form $\underset{\sim}{A}O$ as well as in the form $\underset{\sim}{O}D$. Using only the letters already mentioned, write down the following vectors in a similar form:

(a) **p**+**q**; (b) **q**−**p**; (c) 2**p**−**q**.

(iii) The point X is such that $\underset{\sim}{B}X = \frac{1}{2}(3\mathbf{p}-\mathbf{q})$. On your diagram mark clearly the position of X.   (MEI)

3 Glider P is travelling due north at 60 km h^{-1} and a glider Q is travelling north-west at 40 km h^{-1}. Find the velocity of P relative to Q.

4 An aircraft heads due east at a speed of 200 km h^{-1}, but a wind blowing at a speed of 80 km h^{-1} in the direction of the bearing 220° carries it off south of its course. Find the bearing on which the aircraft actually travels.

Find the bearing on which the aircraft should be heading in order to travel due east allowing for the wind. In this case find how long the journey will take, given that the aircraft's destination is 800 km from its starting-point.   (MEI)

5 A motorboat which travels at 20 km h^{-1} heads due south for a marker buoy which lies 40 km due south of the boat's starting point.

Unfortunately a current running at 10 km h^{-1} from the south-west carries the boat off course. Find:

(a) the distance and bearing of the buoy from the boat after 1 hour;

(b) the course the boat should have taken in order to travel due south, and the time it would then have taken to reach the buoy.   (MEI)

6 (a) A light aircraft is being flown on a certain course. The wind velocity is given by the vector $\begin{bmatrix} 30 \\ 20 \end{bmatrix}$ km h^{-1}, and the velocity of the aircraft relative to the air is $\begin{bmatrix} 50 \\ 60 \end{bmatrix}$ km h^{-1}. Find the column vector which gives the velocity of the aircraft relative to the ground. What is the speed and the direction of the aircraft?

(b) The wind velocity now decreases to half its former value, and at the same time the pilot wishes to change course so that he is flying due north with a ground speed of 100 km h^{-1}. Find his new air velocity.

(c) On another occasion the pilot wishes to fly on a bearing of 100°. The wind velocity is $\begin{bmatrix} -10 \\ 30 \end{bmatrix}$ km h^{-1}, and his airspeed is to be 80 km h^{-1}. Find the direction in which he should fly, and give his air velocity in column-vector form. (MEI)

7 A man wishes to cross a river, travelling as directly across as possible, in a motorboat which travels at a steady speed of 4.0 m s^{-1}. By calculation or by drawing:

(a) find at what angle to the bank, to the nearest degree, he should head if the tide is flowing at 3.0 m s^{-1};

(b) show, with reasons, that if the tide is flowing at 5.0 m s^{-1} the best he can do is to head at an angle of about 37° to the bank.

Verify that it will take him about 10% longer to cross in case (b) than in case (a), and calculate how many metres downstream he will land if the banks are 1.2 km apart.

4

Polynomials and curve sketching

1. SKETCHING CURVES

Example 1
Sketch the graph of $y = (x-2)(x+1)$.

This type of graph was investigated thoroughly in Chapter 2. As the quadratic function is in factored form we can immediately say that

$$y = 0 \quad \Leftrightarrow \quad x-2 = 0 \quad \text{or} \quad x+1 = 0$$
$$\Leftrightarrow \quad x = 2 \quad \text{or} \quad x = {}^-1.$$

The graph therefore crosses the x-axis at $(2, 0)$ and $({}^-1, 0)$.

We can discover where the curve lies above the x-axis and where it lies below the x-axis by considering Table 1.

Table 1

Values of x	${}^-3$	${}^-2$	${}^-1$	0	1	2	3	4
Sign of $(x-2)$	$----$	$----$	$----$	$---$	$---$	$0++$	$+++$	$++$
Sign of $(x+1)$	$----$	$----$	$0++$	$+++$	$+++$	$+++$	$+++$	$++$
Sign of $(x-2)(x+1)$	$++++$	$++++$	$0--$	$---$	$---$	$0++$	$+++$	$++$

The graph can now be sketched as in Figure 1.

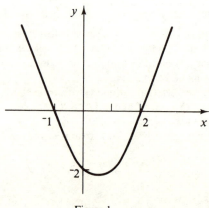

Figure 1

Note that we have 'sketched' the graph and not 'plotted' it. In many cases in mathematics we need to know the shape of a graph without requiring the accuracy of a 'plot'. In this case we plotted three points. The two points where the x-axis is crossed were easily obtained because of the factorised form of the quadratic function. The point where the graph crosses the y-axis is easily found by putting $x = 0$. These three points form a framework for the sketch. Table 1 helps us to complete the sketch.

Note that the sketches in Figure 2 would also fit the same conditions.

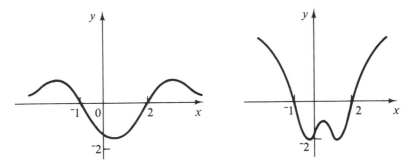

Figure 2

How do we know we have chosen the correct one?

Example 2

Sketch the graph of the function $y = (x+2)(x-1)(x-2)$ and write down the solution set of the *inequality* $(x+2)(x-1)(x-2) > 0$.

Proceeding as in Example 1,

$$y = 0 \quad \Leftrightarrow \quad x+2 = 0 \quad \text{or} \quad x-1 = 0 \quad \text{or} \quad x-2 = 0$$
$$\Leftrightarrow \quad x = {}^-2 \quad \text{or} \quad x = 1 \quad \text{or} \quad x = 2.$$

The graph therefore crosses the x-axis at $({}^-2, 0)$, $(1, 0)$ and $(2, 0)$. When $x = 0$, $y = 4$, hence the graph crosses the y-axis at $(0, 4)$. We obtain Table 2.

Table 2

Values of x	$^-3$	$^-2$	$^-1$	0	1	2	3	4
Sign of $x+2$	$----- $	0 +++	+++	+++	+++	+++	+++	+
Sign of $x-1$	$------ $	$----- $	$---- $	$--- $	0 ++	+++	+++	+
Sign of $x-2$	$------ $	$----- $	$---- $	$--- $	$--- $	0 ++	+++	+
Sign of $(x-2)(x-1)(x-3)$	$----- $	0 +++	+++	+++	+++	0 $--$	0 ++	+++	+

We can now complete the sketch (Figure 3).

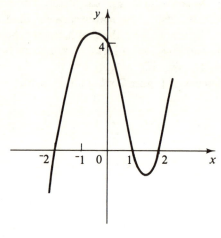

Figure 3

Looking at the table of signs or the graph we see that

$$(x+2)\,(x-1)\,(x-2) > 0 \quad \Leftrightarrow \quad x \in \{x : {}^{-}2 < x < 1\} \cup \{x : x > 2\}.$$

Note that, as in Example 1, other sketches would fit the same framework; for example, those of Figure 4.

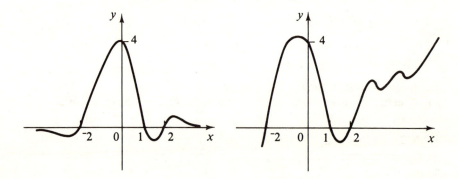

Figure 4

How do we know these sketches are not correct?

Example 3

A certain function is zero only when $x = {}^-2$, $x = {}^-1$ and $x = 3$ and positive only when $x < {}^-2$ and for values of x between $^-1$ and 3. Sketch a simple possible graph for the function.

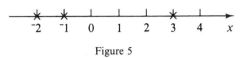

Figure 5

From the zeros given we know that the graph crosses the x-axis at the points marked × in Figure 5.

The other information tells us when the curve is above the x-axis and enables us to complete the sketch as shown in Figure 6.

Notice that there is no point in putting any scale on the y-axis.

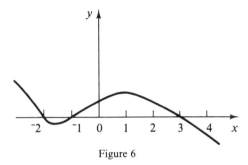

Figure 6

Exercise A

In each of questions 1–5 and 6–10, sketch the graph of the function and write down the solution set of the inequality given.

1   $y = (x-3)(x-1)$, $(x-3)(x-1) > 0$.
2   $y = (x+1)(2x-3)$, $(x+1)(2x-3) \geqslant 0$.
3   $y = (x-2)^2$, $(x-2)^2 < 0$.
4   $y = (x-1)(x+2)(x+3)$, $(x-1)(x+2)(x+3) > 0$.
5   $y = (3x-1)(x+2)(3-x)$, $(3x-1)(x+2)(3-x) < 0$.

6   $y = (x-2)(x-5)$, $(x-2)(x-5) > 0$.
7   $y = (2x-1)(x+2)$, $(2x-1)(x+2) \leqslant 0$.
8   $y = (x-4)(x+2)(3-x)$, $(x-4)(x+2)(3-x) \leqslant 0$.
9   $y = (2x-1)(x-3)^2$, $(2x-1)(x-3)^2 > 0$.
10   $y = (x-2)^3$, $(x-2)^3 < 0$.

In each of the following questions, sketch simple curves which could be graphs of functions satisfying the given information.

11   Zero only when $x = {}^-1$ and 2; positive only for values of x between $^-1$ and 2.
12   Zero only when $x = 1$, 2 or 3; positive only when x is less than 1 or between 2 and 3.
13   Zero only when $x = {}^-2$, $\frac{1}{2}$ or 1; negative only when x lies between $^-2$ and $\frac{1}{2}$ or is greater than 1.

14   Zero only when $x = 0$ or 5; negative only between these two values.
15   Zero only when $x = {}^-1$, 0 or 1; positive only when x lies between $^-1$ and 0 or is greater than 1.
16   Zero only when $x = {}^-1$ and 4; positive only when x is greater than 4.

2. POLYNOMIALS

If we take the expression $(x+2)(x-1)(x-2)$ and multiply out we obtain the following:

$$\begin{aligned}
(x+2)(x-1)(x-2) &= (x+2)(x^2-3x+2) \\
&= x(x^2-3x+2)+2(x^2-3x+2) \\
&= x^3-3x^2+2x+2x^2-6x+4 \\
&= x^3-x^2-4x+4.
\end{aligned}$$

Notice how the 'like terms', for example $^-3x^2$ and $2x^2$, have been collected together.

Sometimes a careful use of the commutative law will shorten the working. For example, in this case, $(x+2)(x-1)(x-2)$ can be written as

$$\begin{aligned}
(x-1)&(x-2)(x+2) \\
&= (x-1)(x^2-4) \quad \text{(multiplying the second pair of factors first)} \\
&= x^3-x^2-4x+4, \text{ as before.}
\end{aligned}$$

An expression such as x^3-x^2-4x+4 is called a *polynomial* in x. Other examples of polynomials are

$$(a) \ 3+\tfrac{7}{8}x-7x^2+x^4, \quad (b) \ 5t^7-\tfrac{1}{3}t^4-6, \quad (c) \ x^2-5x+2, \quad (d) \ 2+5x.$$

Note that (b) is a polynomial in t.

A polynomial is constructed by taking a symbol (usually x though not necessarily so) and forming sums, products and differences with x and given numbers, subject to the rules of ordinary algebra. The powers of x can be sorted into order (either ascending or descending) and the expression written in its simplest form.

In the polynomial $3+\tfrac{7}{8}x-7x^2+x^4$ we say that the *coefficient* of x is $\tfrac{7}{8}$ and the coefficient of x^2 is $^-7$. What is the coefficient of x^4? What is the coefficient of x^3? 3 is called the constant term since it is independent of any value given to the symbol used.

The *degree* of the polynomial is the index of the highest power appearing. So the degree of (a) above is 4 and the degree of (b) is 7. What are the degrees of (c) and (d)?

Two polynomials in x are *equal* if they have exactly the same coefficients and therefore the same degree; in other words, if they are identical.

In Chapter 2 we met polynomials of degree 2, or *quadratic* polynomials. The polynomial at the start of this section is of degree 3 and is often referred to as a *cubic* polynomial. Polynomials of degree 1 (e.g. $2+5x$) are known as *linear* polynomials.

Exercise B

In each of questions 1–5 and 6–10, state (*a*) the degree, (*b*) the constant term, (*c*) the coefficients of the terms indicated.

1 x^2+5x+6; x^2, x.
2 x^3-7x+1; x^2, x.
3 x^5-6x^3+4x; x^2, x^3.
4 $5+x^2-3x^4$; x^2, x^4.
5 $t^5+4t^3-5t^2+t-3$; t^3, t

6 x^2-7x-2; x^2, x.
7 x^3+4x^2-7x; x^2, x.
8 x^7-x^4+19; x^2, x^4.
9 $13-17x-19x^2-23x^3$; x, x^2.
10 $3+5z-2z^2+9z^3$; z^2, z^3

In questions 11–13 and 14–16, multiply out the expressions to form cubic polynomials.

11 $(x-2)(x+1)(x+3)$.
12 $(x+1)(2-x)(3x+1)$.
13 $(2x-1)(1-x)(4-3x)$.

14 $(x-3)(x-1)(x+4)$.
15 $(x-4)(x-5)(6x+7)$.
16 $(3-x)(3-x+1)(x+3)$.

In questions 17–21, multiply out the expressions in the two ways indicated.

17 $[(3x+1)(x-2)](x+3)$; $(3x+1)[(x-2)(x+3)]$.
18 $(x-3)[(x+3)(2x+1)]$; $[(x-3)(x+3)](2x+1)$.
19 $(x+1)[(x-3)(x-1)]$; $(x-3)[(x+1)(x-1)]$.
20 $(t+5)[(t+7)(t-5)]$; $[(t+5)(t-5)](t+7)$.
21 $x[(x-2)(x+3)]$; $[x(x-2)](x+3)$.

In questions 22–27 multiply out the expressions, using the associative and commutative laws where possible to simplify the working.

22 $(x-1)(x+1)(x+3)$.
23 $(x-2)(2x+1)(x+2)$.
24 $(x-1)(x+2)(3x-5)$.
25 $(2x+3)(x+1)(2x-3)$.
26 $(x-2)(3x+1)(x-2)$.
27 $(x-3)(x+1)(x+3)(x-1)$.

3. ADDITION AND SUBTRACTION OF POLYNOMIALS

We can add and subtract two polynomials P and Q to form their sum (written $P+Q$) and difference (written $P-Q$) respectively. In each case we combine the 'like terms'.

Example 4
Find $P+Q$ and $P-Q$ when $P=2+5x^2+x^3-9x^4$ and $Q=x^5+4x^4-3x^2-x+1$.

We begin by writing both polynomials either in ascending or descending order, choosing whichever we wish. Thus

$$P+Q=(2+5x^2+x^3-9x^4)+(1-x-3x^2+4x^4+x^5)$$
$$=(2+1)+(^-x)+(5x^2-3x^2)+x^3+(^-9x^4+4x^4)+x^5$$
$$=3-x+2x^2+x^3-5x^4+x^5.$$
$$P-Q=(2+5x^2+x^3-9x^4)-(1-x-3x^2+4x^4+x^5)$$
$$=1+x+8x^2+x^3-13x^4-x^5.$$

Exercise C

In questions 1–3 and 4–6, add the two polynomials and also subtract the second from the first.

1 $2-6x-x^3$; x^3+5x^2+x-7.

2 $3x^5-2x^3+8x^2+5x-7$; $7x^4-9x^3-3x^2+2x+8$.

3 $5t^3+2t+1$; $3t^2+6t+5$.

---

4 $7-2x+5x^2+3x^4$; $5x^4-7x^3+3x+6$.

5 $t-4t^2+2t^3$; $4-7t^2+t^3$.

6 z^4+2z^2-6; $1-z+z^2+3z^3+z^4$.

---

7 If $P=5x^4-3x^2+2x-1$ and $Q=2x^3+5x+4$, find $P+Q$ and $P-Q$. Hence find $(P+Q)+(P-Q)$ and verify that it is equal to $2P$.

8 Find the sum of $6z^5-5z^4+2z^3$ and $2z^3-3z^2+2z-1$. Find also the result when the second polynomial is subtracted from the first.

9 Find the sum of $9x^4-4x^3+7x^2-x+17$, $7x^4+x^3+5x+3$ and $14-6x-5x^3+2x^4$. Find also the result when the first polynomial is subtracted from the sum of the other two.

4. MULTIPLICATION OF POLYNOMIALS

Suppose $P=x^2-2x+3$ and $Q=1+x-4x^3$: To find the *product* of P and Q (written $P \times Q$), we first write both polynomials in the same order (ascending or descending). We can set out the working in several ways, the most useful of which are the two following.

$$
\begin{aligned}
P \times Q &= (3-2x+x^2)\,(1+x-4x^3) \\
&= 3(1+x-4x^3)-2x(1+x-4x^3)+x^2(1+x-4x^3) \\
&= 3+3x-12x^3-2x-2x^2+8x^4+x^2+x^3-4x^5 \\
&= 3+x-x^2-11x^3+8x^4-4x^5.
\end{aligned}
$$

Alternatively:

	1	x	$0x^2$	$^-4x^3$
3	3	$3x$	$0x^2$	$^-12x^3$
^-2x	^-2x	$^-2x^2$	$0x^3$	$8x^4$
x^2	x^2	x^3	$0x^4$	$^-4x^5$

$$
\begin{aligned}
P \times Q &= 3+(-2+3)\,x+(1-2+0)\,x^2+(1+0-12)\,x^3+(0+8)\,x^4-4x^5 \\
&= 3+x-x^2-11x^3+8x^4-4x^5.
\end{aligned}
$$

Notice how terms are collected *diagonally* and notice also the column with zero coefficients since x^2 does not appear in Q.

Sometimes we are only interested in the coefficient of a particular term in the product. We can pick this out by finding the appropriate entries in the table above.

Example 5
Find the term in x^3 in the product of $x^2 - 5x + 2$ and $x^3 + 3x^2 - 4x + 6$.

We begin the 'table' arrangement for multiplication:

	x^3	$3x^2$	^-4x	6
x^2				
^-5x				
2				

Looking along the first row we see that x^3 will arise when we multiply x^2 and ^-4x.

	x^3	$3x^2$	^-4x	6
x^2			$^-4x^3$	
^-5x				
2				

We know that like terms appear diagonally, so we can now put in all terms in x^3 and find their sum.

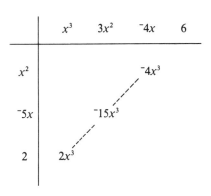

	x^3	$3x^2$	^-4x	6
x^2			$^-4x^3$	
^-5x		$^-15x^3$		
2	$2x^3$			

giving $(2 - 15 - 4) x^3 = {}^-17x^3$.

Exercise D

In questions 1–3 and 4–6, multiply together the two polynomials given.

1 $x^2 - 5x + 3$; $2x - 5$.   **2** $1 + 3x + x^2$; $x^2 - 4x + 2$.

3 $t^3 - 2t^2 + 5t$; $4 - t - t^2$.

4 $2x^2 - 5x + 6$; $3x + 1$.   **5** $x^3 - 7x + 2$; $2x + x^2 - x^3$.

6 $z^4 - 2z + 1$; $3z^2 + 2$.

7 In each of the following questions find, in the product of the two polynomials, the coefficient of the term given, carrying out as little working as necessary.

   (*a*) $x^3 - 5x^2 + 2x + 3$; $x^2 + 2x - 5$: x^4:

   (*b*) $x^2 - x - 1$; $x^4 - 4x^2 + 3x + 2$: x^2.

   (*c*) $2 + 4t - t^3$; $t^5 + 3t^4 - 2t^2$: t^5.

   (*d*) $2x + 1$; $3x^3 - 5x^2 + 4x - 3$: x^3.

8 Multiply out the following:

   (*a*) $(x+1)^2$;   (*b*) $(x+1)^3$;   (*c*) $(x+1)^4$;   (*d*) $(x+1)^5$.

9 Multiply out the following:

   (*a*) $(2x-1)^4$;   (*b*) $(3-x)^3$;   (*c*) $(3x+2)^3$.

10 Multiply out:

   (*a*) $(a+b)^2$;   (*b*) $(a+b)^3$.

5. PASCAL'S TRIANGLE

The pattern of coefficients you should have found in Exercise D question 8 is

$$
\begin{array}{ccccccc}
 & & 1 & 2 & 1 & & \\
 & 1 & 3 & 3 & 1 & & \\
1 & 4 & 6 & 4 & 1 & & \\
1 & 5 & 10 & 10 & 5 & 1 &
\end{array}
$$

Looking at this pattern, can you guess the results of multiplying out $(x+1)^6$ and $(x+1)^7$?

The coefficients form part of a well-known pattern of numbers known as *Pascal's triangle*.

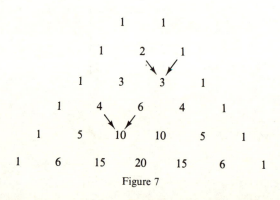

Figure 7

Each number inside the table is the sum of the two numbers directly above and either side of it. For example in Figure 7, $2+1 = 3$ and $6+4 = 10$.

If we take the expansion of $(x+1)^4$ and multiply by $(x+1)$,

$$
\begin{array}{c|cccccc}
 & x^4 & + & 4^3 & + & 6x^2 & + & 4x & + & 1 \\
\hline
 & & & & & & & & & \\
x & x^5 & + & 4x^4 & + & 6x^3 & + & 4x^2 & + & x \\
 & & & & & & & & & \\
1 & x^4 & + & 4x^3 & + & 6x^2 & + & 4x & + & 1 \\
\end{array}
$$

we obtain

$$(x+1)^5 = x^5 + (1+4)x^4 + (4+6)x^3 + (6+4)x^2 + (4+1)x + 1.$$

This shows why the coefficients are given by Pascal's triangle:

$$1+4 = 5, \quad 4+6 = 10, \quad 6+4 = 10, \quad 4+1 = 5,$$

exactly as the 6th row of Figure 7 was obtained from the 5th row.

Your answers to question 10 of Exercise D should have been

$$(a+b)^2 = a^2 + 2ab + b^2,$$
$$(a+b)^3 = a^3 + 3a^2b + 3ab^2 + b^3.$$

In these two examples the numbers 1, 2, 1 and 1, 3, 3, 1 come from Pascal's triangle and we see that the expressions are symmetrical in a and b. However, if we replace b by x, we get, for example,

$$(a+x)^3 = a^3 + 3a^2x + 3ax^2 + x^3,$$

so that, if we think of this as a polynomial in x, the coefficients include powers of a as well. We notice that the powers of a decrease as the powers of x increase, in a symmetrical way. Hence, using the line 1, 4, 6, 4, 1 from Pascal's triangle we can write

$$(a+x)^4 = a^4 + 4a^3x + 6a^2x^2 + 4ax^3 + x^4.$$

We are now in a position to write down the 'expansions' of similar expressions very quickly.

Example 6

Expand as polynomials in x: (a) $(2+x)^3$; (b) $(3-x)^4$; (c) $(2x+1)^5$.

(a) $(2+x)^3 = 2^3 + 3(2)^2 x + 3(2)x^2 + x^3$
$= 8 + 12x + 6x^2 + x^3.$

(b) $(3-x)^4 = 3^4 + 4(3)^3(-x) + 6(3)^2(-x)^2 + 4(3)(-x)^3 + (-x)^4$
$= 81 - 108x + 54x^2 - 12x^3 + x^4.$

(c) $(2x+1)^5 = (2x)^5 + 5(2x)^4 \times 1 + 10(2x)^3 \times 1^2 + 10(2x)^2 \times 1^3 + 5(2x) \times 1^4 + 1^5$
$= 32x^5 + 80x^4 + 80x^3 + 40x^2 + 10x + 1.$

Example 7

Expand $(2+x)^5$ in ascending powers of x. By putting x equal to 0.01, find $(2.01)^5$ giving your answer to 4 decimal places.

$$(2+x)^5 = 2^5 + 5 \times 2^4 x + 10 \times 2^3 x^2 + 10 \times 2^2 x^3 + 5 \times 2x^4 + x^5$$
$$= 32 + 80x + 80x^2 + 40x^3 + 10x^4 + x^5.$$

When $x = 0.01$,

$$(2+0.01)^5 = 32 + 80 \times 0.01 + 80 \times 0.0001 + 40 \times 0.000001$$
$$+ 10 \times 0.000\,000\,01 + 0.000\,000\,000\,1$$
$$= 32 + 0.8 + 0.008 + 0.000\,04 + 0.000\,000\,1 + 0.000\,000\,000\,1$$
$$= 32.808\,040\,100\,1$$
$$= 32.8080 \text{ to 4 d.p.}$$

Notice how the terms become very small so that to find the answer correct to 4 decimal places only the first three terms are relevant.

Exercise E

1   Continue Pascal's triangle in Figure 7 for two more rows. Hence write down as polynomials:
   (*a*) $(1+x)^6$;   (*b*) $(1+x)^7$;   (*c*) $(1+x)^8$.

2   Write down as polynomials in x:
   (*a*) $(a+x)^5$;   (*b*) $(a+x)^6$.

3   Using Pascal's triangle, write down and simplify as polynomials the expansions of:
   (*a*) $(4+x)^3$;   (*b*) $(2-x)^3$;   (*c*) $(3x+1)^4$;   (*d*) $(1-x)^5$;   (*e*) $(3-2x)^4$.

4   Expand $(2-x)^4$ as a polynomial in x. Hence, putting $x = 0.1$, find the exact value of 1.9^4 without using a calculator.

5   Expand $(3+x)^5$ as a polynomial in x and by putting $x = 0.02$ find the value of 3.02^5 correct to 4 decimal places. How many terms of the polynomial are relevant to your answer?

6   Expand $(2-x)^6$ as a polynomial in x. Use the first three terms to find an approximate value of 1.97^6. Work out also the fourth term and state the accuracy of your first answer.

7   Expand $(1-3x)^5$ as a polynomial in x and by giving x a suitable value find the value of 0.994^5 correct to 4 significant figures.

6. CUBIC EQUATIONS

In Section 1 we met expressions of the form

$$y = (x+2)(x-1)(x-2).$$

It was a simple matter to solve the equation $y = 0$ for

$$y = 0 \quad \Leftrightarrow \quad (x+2)(x-1)(x-2) = 0$$
$$\Leftrightarrow \quad x+2 = 0 \quad \text{or} \quad x-1 = 0 \quad \text{or} \quad x-2 = 0$$
$$\Leftrightarrow \quad x = {}^-2 \quad \text{or} \quad x = 1 \quad \text{or} \quad x = 2.$$

It was easy to find these solutions because the equation was already in factorised form. It would not have been so easy if we have been given the equation in the alternative form $x^3 - x^2 - 4x + 2 = 0$. In this section we shall investigate a possible method for solving such *cubic equations*.

Example 8
Solve the equations:

(a) $(x-3)(2x^2+7x-4) = 0$; (b) $(2x+1)(x^2+3) = 0$;
(c) $x^3 - 9x = 0$; (d) $(x+4)^4 = 0$.

(a) In this case the expression has already been partially factorised. Noting that $2x^2 + 7x - 4 = (2x-1)(x+4)$ we can write

$$(x-3)(2x^2+7x-4) = 0 \iff (x-3)(2x-1)(x+4) = 0$$
$$\iff x-3 = 0 \quad \text{or} \quad 2x-1 = 0 \quad \text{or} \quad x+4 = 0$$
$$\iff x = 3 \quad \text{or} \quad x = \tfrac{1}{2} \quad \text{or} \quad x = {}^-4.$$

(b) Here we cannot proceed as in (a) because $x^2 + 3$ will not factorise. In fact $x^2 + 3 = 0 \iff x^2 = {}^-3$ which has no real solution. Hence the only solution is when $2x+1 = 0$, that is, $x = {}^-\tfrac{1}{2}$.

(c) $$x^3 - 9x = x(x^2 - 9)$$
$$= x(x-3)(x+3).$$

Hence $\quad x^3 - 9x = 0 \iff x(x-3)(x+3) = 0$
$$\iff x = 0 \quad \text{or} \quad x-3 = 0 \quad \text{or} \quad x+3 = 0$$
$$\iff x = 0 \quad \text{or} \quad x = 3 \quad \text{or} \quad x = {}^-3.$$

(d) $(x+4)^4 = 0 \iff x+4 = 0$ or $x+4 = 0$ or $x+4 = 0$ or $x+4 = 0$,
since all four factors are $(x+4)$. Thus

$$(x+4)^4 = 0 \iff x = {}^-4.$$

We usually say that $(x+4)^4 = 0$ has four equal roots.

Exercise F

Solve the following equations:

1 $(x-3)(2x+5) = 0$.
2 $(2x-1)^2 = 0$.
3 $x^2 - 7x = 0$.
4 $(x-1)(x-4)(x+3) = 0$.
5 $(2x-1)(x^2-4) = 0$.
6 $(x+2)(x^2+3x-10) = 0$.
7 $x(x^2-3) = 0$.
8 $x^3 - 5x^2 + 6x = 0$.

9 $(x+2)(3x-1) = 0$.
10 $(x+2)^2 = 0$.
11 $2x^2 + 3x = 0$.
12 $(x+2)(2x-3)(x+7) = 0$.
13 $(x+2)(x^2-9) = 0$.
14 $(2x-1)(x^2+5x+6) = 0$.
15 $2x(x^2+1) = 0$.
16 $2x^3 + 3x^2 - 2x = 0$.

7. FACTORISATION

Most of the equations in Exercise F were fairly straightforward because the polynomials were already partially factorised. What would be the procedure with the equation $2x^3 + 15x^2 + 22x - 15 = 0$? If we could guess one of the factors of the polynomial we could then proceed as in Example 8(a). For the moment let us assume that we have (correctly!) guessed that $x+3$ is a factor of $2x^3 + 15x^2 + 22x - 15$. Then we can reconstruct the 'table' for multiplication. We can begin

$$
\begin{array}{c|c}
 & \\
\hline
x & 2x^3 \\
3 & \\
\end{array}
$$

and we see that to get the $2x^3$, we must have $2x^2$ above it;

$$
\begin{array}{c|c}
 & 2x^2 \\
\hline
x & 2x^3 \\
3 & \\
\end{array}
\quad \rightarrow \quad
\begin{array}{c|c}
 & 2x^2 \\
\hline
x & 2x^3 \\
3 & 6x^2 \\
\end{array}
$$

and then, multiplying the $2x^2$ by the 3 at the left gives us $6x^2$ in the table as shown.

We have $15x^2$ in total and we know that like terms appear on diagonals, so the next two steps are

$$
\begin{array}{c|cc}
 & 2x^2 & \\
\hline
x & 2x^3 & 9x^2 \\
3 & 6x^2 & \\
\end{array}
\quad \rightarrow \quad
\begin{array}{c|cc}
 & 2x^2 & 9x \\
\hline
x & 2x^3 & 9x^2 \\
3 & 6x^2 & \\
\end{array}
$$

where the $9x^2 = 15x^2 - 6x^2$, and the $9x$ multiplied by the x at the left gives $9x^2$. We continue

$$
\begin{array}{c|cc}
 & 2x^2 & 9x \\
\hline
x & 2x^3 & 9x^2 \\
3 & 6x^2 & 27x \\
\end{array}
\quad \rightarrow \quad
\begin{array}{c|ccc}
 & 2x^2 & 9x & {}^-5 \\
\hline
x & 2x^3 & 9x^2 & {}^-5x \\
3 & 6x^2 & 27x & \\
\end{array}
$$

and, as a final check, we see that $3 \times (^-5)$ gives $^-15$, the correct entry for the bottom right hand corner of the table. We can now write

$$2x^3 + 15x^2 + 22x - 15 = (x+3)(2x^2 + 9x - 5)$$

and then we have, since $2x^2 + 9x - 5 = (2x-1)(x+5)$,

$$2x^3 - 15x^2 + 22x - 15 = 0 \quad \Leftrightarrow \quad (x+3)(2x-1)(x+5) = 0$$
$$\Leftrightarrow \quad x+3 = 0 \quad \text{or} \quad 2x-1 = 0 \quad \text{or} \quad x+5 = 0$$
$$\Leftrightarrow \quad x = {}^-3 \quad \text{or} \quad x = \tfrac{1}{2} \quad \text{or} \quad x = {}^-5.$$

Notice that, in reconstructing the *multiplication* 'table', we have *divided* $2x^3 + 15x^2 + 22x - 15$ by $x+3$.

Exercise G

1 Divide $x^3 + 4x^2 - x - 4$ by $x - 1$.

2 Divide $2x^3 - 3x^2 - 7x + 3$ by $2x + 3$.

3 Factorise $x^3 + 3x^2 - 4x - 12$ given that $x - 2$ is a factor. Hence solve the equation $x^3 + 3x^2 - 4x - 12 = 0$.

4 Factorise $4x^3 - 8x^2 - 11x - 3$ given that $2x + 1$ is a factor. Hence solve the equation $4x^3 - 8x^2 - 11x - 3 = 0$.

5 Divide $x^3 - 2x^2 + 2x + 5$ by $x + 1$.

6 Divide $3x^3 - x^2 + 15x - 5$ by $3x - 1$.

7 Factorise $2x^3 + 7x^2 + 2x - 3$ given that $2x - 1$ is a factor. Hence solve the equation $2x^3 + 7x^2 + 2x - 3 = 0$.

8 Factorise $3x^3 - x^2 + 6x - 2$ given that $3x - 1$ is a factor. Hence solve the equation $3x^3 - x^2 + 6x - 2 = 0$.

9 Given that 1 is a root of the equation $x^3 + 4x^2 - x - 4 = 0$, find the other two roots. (*Hint.* If 1 is a root of the equation then $x - 1$ will be a factor of the polynomial on the left-hand side when the equation is written in zero form.)

10 Given that $\frac{1}{2}$ is a root of the equation $2x^3 - x^2 - 10x + 5 = 0$ find all the roots of the equation. (*Hint.* The factor to use is $2x - 1$.)

11 Given that $^-2$ is a root of the equation $x^3 - 12x - 16 = 0$, find all the roots of the equation.

8. THE FACTOR THEOREM

Questions 9, 10 and 11 of Exercise G give us a clue to our 'inspired' guess in the previous section. In question 9, knowing that 1 is a root of the equation tells us that $x - 1$ must be a factor of the polynomial. Division by this factor followed by factorisation of the quadratic quotient gives us all the solutions.

$$x^3 + 4x - x - 4 = (x - 1)(x^2 + 5x + 4)$$
$$= (x - 1)(x + 1)(x + 4);$$
$$x^3 + 4x - x - 4 = 0 \iff (x - 1)(x + 1)(x + 4) = 0$$
$$\iff x = 1 \quad \text{or} \quad x = {}^-1 \quad \text{or} \quad x = {}^-4.$$

Example 9
Solve the equation $x^3 + 5x^2 - 2x - 24 = 0$.

We want to find factors of the polynomial on the left of the equation and we know that the $^-24$ is formed by multiplying together all the terms in the factors which do not contain x. So if we list all the possible factors of 24 ($\pm 1, \pm 2, \pm 3,$ $\pm 4, \pm 6, \pm 12, \pm 24$) we can see that *possible* factors of the left-hand side are $x - 1, x + 1; x - 2, x + 2;$ etc. But if, for example, $x - 1$ is a factor of the left-hand side, we know that 1 is a root of the equation and, if we write

$$P(x) = x^3 + 5x^2 - 2x - 24,$$

that will mean $P(1) = 0$. However,

$$P(1) = 1 + 5 - 2 - 24 = {}^-20 \neq 0,$$

so $x - 1$ is not a factor of the left-hand side.

We continue to try other possibilities. $P(^-1)$ is obviously no use, but

$$P(2) = 8 + 20 - 4 - 24 = 0.$$

That means that 2 is a root of the equation and so $x - 2$ is a factor of the left-hand side. We can now complete the solution:

$$x^3 + 5x^2 - 2x - 24 = (x-2)(x^2 + 7x + 12)$$
$$= (x-2)(x+3)(x+4).$$

Thus

$$x^3 + 5x^2 - 2x - 24 = 0 \iff x - 2 = 0 \quad \text{or} \quad x + 3 = 0 \quad \text{or} \quad x + 4 = 0$$
$$\iff x = 2 \quad \text{or} \quad x = {}^-3 \quad \text{or} \quad x = {}^-4.$$

(Note that some care is required with signs: if we had tried $P(^-3)$ we should get $P(^-3) = 0$ and the corresponding factor is $x + 3$, not $x - 3$.)

This illustrates the *factor theorem* which states that if a is a root of the polynomial equation in x, $P(x) = 0$, then $(x - a)$ is a factor of $P(x)$ and, conversely, if $x - a$ is a factor of the polynomial in x, $P(x)$, then a is a root of the equation $P(x) = 0$.

Note that we always work with the polynomial equation in zero form.

Exercise H

Solve the following equations:

1   $x^3 + 3x^2 - x - 3 = 0.$ 2 $x^3 - 2x^2 + 5x - 10 = 0.$

3   $x^3 + x^2 - 8x - 12 = 0.$ 4 $2x^3 - 3x^2 - 11x + 6 = 0.$

5   $x^3 - 7x - 6 = 0.$ 6 $x^3 + 2x^2 - 4x - 8 = 0.$

7   $2x^3 - 7x^2 - 5x + 4 = 0.$ 8 $6x^3 + 7x^2 - 9x + 2 = 0.$

9. SUMMARY

(1) $x^3 - 2x^2 - 5x + 6$ is an example of a polynomial in x. The *degree* is the highest power of x (in this case 3) and, for example, $^-2$ is the *coefficient* of x^2.

(2) If $x^3 - 2x^2 - 5x + 6$ is written as $(x-1)(x+2)(x-3)$ we say that it has been *factorised* and $x-1$, $x+2$, $x-3$ are called *factors*. The *roots* of the equation $(x-1)(x+2)(x-3) = 0$ are 1, $^-2$, 3.

(3) If $P(x)$ is a polynomial written in factorised form, the graph of $y = P(x)$ can be sketched:

(*a*) by finding where it crosses the x-axis from the factors of $P(x)$;

(*b*) by considering the sign of P for various values of x.

(4) The appropriate row of Pascal's triangle gives the coefficients when $(a+b)^n$ is multiplied out or *expanded*.

(5) The factor theorem states that, if a is a root of the polynomial equation in x, $P(x) = 0$, then $(x-a)$ is a factor of $P(x)$ and, conversely, if $x - a$ is a factor of the polynomial in x, $P(x)$, then a is a root of the equation $P(x) = 0$.

Miscellaneous Exercise

1   In each case below,
    (i) sketch the graph of $y = P(x)$;
    (ii) find the roots of the equation $P(x) = 0$;
    (iii) multiply out $P(x)$.
        (a) $P(x) = (x-3)(x-2)(x-1)$;
        (b) $P(x) = (x-2)x(x+2)$;
        (c) $P(x) = (2x-1)(x+1)(3x-2)$;
        (d) $P(x) = (x-2)^2(x+1)$;
        (e) $P(x) = (x-1)(x^2+2)(x+1)$;
        (f) $P(x) = (2-x)(x+1)$;
        (g) $P(x) = x(x+1)(x+2)(x-1)$.

2   Write down the solution set of the given inequality for the corresponding polynomial in question 1:
    (a) $P(x) > 0$; (b) $P(x) \leqslant 0$; (c) $P(x) > 0$; (d) $P(x) \leqslant 0$; (e) $P(x) > 0$;
    (f) $P(x) < 0$; (g) $P(x) > 0$.

3   Expand the following:
    (a) $(p+q)^3$; (b) $(a+x)^5$; (c) $(1-x)^4$; (d) $(1+2x)^3$; (e) $(3-4x)^4$.

4   By expanding $(1-2x)^5$ in ascending powers of x and then putting $x = 0.01$ find the value of 1.02^5 to 3 decimal places.

5   Use the expansion of $(1-3x)^7$ to find the value of 0.997^7 to 4 decimal places, writing down only the relevant terms.

6   In each of the following equations you are given one root. Find the other roots.
    (a) $x^2 - 5x - 6 = 0$; $^-1$.   (b) $x^3 - 3x^2 + 4x - 4 = 0$; 2.
    (c) $x^4 - 13x^2 + 36 = 0$; $^-2$.   (d) $x^4 + 2x^3 - 13x^2 - 14x + 24 = 0$; 1.

7   For each of the following polynomials, find all the factors and solve the equation $P(x) = 0$.
    (a) $P(x) = x^4 - 5x^2 + 4$;   (b) $P(x) = x^3 + x^2 - 6x$;
    (c) $P(x) = 9x^3 + 9x^2 - 16x - 16$; (d) $P(x) = x^4 - 2x^3 - 13x^2 + 14x + 24$.

8   Sketch the graph of $y = x(x-3)(x-4)$. Use your result to sketch the graphs of:
    (a) $y^2 = x^2(x-3)^2(x-4)^2$; (b) $y^2 = x(x-3)(x-4)$.

9   Sketch the graphs of:
    (a) $y = x^2(x-1)(x-2)$; (b) $y = x(x-1)(x-2)^2$.

5

Differentiation and gradients

1. DISPLACEMENT AND VELOCITY

In Chapter 1 we saw that the velocity of an object could be obtained by differentiating the displacement function, and our numerical work suggested the following pattern of derived functions (Table 1).

Table 1

Function	Derived function
$s: t \to t$	$s': t \to 1$
$s: t \to t^2$	$s': t \to 2t$
$s: t \to t^3$	$s': t \to 3t^2$
$s: t \to t^4$	$s': t \to 4t^3$

In general, we suspect that the derived function of

$$s: t \to t^n$$

is

$$s': t \to nt^{n-1}.$$

We also saw how to apply these results to polynomials. None of these results was actually proved – indeed we only showed that the derivative of $5t^2$ was $10t$ when $t = 1, 2,$ or 3. With the use of Pascal's triangle (Chapter 4, Section 5) the general result can be proved; proofs of the main results are given on page 89.

The pattern also extends to negative powers of t, in particular to $s: t \to \dfrac{1}{t} = t^{-1}$.

In this case, $s': t \to -t^{-2} = -\dfrac{1}{t^2}$. (See p. 89.)

Example 1
A stone is thrown upwards in such a way that its height above the ground is $(20t - 4.9t^2)$ metres after t seconds.
(a) With what velocity was it thrown?
(b) What is its velocity after 2 seconds?
(c) What is its greatest height?

From
$$s(t) = 20t - 4.9t^2 \text{ we can deduce that}$$
$$v(t) = s'(t) = 20 - 9.8t.$$

(a) $v(0) = 20$. So the stone was thrown with a velocity of 20 ms^{-1} upwards.

76

(b) $v(2) = 20 - 9.8 \times 2 = 0.4$. After 2 seconds the velocity of the stone is 0.4 m s^{-1} upwards.

(c) The stone reaches its greatest height when $v = 0$ (the moment when it stops going upwards and begins to descend).

$$v = 0 \quad \Leftrightarrow \quad 20 - 9.8t = 0 \quad \Leftrightarrow \quad t = \frac{20}{9.8} \approx 2.04.$$

So it reaches its maximum height after about 2.04 s. To find the height, we calculate $s(2.04)$.

$$s(2.04) = 20 \times 2.04 - 4.9 \times 2.04^2 \approx 20.4.$$

The maximum height reached is 20.4 m.

Exercise A

Write down the derived functions of:

1 $s : t \to t^9$.

2 $s : t \to 4t^3$.

3 $s : t \to 5t^7 + 6t^2$.

4 $s : t \to (3t + 1)^2$. (*Hint.* Multiply out first.)

5 $s : t \to (2t)^3 - 3t$.

6 $s : t \to t^4 + 2t^3 + 7$.

7 $s : t \to 7t^{-1}$.

8 $s : t \to 3t^{-2} + 7$.

9 $s : t \to \dfrac{2}{t^2}$.

10 $s : t \to \dfrac{2}{t} + \dfrac{2}{t^3}$.

11 $s : t \to t^{11}$.

12 $s : t \to 2t^5$.

13 $s : t \to 3t^5 + 5t^3$.

14 $s : t \to (5t - 1)^2$.

15 $s : t \to (6t)^3 + 7t^{-1}$.

16 $s : t \to 3t^4 + 4t^3 + 7t + 9$.

17 $s : t \to 4t^9 + t^{-2} - 5$.

18 $s : t \to \dfrac{8}{t} - \dfrac{4}{t^2}$.

19 $s : t \to t + \dfrac{1}{t}$.

20 $s : t \to t^2 - t^{-2}$.

21 A cricket ball thrown vertically upwards from a height of 2 m, with initial speed of 15 m s^{-1}, reaches a height of $2 + 15t - 5t^2$ metres after t seconds. What is its velocity after (*a*) 1 second, (*b*) 2 seconds?
     What is its maximum height?

22 A brick is dropped down a well and after t seconds it has fallen $4.9t^2$ metres. How long does it take to reach a speed of 20 m s^{-1} and how deep would the well have to be for it to reach this speed?

23 A balloon held on the end of a piece of light string is released and rises vertically so that its height after t minutes is $3 + 2t^2$ metres. What is its velocity after 2 minutes?
     If the balloon bursts when its speed is 40 metres per minute, what is its height when it bursts?

24 On the planet Omega a stone thrown vertically upwards from a height of 1.5 metres with initial speed of 10 m s^{-1}, reaches a height of $1.5 + 10t - 3t^2$ after t seconds. What is its velocity (*a*) after 1.5 seconds, (*b*) after 2 seconds?
     What is its maximum height?

2. RATES OF CHANGE

So far we have differentiated displacement functions to obtain the corresponding velocity functions. This is a particular example of a more general procedure. Given a function $f: x \rightarrow f(x)$, the derived function f' tells us the instantaneous rate of change of $f(x)$ compared with x.

Example 2
It is thought that the temperature in °C of a kettle t minutes after it is switched off is given by

$$\theta(t) = 32 + \frac{60}{t} \quad \text{for} \quad t \geqslant 1.$$

How fast is the kettle cooling after 5 minutes?

$$\theta'(t) = \frac{-60}{t^2} \quad \text{and so} \quad \theta'(5) = \frac{-60}{25} = {}^-2.4.$$

$\theta'(t)$ gives the rate of change of θ with t, so we deduce that after 5 minutes the kettle is cooling at a rate of 2.4 °C min^{-1}. (Notice that the negative sign tells us that the temperature is decreasing, not increasing.)

A particular case of a rate of change is acceleration, which is the rate of change of velocity with time. That is, acceleration, $a(t) = v'(t)$. Since $v(t) = s'(t)$, it is natural to write $a(t) = s''(t)$, indicating that s has to be differentiated twice to obtain a. In general, $f''(x)$ is called the *second derivative* of $f(x)$.

Example 3
The height of a stone is $(20t - 4.9t^2)$ metres at a time t seconds after it is thrown. What is its acceleration?

From $\qquad\qquad\qquad s(t) = 20t - 4.9t^2$

we obtain $\qquad\qquad v(t) = s'(t) = 20 - 9.8t$

and $\qquad\qquad\qquad a(t) = s''(t) = {}^-9.8.$

So the acceleration is 9.8 m s^{-2} downwards at all times.

Exercise B

1   The volume V of a cube of side x is given by $V: x \rightarrow x^3$. What is the rate of change of volume with respect to x?

2   The illumination of the page of a book which is x metres from a source of light is given by

$$I: x \rightarrow \frac{10}{x^2}$$

Find the rate of change of I with respect to x.

---

3   The area of a circular ink-blot of radius r is given by $A = \pi r^2$.
Find the rate of change of the area with respect to the radius.

4   If the cost of sailing a ship one nautical mile is $10 + 3V^2 + \dfrac{5}{V}$, where V is the speed

in knots (i.e. in nautical miles per hour), find the rate of change of the cost with respect to V.

5 The x-coordinate of a particle after t seconds is given by
$$x: t \rightarrow 3t - 5t^2,$$
where the units on the x-axis are metres. Find the velocity and acceleration functions.

6 The distance in metres travelled by a car along a straight road in t seconds is given by
$$s: t \rightarrow 3t + \tfrac{1}{4}t^2 \quad (0 > t > 10).$$
Find the velocity and acceleration functions. What is the acceleration when $t = 8$?

7 After t seconds, the x-coordinate of a particle moving along the x-axis is given by
$$x: t \rightarrow 4t - 3t^3,$$
where the units on the x-axis are metres. Find the x-coordinate after 2 seconds, and the velocity of the particle at the start and after $\tfrac{1}{2}$ second, and also after 2 seconds. When is its acceleration zero? What is the average velocity during the first 2 seconds?

8 If the x-coordinate of a particle after t seconds is given by
$$x: t \rightarrow t^4 - t,$$
find the velocity and acceleration functions.

9 The height, in metres, of a stone thrown vertically upwards from the surface of a planet is $(2 + 16t - 2t^2)$ after t seconds. Calculate the velocity of the stone after 3 seconds and the acceleration due to gravity.

10 After t seconds, the y-coordinate of a particle moving along the y-axis is given by
$$y: t \rightarrow 1 - 3t^2 + t^3,$$
where the units on the y-axis are metres. Find the y-coordinate after 1 second, the velocity of the particle initially and after two seconds. When is its acceleration zero? What is the average velocity during the first two seconds?

11 A particle P is travelling along a straight line so that its distance from the starting-point, x m, after time t seconds is given by the function $x: t \rightarrow 8t - \tfrac{1}{2}t^2$. Calculate the velocity and acceleration functions, and find the distance travelled by P when it first comes to rest. What is its acceleration at this moment?

12 The coordinates of a particle, t seconds after being projected upwards from the top of a cliff at $60°$ to the vertical with a speed of 10 m s^{-1}, are given approximately by
$$x: t \rightarrow 5t$$
$$y: t \rightarrow 3.5t - 5t^2.$$
By considering when the derived function of $y: t \rightarrow 3.5t - 5t^2$ is zero, find the greatest height reached. Hence find the coordinates of the particle at this highest point.

13 The height of a model aeroplane t seconds after taking off is given in metres by
$$h: t \rightarrow t^3 - 6t^2 + 9t$$
until it reaches the ground again. Find (a) the initial time for which it ascends; (b) for how long it descends.
Does it eventually crash?

3. GRADIENTS

Figure 1 shows the graph of the distance of a train from a station t minutes after it reaches its maximum speed. The speed of the train is given by the gradient of the graph: the gradient is 3, so the speed of the train is 3 km min^{-1}. Since the graph crosses the s-axis at 4, the equation of the graph is $s = 3t + 4$.

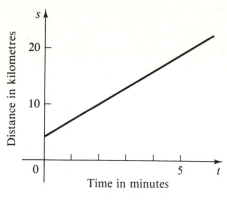

Figure 1

From $s(t) = 3t + 4$ it follows that $s'(t) = 3$, the speed of the train in km min^{-1} and the gradient of the graph.

This correspondence between differentiation and finding the gradient of the graph holds for curves as well as for straight lines. Figure 2 shows the graph of $s(t) = 5t^2$, the distance fallen by a stone in t seconds. The velocity after 3 seconds can be found from the graph by drawing the tangent at P (3, 45) and finding its gradient, which is 30. We also have $s'(t) = 10t$ and so $v(3) = s'(3) = 30$.

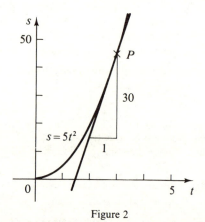

Figure 2

In general the derivative, $f'(x)$, gives the gradient of the graph of f.

Example 4
Find the equation of the tangent to the graph of $f: x \to 3x^3 + x$ at the point (1, 4).
$$f(x) = 3x^3 + x \Rightarrow f'(x) = 9x^2 + 1.$$
The gradient of the curve at $x = 1$ is $f'(1) = 10$.

So the tangent has gradient 10 and passes through (1, 4). (See Figure 3.) Its equation is of the form $y = 10x + b$, where b is a constant to be found. Since it passes through (1, 4),
$$4 = 10 \times 1 + b \quad \text{and so} \quad b = {}^-6.$$
The equation of the tangent is $y = 10x - 6$.

$f'(x)$ gives us the gradient of the graph; $f''(x)$ tells us how the gradient is changing. This is illustrated by the next example.

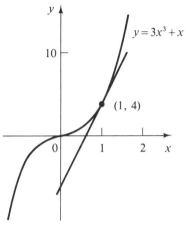

Figure 3

Example 5
If $f(x) = 5x^2 - x^3$, find $f(2)$, $f'(2)$ and $f''(2)$ and sketch the graph near $x = 2$.
$$f(x) = 5x^2 - x^3 \Rightarrow f'(x) = 10x - 3x^2 \Rightarrow f''(x) = 10 - 6x.$$
So $f(2) = 20 - 8 = 12$, $f'(2) = 20 - 12 = 8$ and $f''(2) = 10 - 12 = {}^-2$.
The graph passes through (2, 12) with gradient 8. Since $f''(2)$ is negative, the gradient is decreasing and so the graph is as shown in Figure 4.

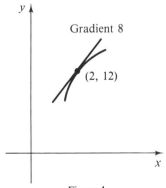

Gradient 8

(2, 12)

Figure 4

Exercise C

1 Find the gradient of $f: x \to 1 + x^2 + x^5$ at the point (1, 3).
2 Find the gradient of $f: x \to x^3 + x$ at the point (2, 10).

3   Find the gradient of $g:x \to x - \frac{1}{3}x^2$ at the point (9, ⁻18).

4   Find the gradient of $p:z \to z + z^{-1}$ when $z = 2$.

5   Find the gradient of $m:q \to q^2 + 8q$ when $q = 4$.

6   Find the gradient of $f:x \to 3 + x^3 + 4x^2$ at the point (1, 8).

7   Find the gradient of $f:x \to x^{-1} - x^2$ at the point (⁻1, ⁻2).

8   Find the gradient of $z:x \to x - \frac{3}{16}x^3$ when $x = 4$.

9   Find the gradient of $v:l \to \frac{1}{3}\pi l^3$ when $l = 2$.

10 Find the gradient of $t:z \to 2z^3 + 10$ when $z = 2$.

11 The graph of the function $f:x \to x^2 - 3x - 5$ goes through the points (5, 5), (4, ⁻1) and (2, ⁻7). Find its gradient at each of these points.

12 Calculate the gradient of the graph of the function $h:z \to z^3 - 5z + 3$ at the points where $z = 3, 1, 0, ⁻2$.

13 Calculate the values of $f'(4), f''(4), f'(⁻1)$ and $f''(⁻1)$ for the function $f:x \to x^3 + 3x - 4$. Sketch the graph near $x = 4$ and near $x = ⁻1$.

14 f is the function $f:x \to 2x^2 - 7x + 3$. Find the value of x when the gradient is 5.

15 For what values of x does the graph of the function $f:x \to 3x^3 - x + 7$ have gradient zero?

16 Find the values of x for which the graph of the function $f:x \to 2x^3 + 3x^2 - 36x - 20$ has zero gradient.

17 Find the gradient of the graph of the function $f:x \to x^3 - 5x^2 + 6x - 3$ at the points where $x = ⁻1$, $x = 1$ and $x = 3$.

18 Calculate the gradient of the graph of the function $k:w \to w^4 - 5w^2 + 1$ at the points where $w = ⁻2, ⁻1, 0, 2$.

19 Calculate the values of $f'(2), f''(2), f'(0)$ and $f''(0)$ for the function $f:x \to x^3 - 4x^2 + 1$. Sketch the graph near $x = 2$ and near $x = 0$.

20 f is the function $f:x \to 3x^2 - x + 1$. Find the value of x when the gradient is ⁻7.

21 For which values of x does the graph of the function $f:x \to x^5 - 5x + 2$ have gradient zero?

22 Find the values of x for which the graph of the function $f:x \to 2x^3 + 3x^2 - 12x - 20$ has zero gradient.

In questions 23–28 and 29–34, find the equation of the tangent to the graph of the function at the given point.

23 $f:x \to x^2$ at (2, 4).

24 $f:x \to 3x - x^2$ at (2, 2).

25 $f:x \to (2x - 1)^2$ at (2, 9).

26 $f:x \to x^2 - x^3$ at (2, ⁻4).

27 $f:x \to 4x^2 - \dfrac{1}{x^3}$ at ($\frac{1}{2}$, ⁻7).

28 $f:x \to x^{-2} + x^{-1}$ when $x = 4$.

29 $f:x \to 2x^2$ at (2, 8).

30 $f:x \to 9x - x^3$ at (1, 8).

31 $f:x \to (x + 2)^2$ at (⁻1, 1).

32 $f:x \to \dfrac{1}{x} - \dfrac{1}{x^2}$ at (⁻1, ⁻2).

33 $f:x \to x^3 - 3x^4$ when $x = 4$.

34 $f:x \to x^{-2} + x^{-3}$ when $x = 2$.

4. STATIONARY POINTS

In Example 1 we found the maximum value of $s(t)$ by finding when $s'(t) = 0$. Differentiation is often used in this way to find the maximum (or minimum) value of a function.

Figure 5 shows the graph of $f : x \rightarrow x^3 - 9x^2 + 15x$.

$$f'(x) = 3x^2 - 18x + 15$$
$$= 3(x^2 - 6x + 5)$$
$$= 3(x - 1)(x - 5).$$

So $f'(x) = 0 \iff x = 1$ or $x = 5$.

From the graph it is clear that $(1, 7)$ is a maximum point, and $(5, {}^{-}25)$ is a minimum point. Notice that these points do not give the greatest and least values of the function – for example $f(8) = 56$, which is greater than $f(1)$, and $f({}^{-}2) = {}^{-}74$, which is less than $f(5)$ – but give 'local' maximum and minimum values. It is in this sense that we shall use the words maximum and minimum.

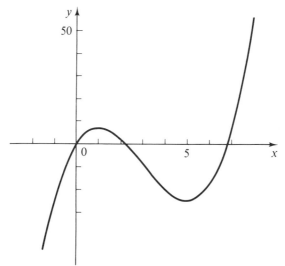

Figure 5

Points where $f'(x) = 0$ are called *stationary points*. They can be of three kinds – a minimum, a maximum or a stationary point of inflexion (points A, B, C in Figure 6). They can be distinguished by considering the sign of the gradient on each side of the point, as follows:

To the left of A, the gradient is negative; to the right it is positive. This is characteristic of a minimum point.

To the left of B, the gradient is positive; to the right it is positive. This is characteristic of a maximum point.

If a stationary point is a point of inflexion, the gradient does not change sign in this way. For example, the gradient is positive on both sides of C in Figure 6.

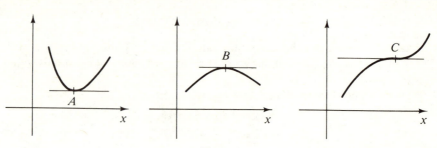

Figure 6

You should recall from Chapter 2 that maximum and minimum points are also called *turning-points*.

Example 6
Find the turning-points on the graph of $f : 4 + 3x - x^3$, and distinguish between them.

$$f(x) = 4 + 3x - x^3 \Rightarrow f'(x) = 3 - 3x^2.$$

So $f'(x) = 0 \Leftrightarrow 3 = 3x^2 \Leftrightarrow x^2 = 1 \Leftrightarrow x = 1$ or $^-1$.

$$f(1) = 4 + 3 - 1 = 6 \quad \text{and} \quad f(^-1) = 4 - 3 + 1 = 2,$$

so the turning-points are at $(1, 6)$ and $(^-1, 2)$. To distinguish between them, we examine the sign of $f'(x)$ (Tables 2 and 3).

Table 2

Value of x	$x < {}^-1$	$x = {}^-1$	$^-1 < x < 1$
Sign of $f'(x)$	$-$	0	$+$
Gradient	╲	$-$	╱

Table 3

Value of x	$^-1 < x < 1$	$x = 1$	$x > 1$
Sign of $f'(x)$	$+$	0	$-$
Gradient	╱	$-$	╲

Therefore when $x = {}^-1$ there is a minimum value of $f(x)$, and when $x = 1$ there is a maximum value of $f(x)$. The turning-points are:

$(^-1, 2)$ minimum; $(1, 6)$ maximum.

The graph can be sketched as in Figure 7.

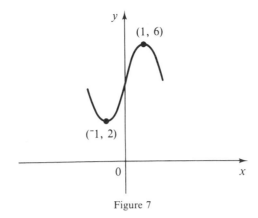

Figure 7

Exercise D

Find and identify the maximum and minimum points of the graph of the functions in the following:

1 (a) $f:x\rightarrow 3x^2$;

(b) $f:x\rightarrow 2x^3-12x^2+5$.

2 (a) $f:x\rightarrow (x+1)^2-4x$;

(b) $f:t\rightarrow t+\dfrac{1}{t}$.

3 (a) $f:x\rightarrow 2x^4$;

(b) $f:x\rightarrow 4x^3-3x+1$.

4 (a) $f:x\rightarrow (x+1)^3-3(x+1)^2$;

(b) $h:z\rightarrow 4z^2+\dfrac{1}{z}$.

5 (a) $f:x\rightarrow x^2-6x+2$;

(b) $g:z\rightarrow z^3-3z-1$.

6 (a) $f:x\rightarrow x^2(3-x)$;

(b) $h:t\rightarrow t+t^{-1}+3$.

7 (a) $f:x\rightarrow 5x^2$;

(b) $z:y\rightarrow y(y^2-12)$.

8 (a) $f:t\rightarrow (t+1)^2-3t$;

(b) $p:q\rightarrow 4q^2-\dfrac{1}{q}$.

9 Find the stationary value of the function $f:x\rightarrow 16x+x^{-2}$, and determine whether this is a maximum or a minimum value.

10 Find the maximum and minimum points on the graph of the function
$$f:x\rightarrow 10x^6+24x^5-45x^4$$
and distinguish between them.   (SMP)

11 Find the maximum and minimum values of the function
$$f:x\rightarrow 2x^3+3x^2-12x+4,$$
and distinguish between them. Sketch the graph of this function for the domain
$$-3\leqslant x\leqslant 2.$$ (SMP)

12 Prove that the maximum value of
$$23000-2400x+90x^2-x^3$$
is 7000. State the value of x for which it occurs and calculate the minimum value. Sketch a graph of this function for $x\geqslant 0$ and $y\geqslant 0$, labelling the coordinates of these turning-points; also state the greatest and least values of the function in this region.   (SMP)

5. APPLICATIONS

Example 7

An open rectangular tank is to have a square base and a minimum capacity of 4000 m³. What is the smallest possible area of sheet metal from which it can be made?

Suppose that the base is of side x metres. Then the area of the base is x^2 m², and since the volume is 4000 m³ the height of the tank must be $\dfrac{4000}{x^2}$ m. The area of each side of the tank is

$$x \times \frac{4000}{x^2} \text{ m}^2 = \frac{4000}{x} \text{ m}^2.$$

The total area, A m², of the base and the four sides is given by

$$A = x^2 + 4 \times \frac{4000}{x}$$

$$= x^2 + \frac{16000}{x}.$$

It is this function we have to minimise.

$$A'(x) = 2x - \frac{16000}{x^2}.$$

$$A'(x) = 0 \quad \Leftrightarrow \quad 2x = \frac{16000}{x^2}$$

$$\Leftrightarrow \quad x^3 = 8000$$

$$\Leftrightarrow \quad x = 20.$$

Table 4

Value of x	$x < 20$	$x = 20$	$x > 20$
Sign of $A'(x)$	−	0	+
Gradient	↘	−	↗

So from Table 4, we see that $A(x)$ has a minimum at $x = 20$.

$$A(20) = 20^2 + \frac{16000}{20} = 400 + 800 = 1200.$$

The minimum area of metal needed is 1200 m².

Exercise E

1   A farmer builds a fence along three sides of a rectangle in order to complete a sheepfold, using an existing fence as the fourth side. If the length of existing fence used is $2x$ metres, and the total length of new fencing is 100 m, show that the area of the sheepfold is given by $A:x \to 100x - 2x^2$. Use calculus to find the maximum area of the sheepfold.

2   If the number of litres of petrol, P, used in going 1 km at V km h^{-1} is given by

$$P:V \to \frac{V^2}{90\,000} - \frac{V}{900} + \frac{1}{4},$$

find the rate of change of P with respect to V, and calculate the value of V which gives the minimum petrol consumption per kilometre.

3   A box is to be made from a sheet of cardboard 1 metre square by cutting away squares from each corner, and then folding along the dotted lines (see Figure 8). Find the size of the squares for the volume to be greatest. (Take x metres to be the side of the shaded square and then the base will be of side $(1 - 2x)$ metres.)

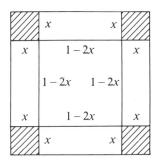

Figure 8

4   A rectangular parcel has a square cross-section of side x m and length $(4 - 4x)$ m (Figure 9). Calculate the maximum volume of the parcel.

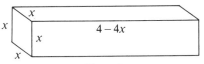

Figure 9

5   The cost of manufacturing a batch of sprigets is a fixed sum of £100, plus £5 per spriget, together with handling costs per spriget which are directly proportional (with constant of proportionality k) to the number of sprigets per batch.
    Show that the cost £C per spriget is given by the function

$$C:N \to 100N^{-1} + kN + 5,$$

where N is the number in a batch.
    Given that $k = 0.04$, find the value of N for which the cost is least and calculate the least cost.

6   What is the smallest area of thin sheet metal needed to make a cylindrical closed tin of capacity 500 ml?

7   A closed rectangular box is made of very thin sheet metal, and its length is three times its width. If the volume of the box is 288 cm³, show that its surface area is given by

$$A : x \to \frac{768}{x} + 6x^2,$$

where x cm is the width of the box.

Find by differentiation the dimensions of the box of least surface area. (You should give some justification for claiming that it is the least, and not the greatest.)

(SMP)

8   In order that a parcel may be sent through the post it must not be longer than one metre nor have the sum of its girth (that is, the distance round) and its length more than 2 metres. What is the greatest permissible volume of the parcel, (*a*) if it has a square cross-section, (*b*) if it is cylindrical?

9   A solid consists of a cylinder of height 12 cm and base radius x cm, from which has been hollowed out the hemisphere of radius x cm having as base one end of the cylinder. The volume V of the solid is thus

$$(12\pi x^2 - \tfrac{2}{3}\pi x^3) \text{ cm}^3.$$

If x can vary but cannot exceed 12, prove that when the hemispherical hollow is as large as possible, the volume of the solid is as large as possible.

10   A pastry cutter is in the shape of a sector of a circle. The perimeter of the cutter is 48 cm. Show that the area it cuts out is $24r - r^2$ cm², where r cm is the radius of the cutter. Hence find the greatest area which such a cutter, with the given perimeter, can have and show that for maximum area the angle of the sector is approximately 115°.

(SMP)

11   A rectangular parcel of square cross-section is such that the sum of its length and girth is 2 m. If its length is x m, show that its volume, V m³, is given by

$$16V = x(2-x)^2.$$

Find the maximum volume of a parcel of this shape, and prove that your result is, in fact, a maximum. (The girth of the parcel is the perimeter of its square cross-section.)

(O & C Additional Maths)

12   An open rectangular tank with a square base is to be constructed with fixed surface area A m² and its volume is to be as large as possible. Prove that the height must be half the side of the square base.

6. SUMMARY

The derived function of $s : t \to t^n$ is $s' : t \to nt^{n-1}$.

The derivative $f'(x)$ gives the rate of change of $f(x)$ with x and the gradient of the graph of $x \to f(x)$. The second derivative $f''(x)$ gives the rate at which the gradient is changing. If $s(t)$ is a displacement function, $s''(t)$ gives the acceleration.

The tangent to the graph of $y = f(x)$ at the point where $x = a$ has gradient $f'(a)$. Its equation can be found from its gradient and the fact that it passes through $(a, f(a))$.

When $f'(x) > 0$, $f(x)$ is increasing, and when $f'(x) < 0$, $f(x)$ is decreasing. When $f'(x) = 0$, the graph of f has a stationary point, which may be a maximum or minimum (turning-points) or a point of inflexion. The three kinds of stationary points can be distinguished by examining the sign of $f'(x)$ on either side of the point.

7. DIFFERENTIATION FROM FIRST PRINCIPLES

(a) $s(t) = t^n$ where n is a positive integer:

The average speed over the time interval from t to $t+h$ is

$$\frac{s(t+h)-s(t)}{h} = \frac{(t+h)^n - t^n}{h}.$$

From Pascal's triangle, we have

$$(t+h)^n = t^n + nt^{n-1}h + \text{terms with } h^2 \text{ or higher powers of } h$$

So $$\frac{s(t+h)-s(t)}{h} = \frac{t^n + nt^{n-1}h + (\text{terms in } h^2 \text{ or higher powers of } h) - t^n}{h}$$

$$= nt^{n-1} + \text{terms in } h \text{ or higher powers of } h.$$

As h approaches zero, this approaches nt^{n-1}, and so the derivative, $s'(t)$, is nt^{n-1}.

(b) $s(t) = \dfrac{1}{t}$:

$$\frac{s(t+h)-s(t)}{h} = \frac{1}{h}\left(\frac{1}{t+h} - \frac{1}{t}\right) = \frac{1}{h}\left(\frac{t}{(t+h)t} - \frac{t+h}{(t+h)t}\right)$$

$$= \frac{1}{h}\left(\frac{t-(t+h)}{(t+h)t}\right)$$

$$= \frac{1}{h} \times \frac{-h}{(t+h)t}$$

$$= \frac{-1}{(t+h)t}$$

This approaches $\dfrac{-1}{t^2}$ as h approaches zero, so

$$s'(t) = \frac{-1}{t^2}.$$

(c) The sum of two functions

If $$s(t) = f(t) + g(t),$$

then $$\frac{s(t+h)-s(t)}{h} = \frac{f(t+h) + g(t+h) - (f(t) + g(t))}{h}$$

$$= \frac{f(t+h)-f(t)}{h} + \frac{g(t+h)-g(t)}{h}.$$

From this we can deduce that

$$s'(t) = f'(t) + g'(t).$$

(*d*) Multiplication by a number

If
$$s(t) = kf(t),$$

then
$$\frac{s(t+h)-s(t)}{h} = \frac{kf(t+h)-kf(t)}{h}$$

$$= k\left(\frac{f(t+h)-f(t)}{h}\right)$$

from which we deduce that
$$s'(t) = kf'(t).$$

Miscellaneous exercise

1 Find the velocity and acceleration of a particle after 3 seconds if it moves along the x-axis so that its x-coordinate after t seconds is given by

$$x:t \to 3t^3 + 5t^2,$$

the units being metres.

2 The position, x, of a particle at time t is given by $x = f(t)$, where $f:t \to t^{-1}+t^2$. Find

     (*a*) $\dfrac{f(5)-f(2)}{3}$,   (*b*) $\dfrac{f(2+h)-f(2)}{h}$.

What do (*a*) and (*b*) represent?

Find the limit of (*b*) as h tends to zero, and say what this represents.

3 Find the derived functions of the following:

     (*a*) $f:x \to (x+1)x$;   (*b*) $f:x \to \dfrac{1}{x}+x$;

     (*c*) $f:x \to (x+1)^2$;   (*d*) $f:x \to \dfrac{1}{x^3}-\dfrac{1}{x^2}$.

4 (*a*) Find the equation of the tangent to $f:x \to x+x^3$ at the point $(2, 10)$.
Repeat part (*a*) for:
(*b*) $f:x \to 1+x+x^2+x^3$ at $x = 3$;
(*c*) $f:x \to x^{-1}$ at $x = 4$ and $x = {}^{-}4$.

5 Find the equation of the tangents at $(0, 1)$ and $(1, {}^{-}1)$ to the graph of $f:x \to 3x^4+x^2-6x+1$. At which other points of the curve does the tangent have a gradient of 8?

6 Sketch the graph and find the greatest value of the function

$$f:x \to (x-1)(2-x).$$

7 The perimeter of a rectangle is 12 cm. What is its greatest possible area?

8 *ABCD* is a square field and the length of a diagonal is $2a$ kilometres. A man starts to walk from A straight across to the opposite corner C at a speed of 4 km h^{-1}. At the same instant a second man starts to walk from B straight across to D at a speed of 3 km h^{-1}. Show that after t hours ($t \leqslant \frac{1}{2}a$) their distance apart is

$$\{(a-3t)^2+(a-4t)^2\}^{\frac{1}{2}}.$$

Prove that their least distance apart is $\frac{1}{5}a$. (*Hint.* Consider the square of the distance.)

                                                             (O & C General Maths)

9 Find the maximum and minimum values of the expression
$$3+6x-2x^3,$$
and distinguish between them.

Sketch the graph of the function $f:x \to 3+6x-2x^3$ for $^-2 \leqslant x \leqslant 2$.

Calculate the equation of the tangent to the graph at $x = 0$ and prove that there is no other tangent to the graph which is parallel to this tangent.

Find the equations of the tangents which are parallel to the x-axis.

10 The area of a rectangular cattle pen is to be made as large as possible. 60 metres of fencing are available. What is the greatest possible area that can be enclosed? Answer the same problem if on one side of the rectangle a hedge can be used as a boundary.

11 Find the equation of the tangent to the parabola $y = \frac{1}{4}x^2$ at $P(2k, k^2)$, and show that it meets the y-axis as far below the x-axis as P is above it.

12 A nut of mass $\frac{1}{4}$ kg is screwed at a distance of x m from the centre of a rod of length 3 m and mass 4 kg. The whole system is set in motion, oscillating about the centre of the rod in a vertical plane. It can be shown that the square of the period of small oscillations is
$$\pi^2(3+\tfrac{1}{4}x^2)/2x.$$
Find the minimum period.

13 The functions
$$f:x \to x^3 - 3x^2 + 3x$$
and $\qquad g:x \to x^3 - 2x^2 + x + k, \quad$ where k is a constant,

have a common turning-point (i.e. there is a value of a such that $f(a) = g(a)$ and $f'(a) = g'(a) = 0$). Find the values of a and k.

14 A rectangle is inscribed in a semicircle of radius a, so that one of its sides, of length $2x$, lies along the diameter of the semicircle. Prove that the area of the rectangle is $2x(a^2 - x^2)^{\frac{1}{2}}$.

By considering the turning values of the function $A:x \to x^2(a^2 - x^2)$, find the value of x which makes the area of the rectangle a maximum, and find the maximum value of the area. (O & C Additional Maths)

Revision exercise 1

1   An object moves in such a way that its displacement (s m) from the origin after t seconds is given by $s = \sqrt{t}$.
    (a) Write down the value of $s(9)$.
    (b) Copy and complete Table 1.

Table 1

h	$9+h$	$\{s(9+h)-s(9)\}/h$
⁻0.1	8.9	
⁻0.01	8.99	
⁻0.001		
0.1	9.1	
0.01	9.01	
0.001		

Hence estimate as accurately as you can the speed of the object when $t = 9$.

2   Sketch the graph of $y = (x+1)^2 - 3$, showing the coordinates of the minimum point.
    Calculate the coordinates of the points where the graph intersects (a) the y-axis, (b) the x-axis, (c) the line $y = 4$. Show these points on your diagram, together with their coordinates.

3   Use the expansion of $(1-2x)^8$ to calculate the value of $(0.98)^8$ correct to 3 significant figures.

4   The vectors **a** and **b** are $\begin{bmatrix} 3 \\ 4 \end{bmatrix}$ and $\begin{bmatrix} 5 \\ 0 \end{bmatrix}$ respectively. Write down in the same form the vectors $\frac{1}{2}(\mathbf{a}+\mathbf{b})$ and $\frac{1}{2}(\mathbf{a}-\mathbf{b})$.
    Draw on squared paper a diagram representing the vectors **a** and **b** by directed line-segments $\widetilde{OA}$ and $\widetilde{OB}$, where O is the origin. Complete the parallelogram $OACB$ and insert the diagonals OC and AB. The diagonals intersect at P.
    Write down the directed line-segment $\widetilde{OC}$ in terms of the vectors **a** and **b**. Write down also the directed line-segments which represent the vectors $\frac{1}{2}(\mathbf{a}+\mathbf{b})$ and $\frac{1}{2}(\mathbf{a}-\mathbf{b})$ respectively.
    Calculate the lengths of the vectors **a** and **b**, and give a geometrical reason why the vectors $\frac{1}{2}(\mathbf{a}+\mathbf{b})$ and $\frac{1}{2}(\mathbf{a}-\mathbf{b})$ are perpendicular to each other.   (MEI)

5   Sketch the graph of $y = x^3 - x^2$. Find the x-coordinates of the two points on the curve where the tangent to the curve has gradient 1.
    One of these points is a point P where the curve intersects the x-axis. Find the equation of the tangent at P, and show that this tangent intersects the curve again at a point Q whose x-coordinate satisfies the equation $x^3 - x^2 - x + 1 = 0$.
    Show that $(x-1)$ is a factor of the left-hand side of this equation and hence find the coordinates of Q.

6 Write the expression $3+4x-x^2$ in the form $a-(x-b)^2$.
   Sketch the graph of $y = 3+4x-x^2$.
   Find the coordinates of the two points where the graph intersects the x-axis.

7 Show that $(x+1)$ is a factor of the polynomial $3x^3+5x^2+x-1$.
   Solve the equation $3x^3+5x^2+x = 1$.
   Sketch the graph of $y = 3x^3+5x^2+x-1$.

8 A particle moves along a straight line in such a way that its displacement (s m) from
   the origin after t seconds is given by the formula

   $$s = 4t^3 - 3t^2 - 18t.$$

   (a) What is the velocity when $t = 0$?
   (b) When does the velocity first become zero? What is then the acceleration?
   (c) When is the acceleration zero? What is the velocity at that time and how far
   is the particle from the origin?

9 A cylindrical tin has radius r cm and height h cm. The total outside surface area
   (including the ends) is 24π cm². Show that

   $$h = \frac{12}{r} - r.$$

   Write down an expression for the volume V cm³ of the tin in terms of r. Hence
   find the value of r which makes V a maximum, and calculate this maximum volume.

10 Referred to x- and y-axes in the normal way, $\mathbf{a} = \begin{bmatrix} 3 \\ 4 \end{bmatrix}$ and $\mathbf{b} = \begin{bmatrix} 2 \\ 1 \end{bmatrix}$.

   (a) Show on a diagram on squared paper the points O, A, P, Q, R, where O is
   the origin and $OA = \mathbf{a}$, $OP = \mathbf{a}+\mathbf{b}$, $OQ = \mathbf{a}+3\mathbf{b}$, $OR = \mathbf{a}-2\mathbf{b}$.
   (b) If S is the point $(15, 10)$, express OS in the form $\mathbf{a}+k\mathbf{b}$.
   (c) Calculate the length of the vector $\mathbf{b}$ and also the distance AS, leaving both
   your answers as square roots. Give a geometrical reason for the relation
   between your two answers to this part of the question and your value of k
   in (b).

11 Copy and complete Table 2, and hence estimate as accurately as you can the gradient
   of the curve $y = \tan x°$ when $x = 45$.

   Table 2

h	$45+h$	$\{\tan(45+h)° - \tan 45°\}/h$
$^-0.1$	44.9	
$^-0.01$		
$^-0.001$		
0.1		
0.01		
0.001		

12 Find the coordinates of the points where the graph of

   $$y = 2x^3 + 3x^2 - 12x + 7$$

   intersects the x-axis.
   Sketch the curve and find the coordinates of its turning-points.

13 Solve the equation $1+4x-x^2 = 0$.
Sketch the graph of $y = 1+4x-x^2$ and find the coordinates of the turning-point. The equation $1+4x-x^2 = k$ has just one solution. Find the value of k.

14 On squared paper draw the usual x- and y-axes; place the origin near the centre of the left-hand edge of the paper and label each axis, using a scale of 2 cm to represent 10 units.

(a) A powerboat leaves a buoy O and heads south at 40 km h^{-1}. However, a current running from due west at 10 km h^{-1} carries the boat off course. Represent these velocities as vectors on your diagram and find the bearing on which the powerboat actually travels.

(b) A helicopter which travels through the air at 50 km h^{-1} leaves the buoy O at the same time as the powerboat. The pilot wishes to travel due east, but she has to allow for a wind blowing from the north at 14 km h^{-1} so she heads the helicopter slightly north of east just sufficiently to ensure that she actually travels due east. Represent these velocities on your diagram and find the speed at which the helicopter travels eastwards.

(c) Find how far apart the powerboat and helicopter are after three hours.

(MEI)

15 An open tank is to be constructed with a square base and four rectangular sides. The capacity is to be 32 m^3. Find the least area of sheet metal required to construct it.

16 (a) Solve the equation $x^2-x = 5$.
(b) Sketch the graphs of $y = x^2-x-5$ and $y = 3/x$.
(c) Show that these graphs intersect at $P(3, 1)$.
(d) Find the equations of the tangents to the two curves at P. These tangents intersect the x-axis at Q and R. Find the coordinates of Q and R.
(e) Find the area of the triangle PQR.

17 A train starts at station P and next stops at station Q. On the journey its distance x km from P is given by the formula

$$x = 90t^2 - 45t^3$$

where t hours is the time elapsed since it left P. Find (a) the time taken to travel from P to Q, (b) the distance PQ, (c) the maximum speed on the journey.

18 An aeroplane which has an airspeed of 350 km h^{-1} is to fly from an airport A to another one B which is 500 km due east of A. There is a steady wind of 50 km h^{-1} blowing from the north. What course should the pilot set, and how long will the flight take?

For the return journey from B to A the pilot sets a course due west, failing to allow for the wind which is now blowing steadily at 50 km h^{-1} from the north-east. The airspeed of the aeroplane is still 350 km h^{-1}. Find the direction in which the plane actually travels.

(MEI)

19 The triangle ABC (Figure 1) is isosceles and right-angled at A. $BC = 18$ cm. $PQRS$ is a rectangle. $PQ = 2r$ cm and $QR = h$ cm.
(a) Write down a formula connecting r and h.
(b) Express the area of $PQRS$ in terms of r only.

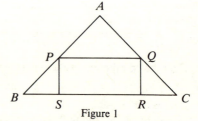

Figure 1

(c) Hence find the maximum possible area of $PQRS$.

(d) Interpreting the diagram as a front elevation of a circular cylinder inscribed in a cone, find the maximum possible volume of the cylinder.

20 (a) Multiply out: (i) $(2x-1)(2x+1)(2x-3)$; (ii) $(2x+1)^3$.

(b) Simplify $(2x-1)(2x+1)(2x-3)-(2x+1)^3$.

21 Express $4x^2-20x+9$ in the form $(2x-p)^2-q$.

Hence solve the equation $4x^2-20x+9 = 0$.

Write down the factors of $4x^2-20x+9$.

22 Show that the point $(5, 6)$ lies on the curve $y = 2x^2-7x-9$ and find the equation of the tangent to the curve at that point. Find also the point on the curve where the tangent is parallel to the line $y = x$. Sketch the curve.

23 A ship P is sailing due south at 6 knots and a ship Q is sailing south-east at 18 knots. Find the velocity of Q relative to P. If initially ship Q was 10 nautical miles north of ship P, find the least distance apart of the two ships.

24 The height of a rocket (h km) above the launch-pad t minutes after blast-off is given by
$$h = \tfrac{1}{4}t(36-24t+10t^2-t^3).$$
Find the maximum velocity and the maximum acceleration.

25 A running track consists of two parallel straight sections of length x m and two semicircular sections of radius r m. The total length of the track is to be 400 m. Write down a relation between x and r.

The area of the rectangular region enclosed by the straight sections (excluding the semicircular ends) is to be a maximum. Find the required values of r and x.

6

Integration

1. INVERSE DIFFERENTIATION

Captain Welsh is trying to land his spaceship on Io. At the moment when he is 15 m above the surface, and falling at 8 m s^{-1}, he fires his rocket engines, so that t seconds later the height of the ship above the surface is given by $s(t) = 1.5t^2 - 8t + 15$. Can you tell what happened?

Differentiation gives the velocity and acceleration as follows:

$$s(t) = 1.5t^2 - 8t + 15 \implies v(t) = s'(t) = 3t - 8$$
$$\implies a(t) = s''(t) = 3.$$

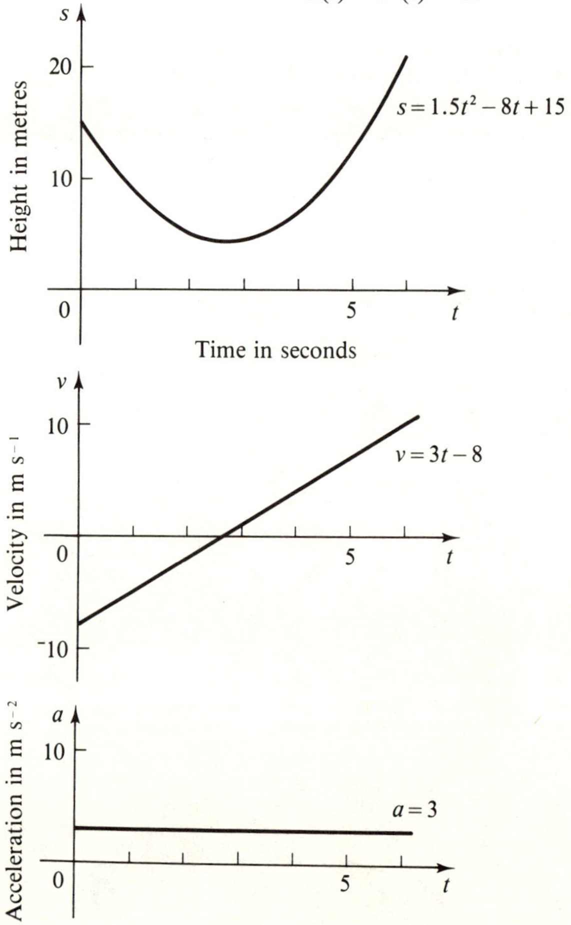

Figure 1

The graphs of s, v and a are shown in Figure 1 and tell the sad story: Captain Welsh fired his engines too hard and left Astronaut Barley, who was waiting below, marooned. Notice that $s(0) = 15$, $v(0) = {}^-8$ (consistent with the information in the second sentence of this section) and that the acceleration was a constant 3 m s^{-2}.

What would have happened if the engines had been fired more gently, to produce a constant acceleration of 2 m s^{-2} instead? That is, can we deduce $v(t)$ and $s(t)$ if $a(t) = 2$?

It is tempting to suggest that, since the derivative of $2t$ is 2, $a(t) = 2 \Rightarrow v(t) = 2t$. But there are many other functions which have a derivative of 2, such as $2t+1$, $2t-7$, $2t+29$, etc. What we *can* say is that $v(t)$ and $2t$ are changing at the same rate, since they have the same derivative, and that therefore, over any period of time, they will change by equal amounts. We write this
$$[v(t)] = [2t],$$
where the square brackets are used as a shorthand for 'the change in'. If we are interested in the change in v between $t = 6$ and $t = 10$, for example, we can write
$$[v(t)]_6^{10} = [2t]_6^{10}$$
and so
$$v(10) - v(6) = 20 - 12 = 8.$$

In this case we know that $v(0) = {}^-8$, and we can use this to find $v(t)$, as follows:
$$[v(t)]_0^t = [2t]_0^t$$
$$v(t) - v(0) = 2t - 0$$
$$v(t) - ({}^-8) = 2t$$
$$v(t) = 2t - 8.$$

We then use the same technique to find $s(t)$.
$$v(t) = 2t - 8 \quad \Rightarrow \quad [s(t)]_0^t \qquad = [t^2 - 8t]_0^t$$

(because the derivative of $t^2 - 8t$ is $2t - 8$)
$$\Rightarrow \quad s(t) - s(0) = t^2 - 8t - 0$$
$$\Rightarrow \quad s(t) - 15 \ = t^2 - 8t$$
$$\Rightarrow \quad s(t) \qquad = t^2 - 8t + 15.$$

Since $s(3) = 0$ and $v(3) = {}^-2$, the spaceship would have landed after 3 seconds with a touch-down speed of 2 m s^{-1}.

Example 1
When the gun goes for the start of a race a dinghy is 4 m behind the line and travelling at 3 m s^{-1}. If its acceleration t seconds later is given by $a(t) = 0.6t + 0.8$, find its speed and position 5 seconds later.
$$a(t) = 0.6t + 0.8$$
$$\Rightarrow \quad [v(t)]_0^t = [0.3t^2 + 0.8t]_0^t \qquad \text{by inverse differentiation}$$
$$\Rightarrow v(t) - 3 = 0.3t^2 + 0.8t - 0 \qquad \text{since } v(0) = 3$$
$$\Rightarrow \quad v(t) = 0.3t^2 + 0.8t + 3$$
$$\Rightarrow \quad [s(t)]_0^t = [0.1t^3 + 0.4t^2 + 3t]_0^t \qquad \text{by inverse differentiation}$$
$$\Rightarrow \quad s(t) = 0.1t^3 + 0.4t^2 + 3t - 4 \quad \text{since } s(0) = 4.$$

Hence $v(5) = 14.5$ and $s(5) = 33.5$.
So after 5 seconds it is 33.5 m in front of the line and travelling at 14.5 m s^{-1}.

Always check, by differentiating, that you have got the inverse process correct. The derivative of $s(t)$ should be $v(t)$, and the derivative of $v(t)$ should be $a(t)$.

So far we have been using $v(0)$ and $s(0)$, but this is not essential: if the values of $v(t)$ and $s(t)$ are known at some other time then that information can be used, as the following example illustrates.

Example 2

Three seconds after passing through a station a train crosses a level-crossing at 21 m s^{-1}. If its acceleration, t seconds from the station, is given by $a(t) = 0.2t + 1$, find its speed through the station and the distance between the station and the crossing.

$$a(t) = 0.2t + 1$$
$$\Rightarrow \quad [v(t)]_3^t = [0.1t^2 + t]_3^t \qquad \text{by inverse differentiation}$$
$$\Rightarrow v(t) - 21 = (0.1t^2 + t) - (0.9 + 3) \quad \text{since } v(3) = 21$$
$$\Rightarrow \quad v(t) = 0.1t^2 + t + 17.1$$
$$\Rightarrow \quad [s(t)]_0^3 = \left[\frac{0.1}{3}t^3 + \tfrac{1}{2}t^2 + 17.1t\right]_0^3 = (0.9 + 4.5 + 51.3) - 0$$
$$\Rightarrow \quad [s(t)]_0^3 = 56.7.$$

The train has travelled 56.7 m between $t = 0$ and $t = 3$, so the distance from station to level-crossing is 56.7 m.

Also, $v(0) = 17.1$, so the train passed through the station at 17.1 m s^{-1}.

The inverse of differentiation is called *integration*.

Example 3

If $v(t) = t^4$, integrate to find $s(t)$.

In this case, we have no information about $s(t)$ at any particular time, so all we can do is to write down the most general expression whose derivative is t^4.

$$v(t) = t^4 \;\Rightarrow\; s(t) = \tfrac{1}{5}t^5 + k \text{ for some number } k.$$

Exercise A

1   Integrate to find $[s(t)]_1^3$ when $v(t)$ is
    (a) $3t^2$; (b) $2t + 3$; (c) $4 - 5t^3$; (d) $2t^4 + 4t^7$.
    Write down $s(t)$ in each case.

2   A runner moves so that her velocity is given by $v(t) = 5 + 0.4t$.
    (a) How far does she move in the first three seconds?
    (b) How long does it take her to cover the first 30 m?

3   A stone drops from rest with constant acceleration 10 m s^{-2}. Find $v(t)$ and $s(t)$, its velocity and distance fallen after t seconds. How far does it travel in the sixth second, and how fast is it falling at the end of this time?

4   A train travelling at 32 m s^{-1} brakes so that its acceleration is given by $a(t) = {}^-0.2t$.
    (a) Find its velocity after t seconds.
    (b) Find how long it takes to stop and the distance required to do so.

5   An arrow travels so that $v(t) = 100 - 4t^2$. How far does it go in the first 3 seconds?

6 A car moves from rest so that $a(t) = 8 - t$ for the first 5 seconds, but $a(t) = {}^-3$ thereafter. How far does it travel before stopping?

7 Integrate to find $s(t)$ if $v(t)$ is:
   (a) t^5; (b) $3t^4 - 2t^3$; (c) $t^2 + t + 1$.

8 Integrate to find $[v(t)]_2^5$ when $a(t)$ is
   (a) $4t^3$; (b) $6t - 5$; (c) $2 - 7t^2$; (d) $4.2t^5 + 6.3t^6$.

9 A rocket rises from rest with constant acceleration 5 m s^{-2}. Find $v(t)$ and $h(t)$, its velocity and height after t seconds. How far does it travel in the fourth second? What is its height after four seconds, and how fast is it then travelling?

10 A cyclist moves so that his velocity at time t seconds is given by $v(t) = 5.9 + 0.6t$. How long does it take him to cover the first 56 metres?

11 A car travelling at 18 m s^{-1} brakes so that its acceleration is given by $a(t) = {}^-4t$. Write down its velocity after t seconds and deduce how many seconds it takes to stop and how far the car has travelled in that time.

12 A bullet travels so that its speed at time t seconds is given by $v(t) = 400 - 25t^2$: After how many seconds does it stop? How far has it travelled in that time?

13 A boat moves from rest so that its acceleration is given by $a(t) = 0.6 + t$ for the first four seconds, but $a(t) = {}^-0.2$ thereafter. How far does it travel before stopping?

14 Integrate to find $s(t)$ if $v(t)$ is:
   (a) t^7; (b) $4t - 7$; (c) $3t^5 - 4t^{11}$.

2. DISPLACEMENT AND DISTANCE

A car moves so that $v(t) = 2t - 3$. Can we find how far it travels in the first five seconds? We take its initial position as origin. Its change in position is given by $[s(t)]_0^5 = [t^2 - 3t]_0^5 = 10$. So it appears to have travelled 10 metres. However $[s(t)]_0^3 = 0$, so it appears not to have moved at all in the first three seconds. Can this be true?

In fact the car moves backwards before going forwards, passing through its original position after three seconds. It has travelled several metres in these three seconds although its overall displacement is zero, since $[s(t)]_0^3 = 0$. Similarly, the fact that $[s(t)]_0^5 = 10$ means that its displacement from its original position is 10 metres after five seconds. The actual distance that it has travelled in this time is considerably greater.

In general $[s(t)]_0^t = [t^2 - 3t]_0^t$

and $s(t) = t^2 - 3t$ since $s(0) = 0$.

Also $v(t) = 2t - 3 < 0 \Leftrightarrow t < 1.5$.

So the car moves backwards until $t = 1.5$, when its velocity is zero and its position is $s(1.5) = {}^-2.25$. When $t = 3$ it is back at the origin, having travelled 4.5 m, and when $t = 5$ it is 10 metres away, having travelled a total distance of 14.5 m.

Displacement from $t = 0$ to 5

$$= \frac{\text{displacement from}}{t = 0 \text{ to } t = 1.5} + \frac{\text{displacement from}}{t = 1.5 \text{ to } t = 3} + \frac{\text{displacement from}}{t = 3 \text{ to } t = 5}$$

$$= \quad [s(t)]_0^{1,\,5} \quad + \quad [s(t)]_{1,\,5}^3 \quad + \quad [s(t)]_3^5$$

$$= \quad ^-2.25 \quad + \quad 2.25 \quad + \quad 10$$

$$= 10.$$

But the distance travelled

$$= \quad 2.25 \quad + \quad 2.25 \quad + \quad 10$$

$$= 14.5.$$

Example 4

The velocity of a fly for a while is given by $v(t) = t^2 - 4t + 3$. Find its displacement in the second second and the distance travelled in the first five seconds.

$$[s(t)] = [\tfrac{1}{3}t^3 - 2t^2 + 3t] \text{ by integration}$$

and

$$[s(t)]_1^2 = s(2) - s(1)$$
$$= (\tfrac{1}{3} \times 8 - 2 \times 4 + 3 \times 2) - (\tfrac{1}{3} - 2 + 3) = \tfrac{2}{3} - \tfrac{4}{3} = ^-\tfrac{2}{3}.$$

So the displacement in the second second is $^-\tfrac{2}{3}$ m. The fly moves 'backwards'.

$$v(t) = 0 \Rightarrow t^2 - 4t + 3 = 0$$
$$\Rightarrow (t-1)(t-3) = 0$$
$$\Rightarrow t = 1 \text{ or } t = 3.$$

The fly is stationary when $t = 1$ and when $t = 3$. Between these times it is moving with negative velocity; at other times it is moving with positive velocity. So, considering its position at these particular times, we obtain:

$$\frac{\text{Displacement}}{\text{from } t = 0 \text{ to } t = 5} = [s(t)]_0^1 + [s(t)]_1^3 + [s(t)]_3^5$$

$$\text{Distance} \quad = \quad \tfrac{4}{3} \quad - \quad \tfrac{4}{3} \quad + \quad \tfrac{20}{3} \quad = \tfrac{20}{3}.$$
$$= \quad \tfrac{4}{3} \quad + \quad \tfrac{4}{3} \quad + \quad \tfrac{20}{3} \quad = \tfrac{28}{3}.$$

So the displacement after 5 seconds is 6.7 m, but the distance travelled is 9.3 m to 2 significant figures.

Exercise B

1   Sketch a graph of $v(t) = 5 - 2t$. Find the displacements at $t = 3$ and at $t = 5$ and the distance travelled in the first five seconds.

2   Sketch a graph of $v(t) = t^2 - 6t + 8$. Find the displacements at $t = 2$ and at $t = 5$ and the distance travelled in the first five seconds.

3   Given that $v(t) = t^2 - 6t + 7$ find, correct to 2 significant figures, when the velocity is zero and when it is negative. Hence find the displacement and distance travelled in the first five seconds.

4   $v(t) = (t-1)(t-2)(t-4)$. State when the velocity is negative. Hence calculate the displacement and distance travelled in the first five seconds.

5   $v(t) = (t-2)(t-3)$. Calculate $[s(t)]_2^3$ and $[s(t)]_3^2$.
    State in words what these results give.

6   Give the value of $[v(t)]^b_a$ and $[v(t)]^a_b$: (a) when $v(t) = t^2$; (b) when $v(t) = 3-t$. Make a general statement about the relationship between $[v(t)]^b_a$ and $[v(t)]^a_b$.

7   Sketch a graph of $v(t) = 8-2t$. Find the displacements at $t = 2$ and at $t = 4$ and the distance travelled in the first four seconds.

8   Sketch a graph of $v(t) = t^2 - 5t + 4$. Find the displacements at $t = 4$ and at $t = 6$ and the distance travelled in the first six seconds.

9   Given that $v(t) = t^2 - 4t + 2$ find, correct to 2 significant figures, when the velocity is zero and when it is negative. Hence find the displacement and distance travelled in the first four seconds.

10 $v(t) = (t-1)(t-3)(t-4)$. State when the velocity is negative. Hence calculate the displacement and distance travelled in the first six seconds.

11 $v(t) = (t-1)(t-4)$. Calculate $[s(t)]^4_1$ and $[s(t)]^1_4$. State in words what these results give.

12 Give the value of $[s(t)]^b_a$ and $[s(t)]^a_b$: (a) when $s(t) = t^3$; (b) when $s(t) = -t$. Make a general statement about the relationship between $[s(t)]^b_a$ and $[s(t)]^a_b$.

3. AREA

We know that when $s(t)$ is differentiated we get $v(t)$. On a graph of $s(t)$ the gradient gives $v(t)$. So differentiation corresponds to finding the gradient of a graph. This is true not only for $s(t)$ but for any other function.

Can you see to what the inverse process, integration, corresponds? We know that when $v(t)$ is integrated we get $[s(t)]$. On a graph of $v(t)$ the area gives $[s(t)]$. So integration corresponds to finding the area under a graph.

In fact, for any function, integration corresponds to finding the area under its graph, and we shall outline a proof of this result in Section 5. But first we shall confirm in some simple cases that the area between the graph of $v(t)$ and the t-axis does indeed give $[s(t)]$.

This is particularly simple to see in the special case when the velocity is constant, 3 m s^{-1} for example. In the first 2 seconds the displacement will be 6 metres which also corresponds to the area from $t = 0$ up to $t = 2$ between the graph of $v(t)$ and the axis, shown shaded in Figure 2. Note that 3 integrated is $[3t]$ and $[3t]^2_0 = 6$.

The next simplest case is when the velocity increases steadily, for example $v(t) = 2t - 3$. The velocity is

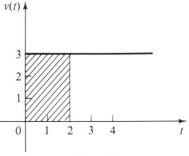

Figure 2

negative for the first 1.5 seconds. The corresponding distance is given by the triangular region shaded horizontally in Figure 3, the area of which is $1.5 \times \frac{3}{2} = 2.25$. From $t = 1.5$ to $t = 5$ the distance is given by the triangular region, shaded vertically, whose area is $3.5 \times \frac{7}{2} = 12.25$. The total shaded area

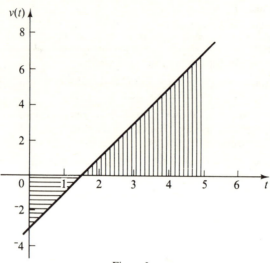

Figure 3

is $2.25 + 12.25 = 14.5$. Note the displacement is $^-2.25 + 12.25 = 10$, the area below the axis being taken as negative. Comparison with Section 2 confirms that integration does give the value of these areas.

We can now use integration to calculate areas under curves.

Example 5
Find the area between the t-axis and the graph of $y = t^3 + 2$ bounded by $t = 1$ and $t = 3$.

This area is shown shaded in Figure 4.

Area $= t^3 + 2$, integrated,

$\quad = [\frac{1}{4}t^4 + 2t]_1^3$

$\quad = 26.25 - 2.25 = 24.$

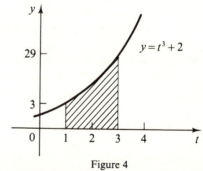

Figure 4

It is useful to have a symbol for integration and an old-fashioned $S - \int -$ for summing areas, is used, together with the symbol dt. (The significance of this symbol is explained in Section 5.)

So in Example 5 we would write:

$$\text{Area} = \int_1^3 (t^3 + 2)\, dt = [\tfrac{1}{4}t^4 + 2t]_1^3 = 24.$$

Care has to be taken when part of the area is below the axis. Integration will give a negative value for such a region but, as with displacement and distance, the positive value has to be taken for the area.

The integral sign, $\int$, can also be used without the *limits* (1 and 3 in this case). So when in Example 3 we integrated t^4 to obtain $\frac{1}{5}t^5+k$, we could have written

$$\int t^4\,dt = \tfrac{1}{5}t^5+k.$$

Example 6

Find the area between the t-axis and the graph of $y = t^2-4t+3$ from $t = 0$ to $t = 5$.

The graph is seen to be below the axis from $t = 1$ to $t = 3$ (Figure 5).

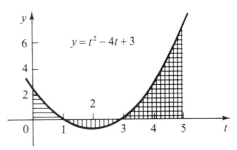

Figure 5

Shaded horizontally: $\displaystyle\int_0^1 (t^2-4t+3)\,dt = [\tfrac{1}{3}t^3-2t^2+3t]_0^1 = \tfrac{4}{3}-0 = \tfrac{4}{3}.$

Shaded vertically: $\displaystyle\int_1^3 (t^2-4t+3)\,dt = [\tfrac{1}{3}t^3-2t^2+3t]_1^3 = 0-\tfrac{4}{3} = \tfrac{-4}{3}.$

Cross shaded: $\displaystyle\int_3^5 (t^2-4t+3)\,dt = [\tfrac{1}{3}t^3-2t^2+3t]_3^5 = \tfrac{20}{3}-0 = \tfrac{20}{3}.$

Total shaded area $= \tfrac{4}{3}+\tfrac{4}{3}+\tfrac{20}{3} = \tfrac{28}{3}.$

Note that $\displaystyle\int_0^5 (t^2-4t+3)\,dt = [\tfrac{1}{3}t^3-2t^2+3t]_0^5 = \tfrac{20}{3}-0 = \tfrac{20}{3}.$

This is what we should get by considering the region below the t-axis to be negative. Compare this example with Example 4.

Example 7

Calculate the area between the x-axis and the graph of $y = x^2+2x$ from $x = 1$ to $x = 4$.

$$\text{Area} = \int_1^4 (x^2+2x)\,dx = [\tfrac{1}{3}x^3+x^2]_1^4 = \tfrac{112}{3}-\tfrac{4}{3} = 36.$$

Notice that dt has become dx since x is the variable. We should get exactly the same result with t as variable:

$$\int_1^4 (t^2+2t)\,dt = [\tfrac{1}{3}t^3+t^2]_1^4 = \tfrac{112}{3}-\tfrac{4}{3} = 36.$$

Exercise C

1   On a graph of the curve $y = t^2$, shade the region corresponding to $\int_1^3 t^2 \, dt$ and calculate its value.

2   Draw a graph of $y = 3x^2 + 2$, shading the region between the curve and x-axis bounded by $x = 1$ and $x = 4$. Estimate the area of this region and calculate the exact value.

3   On a graph of the line $y = 2t + 3$ shade the trapezium between the line and the t-axis bounded by $t = 1$ and $t = 5$. Use geometry to find the area of this trapezium and verify your result by integration.

4   On a graph of $y = 6 - 2x$ show the regions with areas $\int_0^3 (6 - 2x) \, dx$ and $\int_3^5 (6 - 2x) \, dx$, giving the value of each of these expressions. Hence find the area between the graph and the x-axis from $x = 0$ to $x = 5$. Explain why this is not the same as $\int_0^5 (6 - 2x) \, dx$.

5   State the values of x for which the graph of $y = x^2 - 5x + 4$ is below the x-axis. Hence calculate the total area between this curve and the x-axis from $x = 0$ to $x = 6$.

6   Find, correct to 2 significant figures the area between the graph of $y = 3x^2 - 4x + 5$ and the x-axis from $x = 0$ to $x = 5$.

7   Calculate the values of $\int_1^3 f(x) \, dx$, $\int_3^4 f(x) \, dx$, $\int_1^4 f(x) \, dx$:
   (a) for $f(x) = 6x$;   (b) for $f(x) = 6 - 2x$.
   What is the relationship between your three answers in each case?

8   Find: (a) $\int (6x^2 - 4) \, dx$;   (b) $\int (2x^3 + x) \, dx$.

---

9   On a graph of the curve $y = t^3$, shade the region corresponding to $\int_2^5 t^3 \, dt$ and calculate its value.

10 Draw a graph of $y = 6x^2 + 5$, shading the region between the curve and x-axis bounded by $x = 1$ and $x = 3$. Estimate the area of this region and calculate its exact value.

11 On a graph of the line $y = 4t + 5$, shade the trapezium between the line and the t-axis bounded by $t = 1$ and $t = 4$. Use geometry to find the area of this trapezium and verify your result by integration.

12 On a graph of $y = 8 - 2x$ show the regions with areas $\int_0^4 (8 - 2x) \, dx$ and $\int_4^6 (8 - 2x) \, dx$, giving the value of each of these expressions. Hence find the area between the graph and the x-axis from $x = 0$ to $x = 6$. Explain why this is not the same as $\int_0^6 (8 - 2x) \, dx$.

13 State the values of x for which the graph of $y = x^2 - 5x + 6$ is below the x-axis. Hence calculate the total area between this curve and the x-axis from $x = 0$ to $x = 5$.

14 Find, correct to 2 significant figures, the area between the graph of $y = 2x^2 - 3x - 4$ and the x-axis from $x = 1$ to $x = 5$.

15 Calculate the values of $\int_1^4 f(x) \, dx$, $\int_4^6 f(x) \, dx$, $\int_1^6 f(x) \, dx$:
   (a) for $f(x) = 4x$;   (b) for $f(x) = 8 - 2x$.
   What is the relationship between your three answers in each case?

16 Find: (a) $\displaystyle\int(x-9x^2)\,dx$; (b) $\displaystyle\int(1+x+\tfrac{1}{2}x^2)\,dx$.

4. SUMMARY

Inverse differentiation is called integration.

When acceleration $a(t)$ is integrated it gives change of velocity $[v(t)]$.

When velocity $v(t)$ is integrated it gives change of position $[s(t)]$.

$[s(t)]_a^b$ means $s(b)-s(a)$. This gives the displacement from the position when $t = a$ to the position when $t = b$. The distance travelled may well be greater than the displacement. It is found by considering the positive and negative displacements separately.

Integration of a function $f(x)$ gives areas between the graph of the function and the x-axis. An area below the axis is taken to be negative.

5. INTEGRATION AND AREA

We give an outline proof that integration gives the area under a curve. $A(x)$ denotes the area under the graph of $y = f(x)$ up to the point x. When x is increased by an amount h the area will be $A(x+h)$ and the corresponding increase in area, shown shaded, will be $A(x+h)-A(x)$. The average height of this shaded area is given by

$$\frac{(A(x+h)-A(x))}{h}$$

so $\dfrac{A(x+h)-A(x)}{h} \approx f(x).$

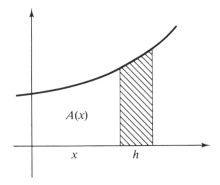

Figure 6

For all reasonable curves, this approximation becomes more and more accurate the smaller h becomes. As h approaches zero, the limit of $\dfrac{A(x+h)-A(x)}{h}$ is $f(x)$.

So the derived function of the area gives the height of the curve.

Thus by integration of the function we can find areas under the graph of the function.

In Figure 6, h can be thought of as an increase in x. The symbol δx is also used, the δ (delta) being used to mean 'increase in'. With this notation, the shaded area is approximately $f(x)\,\delta x$, and the area under the graph can be found approximately by summing many such strips. We could write

$$A(x) = \mathsf{S}\,f(x)\,\delta x$$

and it is from this notation that the notation $\displaystyle\int f(x)\,dx$ is derived.

Miscellaneous exercise

1 A point moves along a line in such a way that its velocity v metres per second at time t seconds is given by
$$v(t) = 1 + 5t - t^2.$$
   (a) For how long is the velocity greater than 5 m s^{-1}?
   (b) How far does the point move between $t = 0$ and $t = 2$?   (SMP)

2 Draw a graph of $y = \dfrac{12}{x+1}$ for $0 \leqslant x \leqslant 5$.

   (a) By using the trapezium rule, or otherwise, estimate to two significant figures the area between this portion of the curve and x-axis. Hence estimate the value of
$$\int_0^5 \frac{1}{x+1}\, dx.$$

   (b) Also estimate the derivative of $1/(x+1)$ at $x = 3$, stating your method clearly.   (SMP)

3 Sketch the curve $y = (x-3)^2$ for $0 \leqslant x \leqslant 5$.
   Find by integration the area enclosed between the x-axis, the line $x = 1$, and part of the curve.   (SMP)

4 Calculate the area enclosed by the graphs of $y = x^2 + 3$ and $y = 4x$.   (SMP)

5 Evaluate $\displaystyle\int_0^2 (x^2 + 1)\, dx$.

   Explain, with the aid of a sketch, how you could deduce the value of $\displaystyle\int_{-2}^2 (x^2 + 1)\, dx$.

   Show, preferably by means of another sketch, that $\displaystyle\int_{-2}^2 x(x^2 + 1)\, dx = 0$.   (SMP)

6 Approximations to the area under a curve between given ordinates can be obtained from the value of five equally spaced ordinates $\frac{1}{2}w$ apart by either of the following formulae:

   (i) $w\left\{\dfrac{y_0 + y_2}{2} + \dfrac{y_2 + y_4}{2}\right\}$, the trapezium rule.

   or

   (ii) $w\{y_1 + y_3\}$, the mid-ordinate rule.

   Explain why, in the case of the curve sketched in Figure 7, the value of T given by

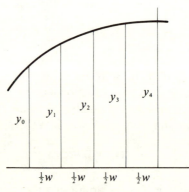

Figure 7

(i) is too small and the value of M given by (ii) is too large. Sketch a curve for which the reverse is true.

Obtain by an exact method the area enclosed by the curve $y = x^2$, the ordinates at $x = 1$, $x = 3$ and the x-axis. Verify in this case that the value $\frac{1}{3}(2M + T)$ is exact. (Take $w = 1$.) (SMP)

7 Calculate $\displaystyle\int_{-3}^{1} (4 - x^2)\,dx$ and show on a sketch the area which this represents.

Show that $\displaystyle\int_{a}^{1} (4 - x^2)\,dx = 0 \Rightarrow a^3 - 12a + 11 = 0$. Write down one root of this equation and show that the other roots satisfy the equation $a^2 + a - 11 = 0$.

Calculate these two roots correct to one place of decimals. Hence indicate on your graph two areas below the x-axis which are together numerically equal to that bounded by the curve above the x-axis. (SMP)

7

Oscillations

1. THE BIG WHEEL

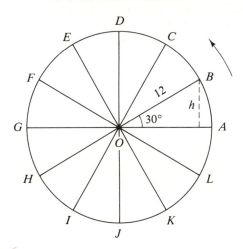

Figure 1

The Big Wheel at a fair has a radius of 12 m and turns anticlockwise. A passenger starts a trip on the wheel at A, level with the centre O of the wheel. What is her height above the level of O when the wheel has turned through 30°, so that she is at B?

This is a simple trigonometrical calculation. If her height above O is h metres, then $h = 12 \sin 30° = 12 \times 0.5 = 6$. In the same way, when the wheel has turned through 60° and the passenger is at C her height is given by $h = 12 \sin 60° = 12 \times 0.866 = 10.4$ to 3 s.f.

In general, if the wheel has turned through an angle $\theta°$, the passenger's height above O is given by the formula

$$h = 12 \sin \theta°.$$

You should recall that, for example, $\sin 150°$ has the same value as $\sin 30°$ — when the wheel has turned through 150° the passenger is at the same height as she was when the wheel had turned through 30°.

Similarly, we can see that when $\theta = 210$, the passenger is at H, and she is as far *below* O as she was *above* O when she was at F or B. Thus for $\theta = 210$, $h = {}^-6$, and we deduce that $\sin 210° = {}^-0.5$. The value of $\sin 210°$ has the same magnitude as $\sin 30°$, but is negative. It is clear that the value of h is negative throughout the interval when $180 < \theta < 360$, and so the value of $\sin \theta$ is also negative throughout this interval.

When the wheel has turned through 360°, we say it has completed a *cycle*. If it now continues on its second cycle the passenger will be back at B when the wheel has turned through 390°. The height h is again positive and we deduce that $\sin 390° = \sin 30°$.

Exercise A

(Do not use a calculator in this exercise.)

1   Given that $\sin 30° = 0.5$ and $\sin 60° = 0.87$ to 2 significant figures, use the diagram of the Big Wheel to write down values for the following:
   (a) $\sin 120°$;   (b) $\sin 300°$;   (c) $\sin 330°$;
   (d) $\sin 420°$;   (e) $\sin 240°$;   (f) $\sin 570°$.

2   Given that $\sin 20° = 0.342$ and $\sin 70° = 0.940$ to 3 significant figures, use the symmetry of the Big Wheel to write down values for the following:
   (a) $\sin 110°$;   (b) $\sin 160°$;   (c) $\sin 340°$;
   (d) $\sin 470°$;   (e) $\sin 250°$;   (f) $\sin 650°$.

3   Given that $\sin 40° = 0.64$ and $\sin 50° = 0.77$ to 2 significant figures, use the symmetry of the Big Wheel to write down values for the following:
   (a) $\sin 140°$;   (b) $\sin 220°$;   (c) $\sin 230°$;
   (d) $\sin 500°$;   (e) $\sin 310°$;   (f) $\sin 400°$.

4   Given that $\sin 10° = 0.17$ and $\sin 80° = 0.98$ to 2 significant figures, use the symmetry of the Big Wheel to write down values of the following:
   (a) $\sin 100°$;   (b) $\sin 170°$;   (c) $\sin 440°$;
   (d) $\sin 190°$;   (e) $\sin 350°$;   (f) $\sin 710°$.

5   Use the diagram of the Big Wheel to deduce values for;
   (a) $\sin 90°$;   (b) $\sin 180°$;   (c) $\sin 270°$;
   (d) $\sin 360°$.

6   The wheel accidentally starts to turn backwards. Write down values of:
   (a) $\sin (^-30)°$;   (b) $\sin (^-180)°$;   (c) $\sin (^-210)°$;
   (d) $\sin (^-270)°$.

2. THE SINE CURVE

Figure 2 shows the graph of the height of the passenger on the Big Wheel above the centre, that is, the graph of $h = 12 \sin \theta°$. This has, of course, the familiar shape of the sine curve. As we expect, the value of h varies between ± 12, since $\sin \theta$ varies between ± 1. The number 12 is called the *amplitude* of the motion. It represents the maximum distance of the passenger above or below the centre of the motion.

Suppose that the lowest point of the Big Wheel was 2 metres above the ground, and that we were measuring the height of the passenger above the ground instead of above the centre of the wheel. Our formula for h would now be

$$h = 12 \sin \theta° + 14,$$

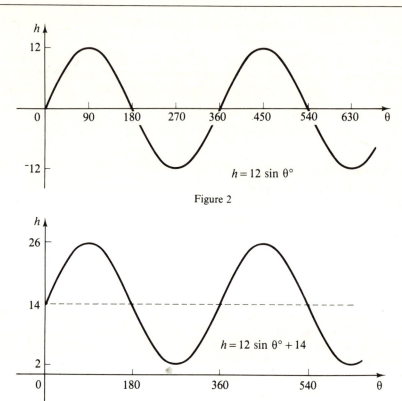

Figure 2

Figure 3

and its graph would look like Figure 3. The amplitude of the motion is still 12, but the graph of $h = 12 \sin \theta°$ has undergone a translation of $\begin{bmatrix} 0 \\ 14 \end{bmatrix}$.

Figure 4 shows the graphs of $h = 12 \sin \theta°$ and $h = \sin \theta°$ (that is, a sine curve with amplitude 1 unit) on the same diagram. It is clear that one graph is the image of the other under a one-way stretch of scale factor 12 in the direction of the h-axis.

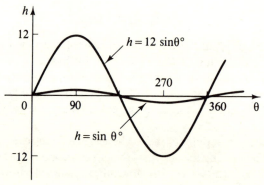

Figure 4

We may regard the graph of $h = \sin \theta°$ as the 'standard' sine curve, from which other sine curves can be obtained by a combination of translations, reflections and one-way stretches. The graph of $h = \sin \theta°$ in Figure 4 is too 'thin' to be of much use, but Figure 5 shows the same curve using a larger scale on the vertical axis. Notice that the graph has been extended to include points corresponding to negative values of θ.

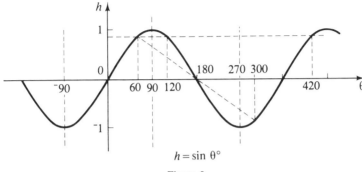

$$h = \sin \theta°$$

Figure 5

Figure 5 indicates various axes of symmetry of the graph, such as the lines $\theta = 90, {}^-90, 270$, and also some centres of half-turn symmetry, such as the point $(180, 0)$.

It is helpful to relate these to the picture of the Big Wheel; for example:
  (i) $\sin 60°$ and $\sin 120°$ have the same value, because the height of the passenger at a point 30° *before* reaching the top of the wheel is the same as her height 30° *after* reaching the top;
 (ii) similarly, $\sin 420° = \sin 60°$ because when $\theta = 420$ the passenger has completed one cycle and is back at the '60°' position;
(iii) $\sin 300° = {}^-\sin 60°$ because when $\theta = 300$ the passenger is the same distance *below* the centre as she was *above* the centre when $\theta = 60$.

The symmetries of the graph can also be used to solve equations, as the following example illustrates.

Example 1
Find all the solutions (to the nearest integer) in the interval $^-360 < \theta < 360$ of the equation $\sin \theta° = 0.77$.

Since $\sin^{-1} 0.77 = 50°$ to the nearest degree, we know that 50 is one solution. We can then deduce other solutions by reference to the graph:

$\theta = 130$ (reflection in $\theta = 90$)

$\theta = {}^-230$ $\left(\text{reflection in } \theta = {}^-90 \text{ or translation of } \begin{bmatrix} ^-360 \\ 0 \end{bmatrix} \right.$

$\left. \text{applied to the solution } \theta = 130 \right)$

$\theta = {}^-310$ $\left(\text{translation of } \begin{bmatrix} ^-360 \\ 0 \end{bmatrix} \right).$

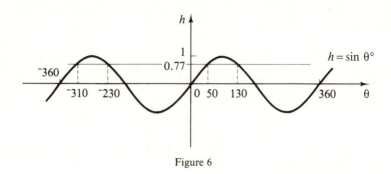

Figure 6

The four solutions, $\theta = {}^-310, {}^-230, 50, 130$, are illustrated in Figure 6.

Exercise B

1 On the same diagram sketch the graphs of:
   (a) $h = \sin \theta°$;   (b) $h = 2 \sin \theta°$;   (c) $h = \frac{1}{2} \sin \theta°$;
   (d) $h = \sin \theta° + 2$;   (e) $h = 2 \sin \theta° + 2$.

2 Use tables or a calculator to find, to the nearest degree, an angle θ such that $\sin \theta° = 0.8$. By using the symmetry of the graph of $h = \sin \theta°$, write down three other solutions of the equation $\sin \theta° = 0.8$ lying in the interval $0 < \theta < 720$.

3 Find (to the nearest integer) four solutions of the equation $\sin \theta° = 0.72$ which lie in the interval $^-360 < \theta < 360$.

4 Use tables or a calculator to find (to the nearest degree) the value of $\sin^{-1} 0.56$. Use the half-turn symmetry of the graph of $h = \sin \theta°$ to find two solutions of the equation $\sin \theta° = {}^-0.56$ in the interval $^-180 < \theta < 180$.

5 In the example of the Big Wheel, use the formula $h = 12 \sin \theta°$ to find (to the nearest integer) the two values of θ in the interval $0 < \theta < 360$ when the passenger is 8 metres above the centre of the wheel.

6 Using the formula $h = 12 \sin \theta° + 14$, find two values of θ in the interval $0 < \theta < 360$ when the passenger is 5 metres above the ground.

7 On the same diagram sketch graphs of:
   (a) $y = \sin x°$;   (b) $y = 3 \sin x°$;   (c) $y = {}^-\sin x°$;
   (d) $y = \sin x° - 1$;   (e) $y = 3 \sin x° - 1$.

8 Find $\sin^{-1} 0.68$. Hence write down two solutions of the equation $\sin \theta° = 0.68$ in the interval $0 < \theta < 360$.

9 Find four solutions of the equation $\sin \theta° = 0.454$ in the interval $0 < \theta < 720$.

10 Find two solutions of the equation $\sin \theta° = {}^-0.848$ in the interval $^-180 < \theta < 180$.

11 On an even bigger Big Wheel, the height of a passenger above the centre when the wheel has turned through $\theta°$ is given by the formula $h = 15 \sin \theta°$. Find two values of θ in the interval $0 < \theta < 360$ when the passenger is 7 metres above the centre of the wheel.

12 On this Big Wheel, the formula for the height of the passenger above the ground is $h = 15 \sin \theta° + 16$. Find two values of θ in the interval $0 < \theta < 360$ when the passenger is 14 metres above the ground.

3. ANGULAR VELOCITY

We return once more to consideration of the Big Wheel. Suppose it is rotating anticlockwise at a steady rate of 10° per second. This measurement is called the *angular velocity* of the wheel. Since the wheel has to turn through 360° to complete one cycle, the time taken to do this will be $\frac{360}{10} = 36$ seconds. This time (36 seconds) is called the *period* of the motion; the period of a motion like this is the time taken to complete one cycle.

If the angular velocity is 10° per second, we can deduce that after t seconds the wheel will have turned through $10t$ degrees; as a formula we may write

$$\theta = 10t.$$

We may now go back to the formula for the height of a passenger above the centre of the wheel, namely

$$h = 12 \sin \theta°$$

and rewrite it as $\qquad h = 12 \sin (10t)°.$

It is important not to allow the 'degrees' symbol to cause confusion. The symbol t represents a *time* (here measured in seconds); it is the expression $(10t)$ which represents the angle of rotation and which is here measured in degrees.

Figure 2 showed the graph of $h = 12 \sin \theta°$. Figure 7 shows the same graph, but the horizontal axis has been labelled both with values of θ and values of t, so that the diagram can also be interpreted as a graph of $h = 12 \sin (10t)°$.

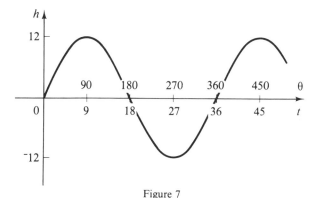

Figure 7

Notice that if we consider the θ scale the graph starts to 'repeat itself' when $\theta = 360$, but if we consider the t scale the graph starts to repeat itself when $t = 36$, because the *period* of the motion is 36 seconds.

If the wheel were turning very slowly, at an angular velocity of only 1° per second, the equation for the passenger's height above the centre would be

$$h = 12 \sin t°.$$

Figure 8 shows the graph of $h = 12 \sin (10t)°$ and $h = 12 \sin t°$ on the same diagram. Notice how the graph of $h = 12 \sin (10t)°$ completes *ten* cycles during the time that the graph of $h = 12 \sin t°$ takes to complete *one* cycle. Of course,

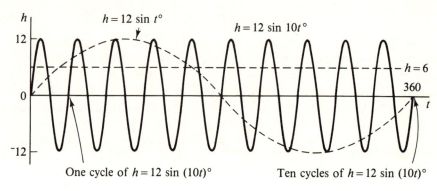

One cycle of $h = 12 \sin (10t)°$ Ten cycles of $h = 12 \sin (10t)°$

Figure 8

this is because in the former case the wheel is turning ten times as fast. The graph of $h = 12 \sin (10t)°$ is the image of the graph of $h = 12 \sin t°$ under a one-way stretch of scale factor $\frac{1}{10}$ parallel to the t-axis.

Example 2

A ride on the Big Wheel lasts six minutes. On how many occasions is the passenger 6 metres above the centre (a) when the wheel is rotating at 1° per second, (b) when it is rotating at 10° per second? Find the corresponding values of t.

Since the ride lasts six minutes (360 seconds) we require solutions to the equations:

(a) $6 = 12 \sin t°$;   (b) $6 = 12 \sin (10t)°$

in the interval $0 \leqslant t \leqslant 360$.

The line $h = 6$ intersects one curve twice in that interval, and the other curve 20 times in the same interval (see Figure 8), so the answers are:

(a) 2;   (b) 20.

In case (a) we require solutions of $\sin t° = 0.5$. Since $\sin^{-1} 0.5 = 30°$ one solution is $t = 30$, and from the symmetry of the graph we deduce that another solution is $t = 150$. These are the only solutions in the interval $0 \leqslant t \leqslant 360$ and so the required values of t are 30 and 150.

In case (b) it is convenient to substitute $\theta = 10t$, thus altering the equation $6 = 12 \sin (10t)°$ to $6 = 12 \sin \theta°$. We already know that the only two solutions in the interval $0 \leqslant \theta \leqslant 360$ (that is, the first cycle of the graph) are

$$\theta = 30 \quad \text{and} \quad \theta = 150$$

corresponding to $t = 3 \quad \text{and} \quad t = 15.$

There are two solutions in each cycle. Since each cycle takes 36 seconds, other solutions are

$$t = 3 + 36, \, 3 + 72, \, 3 + 108, \, 3 + 144, \text{ etc.}$$
$$t = 15 + 36, \, 15 + 72, \, 15 + 108, \, 15 + 144, \text{ etc.}$$

The complete set of solutions in the interval $0 \leqslant t \leqslant 360$ is {3, 15, 39, 51, 75, 87, 111, 123, 147, 159, 183, 195, 219, 231, 255, 267, 291, 303, 327, 339}.

Example 3

Solve $\sin 3t° = 0.84$, giving all solutions (to the nearest integer) in the interval $^-360 \leqslant t \leqslant 360$.

First we observe that the graph of $h = \sin 3t°$ completes three cycles during the time that the graph of $h = \sin t°$ completes one, so that $\sin 3t°$ has period $\frac{1}{3} \times 360$, that is, 120.

We then find the solutions in the first cycle:

$$\sin^{-1} 0.84 = 57°, \text{ to the nearest degree,}$$

so
$$3t = 57 \text{ or } 180 - 57 \text{ in the first cycle,}$$

$$t = 19 \text{ or } 41.$$

All other solutions are obtained by adding or subtracting multiples of the period. We only retain those in the interval required, $^-360 \leqslant t \leqslant 360$:

$$t = 19, 19 - 120, 19 - 240, 19 - 360, 19 + 120, 19 + 240,$$

$$t = 41, 41 - 120, 41 - 240, 41 - 360, 41 + 120, 41 + 240.$$

The solution set is

$$\{^-341, ^-319, ^-221, ^-199, ^-101, ^-79, 19, 41, 139, 161, 259, 281\}.$$

Exercise C

1   The angular velocity of a Big Wheel of radius 12 m is 4° per second.
    (a) What is its period?
    (b) The ride lasts six minutes. On how many occasions will the passenger be 8 metres above the centre of the wheel?
    (c) Find the corresponding values of t.

2   Repeat question 1 when the angular velocity is 6° per second.

3   Solve the following equations, giving all solutions in the interval $0 \leqslant t \leqslant 360$:
    (a) $\sin 2t° = 0.72$;   (b) $\sin 3t° = ^-0.56$;   (c) $\sin \frac{1}{2}t° = 0.8$.

4   A bicycle wheel is spinning at 120° per second. The height (h cm) of the valve above the centre at time t second is given by the formula

$$h = 35 \sin 120t°.$$

    (a) What is the period of the motion?
    (b) Find the times in the interval $0 \leqslant t \leqslant 10$ when the valve is 7 cm below the level of the centre.

5   Repeat question 4 when the angular velocity is 90° per second.

6   Solve the following equations, giving all solutions in the interval $0 \leqslant t \leqslant 180$:
    (a) $\sin 5t° = 0.4$;   (b) $\sin 8t° = ^-0.68$;   (c) $\sin \frac{1}{3}t° = 0.454$.

4. SINUSOIDAL MOTION

The work in this chapter so far has been derived from the fact that, provided we measure from a suitable origin, the height of a point fixed on the circumference of a circle of radius a which is rotating at a constant angular velocity of $k°$ per second is given by the formula

$$h = a \sin kt°;$$

the *amplitude* of the motion is a (the maximum distance the point can reach from the centre of the motion) and the *period* of the motion is $\frac{360}{k}$ seconds (the time taken to complete one cycle).

There are many other situations where some physical quantity fluctuates in a regular manner, repeating its changes periodically at equal intervals of time. In many of these the variation, plotted against time, approximates more or less closely to the form of the sine curve, in which case it is said to be *sinusoidal*, or the motion is said to be *simple harmonic*. Here are some examples.

(*a*) The height, h, of the tide above mean tide-level varies with time according to the approximate formula

$$h = a \sin kt°,$$

where a is the maximum height above or below mean tide-level.

(*b*) If an object hanging motionless on the end of an elastic spring is given a small vertical push, its distance from its original position varies with time according to the formula

$$x = a \sin kt°.$$

(*c*) When a coil of wire rotates in a magnetic field a current i amps is produced in the coil, which is not constant but varies with time according to the formula

$$i = a \sin kt°.$$

This is the familiar domestic alternating current.

(*d*) If a tuning fork is struck, the distance between the prongs varies with time sinusoidally; so does the pressure of air at a given point in an organ pipe, or the displacement of a definite point on a vibrating violin string.

There are two important points to remember when using the expression $A \sin kt°$ in each of the above cases:

(i) the use of the 'degrees' symbol ° does not mean that an *angle* is relevant. The symbol t represents a time, and the ° symbol is probably best thought of as a reminder to ensure that the calculator is set to operate in degrees before using the sine function key;

(ii) in the example of the Big Wheel the symbol k represented an angular velocity, but in the cases listed above there is no 'angular velocity' involved; the quantity we usually know in such situations is the period, P. Since $\sin kt°$ has period $\frac{360}{k}$, we have

$$P = \frac{360}{k} \text{ and hence } k = \frac{360}{P}.$$

In some situations, such as examples (*c*) and (*d*) above, the period is a very short interval of time, perhaps $\frac{1}{100}$ of a second or less. If one cycle takes $\frac{1}{100}$ s, then 100 cycles will be completed in 1 second. Here we would say that the *frequency* is 100 cycles per second.

A frequency of 1 cycle per second is called 1 hertz; you will know that radio frequencies, for example, are given in kilohertz (kHz). If a sinusoidal motion has a period p seconds, then its frequency is $\dfrac{1}{p}$ Hz.

Example 4
An object hanging motionless at the end of an elastic spring is pushed down so that it oscillates to a maximum of 6.0 cm above and below its equilibrium position. It completes one cycle every 5 seconds. Find:

(a) the displacement after 3 seconds;

(b) the time elapsed on the first two occasions that it is 4.0 cm below its starting position.

The amplitude of the motion is 6.0 cm. The period is 5 seconds, so we calculate the value of k in the expression $A \sin kt°$ from the fact that $\frac{360}{k} = 5$; this gives $k = 72$. Hence, if d cm denotes the displacement downwards at time t seconds,

$$d = 6 \sin 72t°.$$

(a) When $t = 3$, $d = 6 \sin 216° = 6 \times (^-0.588)$
$$= ^-3.5 \text{ to 2 s.f.}$$

After 3 seconds it is 3.5 cm above its original position.

(b) When $d = 4$, then $4 = 6 \sin 72t°$
$$\Rightarrow \quad \sin 72t° \approx 0.6667.$$

Now $\sin^{-1} 0.667 \approx 41.8°$, so that (using the symmetry of the sine curve) the first two values of $72t$ are 41.8 and 138.2, and thus the first two values of t are 0.58 and 1.92 approximately. These are the numbers of seconds elapsed on the first two occasions when the object is 4.0 cm below its starting position (once on the 'way down' and once on the 'way up').

Exercise D

1   The *frequency* of an alternating current is 50 cycles per second. What is its period? If the formula for the current i is given by $i = A \sin kt°$, what is the value of k?

2   A pendulum which is initially hanging vertically is set in motion so that it swings out to a maximum angle of 15°. Its period is 3 seconds. Assuming that the motion is simple harmonic, write down the formula connecting its angular displacement $\theta°$ from the vertical with the time t seconds which has elapsed since it was set in motion. Calculate its angular displacement after

(a) 0.75 seconds, (b) 1.5 seconds, (c) 2.25 seconds, (d) 1 second.

Calculate also the first time at which its displacement is 6° from the vertical.

3   At a certain port the tide has a greatest height of 4 metres above mean tide-level and its period is 12 hours. Write down a formula connecting its height (h metres) above mean tide-level with the time (t hours) which has elapsed since a moment when

the tide was at its mean, and rising. If mean tide was at 7 a.m. on a certain day, when was the next (*a*) high tide, (*b*) low tide?

Calculate the height of the tide at noon on that day.

A boat cannot leave that port when the tide is more than 1 metre below mean tide-level. What is the latest time before 4 p.m. that the boat can leave port?

4 A toyshop display consists of a plastic duck hanging on the end of a long spring. It is initially at rest, but is then given a sharp tap so that it descends a maximum of 50 cm and then oscillates sinusoidally, making one complete cycle every two seconds. If *x* cm is its depth below its original position *t* seconds after the tap, write down a formula connecting *x* and *t*.

After how long does it
(*a*) return to its original level,
(*b*) first arrive at its lowest point,
(*c*) first arrive at its highest point,
(*d*) first arrive at a point 25 cm below its original point,
(*e*) next arrive at a point 25 cm below its original point,
(*f*) first arrive at a point 25 cm above its original point?

5 A pendulum which is initially hanging vertically is set in motion so that it swings out to a maximum angle of 10°. Its period is 1 second. Write down a formula connecting its angular displacement $\theta°$ from the vertical with the time *t* seconds which has elapsed since it was set in motion. What is its angular displacement after
(*a*) 0.5 seconds,   (*b*) 1 second,   (*c*) 1.2 seconds,   (*d*) 3.4 seconds?
Calculate the first two times at which its angular displacement is 3°.

6 On a certain beach the water is 5 metres from the promenade at high tide and 65 metres from the promenade at low tide. Assuming that the period of the tide is 13 hours and that the distance of the water from mean tide-level varies sinusoidally with time, write down a formula for the distance *d* metres of the water above mean tide-level *t* hours after it was at that level.

If low tide was at 9 a.m., how far from the promenade will the water be at 12.15 p.m.?

A holidaymaker settles in a deckchair early in the morning at a point 15 metres from the promenade. At what time will the tide reach him?

5. SHIFTING THE ORIGIN

A sinusoidal motion will have an equation $x = A \sin kt°$ only if the quantity *x* is measured from a moment when the oscillation is at its centre. We have already met, in Section 2, a situation where *x* (or, in that case, *h*) was not measured from the centre of the motion. There we measured the height of a passenger on the Big Wheel with reference to the *ground* rather than the centre of the wheel, using the formula $h = 12 \sin \theta° + 14$ instead of the formula $h = 12 \sin \theta°$. This was an example of shifting the origin, or, more accurately, shifting the sine curve by a translation $\begin{bmatrix} 0 \\ 14 \end{bmatrix}$. In the same way, measuring the time from a moment other than the centre of the motion will translate the sine curve parallel to the *horizontal* axis.

Example 5

The Big Wheel, of radius 12 metres, is rotating with an angular velocity of 10° per second. The boarding point for passengers is at X, where OX makes an angle 60° below the horizontal. Write down a formula for the height of the passenger above the boarding point t seconds after the wheel has started. (See Figure 9.)

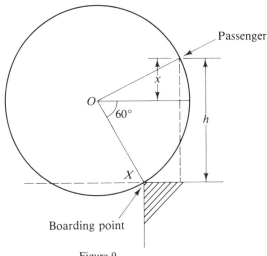

Figure 9

It is best to start the solution by measuring the height x metres above the *centre* of the wheel. If time were also to be measured from the moment when the passenger is level with the centre we would have the familiar formula $x = 12 \sin 10t°$. However, when the passenger reaches this point the wheel has already turned through 60°, so 6 seconds have already elapsed. We require x to have the value zero when $t = 6$; this is easily achieved by rewriting the formula as

$$x = 12 \sin 10(t-6)°.$$

The expression $(t-6)$ appears in place of t because the first 6 seconds are used to bring the passenger up to the level of O.

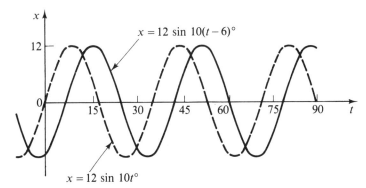

Figure 10

The centre of the wheel is $12 \sin 60°$ ($= 10.4$ approximately) metres above the boarding point, so that the height h metres above the boarding point is given by $h = x + 10.4$; thus the required formula connecting h and t is

$$h = 12 \sin 10(t-6)° + 10.4.$$

Check that when $t = 0$, $h = 0$, as required.

Figure 10 shows the graph of $x = 12 \sin 10(t-6)°$. Notice that it is the image of the graph of $x = 12 \sin 10t°$ under a translation $\begin{bmatrix} 6 \\ 0 \end{bmatrix}$.

Example 6

Solve the equation $12 \sin 10(t-6)° = 1$, for $0 \leqslant t \leqslant 60$.

Let $\theta = 10(t-6)$, so that $12 \sin \theta° = 1$. Then $\sin \theta = \frac{1}{12} \approx 0.08333$. $\sin^{-1} 0.08333 \approx 4.8°$, so one solution for θ is 4.8. By the symmetry of the sine curve, the other solution in the first cycle is 175.2, so two solutions for t can be found from

$$10(t-6) = 4.8$$

and

$$10(t-6) = 175.2,$$

giving $t = 6.48$ and $t = 23.52$.

The period of the motion is $\frac{360}{10} = 36$ seconds, so there will be other values of t which are 36 seconds 'further on' than these, namely 42.48 and 59.52. The next one will be 78.48, which is outside the interval $0 \leqslant t \leqslant 60$. The required solutions are (to 2 decimal places):

$$t = 6.48, \ 23.52, \ 42.58, \ 59.52.$$

Exercise E

1   In each part, sketch both graphs on the same diagram. Give three values of t at which each graph cuts the t-axis.
    (a) $h = \sin t°$, $h = \sin (t-20)°$;   (b) $y = \sin t°$, $y = \sin (t+40)°$;
    (c) $y = \sin 2t°$, $y = \sin 2(t+30)°$;   (d) $x = \sin 3t°$, $x = \sin (3t+90)°$.

2   Solve the following equations, giving answers in the interval $0 \leqslant x \leqslant 360$ to the nearest integer:
    (a) $\sin (x-20)° = 0.788$;   (b) $\sin 3(x-20)° = 0.866$;
    (c) $\sin \frac{1}{2}(x+40)° = 0.225$;   (d) $\sin (2x+40)° = {}^-0.891$.

3   (a) A child is playing on a swing. The amplitude of the swing is $20°$ and the period 3 seconds. Write down a formula for the child's angular displacement, $\theta°$, from the vertical t seconds after her father gave her the first 'push' when the swing was vertical. (Assume simple harmonic motion.)
    (b) If instead her father pulls the swing out to a point $20°$ from the vertical and then lets go, how long will the swing take to descend to the vertical position? What is the formula for the motion in this case? When is the first time that the swing is displaced $10°$ from the vertical (i) on the way down, (ii) on the way up the other side?

4   In each part, sketch both graphs on the same diagram. Give three values of t at which each graph cuts the t-axis.
    (a) $y = \sin t°$, $y = \sin (t+40)°$;   (b) $h = \sin t°$, $h = \sin (t-10)°$;
    (c) $y = \sin 3t°$, $y = \sin 3(t+30)°$;   (d) $x = \sin \frac{1}{3}t°$, $x = \sin (\frac{1}{3}t-10)°$.

5   Solve the following equations, giving answers to the nearest integer in the interval.
$^-180 \leqslant x \leqslant 180$:
    (a) $\sin 2(x+50)° = 0.5$;   (b) $\sin \frac{1}{2}(x-25)° = 1$;
    (c) $\sin (\frac{1}{2}x-30)° = 0.122$;   (d) $\sin (2x+40)° = ^-0.891$.

6   On a certain beach the tide has an amplitude of 6 metres and a period of 12.5 hours.
Write down a formula for the height (h metres) of the tide above mean tide-level
t hours after midnight on Monday night if the last mean tide on Monday was
11.45 p.m. and the tide was then rising.
    Calculate the times when $h = 1$ on Tuesday morning.

6. THE COSINE FUNCTION

The graph of $y = \cos x°$ is the image of the graph of $y = \sin x°$ under the
translation $\begin{bmatrix} ^-90 \\ 0 \end{bmatrix}$ (see Figure 11). Thus we have the relation

$$\cos x° = \sin (x+90)°.$$

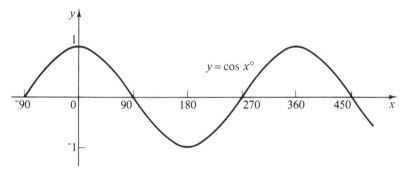

Figure 11

Notice that the cosine function graph (a) 'starts' at $(0, 1)$ and (b) initially
decreases as x increases. This means that the cosine function is particularly
convenient to describe sinusoidal variation when we measure time, not from the
moment when the object concerned is at the 'centre' of its motion, but from
an extremity. As an example, consider the diagram of the Big Wheel in Section 1:
suppose that, instead of measuring the height of the passenger above the centre,
we are interested in the distance that the passenger finds herself to the right of
the centre of the wheel at any time. If this distance is x metres, it is clear that
initially x has its greatest value, 12, and that x decreases just after the wheel
starts. This makes the cosine function appropriate to describe the motion. The
required equation is in fact

$$x = 12 \cos kt°,$$

where k depends on the angular velocity.
    Compare this with the formula for the height (h metres) of the passenger above
the centre of the wheel:

$$h = 12 \sin kt°;$$

the periods are the same, and so are the maximum values of h and x. Only the function (cosine instead of sine) is different.

Since the graph of the cosine function is only a translation of the graph of the sine function, the principles which have been used in this chapter so far can be applied equally well to the cosine function. The following exercise illustrates this point.

Exercise F

1 An object hanging on a spring is pulled down a distance of 8.0 cm and then released. Its period is 4 seconds. Write down a formula connecting its displacement d cm from its original position with the time t seconds which has elapsed since it was released.

2 (a) Write down three vertical axes of symmetry of the graph $y = \cos x°$.
   (b) Write down the coordinates of three centres of half-turn symmetry of the graph of $y = \cos x°$.
   (c) Given that $\cos 35° = 0.82$ approximately, write down the values of:
       (i) $\cos (^-35)°$; (ii) $\cos 145°$; (iii) $\cos 215°$; (iv) $\cos 325°$.

3 Sketch the following graphs:
   (a) $y = \cos 4x°$;   (b) $y = 4 \cos x°$;   (c) $y = 4 + \cos x°$;
   (d) $y = {}^-\cos x°$;   (e) $y = \cos (x - 90)°$; (f) $y = \cos (x + 20)°$;
   (g) $y = \cos (3x - 45)°$; (h) $y = \cos 3(x - 45)°$.

4 Solve the following equations, giving solutions in the interval $0 \leqslant x \leqslant 360$:
   (a) $\cos x° = 0.5$; (b) $\cos x° = {}^-0.788$;   (c) $\cos 2x° = 0.829$;
   (d) $\cos x° = 0$;   (e) $\cos (\frac{1}{2}x + 40)° = 0.891$; (f) $\cos 2(x + 50)° = 0.866$.

5 Solve the following equations, giving solutions in the interval $^-180 \leqslant x \leqslant 180$:
   (a) $\cos x° = 0.225$; (b) $\cos 3x° = {}^-0.5$; (c) $\cos (2t - 10)° = 0.866$.

6 The period of the tide at a certain resort is 12 hours, and the difference between high tide and the mean tide-level is 8 metres.
   Write down a formula connecting the height h of the tide above mean tide-level with the time t hours which has elapsed since the last high tide.

7 Given that $\cos 160° = {}^-0.94$ approximately, write down the values of:
   (a) $\cos 200°$; (b) $\cos 20°$; (c) $\cos (^-160)°$; (d) $\cos 340°$.

8 Sketch the following graphs:
   (a) $y = 2 + \cos x°$; (b) $y = 2 \cos x°$;   (c) $y = \cos 2x°$;
   (d) $y = 2 - \cos x°$; (e) $y = \cos (x - 20)°$; (f) $y = \cos (2x - 20)°$;
   (g) $y = \cos 2(x - 20)°$.

9 Solve the following equations, giving solutions in the interval $0 \leqslant x \leqslant 360$:
   (a) $\cos x° = 0.848$;   (b) $\cos x° = {}^-1$;   (c) $\cos 3x° = {}^-0.682$;
   (d) $\cos \frac{1}{2}x° = 0.766$; (e) $\cos (2x - 50)° = 0$; (f) $\cos (3x - 26)° = {}^-0.515$.

10 Solve the following equations, giving solutions in the interval $^-180 \leqslant x \leqslant 180$:
   (a) $\cos x° = 0.8746$; (b) $\cos 3x° = {}^-0.309$; (c) $\cos (x - 24)° = {}^-0.9848$.

7. THE TANGENT FUNCTION

When the passenger on the Big Wheel is at P, as shown in Figure 12, we know that the gradient of the line OP is $\dfrac{h}{d}$, and that this ratio is called tan $\theta°$.

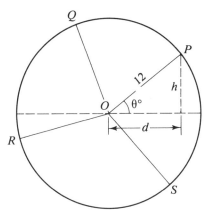

Figure 12

We now proceed in the same way as we did for sin $\theta°$ and *define* tan $\theta°$ for angles outside the interval $0 \leqslant \theta < 90$ as the *gradient of the line joining the centre of the wheel to the passenger*. This means that when $90 < \theta < 180$ and the passenger is at Q, for example, tan $\theta°$ is *negative*. Similarly, tan $\theta°$ is also negative when the passenger is at S. At R, however, when $180 < \theta < 270$, tan $\theta°$ is positive.

In Figure 12 we know that $h = 12 \sin \theta°$ and $d = 12 \cos \theta°$, so that

$$\tan \theta° = \frac{h}{d} = \frac{12 \sin \theta°}{12 \cos \theta°},$$

and we have an alternative definition for tan $\theta°$:

$$\tan \theta° = \frac{\sin \theta°}{\cos \theta°}.$$

Clearly tan $\theta°$ is zero whenever sin $\theta°$ is zero, i.e. when $\theta = 0,\ \pm 180,$ $\pm 360,\ \ldots$ This is consistent with the gradient definition, since for these values of θ the passenger is level with the centre of the wheel and the gradient of the line joining her to the centre is zero.

When the passenger is at the top or bottom of the wheel the line joining her to the centre is vertical, so the slope is undefined – it is 'infinitely large'. These points correspond to $\theta = \pm 90,\ \pm 270,\ \ldots$, when cos $\theta°$ is zero, and therefore the fraction $\dfrac{\sin \theta°}{\cos \theta°}$ is undefined. The graph of tan $\theta°$ is *discontinuous* at these values of θ. When θ is very close to any of these values, cos $\theta°$ will be very close to zero, so that tan $\theta°$ will be numerically very large (Figure 13). Notice how the graph is composed of separate segments, all identical and spaced 180 apart; the

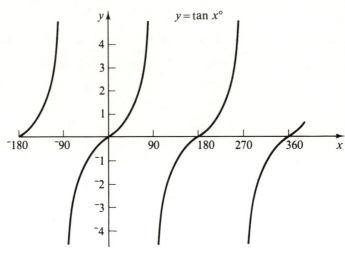

Figure 13

period of the function $f: x \rightarrow \tan x°$ is 180, unlike the functions sine and cosine, whose periods are both 360. Notice also that the graph has no reflective symmetry, though it has half-turn symmetry about the points $(0, 0)$, $(\pm 180, 0)$, etc. and also about the points $(\pm 90, 0)$, $(\pm 270, 0)$, etc.

Exercise G

1 Sketch the graphs of the following:
(a) $y = \tan x° - 2$; (b) $y = 2 \tan x°$; (c) $y = \tan (x + 45)°$;
(d) $y = \tan 2x°$; (e) $y = ^-\tan x°$; (f) $y = \tan (2x + 60)°$.

2 Given that $\tan 35° = 0.7$ approximately, write down the values of:
(a) $\tan (^-35)°$; (b) $\tan 215°$; (c) $\tan 145°$; (d) $\tan 325°$.

3 Solve the following equations, giving solutions in the interval $0 \leqslant x \leqslant 360$:
(a) $\tan x° = ^-0.231$; (b) $\tan x° = 0$; (c) $\tan (x - 30)° = 1$;
(d) $\tan 2x° = 0.9$; (e) $\tan \frac{1}{2}x° = ^-1.6$.

4 Sketch graphs of the following:
(a) $y = \tan x° + 1$; (b) $y = ^-\tan x° + 1$; (c) $y = \tan (^-x + 20)°$;
(d) $y = \tan 3x°$; (e) $y = \tan \frac{1}{2}x°$; (f) $y = \tan (3x - 90)°$.

5 Given that $\tan 71° = 2.9$ approximately, write down the values of
(a) $\tan 251°$; (b) $\tan (^-71)°$; (c) $\tan 109°$; (d) $\tan (^-109)°$.

6 Solve the following equations, giving solutions in the interval $^-180 \leqslant x \leqslant 180$:
(a) $\tan x° = 0.51$; (b) $\tan x° = ^-1.54$; (c) $\tan (x + 20)° = 0.81$;
(d) $\tan 3x° = ^-0.51$; (e) $\tan (2x + 15)° = 1$.

8. SUMMARY

If a quantity x is varying sinusoidally with time, has a maximum value a, and has period P, then the formula connecting x with t is

$$x = a \sin\left(\frac{360}{P}\right) t° \text{ if time is measured from the 'centre' of the motion,}$$

$$x = a \cos\left(\frac{360}{P}\right) t° \text{ if time is measured from an extremity of the motion.}$$

$y = \sin x°$ and $y = \cos x°$ both have period 360; $y = \tan x°$ has period 180. If x is replaced by kx in any of these functions, the period is *reduced* by a factor k.

Given a number k, tables or a calculator yield only *one* solution of an equation such as $\sin x° = k$, $\cos x° = k$, $\tan x° = k$. To find other solutions it is necessary to consider the symmetries of the graphs of $\sin x°$, $\cos x°$ or $\tan x°$.

$$\cos x° = \sin (x+90)°.$$

$$\tan x° = \frac{\sin x°}{\cos x°}.$$

Miscellaneous exercise

1 Sketch the graphs of:
   (a) $y = \cos x°$; (b) $y = \cos (x+90)°$; (c) $y = {}^{-}\cos (x+90)°$.
   Hence express ${}^{-}\cos (x+90)°$ in simpler form.

2 Solve the following equations for values of x in the interval $^{-}180 \leqslant x \leqslant 180$:
   (a) $\sin (3x+30)° = {}^{-}0.7$; (b) $\cos (40-2x)° = 0.3$; (c) $\tan (0.5x+10)° = 2$.

3 Sketch on the same diagram the graphs of $y = \sin x°$ and $y = \cos x°$. Hence find the solutions of the equation $\sin x° = \cos x°$ in the interval $0 \leqslant x \leqslant 360$.

4 Sketch on the same diagram the graphs of $y = \sin 2x°$ and $y = \cos x°$. Hence solve the equation $\sin 2x° = \cos x°$ in the interval $0 \leqslant x \leqslant 360$, given that one solution is $x = 30$.

5 Sketch the graph of $y = \cos 3x°$ for $0 \leqslant x \leqslant 180$. Calculate all the values of x in the interval $0 \leqslant x \leqslant 180$ for which
   (a) $\cos 3x° = \frac{\sqrt{3}}{2}$, (b) $\cos 3x° = {}^{-}\frac{1}{2}$.
   Hence write down the complete solution set of $^{-}\frac{1}{2} \leqslant \cos 3x° \leqslant \frac{\sqrt{3}}{2}$ for $0 \leqslant x \leqslant 180$.
   (SMP)

6 A sand-bar is 7 metres below mean tide-level. The amplitude of the tide is 5 metres and its period 12.5 hours. Write down a formula for the height (h metres) of the tide above the sand-bar t hours after noon on Friday if the tide was at its mean at 10 a.m. that day and rising.
   (a) What was the depth of water over the sand-bar at noon?
   (b) A ship needs a depth of 3 metres over the sand-bar in order to cross it. Between which times on that day can it cross the bar?

7 When a violin string tuned to A is bowed, the string vibrates with a frequency of 440 cycles per second. What is the period of the vibration? If the amplitude is 0.5 mm, write down a formula for the displacement y mm of a particular point P on the string t seconds after it is at a point of maximum displacement.

8

Triangles

Practical calculations in trigonometry depend on the existence of relations between the sides and angles of a figure, and among these relations those that refer to the sides and angles of a triangle are both the simplest and the most important. Any figure made up of line-segments can ultimately be divided into triangles; and a triangle has six elements which can be measured – the lengths of its three sides, and the magnitudes of its three angles.

We know from practical experience that we cannot choose these six elements arbitrarily; in fact once we have chosen a suitable set of three of them, the triangle can be constructed and the other three found. There must therefore be relations which connect them.

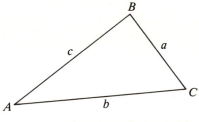

Figure 1

We name the angles of the triangle ABC by the names of the vertices, A, B and C, and we also use the same letters A, B and C to denote the measures of these angles. The lengths of the sides we denote by the small letters a, b, c, chosen so that a is opposite the angle A, and so on. (See Figure 1.)

1. THE AREA OF A TRIANGLE

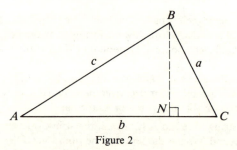

Figure 2

Area of triangle $ABC = \frac{1}{2} \times$ base $\times$ height $= \frac{1}{2} \times AC \times BN$

From triangle ABN (Figure 2), $BN = c \sin A$,

so the area of triangle $ABC = \frac{1}{2}bc \sin A$.

Also, from triangle CBN, $BN = a \sin C$
       and the area of triangle $ABC = \frac{1}{2}ab \sin C$.
It can also be shown that the area of triangle $ABC = \frac{1}{2}ca \sin B$.
So we have

$$\text{area of triangle } ABC = \tfrac{1}{2}bc \sin A = \tfrac{1}{2}ca \sin B = \tfrac{1}{2}ab \sin C.$$

Notice the threefold symmetry of these results and that each expression involves two sides of the triangle and the angle between them.

Example 1
Find the area of the triangle shown in Figure 3.

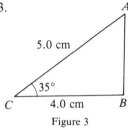

$a = 4.0$, $b = 5.0$, $C = 35°$.
Area $= \frac{1}{2}ab \sin C$ cm²
     $= \frac{1}{2} \times 4 \times 5 \sin 35°$ cm²
     $= 5.7$ cm² to 2 s.f.

Figure 3

Example 2
Find angle A, given that $b = 10$, $c = 11$ and the area of triangle ABC is 36 square units.

From the formula, area $= \frac{1}{2}bc \sin A$,

$$36 = \tfrac{1}{2} \times 10 \times 11 \sin A$$
$$\sin A \approx 0.655$$
$$A = 41° \text{ or } 139° \text{ to the nearest degree.}$$

The two possibilities are illustrated in Figure 4.

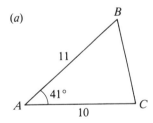
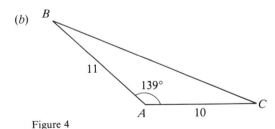

Figure 4

Exercise A

Find the area of the triangles in questions $1-4$, giving your answers to an accuracy of 2 significant figures.

1 $a = 10$ cm, $b = 12$ cm, $C = 48°$.

2 The triangle ABC shown in Figure 5.

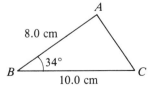

Figure 5

3 $AB = 33$ cm, $AC = 57$ cm, $A = 138.6°$.

4 The triangle PQR shown in Figure 6.

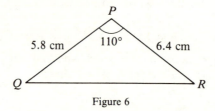

Figure 6

5 In triangle ABC, $a = 4.0$ m, $b = 5.0$ m, and its area is 6.8 m². Find the possible sizes of angle C to the nearest degree.

6 In triangle ABC, $a = 7$, $b = 5$, $c = 3$ and $A = 120°$. Find B and C. Calculate $\dfrac{a}{\sin A}, \dfrac{b}{\sin B}$ and $\dfrac{c}{\sin C}$, and comment on your answers.

Find the areas of the triangles in questions 7–10, giving your answers to an accuracy of 2 significant figures.

7 Triangle PQR, in which $p = 8.0$ cm, $q = 15$ cm, and $R = 124°$.

8 The triangle LMN, shown in Figure 7.

9 The triangle ABC, in which $a = 2.4$ cm, $b = 3.2$ cm and $C = 60°$.

10 The triangle XYZ shown in Figure 8.

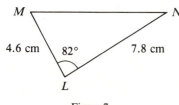

Figure 7

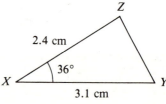

Figure 8

11 In triangle ABC, $a = 7.9$ cm, $c = 5.4$ cm, and the area of the triangle is 12.6 cm². Find the possible sizes of angle B to the nearest degree.

12 Find the area of the triangle EFG shown in Figure 9 and hence find the angles E and G. Calculate $\dfrac{e}{\sin E}, \dfrac{f}{\sin F}$ and $\dfrac{g}{\sin G}$.

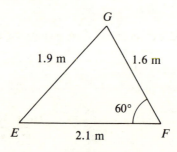

Figure 9

2. THE SINE FORMULA

It is clear that the longest side of a triangle is opposite the largest angle and the smallest side is opposite the smallest angle.

Is it true that the lengths of the sides of a triangle are proportional to the opposite angles?

Suppose we draw a triangle ABC in which $BC = 10$ cm, $B = 20°$, $C = 60°$ and $A = 100°$ (Figure 10). Would AB be 6 cm and AC be 2 cm?

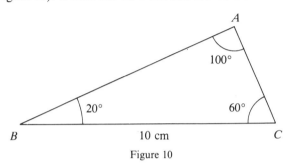

Figure 10

If $BA = 6$ cm and $AC = 2$ cm, then $BA + AC = 8$ cm, which is less than BC, which is impossible. So it cannot be true that the lengths of the sides are proportional to the opposite sides.

However, we can deduce a connection between the sides and the angles opposite them from the area formula which we met in Section 1. We have:

$$\text{area of triangle } ABC = \tfrac{1}{2}bc\sin A = \tfrac{1}{2}ac\sin B = \tfrac{1}{2}ab\sin C$$
$$\Rightarrow \quad 2 \times \text{area of triangle} = bc\sin A = ac\sin B = Ab\sin C$$
$$\Rightarrow \quad \frac{2 \times \text{area of triangle}}{abc} = \frac{\sin A}{a} = \frac{\sin B}{b} = \frac{\sin C}{c}.$$

So the sides are proportional to the *sines* of the opposite angles.

Example 3

In triangle ABC, $A = 100°$, $B = 20°$ and $BC = 10$ cm. Find AC and AB.

$$\frac{\sin A}{a} = \frac{\sin B}{b} \quad \text{so} \quad \frac{\sin 100°}{10} = \frac{\sin 20°}{b}$$

$$b\sin 100° = 10\sin 20°$$

$$b = \frac{10\sin 20°}{\sin 100°} \approx 3.48.$$

$$AC = 3.5 \text{ cm to 2 s.f.}$$

Since $A = 100°$ and $B = 20°$, $C = 60°$ and so

$$\frac{\sin A}{a} = \frac{\sin C}{c} \quad \text{gives} \quad \frac{\sin 100°}{10} = \frac{\sin 60°}{c}.$$

$$c = \frac{10 \sin 60°}{\sin 100°}$$

$$= \frac{10 \times 0.866}{0.985} \approx 8.81.$$

$$AB = 8.8 \text{ cm to 2 s.f.}$$

Example 4

In triangle PQR, $P = 30°$, $p = 8$ and $r = 12$. Find R.

$$\frac{\sin P}{p} = \frac{\sin R}{r} \quad \text{so} \quad \frac{\sin 30°}{8} = \frac{\sin R}{12}$$

$$\sin R = \frac{12 \sin 30°}{8}$$

$$= \frac{12 \times 0.5}{8}$$

$$= 0.75.$$

$$R = 49° \text{ or } 131° \text{ to the nearest degree.}$$

Note that both values of R are possible, as illustrated in Figure 11.

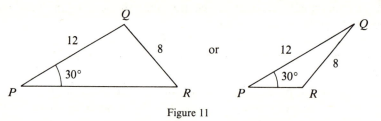

Figure 11

Exercise B

In your answers to questions in this exercise, give all angles to the nearest degree and all lengths to an accuracy of 3 significant figures.

1 Find c in the triangle ABC shown in Figure 12.

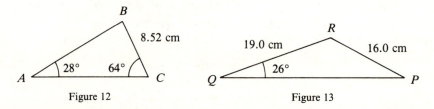

Figure 12   Figure 13

2 Find P in the triangle PQR shown in Figure 13.

3 If, in triangle ABC, $b = 7.24$ cm, $B = 86°$ and $C = 48°$, find c.

4 If, in triangle ABC, $b = 5.31$ m, $A = 55°$ and $B = 83°$, find a and c.

5 If, in triangle PQR, $p = 13.6$, $q = 11.7$ and $P = 42°$, find Q.

6 If, in triangle PQR, $q = 8.34$, $r = 4.23$ and $Q = 79°$, find R and p.

7 Given that, in triangle ABC, $A = 67°$ and $c = 20.0$ cm, find possible values for B if:

(a) $a = 25.0$ cm;   (b) $a = 19.2$ cm;   (c) $a = 18.6$ cm.

Sketch the triangles corresponding to your answers.

8 Find b in the triangle ABC shown in Figure 14.

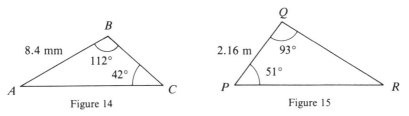

Figure 14   Figure 15

9 Find p in the triangle PQR shown in Figure 15.
10 In triangle LMN, $L = 22°$, $l = 4.29$ cm, $m = 7.12$ cm. Find M.
11 In triangle ABC, $A = 123°$, $c = 19.6$ m, $a = 24.8$ m. Find b and B.
12 If, in triangle PQR, $PQ = 15.2$ cm, $QR = 11.1$ cm and $\angle QPR = 26°$, find $\angle QRP$.
13 Given that, in triangle ABC, $a = 14.8$ m, $B = 62.5°$ and $C = 51°$, find b and c.
14 If, in triangle PQR, $p = 5.46$ cm, $Q = 38°$ and $R = 75°$, find q and r.

3. THE COSINE FORMULA

The size and shape of a triangle are completely determined when two sides and the angle between them are known: for example, a, b and C in a triangle ABC. The sine rule does not enable us to find c or A or B in this situation, since the information does not give the value of $\dfrac{a}{\sin A}$ or $\dfrac{b}{\sin B}$ or $\dfrac{c}{\sin C}$.

Suppose we know a, b and C. We can find c as follows (see Figure 16).

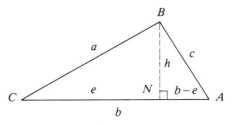

Figure 16

From triangle BCN,
$$a^2 = h^2 + e^2.$$
From triangle ABN,
$$\begin{aligned}
c^2 &= h^2 + (b-e)^2 \\
&= h^2 + (b^2 - 2be + e^2) \\
&= h^2 + e^2 + b^2 - 2be \\
&= a^2 + b^2 - 2be.
\end{aligned}$$

From triangle BCN, $e = a \cos C$

so $c^2 = a^2 + b^2 - 2ab \cos C$.

This is the *cosine formula*.

The same result applies even when one of the angles is obtuse (remember that the cosine of an obtuse angle is negative).

Notice that if $C = 90°$, then $c^2 = a^2 + b^2$ since $\cos 90° = 0$,

and that   if $C < 90°$, then $c^2 < a^2 + b^2$

and   if $C > 90°$, then $c^2 > a^2 + b^2$.

As with the formulae for the area of a triangle, there are three symmetrical forms of the cosine formula:

$$a^2 = b^2 + c^2 - 2bc \cos A;$$
$$b^2 = c^2 + a^2 - 2ca \cos B;$$
$$c^2 = a^2 + b^2 - 2ab \cos C.$$

Example 5

Given that $c = 10$, $a = 7$, and $B = 37°$, find b. (See Figure 17.)

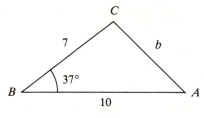

Figure 17

By the cosine formula

$$b^2 = c^2 + a^2 - 2ac \cos B$$
$$= 10^2 + 7^2 - 2 \times 10 \times 7 \times \cos 37°$$
$$= 149 - 111.8.$$

Hence $b = 6.1$ to 2 s.f.

Example 6

Given $A = 114°$, $AC = 5.8$ cm, $AB = 3.9$ cm, find BC. (See Figure 18).

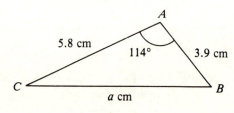

Figure 18

By the cosine formula

$$a^2 = b^2 + c^2 - 2bc \cos A$$
$$= 5.8^2 + 3.9^2 - 2 \times 5.8 \times 3.9 \times \cos 114°$$
$$= 48.85 + 18.40$$
$$= 67.25.$$

Hence $BC = 8.2$ cm.

The three forms of the cosine formula can be rearranged to give the cosines of the angles A, B and C:

$$\cos A = \frac{b^2 + c^2 - a^2}{2bc}; \quad \cos B = \frac{c^2 + a^2 - b^2}{2ca}; \quad \cos C = \frac{a^2 + b^2 - c^2}{2ab}.$$

Example 7
Given that a triangle has sides of
6.0 cm, 7.0 cm and 8.0 cm, find all its
angles. (See Figure 19.)

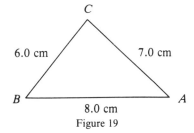

Figure 19

Writing $a = 6$, $b = 7$, $c = 8$, we have

$$\cos A = \frac{b^2 + c^2 - a^2}{2bc} = \frac{7^2 + 8^2 - 6^2}{2 \times 7 \times 8} = 0.6875,$$

so $A = 47°$ to the nearest degree.

Also $$\cos B = \frac{c^2 + a^2 - b^2}{2ca} = \frac{8^2 + 6^2 - 7^2}{2 \times 8 \times 6} = 0.53125,$$

so $B = 58°$ to the nearest degree.

Finally, $$\cos C = \frac{a^2 + b^2 - c^2}{2ab} = \frac{6^2 + 7^2 - 8^2}{2 \times 6 \times 7} = 0.25,$$

so $C = 76°$ to the nearest degree.

Note that rounding the angles to the nearest degree causes the answers to add up to 181°. (Had we given the angles to the nearest tenth of a degree, we would have written down 46.6°, 57.9° and 75.5°.)

Exercise C

1   If, in triangle ABC, $c = 83$ cm, $b = 67$ cm and $A = 103°$, find a.

2   Calculate the length AB in the triangle
shown in Figure 20.

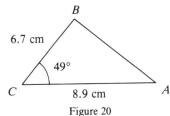

Figure 20

3   Calculate all the angles of a triangle with sides 2, 3 and 4 units, giving your answers to the nearest degree.

4   Two ships are close together in mid-ocean. The *Arethusa* is proceeding at 25 knots on a course bearing 258° and the *Bellerephon* is travelling due south at 21 knots. How far apart are they (*a*) after 1 hour, (*b*) after 2 hours, (*c*) after *t* hours, assuming they maintain their courses and speeds?

5   Points *P* and *Q* have position vectors $\begin{bmatrix} 3 \\ 4 \end{bmatrix}$ and $\begin{bmatrix} 1 \\ 2 \end{bmatrix}$ respectively. Find the lengths of *OP*, *OQ* and *PQ*. Use the cosine formula to find the angles *POQ* and *OPQ*.

6   A ship sails 5.5 km on a bearing of 039°, and then 4.1 km on a bearing of 072°. How far is it then from its starting-point?

7   If, in triangle *ABC*, *a* = 34, *b* = 71 and *C* = 60°, find *c*.

8   Calculate the length *AC* in the triangle *ABC* shown in Figure 21.

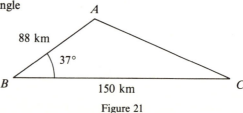

Figure 21

9   Calculate all the angles of a triangle with sides 5.7 cm, 8.3 cm and 11.5 cm.

10   Bilton is 6.7 km due north of Applemouth, and Chirby is 12.4 km from Applemouth on a bearing of 037°. How far is it from Bilton to Chirby?

11   Points *L* and *M* have position vectors $\begin{bmatrix} 5 \\ 12 \end{bmatrix}$ and $\begin{bmatrix} 15 \\ -8 \end{bmatrix}$ respectively. Find the lengths of *OL*, *OM* and *LM*, and use the cosine formula to find angles *OLM* and *MOL*.

12   An aircraft flies 87 km on a track of 057° and then 34 km on a bearing of 068°. How far is it now from its starting-point?

4. SCALAR PRODUCTS

Relative to a tracking station on the ground the position vectors of two aircraft are $\begin{bmatrix} 4 \\ 8 \\ 1 \end{bmatrix}$ and $\begin{bmatrix} -3 \\ 5 \\ 2 \end{bmatrix}$, the units being kilometres. What is the angle between these two vectors?

This question can be answered by considering the triangle with vertices at the tracking station and the two aircraft, and using the cosine formula, as follows.

Take the origin *O* at the tracking station, so that the aircraft are at *A*(4, 8, 1) and *B* (⁻3, 5, 2).

$$OA^2 = 4^2 + 8^2 + 1^2 = 81, \quad OB^2 = (^-3)^2 + 5^2 + 2^2 = 38,$$

$$\underset{\sim}{AB} = \underset{\sim}{OB} - \underset{\sim}{OA} = \begin{bmatrix} -3 \\ 5 \\ 2 \end{bmatrix} - \begin{bmatrix} 4 \\ 8 \\ 1 \end{bmatrix} = \begin{bmatrix} -7 \\ -3 \\ 1 \end{bmatrix}$$

so that $AB^2 = (^-7)^2 + (^-3)^2 + 1^2 = 59.$

Now, from the cosine formula,

$$\cos AOB = \frac{OA^2 + OB^2 - AB^2}{2 \times OA \times OB} = \frac{81 + 38 - 59}{2\sqrt{81}\sqrt{38}} = \frac{30}{\sqrt{81}\sqrt{38}}$$

and hence   angle $AOB = 57°$ to the nearest degree.

Notice where the numbers in the final expression for $\cos AOB$ have come from.

$\sqrt{81}$ is the length of the vector $\begin{bmatrix} 4 \\ 8 \\ 1 \end{bmatrix}$, that is, $\sqrt{(4^2 + 8^2 + 1^2)}$;

$\sqrt{38}$ is the length of the vector $\begin{bmatrix} -3 \\ 5 \\ 2 \end{bmatrix}$, that is, $\sqrt{[(-3)^2 + 5^2 + 2^2]}$.

30 has also been found from these two vectors, but in rather a roundabout way. In fact it could have been found directly as follows:

$$\begin{bmatrix} 4 \\ 8 \\ 1 \end{bmatrix} \cdots \begin{bmatrix} -3 \\ 5 \\ 2 \end{bmatrix} \quad \begin{array}{l} \cdots 4 \times (-3) = -12 \\ \cdots 8 \times 5 = 40 \\ \cdots 1 \times 2 = 2 \end{array}$$

$$\overline{30}$$

In general, it can be shown that if $\theta°$ is the angle between the vectors

$$\mathbf{p} = \begin{bmatrix} x_1 \\ y_1 \\ z_1 \end{bmatrix} \quad \text{and} \quad \mathbf{q} = \begin{bmatrix} x_2 \\ y_2 \\ z_2 \end{bmatrix}, \quad \text{then}$$

$$\cos \theta° = \frac{x_1 x_2 + y_1 y_2 + z_1 z_2}{pq},$$

where p and q are the lengths of the vectors.

Equivalently, we can write

$$pq \cos \theta° = x_1 x_2 + y_1 y_2 + z_1 z_2.$$

This quantity is called the *scalar product* of the two vectors and is written $\mathbf{p.q}$. The equality of these two expressions for $\mathbf{p.q}$ is proved in Section 7.

The formula for $\cos \theta°$ can be written

$$\cos \theta° = \frac{\mathbf{p.q}}{pq}$$

and it is this form we shall use most frequently.

Note that, in two dimensions, $\begin{bmatrix} x_1 \\ y_1 \end{bmatrix} . \begin{bmatrix} x_2 \\ y_2 \end{bmatrix} = x_1 x_2 + y_1 y_2.$

Example 8

Find the angle between the vectors $\begin{bmatrix} 1 \\ 2 \\ 3 \end{bmatrix}$ and $\begin{bmatrix} 4 \\ -1 \\ 2 \end{bmatrix}$.

The lengths of the vectors are

$$(1^2+2^2+3^2) = \sqrt{14}$$

and

$$\sqrt{(4^2+(^-1)^2+2^2)} = \sqrt{21}.$$

Their scalar product is

$$(1\times4)+(2\times{}^-1)+(3\times2) = 4-2+6 = 8.$$

Hence

$$\cos\theta° = \frac{8}{\sqrt{14}\sqrt{21}} \approx 0.467,$$

$$\theta° = 62.1°.$$

The angle between the vectors is 62.1°.

Example 9

p and **q** are vectors of magnitudes 2 and 3 units respectively. Find **p.q** if the angle between them is (*a*) 60°, (*b*) 140°.

(*a*) **p.q** = $pq\cos\theta°$ = $2\times3\times\cos60°$ = 3.
(*b*) **p.q** = $pq\cos\theta°$ = $2\times3\times\cos140°$ = $^-4.6$.

Exercise D

1 Find the scalar products of:

(*a*) $\begin{bmatrix}3\\1\end{bmatrix}$ and $\begin{bmatrix}1\\2\end{bmatrix}$; (*b*) $\begin{bmatrix}3\\1\\4\end{bmatrix}$ and $\begin{bmatrix}3\\3\\6\end{bmatrix}$; (*c*) $\begin{bmatrix}3\\1\\4\end{bmatrix}$ and $\begin{bmatrix}-1\\-1\\-2\end{bmatrix}$.

2 Find the angles between:

(*a*) $\begin{bmatrix}4\\7\end{bmatrix}$ and $\begin{bmatrix}3\\1\end{bmatrix}$; (*b*) $\begin{bmatrix}-1\\-4\end{bmatrix}$ and $\begin{bmatrix}3\\2\end{bmatrix}$.

3 Find the angles between:

(*a*) $\begin{bmatrix}2\\3\\6\end{bmatrix}$ and $\begin{bmatrix}2\\-1\\2\end{bmatrix}$; (*b*) $\begin{bmatrix}1\\1\\1\end{bmatrix}$ and $\begin{bmatrix}1\\-1\\1\end{bmatrix}$.

4 If $\mathbf{d} = \begin{bmatrix}2\\3\\1\end{bmatrix}$, $\mathbf{e} = \begin{bmatrix}-4\\2\\5\end{bmatrix}$ and $\mathbf{f} = \begin{bmatrix}5\\3\\-2\end{bmatrix}$, find:

(*a*) **d.e**; (*b*) **d.f**; (*c*) **e+f**;
(*d*) **d.(e+f)**.

5 If *A*, *B*, *C* have coordinates (3, 1, 4), (4, ⁻1, 2) and (0, 3, ⁻2) respectively:
(*a*) write down the vectors $A\tilde{B}$ and $A\tilde{C}$ as column-matrices;
(*b*) find the lengths of $A\tilde{B}$ and $A\tilde{C}$;
(*c*) find the scalar product $A\tilde{B}.A\tilde{C}$;
(*d*) find the angle *BAC*.

6 Find the scalar products of:

(*a*) $\begin{bmatrix}3\\1\end{bmatrix}$ and $\begin{bmatrix}2\\5\end{bmatrix}$; (*b*) $\begin{bmatrix}3\\1\\-2\end{bmatrix}$ and $\begin{bmatrix}-1\\2\\7\end{bmatrix}$; (*c*) $\begin{bmatrix}2\\5\end{bmatrix}$ and $\begin{bmatrix}-1\\2\end{bmatrix}$.

7 Find the angles between:

(a) $\begin{bmatrix} 6 \\ 1 \end{bmatrix}$ and $\begin{bmatrix} 2 \\ -5 \end{bmatrix}$; (b) $\begin{bmatrix} -7 \\ 4 \end{bmatrix}$ and $\begin{bmatrix} 3 \\ -3 \end{bmatrix}$.

8 Find the angles between:

(a) $\begin{bmatrix} 3 \\ 4 \\ 5 \end{bmatrix}$ and $\begin{bmatrix} -2 \\ 0 \\ 2 \end{bmatrix}$; (b) $\begin{bmatrix} 4 \\ 7 \\ 1 \end{bmatrix}$ and $\begin{bmatrix} 5 \\ -3 \\ 1 \end{bmatrix}$.

9 If $\mathbf{p} = \begin{bmatrix} -1 \\ 1 \end{bmatrix}$, $\mathbf{q} = \begin{bmatrix} -5 \\ 3 \end{bmatrix}$ and $\mathbf{r} = \begin{bmatrix} 3 \\ 2 \end{bmatrix}$, find:

(a) $\mathbf{p.q}$; (b) $\mathbf{p.r}$; (c) $\mathbf{q+r}$;
(d) $\mathbf{p.(q+r)}$.

10 If A, B and C have coordinates $(1, {}^{-}2)$, $(5, {}^{-}1)$ and $(0, 2)$ respectively:

(a) write down the vectors $\underset{\sim}{AB}$ and $\underset{\sim}{AC}$ as column-matrices;
(b) find the lengths of $\underset{\sim}{AB}$ and $\underset{\sim}{AC}$;
(c) find the scalar product $\underset{\sim}{AB}.\underset{\sim}{AC}$;
(d) find the angle BAC.

*11 If $\mathbf{p}$ and $\mathbf{q}$ are perpendicular, what is the value of $\mathbf{p.q}$?

*12 If $\mathbf{p} = \begin{bmatrix} 4 \\ 7 \\ 2 \end{bmatrix}$, find:

(a) the length of $\mathbf{p}$; (b) $\mathbf{p.p}$.
Make a general statement about $\mathbf{q.q}$ for any vector $\mathbf{q}$.

13 $\begin{bmatrix} 3 \\ y \\ 4 \end{bmatrix}$ is perpendicular to $\begin{bmatrix} 5 \\ 3 \\ -6 \end{bmatrix}$. Find y.

14 Vectors $\mathbf{a}$, $\mathbf{b}$, $\mathbf{c}$, $\mathbf{d}$, $\mathbf{e}$ lie in a plane, have magnitudes 5, 3, 4, 3, 2 respectively and directions as shown in Figure 22. Calculate:

(a) $\mathbf{a.b}$; (b) $\mathbf{b.c}$; (c) $\mathbf{d.c}$;
(d) $\mathbf{d.e}$; (e) $\mathbf{a.e}$.

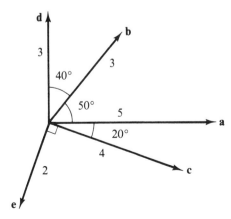

Figure 22

15 Using the same vectors as in question 14, calculate **c.a** and **c.b**. Use the cosine formula to calculate the magnitude and direction of **a**+**b**. Hence calculate **c**.(**a**+**b**). What do you notice?

16 Using the same vectors as in question 14, calculate **d.b** and **d.c**. Use the cosine formula to find the magnitude and direction of **b**+**c**. Hence calculate **d**.(**b**+**c**). What do you notice?

17 Using your answers to questions 15 and 16, calculate (**a**+**b**).(**b**+**c**). Compare your answer with the values of **a.b**, **a.c**, **b.b** and **b.c**. Is it what you would have expected?

18 Using the same vectors as in question 14, calculate **d.a** and **e.c**. Explain what it is about the vectors **d** and **a** and the vectors **e** and **c** which make their scalar products so special.

19 In Figure 23, *OABCDEFG* is a cube of side 5 units.
 (*a*) Write down the coordinates of *C*, *E* and *G*.
 (*b*) Write *CE* and *CG* as column-matrices.
 (*c*) Calculate the angle between the lines *CE* and *CG*.
 (*d*) Use a scalar product to find the angle between *CE* and *AG*.

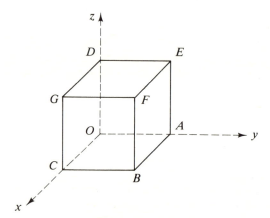

Figure 23

5. PERPENDICULAR VECTORS

Two non-zero vectors are perpendicular if their scalar product is zero (since $\cos 90° = 0$).

If $\mathbf{p} = \begin{bmatrix} x_1 \\ y_1 \end{bmatrix}$ and $\mathbf{q} = \begin{bmatrix} x_2 \\ y_2 \end{bmatrix}$ are perpendicular, then

$$\mathbf{p \cdot q} = x_1 x_2 + y_1 y_2 = 0.$$

For example, $\begin{bmatrix} 5 \\ -2 \end{bmatrix}$ is perpendicular to $\begin{bmatrix} 2 \\ 5 \end{bmatrix}$ since $\begin{bmatrix} 5 \\ -2 \end{bmatrix} \cdot \begin{bmatrix} 2 \\ 5 \end{bmatrix} = 10 - 10 = 0.$

Notice that $\begin{bmatrix} 10 \\ -4 \end{bmatrix}, \begin{bmatrix} -5 \\ 2 \end{bmatrix}, \begin{bmatrix} 50 \\ -20 \end{bmatrix}$ are also perpendicular to $\begin{bmatrix} 2 \\ 5 \end{bmatrix}$.

Example 10

Find t if $\begin{bmatrix} 2 \\ 2 \\ -1 \end{bmatrix}$ and $\begin{bmatrix} 2 \\ 1 \\ t \end{bmatrix}$ are perpendicular.

$$\begin{bmatrix} 2 \\ 2 \\ -1 \end{bmatrix} . \begin{bmatrix} 2 \\ 1 \\ t \end{bmatrix} = 0 \;\Rightarrow\; 4+2-t = 0 \;\Rightarrow\; t = 6.$$

Example 11
Find the equation of the line perpendicular to $y = 3x+2$ passing through the point $(^-3, 4)$.

The line $y = 3x+2$ is sketched in Figure 24. Since its gradient is 3, it is in the direction of the vector $\begin{bmatrix} 1 \\ 3 \end{bmatrix}$. The vector $\begin{bmatrix} ^-3 \\ 1 \end{bmatrix}$ is perpendicular to this, and so the line required is parallel to $\begin{bmatrix} ^-3 \\ 1 \end{bmatrix}$ and therefore has gradient $^-\frac{1}{3}$. Its equation is $y = ^-\frac{1}{3}x+c$, for some number c.

Since the line passes through $(^-3, 4)$,
$$4 = ^-\tfrac{1}{3} \times ^-3 + c,$$
$$c = 3.$$

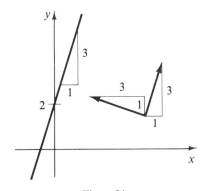

Figure 24

The equation of the line is $y = ^-\frac{1}{3}x+3$.

In general, a line with equation $y = mx+c$ is perpendicular to a line with equation $y = \dfrac{^-1}{m}x+k$.

Exercise E

1   Find the value of k if the vectors $\begin{bmatrix} 2 \\ 1 \end{bmatrix}$ and $\begin{bmatrix} 3 \\ k \end{bmatrix}$ are perpendicular.

2 If the points A, B and C have coordinates $(1, 2, 6)$, $(^-1, 3, 2)$ and $(^-2, 5, 3)$ respectively, prove that the angle ABC is $90°$.

3 Find a and b if the vector $\begin{bmatrix} a \\ b \\ 1 \end{bmatrix}$ is perpendicular to both $\begin{bmatrix} 2 \\ 5 \\ ^-1 \end{bmatrix}$ and $\begin{bmatrix} 1 \\ 4 \\ 1 \end{bmatrix}$.

4 Find the equations of the lines through the given points which are (*a*) parallel to and (*b*) perpendicular to the given lines.
  (i) $(2, 3)$, $3x - y = 5$;
  (ii) $(4, {}^-5)$, $2x + y = 5$.

5 Find the angles between the lines:
  (*a*) $x + y = 2$ and $4x - 3y = 1$;
  (*b*) $2x - y = 5$ and $x + 2y + 3 = 0$.

---

6 Find the value of k if the vectors $\begin{bmatrix} 3 \\ 4 \\ ^-2 \end{bmatrix}$ and $\begin{bmatrix} k \\ 1 \\ k \end{bmatrix}$ are perpendicular.

7 If P is the point with coordinates $(3, 1, 2)$, Q is $(^-2, 2, 1)$ and R is $(1, 5, {}^-11)$, prove that the angle PQR is $90°$.

8 Find t if the vector $\begin{bmatrix} 2 \\ 1 \end{bmatrix} + t \begin{bmatrix} 1 \\ ^-1 \end{bmatrix}$ is perpendicular to the vector $\begin{bmatrix} 1 \\ 3 \end{bmatrix}$.

9 Find the equations of the lines through the given points which are (*a*) parallel to and (*b*) perpendicular to the given lines.
  (i) $(1, {}^-1)$, $7x + 5y + 6 = 0$;
  (ii) $(0, {}^-4)$, $2y = 3x + 5$.

10 Find the angles between the lines:
  (*a*) $5x - 12y + 3 = 0$ and $12x + 5y = 11$;
  (*b*) $x = 3y + 7$ and $x + 2y + 6 = 0$.

6. SUMMARY

Area of a triangle $= \frac{1}{2}$base $\times$ height

$$= \tfrac{1}{2}ab \sin C = \tfrac{1}{2}ac \sin B = \tfrac{1}{2}bc \sin A$$

Sine formula $\dfrac{\sin A}{a} = \dfrac{\sin B}{b} = \dfrac{\sin C}{c}$

Cosine formula $a^2 = b^2 + c^2 - 2bc \cos A$

Scalar product $\mathbf{p}.\mathbf{q} = pq \cos \theta°$ (where $\theta°$ is the angle between $\mathbf{p}$ and $\mathbf{q}$)

$$= x_1 x_2 + y_1 y_2 + z_1 z_2 \text{ if } \mathbf{p} = \begin{bmatrix} x_1 \\ y_1 \\ z_1 \end{bmatrix} \text{ and } \mathbf{q} = \begin{bmatrix} x_2 \\ y_2 \\ z_2 \end{bmatrix}.$$

Two non-zero vectors are perpendicular if their scalar product is zero.

7. SCALAR PRODUCT AND COMPONENTS

We now give an outline proof that if $\mathbf{p} = \begin{bmatrix} x_1 \\ y_1 \\ z_1 \end{bmatrix}$ and $\mathbf{q} = \begin{bmatrix} x_2 \\ y_2 \\ z_2 \end{bmatrix}$ then $\mathbf{p.q}$

(defined as $pq \cos \theta°$) is equal to $x_1 x_2 + y_1 y_2 + z_1 z_2$.

If P and Q are the points with position vectors $\mathbf{p}$ and $\mathbf{q}$ respectively (Figure 25), then

$$OP^2 = p^2 = x_1^2 + y_1^2 + z_1^2 \quad \text{and} \quad OQ^2 = q^2 = x_2^2 + y_2^2 + z_2^2.$$

Also

$$\underset{\sim}{QP} = \begin{bmatrix} x_1 - x_2 \\ y_1 - y_2 \\ z_1 - z_2 \end{bmatrix} \quad \text{so} \quad QP^2 = (x_1 - x_2)^2 + (y_1 - y_2)^2 + (z_1 - z_2)^2.$$

$$= x_1^2 - 2x_1 x_2 + x_2^2 + y_1^2 - 2y_1 y_2 + y_2^2 + z_1^2 - 2z_1 z_2 + z_2^2$$

$$= x_1^2 + y_1^2 + z_1^2 + x_2^2 + y_2^2 + z_2^2 - 2x_1 x_2 - 2y_1 y_2 - 2z_1 z_2$$

$$= p^2 + q^2 - 2(x_1 x_2 + y_1 y_2 + z_1 z_2).$$

But, from the cosine formula,

$$QP^2 = p^2 + q^2 - 2pq \cos \theta°.$$

Hence

$$\mathbf{p.q} = pq \cos \theta = x_1 x_2 + y_1 y_2 + z_1 z_2.$$

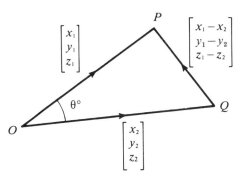

Figure 25

Miscellaneous exercise

1 For the points A, B, C shown in Figure 26,

   (a) show that the scalar product $\underset{\sim}{AB}.\underset{\sim}{AC} = 11$;

   (b) calculate the lengths of AB and AC. Hence calculate the angle CAB to the nearest degree.

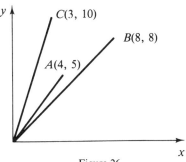

Figure 26

2   Given the points A (1, 2, 4), B (5, 3, 7) and C (3, 7, 4) write down the vectors $\underset{\sim}{AB}$
and $\underset{\sim}{AC}$ in column-matrix form. Calculate
   (a) the scalar product $\underset{\sim}{AB}.\underset{\sim}{AC}$;
   (b) the lengths of AB and AC.
   Hence show by calculation that the angle $BAC \approx 62°$.
   Calculate also the area of the triangle ABC.

3   The vertices of a square-based pyramid have coordinates as follows: A (0, 0, 0),
B (0, 5, 0), C (4, 5, 3), D (4, 0, 3) and V (3, 2, 5).
   (a) Write down the displacements $\underset{\sim}{BC}$ and $\underset{\sim}{DV}$ in column-matrix form and state
       their lengths.
   (b) Using scalar products, calculate the angle between the edges BC and DV.

4   Three vectors **a, b** and **c** are as follows:
   $\mathbf{a} = 2\mathbf{i} - \mathbf{j}, \mathbf{b} = \mathbf{i} + 4\mathbf{j}, \mathbf{c} = 3\mathbf{i} + 2\mathbf{j}.$
   (a) Find the value of s if the vector $\mathbf{a} + s\mathbf{b}$ is parallel to **c**.
   (b) Find the value of t if the vector $\mathbf{a} + t\mathbf{b}$ is perpendicular to **c**.

5   A mast has its base at (0, 0, 0) and its top at T (0, 0, 24). It is supported by guys
from A (10, 0, 0) and B (0, 10, 0) to T and from C (‾6, ‾6, 0) to the point
M (0, 0, 17).
   Calculate the magnitudes of the vectors $\underset{\sim}{AT}$ and $\underset{\sim}{CM}$, and also the angle between
them.

6   (a) Given $\mathbf{a} = \begin{bmatrix} 3 \\ 4 \end{bmatrix}$, find a vector **u** of unit magnitude which is parallel to **a**.

   (b) Given $\mathbf{b} = \begin{bmatrix} 1 \\ -2 \end{bmatrix}$ and $\mathbf{c} = \begin{bmatrix} -1 \\ 3 \end{bmatrix}$, calculate the scalar products (i) **a.b**, (ii) **u.c**.
   Explain with the aid of a diagram why these two quantities are opposite in sign.

7   A tetrahedron is formed with its vertices at the points O (0, 0, 0), A (2, 3, 4),
B (x, y, z) and C (5, 2, ‾1). Show that, if OA is perpendicular to BC, then (x, y, z)
must satisfy the equation $2x + 3y + 4z = 8$.
   If also AC is perpendicular to OB, find another linear equation satisfied by x, y
and z.
   Hence show that OC is perpendicular to AB.

8   Two girls on a level beach wish to work out the height of a kite which they can see
due west of them. Claire walks to C, 50 metres away from Ann at A, and finds the
bearing of the kite is now 060°, and the bearing of Ann is 070°.
   (a) Verify that angle $CPA = 150°$, where P is the point on the beach directly below
       the kite.
   (b) Use the sine formula to confirm that CP is about 34 m.
   (c) Hence calculate the height of the kite if Claire observes its angle of elevation
       to be about 63°.
   (d) If the boy at B flying the kite is 40 m due north of Claire, use the cosine
       formula to confirm that BP is about 37 m.
   (e) Calculate the direct distance from the boy to the kite.

9   A bee, which flies through the air at 6 m s⁻¹, is aiming to reach a tree 600 m due south
of it. It happens that the sun is also due south, so the bee wishes to move directly
towards the sun, but there is a wind blowing in the direction 052°, at 2.5 m s⁻¹.
   Sketch a triangle of velocities, showing the resultant velocity of the bee in the
direction due south, and mark the known values on your diagram. By calculation
show that the bee must aim in a direction 19.2° to the right of the sun.
   Calculate how long the bee will take to reach the tree.

10 In a survey of the Cotanni Plains, a point on Stony Hill has been marked which is precisely 1 km due north of the base at Lone Tree. From this point one can see Black Rock on a bearing of 062°. The bearing of Black Rock from Lone Tree is 042°. Calculate the distance from Black Rock to Lone Tree.

There is a waterhole invisible from Lone Tree from which the bearing and distance of Black Rock are known to be 342° and 2.12 km. How far is the waterhole from Lone Tree?

11 A rectangular block has three edges meeting at right-angles at a corner *A*. $AB = 15$ cm, $AC = 36$ cm and $AD = 20$ cm. It is sawn through the vertices *B*, *C*, *D*, exposing the triangle *BCD*. Calculate the sides of this triangle and the angle *CBD*.

12 A triangle *ABC* is shown in Figure 27.

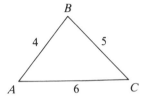

Figure 27

  (*a*) Show by calculation that the largest angle is approximately 83°.
  (*b*) The circle through the vertices *ABC* has centre *O*. Using the fact that the angle *AOC* is twice the angle *ABC*, or otherwise, calculate the diameter of this circle.

13 A helicopter is flying on a course due east with an airspeed of 120 km h⁻¹. The wind is blowing in the direction with bearing 197°. If the windspeed is 30 km h⁻¹, in what direction and with what speed does the helicopter travel relative to the ground?

14 A ship is steaming with constant velocity and is sighted on a bearing 035° at a range of 5 nautical miles. Half an hour later it is sighted from the same point on a bearing of 107° at a range of 6.2 nautical miles. Calculate its velocity, giving its magnitude to the nearest knot and its direction to the nearest degree.

15 A light aircraft with an airspeed of 200 km h⁻¹ has to fly to an airfield 80 km away on a bearing of 140°. If the wind is blowing from the north with a speed of 40 km h⁻¹, what course should the pilot set, and how long will the journey take?

9

The sine and cosine functions

The height, h m, of the tide above mean sea-level at Muchmud-by-the-Sea is modelled by the formula $h = 5 \sin (30t)°$ where t is the number of hours after noon on 1st April. How fast is the tide rising at 2.30 p.m.? To answer questions such as these we need to find the derivative of the sine function.

1. DIFFERENTIATION OF $x \rightarrow \sin x°$

When we draw the graph of the function $x \rightarrow \sin x°$, we normally use unequal scales for the axes, as in Figure 1. It would be very inconvenient to use equal

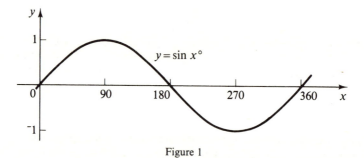

Figure 1

scales because the graph would look rather spread out along the x-axis (Figure 2).

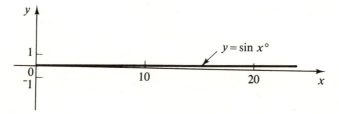

Figure 2

We already know (Chapter 5) that when we differentiate a function f, the derived function f' which we obtain tells us the gradient of the graph of f at any point. So when we look at the graph in Figure 1, we get a false impression of the gradient because of the unequal scales. However, an inspection of the graph can give us some useful information.

144

If $g(x) = \sin x°$, we can deduce at sight from the graph that
$$g'(0) = g'(360) = {}^-g'(180),$$
$$g'(90) = 0,$$
and
$$g'(270) = 0.$$
In the first exercise we investigate the derived function of g more fully.

Exercise A

*1 Let $k(x) = \dfrac{\sin (x+0.1)° - \sin x°}{0.1}$,

i.e. $k(x) \approx g'(x)$, where $g(x) = \sin x°$.

Copy and complete Table 1, but leave the third column blank until question 3.

Table 1

x	$k(x)$	$k(x)/0.01745$
0		
30		
60		
90		

*2 (a) Extend the table of values of $k(x)$ obtained in question 1 for values of x from 120 to 360, by steps of 30. (You may be able to make use of the symmetries of the graph of $y = \sin x°$.)

(b) Using a scale of 1 cm to 30 units for the horizontal axis, and 1 cm to 0.005 units for the vertical axis, use your results to draw the graph of the function $k(x)$.

*3 Calculate the values of $\dfrac{k(x)}{0.01745}$, and put these in the third column of your table, corrected to 3 decimal places.

It appears from the results of Exercise A that the graph of k is familiar: it seems to be in the form of a cosine curve. However it is not the graph of $x \rightarrow \cos x°$ itself, because it has a maximum of only about 0.01745 (at 0 and 360) and a minimum of about $^-0.01745$ (at 180). So $k(x) \approx 0.01745 \cos x°$, i.e. the graph of k is a cosine curve squashed by a scale factor of 0.01745 in a direction parallel to the y-axis.

We could therefore guess that if
$$g(x) = \sin x°,$$
then
$$g'(x) \approx 0.01745 \cos x°.$$

This result can be demonstrated more generally, without restricting ourselves to particular values of x, as we have done above. Figure 3(*a*) shows a circle of unit radius. Now we already know that, if h is small, then
$$g'(x) \approx \frac{\sin (x+h)° - \sin x°}{h}.$$

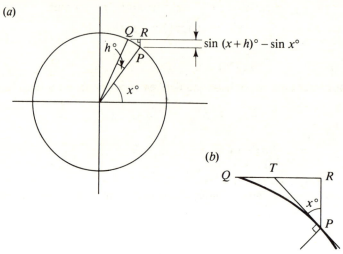

Figure 3

In Figure 3(b), PT is the tangent to the circle at P. Angle TPR is equal to $x°$,

$$\sin (x+h)° - \sin x° = PR,$$

and so

$$\sin (x+h)° - \sin x° = PR = PT \cos x°.$$

Now, if h is small,

$$PT \approx \text{arc } PQ, \text{ so}$$

$$PT \approx \frac{h}{360} \times 2\pi.$$

Hence we obtain

$$\sin (x+h)° - \sin x° = PT \cos x°$$

$$\approx \frac{h}{360} \times 2\pi \times \cos x°$$

$$\approx \frac{h\pi}{180} \cos x°.$$

So

$$g'(x) \approx \frac{\sin (x+h) - \sin x°}{h},$$

$$\approx \frac{\pi}{180} \cos x°.$$

If we evaluate $\frac{1}{180}\pi$ on a calculator, we obtain $\frac{1}{180}\pi = 0.01745...$, which corresponds to the results which we previously obtained. So the derived function has a particularly simple form except for the factor of $\frac{1}{180}\pi$. We shall see later that the number $\frac{1}{180}\pi$ arises in other contexts. We are eventually better off changing matters so that it doesn't appear at all; this turns out to be very easy to do!

2. RADIAN MEASURE

If we want to avoid the appearance of the factor $\pi/180$ when we differentiate $\sin x°$, we need to make the gradient of the curve $180/\pi$ times greater at every point. This can be done very simply by relabelling the x-axis with numbers which are $\pi/180$ of the present values, replacing 90 by $\frac{1}{2}\pi$, 180 by π, etc. This new

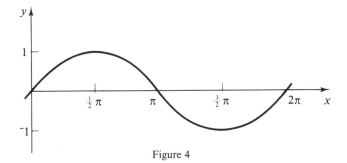

Figure 4

function completes a cycle in an interval of 2π, rather than 360 as before, and so its gradient will be $180/\pi$ times greater than that of the old function. It is now easy to draw the graph using the same scales for each axis. (See Figure 4.)

We shall call this new function simply $x \to \sin x$ (as opposed to $x \to \sin x°$). Since the change of scale has increased the gradient by a factor of $180/\pi$, it follows that for $g(x) = \sin x$, the derived function is given by $g'(x) = \cos x$.

In addition we have devised an alternative way of measuring angle called radian measure. In this system we consider a whole turn to correspond to an angle of 2π units, called *radians*, rather than 360° (Figure 5). This is not as

Figure 5

unnatural as it may seem at first. For a unit circle, 2π is simply its circumference, and so we are adopting a system of measuring angle by arc lengths on a unit circle. Similarly $\frac{1}{2}\pi$ is simply a quarter of the circumference, and this is the radian measure for 90° (Figure 6).

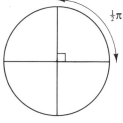

Figure 6

Figure 7

To form an idea of the size of 1 radian, we can see that this is the angle corresponding to a unit arc length on the unit circle. From Figure 7, it is clear that 1 radian is a little less than 60°, since we would have an equilateral triangle if the circular arc was a line-segment of unit length.

In fact, as 2π radians equal 360°, it follows that

$$1 \text{ radian} = \frac{360°}{2\pi}$$

$$\approx 57.29°.$$

To change from radians to degrees, we multiply by $180/\pi$, and therefore to change from degrees to radians we multiply by $\pi/180$. It is important to be able to think in both systems, and it is useful to be able to switch readily from one to the other. Table 2 gives some common radian equivalents of degree measure.

Table 2

Degrees	Radians
0	0
30	$\frac{1}{6}\pi$
45	$\frac{1}{4}\pi$
60	$\frac{1}{3}\pi$
90	$\frac{1}{2}\pi$
120	$\frac{2}{3}\pi$
135	$\frac{3}{4}\pi$
150	$\frac{5}{6}\pi$
180	π

Example 1
Find the equivalent in radian measure of 117°.

$$117° = 117 \times \frac{\pi}{180} \text{ radians}$$

$$= \frac{117}{180} \times \pi \text{ radians}$$

$$= 0.65\pi \text{ radians}.$$

It is often convenient to leave the result as a multiple of π in this way. However, if we complete the calculation, we obtain

$$117° \approx 2.04 \text{ radians}.$$

Exercise B

1 (i) Figure 8 shows a sector AOB of angle $\frac{1}{2}\pi$ radians.

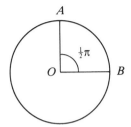

Figure 8

For each of the following, draw a similar diagram to show a sector of angle:
(a) $\frac{1}{4}\pi$ radians; (b) $\frac{2}{3}\pi$ radians; (c) $\frac{3}{4}\pi$ radians;
(d) $\frac{1}{6}\pi$ radians; (e) $\frac{3}{2}\pi$ radians.
(ii) Without using a calculator, give the equivalents in degrees of the angles in (i).

2 Use a calculator to work out the equivalents in degrees to 3 significant figures of the following angles in radians:
(a) 1.5; (b) 2.1; (c) 6.81;
(d) 2.89; (e) 26.1.

3 Work out the equivalents in radians of the following, expressing each answer as a multiple of π:
(a) $22\frac{1}{2}°$; (b) $81°$; (c) $144°$;
(d) $378°$; (e) $594°$.

4 Work out the equivalents in radians of the following:
(a) 127; (b) $482°$; (c) $91.2°$;
(d) $114.2°$; (e) $294.1°$.

5 Give the values, correct to 3 significant figures, of:
(a) $\sin \frac{1}{8}\pi$; (b) $\cos \frac{1}{8}\pi$; (c) $\cos 4$; (d) $\tan \frac{1}{4}\pi$;
(e) $\sin 1$; (f) $\tan 0.499\pi$; (g) $\sin 0.1$;
(h) $\sin 0.01$; (i) $\sin 6.283$.

6 Write down, without using a calculator, the values of:
(a) $\cos \frac{1}{2}\pi$; (b) $\sin \pi$; (c) $\tan \frac{3}{4}\pi$;
(d) $\sin \frac{3}{2}\pi$; (e) $\cos 5\pi$.

7 (i) Draw diagrams, as in question 1, to show angles of:
(a) $\frac{1}{8}\pi$ radians; (b) $\frac{1}{3}\pi$ radians; (c) $\frac{5}{6}\pi$ radians;
(d) $\frac{7}{8}\pi$ radians; (e) $\frac{4}{3}\pi$ radians.
(ii) Without using a calculator give the equivalents in degrees of the angles in (i).

8 Use a calculator to work out the equivalents in degrees to 3 significant figures of the following angles in radians:
(a) 1.3; (b) 2.4; (c) 3.54;
(d) 4.71; (e) 320.4.

9 Work out the equivalents in radians of the following, expressing each answer as a multiple of π:
(a) $67\frac{1}{2}°$; (b) $108°$; (c) $150°$;
(d) $252°$; (e) $405°$.

10 Work out the equivalents in radians of the following:
  (a) 46.1°; (b) 115°; (c) 187°;
  (d) 212.1°; (e) 458°.

11 Give the values, correct to 3 significant figures, of:
  (a) $\sin \frac{2}{3}\pi$; (b) $\tan \frac{1}{8}\pi$; (c) $\cos \frac{4}{3}\pi$;
  (d) $\cos 1$; (e) $\tan(0.501\pi)$; (f) $\sin 2$;
  (g) $\cos 0.01$; (h) $\cos 7.33$.

12 Write down, without using a calculator, the values of:
  (a) $\tan \frac{1}{4}\pi$; (b) $\sin \frac{1}{2}\pi$; (c) $\cos \pi$;
  (d) $\cos \frac{5}{2}\pi$; (e) $\sin 5\pi$.

13 Find $\sin^{-1} 0.932$. Hence write down all the solutions of the equation $\sin x = 0.932$ in the interval $0 < x < 2\pi$.

14 Find four solutions of the equation $\sin x = 0.8$ in the interval $0 < x < 4\pi$.

15 Solve the equation $\cos 2x = {}^-0.15$ in the interval $^-\pi < x < \pi$.

16 On a Big Wheel the height, h metres, of a passenger above the ground is given by $h = 10 \sin x + 11$. Find two values of x in the interval $0 < x < 2\pi$ when the passenger is 9 metres above the ground.

17 Write down, without using a calculator, the values of:
  (a) $\sin^{-1}(1)$; (b) $\cos^{-1}(1)$; (c) $\tan^{-1}(0)$.

18 Find $\cos^{-1}(0.5)$. Hence write down all the solutions of the equation $\cos x = 0.5$ in the interval $0 < x < 2\pi$.

19 Find all the solutions of the equation $\sin x = 0.17$ in the interval $^-2\pi < x < 2\pi$.

20 Solve the equation $\sin \frac{1}{2}x = {}^-0.1$ in the interval $0 < x < 8\pi$.

21 The height above the road surface, h metres, of the valve of a bicycle tyre is given by
$$h = 0.35 - 0.3 \cos x.$$
Find two values of x when the valve is 0.5 metres above the road surface.

22 Write down, without using a calculator, the values of:
  (a) $\sin^{-1}(0)$; (b) $\cos^{-1}(0)$; (c) $\tan^{-1}(1)$.

3. ARC LENGTH AND SECTOR AREA

(a) Length of a circular arc

Example 2
Find the length of the arc AB of the circle shown in the diagram.

The circumference, C cm, of the circle is given by

$$C = 2\pi r,$$

and so, in this case,

$$C = 2\pi \times 3.$$

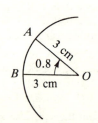

Figure 9

The arc AB is $\dfrac{0.8}{2\pi}$ times the circumference,

$$\text{so arc length } AB = \frac{0.8}{2\pi} \times 2\pi \times 3 \text{ cm}$$

$$= 0.8 \times 3 \text{ cm},$$

$$= 2.4 \text{ cm}.$$

We can see that the calculation turns out to be very simple; we have in effect multiplied the angle (in radians) by the radius.

In general, if the angle is α radians, and the circle is of radius r (Figure 10), we have

$$\text{arc length } AB = \frac{\alpha}{2\pi} \times 2\pi r$$

so arc length $AB = \alpha r$.

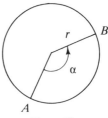

Figure 10

In particular, when the angle is 1 radian, the arc length is r. (Figure 11 shows a sector which is just an enlargement, scale factor r, of that in Figure 7.)

Figure 11

(b) Area of a sector

Figure 12 shows a circle of radius r. The area of the sector AOB, where angle $AOB = \alpha$ radians, is just $\dfrac{\alpha}{2\pi}$ of the entire area.

So the area of sector $AOB = \dfrac{\alpha}{2\pi} \times \pi r^2$.

$$= \tfrac{1}{2}\alpha r^2.$$

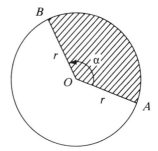

Figure 12

The simple forms of these formulae, when the angle is given in radians, should be compared with the corresponding formulae when the angle is given in degrees (Exercise C, questions 5 and 6).

Exercise C

1 Find the arc length through which you move on a Big Wheel of radius 10 m when it turns through an angle of
   (*a*) 150°,   (*b*) $\frac{2}{9}\pi$ radians,   (*c*) 34.1°,
   (*d*) 2.3 radians.

2 A car's windscreen wiper is of length 45 cm and blade length 30 cm (Figure 13). Find the area of windscreen cleaned if it sweeps through
   (*a*) 80°,   (*b*) 1.5 radians.

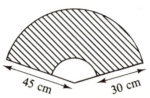

Figure 13

3 (*a*) Taking the Earth's radius as 4000 miles, how far does somebody on the equator move:
   (i) when the Earth rotates about its axis through 60°;
   (ii) when the Earth rotates about its axis through 2.4 radians;
   (iii) during a period of 40 minutes? (Ignore motion relative to the sun.)
   (*b*) (i) Through how many degrees does the Earth rotate in one hour?
   (ii) One 'horad' is the length of time the Earth takes to rotate through 0.25 radians. How many horads are there in one day?

4 Figure 14 shows a section of model racing-car track. *AB* and *PQ* are circular arcs, centre *O*, of radius 15 cm and 30 cm respectively. The area of the curved section of the track is 610 cm². Find the angle α.

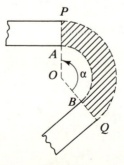

Figure 14

5 (a) Calculate the circumference of the
     circle shown in Figure 15.
   (b) What fraction of this circumference
     is the length of the arc AB?
   (c) Hence find the length of the arc
     AB. Compare your answer with
     Example 2.
   (d) Show that, for an arc of angle $\theta°$, in
     a circle of radius r, the length of the
     arc is $\frac{1}{180}\pi\theta r$.

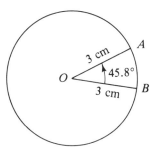

Figure 15

6 Show that a sector of angle $\theta°$ in a circle of radius r has an area $\frac{1}{180}\pi \cdot \frac{1}{2}\theta r^2$. Compare
   this result with that given above when the angle is measured in radians.

4. DIFFERENTIATION OF $x \to \sin kx$

Consider the two functions $g(x) = \sin x$ and $f(x) = \sin 2x$. The graph of f is
obtained from that of g by a stretch parallel to the x-axis of scale factor $\frac{1}{2}$, with
the y-axis invariant (Figure 16).

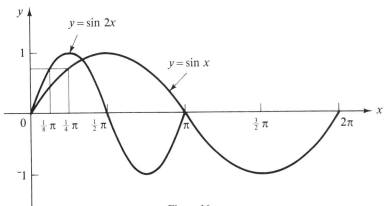

Figure 16

Now for $g(x) = \sin x$, we have
                                $g'(x) = \cos x$.

For instance, if $x = \frac{1}{4}\pi$,
                                $g'(x) = \cos \frac{1}{4}\pi \approx 0.7071.$

If we consider the function f, we can see that when $x = \frac{1}{8}\pi$, the gradient of f is
exactly twice that of g when $x = \frac{1}{4}\pi$, i.e.
$$f'(\tfrac{1}{8}\pi) = 2g'(\tfrac{1}{4}\pi)$$
$$= 2 \cos \tfrac{1}{4}\pi$$
$$\approx 2 \times 0.7071 = 1.4142.$$

In general $f'(x) = 2g'(2x)$

or $f'(x) = 2\cos 2x.$

The above results suggest that

$$f(x) = \sin kx,$$
$$\Rightarrow \quad f'(x) = k\cos kx.$$

This is a special case of a general rule, called the *chain rule*, which will be dealt with more thoroughly in Chapter 23. For the rest of this chapter we shall assume that this property, which we have studied in connection with the sine function, is common to all functions, i.e. that

$$f(x) = g(kx)$$
$$\Rightarrow \quad f'(x) = kg'(kx).$$

Example 3
Differentiate $f : x \to 2\sin 3x$.

The graph of f is obtained from the graph of $g : x \to \sin x$ by stretches with scale factors $\frac{1}{3}$ and 2 parallel to the x- and y-axes respectively. These have the effect of increasing the gradient by factors of 3 and 2.

Hence $f'(x) = 2 \times 3\cos 3x = 6\cos 3x$.

Exercise D

1   (a) Draw on the same axes the graphs of

          (i) $f(x) = \sin x$

     and (ii) $g(x) = \sin 3x$,

     using equal scales for each axis. What transformation maps the graph of f onto the graph of g?
   (b) The point $(\frac{5}{6}\pi, \frac{1}{2})$ is on the graph of f. What is its image under the transformation of part (a)?
   (c) (i) Evaluate $f'(\frac{5}{6}\pi)$.
       (ii) Hence evaluate $g'(\frac{5}{18}\pi)$.
   (d) Copy and complete Table 3.

Table 3

x	$f(3x)$	$g(x)$	$f'(3x)$	$g'(x)$
$\frac{1}{4}\pi$				
$\frac{1}{3}\pi$				
$\frac{1}{2}\pi$				
$\frac{2}{3}\pi$				
$\frac{3}{4}\pi$				

   (e) If $g(x) = \sin 3x$, what is $g'(x)$?

2 Differentiate the following functions:
   (a) $x \to \sin 4x$;   (b) $x \to \sin 7x$;   (c) $x \to \sin \frac{1}{2}x$;

   (d) $x \to \sin \dfrac{\pi}{180}x$;   (e) $x \to 3 \sin 2x$;   (f) $x \to 6 \sin \frac{1}{2}x$.

---

3 (a) Draw on the same axes the graphs of
$$\text{(i) } f(x) = \sin x,$$
   and
$$\text{(ii) } h(x) = \sin 4x,$$
   for values of x from 0 to $\frac{1}{2}\pi$, using equal scales for each axis. What transformation maps the graph of f onto the graph of h?
   (b) The point $(\frac{1}{3}\pi, 0.866)$ is on the graph of f. What is its image under the transformation of part (a)?
   (c) (i) Evaluate $f'(\frac{1}{3}\pi)$.
      (ii) Hence evaluate $h'(\frac{1}{12}\pi)$.
   (d) Copy and complete Table 4.
   (e) What is $h'(x)$?

Table 4

x	$f(4x)$	$h(x)$	$f'(4x)$	$h'(x)$
$\frac{1}{8}\pi$				
$\frac{1}{4}\pi$				
$\frac{3}{8}\pi$				
$\frac{1}{2}\pi$				

4 Differentiate the following functions:
   (a) $x \to \sin 5x$;   (b) $x \to \sin 3x$;   (c) $x \to \sin \frac{1}{4}x$;
   (d) $x \to \sin \frac{1}{10}x$;   (e) $x \to 100 \sin 0.1x$;   (f) $x \to \frac{1}{10} \sin 10x$.

5. DIFFERENTIATION OF $x \to \cos x$

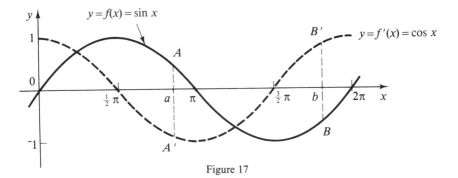

Figure 17

Figure 17 shows the graph of $f(x) = \sin x$, and its derived function $f'(x) = \cos x$. So the gradient at the point A on the graph of f is given by the value of y at the point A' on the graph of f'. Similarly the gradient at B is equal to the value of y at B'.

This relationship between the graphs will still hold if the curves are both translated parallel to the x-axis. Such a translation will affect neither the gradient at a point of the one graph nor the value of y at the corresponding point on the other graph. In particular, if the two curves are given a translation of $\begin{bmatrix} -\frac{1}{2}\pi \\ 0 \end{bmatrix}$, we obtain Figure 18. The graph of f has then become the graph of $g : x \rightarrow \cos x$, and the graph of f' has become the graph of $x \rightarrow {}^-\sin x$. So we deduce that

$$g(x) = \cos x \quad \Rightarrow \quad g'(x) = {}^-\sin x.$$

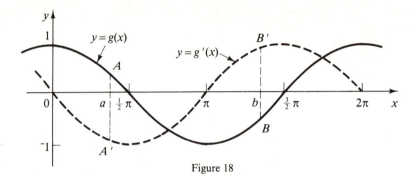

Figure 18

Example 4
Differentiate $f : x \rightarrow 4 \cos 3x$.

The arguments about stretches applied to the graph of $g : x \rightarrow \sin x$ can be used for the graph of $x \rightarrow \cos x$ also. So we can deduce immediately that

$$f'(x) = 4 \times 3 \times {}^-\sin 3x = {}^-12 \sin 3x.$$

6. INTEGRATION OF sin x AND cos x

Example 5

Find
$$\int_0^{\frac{1}{6}\pi} (\cos 2x)\, dx.$$

To find this integral, we need first to find a function which differentiated gives $\cos 2x$.

If we try $f(x) = \sin 2x$ as our function, we get $f'(x) = 2 \cos 2x$, which is twice as big as we require. So the function we need is

$$f(x) = \tfrac{1}{2} \sin 2x,$$

since differentiating this gives

$$f'(x) = \tfrac{1}{2} \times 2 \cos 2x,$$
$$= \cos 2x.$$

So
$$\int_0^{\frac{1}{6}\pi} (\cos 2x)\, dx = \left[\tfrac{1}{2} \sin 2x \right]_0^{\frac{1}{6}\pi},$$
$$= \tfrac{1}{2} \sin \left(\tfrac{1}{3}\pi \right) - \tfrac{1}{2} \sin (0),$$
$$= 0.433 \text{ to 3 s.f.}$$

Generalising from the above example, we may write

$$\int (\cos kx)\, dx = \frac{1}{k} \sin kx + C.$$

Also, if $f(x) = \cos kx$

we have $f'(x) = {}^-k \sin kx.$

So $\int (\sin kx)\, dx = \frac{{}^-1}{k} \cos kx + C.$

Example 6
The speed at which the valve on a bicycle wheel is moving forward is modelled by $v(t) = 6.6 + 6.5 \sin 20t$, where $v(t)$ is the speed in metres per second at time t seconds. Find how far forward it moves between $t = 0.6$ and $t = 0.8$.

$$v(t) = 6.6 + 6.5 \sin 20t$$

$$\Rightarrow \quad \left[s(t) \right]_{0.6}^{0.8} = \int_{0.6}^{0.8} (6.6 + 6.5 \sin 20t)\, dt$$

$$= \left[6.6t - 6.5 \times \tfrac{1}{20} \cos 20t \right]_{0.6}^{0.8}$$

$$= \left[6.6t - 0.325 \cos 20t \right]_{0.6}^{0.8}$$

$$\approx 5.591 - 3.686$$

$$= 1.905.$$

The valve moves forward 1.9 m (to 2 s.f.).

Exercise E

1   Differentiate:
    (a) $x \to \cos 8x$;   (b) $x \to 3 \cos 4x$;   (c) $x \to \sin 2x + \cos 3x$;
    (d) $x \to \cos \tfrac{1}{5}x$;   (e) $x \to 5 \cos \tfrac{1}{10}x$.

2   Find:
    (a) $\int (\cos 3x)\, dx$;   (b) $\int (\sin 10x)\, dx$;   (c) $\int (\cos \pi x + \sin \pi x)\, dx$;
    (d) $\int (2 \sin 3x - 4 \cos 2x)\, dx$.

3   Find the maximum value of the function

$$f(x) = 5 \cos 2x + 12 \sin 2x.$$

4   (a) Evaluate (i) $\int_0^{\frac{1}{2}\pi} (\sin x)\, dx$,   (ii) $\int_{\pi}^{\frac{3}{2}\pi} (\sin x)\, dx$, and sketch the regions whose areas are given by these integrals.
    (b) Without further integration, use the results of part (a) to evaluate
       (i) $\int_0^{\pi} (\cos x)\, dx$,   (ii) $\int_0^{\pi} (\sin x)\, dx$.

5   (a) What transformation maps the curve $y = \cos x$ onto the curve $y = \cos 3x$?
    (b) Evaluate $\int_0^{\frac{1}{3}\pi} (\cos x)\, dx$.

(c) Write down the value of $\int_0^{\frac{1}{6}\pi} (\cos 3x)\,dx$, and then check your answer by integration.

6   The height of the tide at Muchmud-by-the-Sea, in metres above mean sea-level, is modelled by the formula

$$h(t) = 5 \cos \tfrac{1}{6}\pi t$$

where $h(t)$ m is the height t hours after high tide.
   (a) Find $h'(t)$.
   (b) At what rate is the tide rising (i) 8 hours, (ii) 9 hours, (iii) 5 hours, after high tide?

7   Differentiate:
   (a) $x \rightarrow \cos 5x$;   (b) $x \rightarrow 4 \cos 7x$;   (c) $x \rightarrow \cos 3x - \sin 5x$;
   (d) $x \rightarrow \cos \tfrac{1}{12}x$;   (e) $x \rightarrow 9 \cos \tfrac{1}{12}x$.

8   Find:
   (a) $\int (\cos 6x)\,dx$;   (b) $\int (\sin 8x)\,dx$;   (c) $\int (2 \cos 2x + 3 \sin 6x)\,dx$;
   (d) $\int (\tfrac{1}{4} \sin \tfrac{1}{8}x - \tfrac{1}{2} \cos 3x)\,dx$.

9   Find the maximum value of the function

$$g(x) = 8 \sin 3x - 15 \cos 3x.$$

10   (a) Evaluate (i) $\int_0^{\pi} (\sin x)\,dx$,   (ii) $\int_{\pi}^{2\pi} (\sin x)\,dx$, and sketch the regions whose areas are given by these integrals.
   (b) Without further integration, use the results of part (a) to evaluate
   (i) $\int_0^{\frac{1}{2}\pi} (\sin x)\,dx$,   (ii) $\int_{\frac{1}{2}\pi}^{\frac{3}{2}\pi} (\cos x)\,dx$.

11   (a) What transformation maps the curve $y = \sin x$ onto the curve $y = \sin \tfrac{1}{2}x$?
   (b) Evaluate $\int_0^{6\pi} (\sin x)\,dx$.
   (c) Write down the value of $\int_0^{\frac{1}{3}\pi} (\sin \tfrac{1}{2}x)\,dx$, and then check your answer by integration.

12   The displacement of a mass oscillating on a spring is given by

$$x = 10 + \sin \tfrac{1}{4}\pi t.$$

   (a) Find its velocity as a function of time.
   (b) When is it stationary?
   (c) Find its acceleration as a function of time.
   (d) When is the acceleration zero?

7. SUMMARY

For the function $g(x) = \sin x°$, the derived function

is given by $g'(x) = \dfrac{\pi}{180} \cos x°$.

Radian measure: 2π radians $= 360°$;

$$x \text{ degrees} = x \times \frac{\pi}{180} \text{ radians};$$

$$\alpha \text{ radians} = \alpha \times \frac{180}{\pi} \text{ degrees}.$$

Arc length and sector area:

For a circle of radius r, centre O,

arc length $AB = \alpha r$,

area of sector $AOB = \frac{1}{2}\alpha r^2$,

where α is the radian measure of angle AOB.

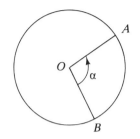

Figure 19

Differentiation of $\sin kx$, $\cos kx$:

$f(x)$	$f'(x)$
$\sin kx$	$k \cos kx$
$\cos kx$	$^-k \sin kx$

Integration of $\sin kx$, $\cos kx$:

$f(x)$	$\int f(x)\,dx$
$\sin kx$	$\dfrac{^-1}{k}\cos kx + C$
$\cos kx$	$\dfrac{1}{k}\sin kx + C$

Miscellaneous exercise

1 The velocity of a particle moving in a straight line is given at time t by $v(t) = 3 \cos \pi t$.
   (a) What is its initial speed?
   (b) Write down the acceleration at time t.
   (c) When does it first stop? Show that its acceleration is then of magnitude 3π.
   (d) State its displacement from the initial position at time t, and find after how long it first returns to its starting point.
   (e) What is its greatest displacement from the initial position? (SMP)

2 The framework for a solar cell for a satellite is made from a length of wire bent into the shape of a sector of a circle of radius r and angle θ, as shown in Figure 20. Write

Figure 20

down, in terms of r and θ, expressions for (a) the length of the wire, (b) the area, A, of the sector.

It is decided that the wire shall be of length 3 metres. Show, by eliminating θ from your expressions (a) and (b), that

$$A = r(3-2r)/2.$$

Hence calculate the greatest area of the cell.   (SMP)

3 Calculate $\displaystyle\int_{1}^{2} \sin x \, dx$ and, on a sketch graph, mark the area corresponding to this.

If $\displaystyle\int_{p}^{p+1} \sin x \, dx = 0$, state a possible value of p.   (SMP)

4 Draw the graph of $y = \sin x$ for $0 \leqslant x \leqslant \pi$, plotting points with x-coordinates at intervals of $\frac{1}{6}\pi$, and labelling them O, A, B, C, D, E, F successively.

Calculate the area of the polygon $OABCDEF$ by dividing it into strips parallel to the y-axis, or otherwise. Hence give an approximation to $\displaystyle\int_{0}^{\frac{1}{2}\pi} \sin x \, dx$.   (SMP)

5 (Throughout this question the domain is $0 \leqslant x \leqslant \pi$.)
Sketch the graph of $f(x) = 9x(2x - \pi)$ and state where it crosses the x-axis.

Verify that the graph of $g(x) = 10\pi^2 \sin x$ meets the graph of $f(x)$ where $x = \frac{5}{6}\pi$, and state the other values of x at which they intersect. Sketch the graph of $g(x)$ on the same axes.

    (a) Calculate (i) the area of $g(x)$ above the x-axis, (ii) the area of $f(x)$ below the x-axis.

    (b) Write down an expression involving definite integrals for the area enclosed between the curves $y = f(x)$ and $y = g(x)$.   (SMP)

6

Figure 21

A length of wire is bent into the shape of a sector of a circle of radius r with arc length one unit (Figure 21).

    (a) Calculate the area enclosed.

    (b) State, in terms of r, what the area would have been if it had been bent into the shape of a square.

    (c) Hence show that the percentage by which the area of the square exceeds the area of the sector is

$$50r - 50 + 25/2r.$$

    (d) Calculate the value of r for which this percentage is least.

    (e) State the corresponding length of wire.   (SMP)

7 A and B are points on the circumference of a circle of radius r with centre O (Figure 22). The angle AOB is θ radians. Express in terms of r and θ,

    (a) x, the area of triangle OAB;

    (b) y, the area of sector OAB;

    (c) z, the area of the segment shaded.

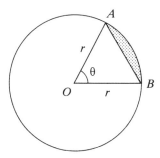

Figure 22

Find the value of θ for which x is a maximum, explaining your method. Explain why z is not a minimum for this value of θ.   (SMP)

8   (a) Given that $f(x) = \sin x°$, express $x°$ in radians and find $f'(x)$.

(b) Show that the gradient of the graph of $y = \sin x°$ at $x = 30$ is $\dfrac{\pi\sqrt{3}}{360}$.

(c) A ladder of length 4 units is leaning against a vertical wall. The ladder makes an angle of $\theta°$ with the horizontal and reaches a distance h units up the wall (Figure 23). Express h in terms of $\theta°$ and evaluate h when $\theta = 30$.

Figure 23

(d) If θ is now changed to 33, use your answers to (b) and (c) to find an approximation for the new value of h, leaving your answer in terms of π.   (SMP)

9   The position of a point moving on the x-axis is given at time t by the equation

$$x = t^2 - \sin t.$$

(a) Find x at (i) time $t = 0$, (ii) time $t = \tfrac{1}{3}\pi$.
(b) Find the average velocity over the time from $t = 0$ to $t = \tfrac{1}{3}\pi$.
(c) Write down a formula for the velocity at time t.
(d) Show that the velocity at time $t = 0$ is -1.
(e) Write down a formula for the acceleration at time t. What will be the least value of the acceleration?   (SMP)

10   The angle θ which a pendulum makes with the vertical t seconds after it is released is given, for $t \geqslant 1$, by $\theta = f(t) = t^{-1} - \cos t$.
(a) Write down the value of $f'(t)$.
(b) State, correct to 2 places of decimals, the six values needed to complete the table given below:

t	1	2	3	4	5	6	7	8	9	10
$1/t^2$	1	0.25		0.06	0.04		0.02	0.02		0.01
$\sin t$	0.84	0.91		-0.76	-0.96		0.66	0.99		-0.54

(c) With the same axes, plot the graphs of sin t and of $1/t^2$ for $1 \leqslant t \leqslant 10$. Take a scale of one centimetre to one unit horizontally and to one tenth of a unit vertically.

(d) Read off to 2 significant figures the values of t at the four points where the graphs intersect.

(e) Hence, by calculating values of θ corresponding to two of the above values of t, show that the greatest deflection in the first 5 seconds is about 76°.

(SMP)

11 Sketch the graphs of

(a) $y = \sin x$,   (b) $y = \sin (x - \frac{1}{2}\pi)$,   (c) $y = {}^-\sin (x - \frac{1}{2}\pi)$.

Hence express ${}^-\sin (x - \frac{1}{2}\pi)$ in a simpler form.

(SMP)

12 Solve the following equations for values of x in the interval ${}^-\pi \leqslant x \leqslant \pi$:

(a) $\sin (2x + \frac{1}{3}\pi) = 0.5$;   (b) $\tan (\frac{1}{3}x - \frac{2}{3}\pi) = \sqrt{3}$;   (c) $3 \cos (\pi - 4x) = 4$.

13 The number of hours of daylight at a point on the Arctic Circle is given approximately by

$$d = 12 - 12 \cos \tfrac{1}{6}\pi \, (t + \tfrac{1}{3})$$

where t is the number of months which have elapsed since 1st January.

(a) Find d (i) on 21st March ($t \approx 2.7$),   (ii) on 21st June ($t \approx 5.7$).

(b) When will there be 5 hours of daylight?

(SMP)

14 (a) Find the period of each of the functions:

(i) $x \to \sin x$;   (ii) $x \to \cos x$;   (iii) $x \to \tan x$;

(iv) $x \to \sin 2x$;   (v) $x \to \cos \frac{1}{3}x$;   (vi) $x \to \tan \pi x$.

(b) A boat anchored in a harbour in rough weather bobs up and down sinusoidally with period $\frac{5}{2}\pi$ seconds. If the amplitude of the motion is 0.8 metre, write a formula for the vertical displacement above the calm water-level as a function of time.

Calculate the boat's displacement after (i) 1 second, (ii) π seconds.

What is the frequency of the oscillation?

(SMP)

10

Kinematics

1. POSITION VECTORS

Kinematics is the study of motion. We are familiar with many examples of objects in motion – for example, a fly in a room, a car on a road, a ship on the sea, a ball thrown in the air, a satellite orbiting the earth, etc. – and so this is a topic with many immediate applications.

In order that we can study the motion of an object, we must first be able to find its *position* at any time and to do this we consider the object to be moving relative to a *fixed frame of reference*.

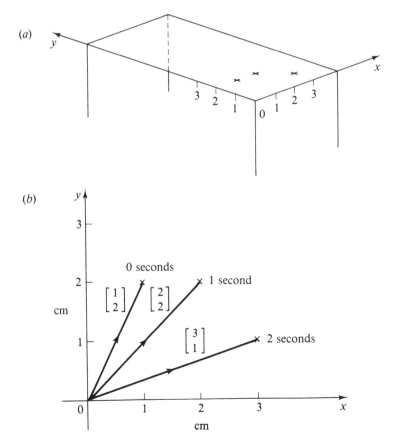

Figure 1

Example 1

A fly is crawling across a table. With respect to Cartesian coordinates fixed in one corner its positions after 0, 1, 2 seconds are $(1,2)$, $(2,2)$, $(3,1)$ respectively, where the scale units are centimetres. Express the position of the fly in vector form. (See Figure 1(*a*).)

We can write these positions in terms of *position vectors*. If we take 1 cm as the unit of length, the positions of the fly are given by the position vectors $\begin{bmatrix} 1 \\ 2 \end{bmatrix}$, $\begin{bmatrix} 2 \\ 2 \end{bmatrix}$, $\begin{bmatrix} 3 \\ 1 \end{bmatrix}$, respectively. (See Figure 1(*b*).)

We can find the vector *displacement* of the fly as follows:

From 0 to 1 second the displacement is $\begin{bmatrix} 2 \\ 2 \end{bmatrix} - \begin{bmatrix} 1 \\ 2 \end{bmatrix} = \begin{bmatrix} 1 \\ 0 \end{bmatrix}$.

From 1 to 2 seconds the displacement is $\begin{bmatrix} 3 \\ 1 \end{bmatrix} - \begin{bmatrix} 2 \\ 2 \end{bmatrix} = \begin{bmatrix} 1 \\ -1 \end{bmatrix}$.

Position vectors may be used just as readily to give the position of an object in three dimensions. Suppose that the fly of Example 1 had jumped 2 cm above the table after 2 seconds. Defining $\begin{bmatrix} 1 \\ 0 \\ 0 \end{bmatrix}$, $\begin{bmatrix} 0 \\ 1 \\ 0 \end{bmatrix}$ to represent 1 cm along the two edges of the table and $\begin{bmatrix} 0 \\ 0 \\ 1 \end{bmatrix}$ to represent 1 cm upwards perpendicular to the table, the positions of the fly are given by the position vectors $\begin{bmatrix} 1 \\ 2 \\ 0 \end{bmatrix}$, $\begin{bmatrix} 2 \\ 2 \\ 0 \end{bmatrix}$, $\begin{bmatrix} 3 \\ 1 \\ 2 \end{bmatrix}$.

Note that the units of length used in position vectors must be stated or given in form such as 'a displacement of $\begin{bmatrix} 4 \\ 7 \\ 3 \end{bmatrix}$ m.'

2. CONSTANT VELOCITY

In the examples above we do not know how the fly moved between the positions given and so we cannot study its motion in any detail. The following example gives more information, allowing deeper analysis.

Example 2

The position vector of a ship t hours after noon is given by $\begin{bmatrix} 40t+10 \\ 100-30t \end{bmatrix}$, where the axes point east and north and distance is in km. Discuss the ship's motion.

In this case we can plot the exact course as the position of the ship is known at all times. We shall write the position vector of the ship after t hours as $\mathbf{p}(t)$, thus $\mathbf{p}(0) = \begin{bmatrix} 10 \\ 100 \end{bmatrix}$, $\mathbf{p}(1) = \begin{bmatrix} 50 \\ 70 \end{bmatrix}$ etc. Drawing these position vectors allows us to find the path of the ship. Reference to Figure 2 indicates that the ship is travelling in a straight line.

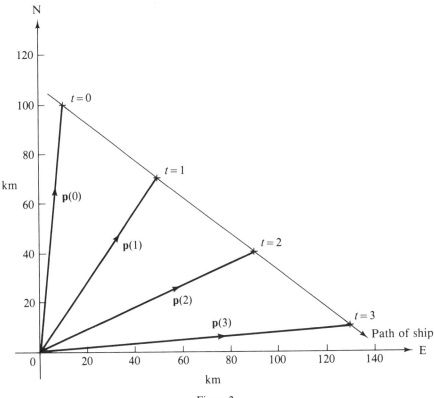

Figure 2

We can again calculate the displacement vectors:
From 0 to 1 hour the displacement is

$$\begin{bmatrix} 50 \\ 70 \end{bmatrix} - \begin{bmatrix} 10 \\ 100 \end{bmatrix} = \begin{bmatrix} 40 \\ -30 \end{bmatrix}.$$

From t to $t+1$ hours the displacement is

$$\begin{bmatrix} 40(t+1)+10 \\ 100-30(t+1) \end{bmatrix} - \begin{bmatrix} 40t+10 \\ 100-30t \end{bmatrix} = \begin{bmatrix} 40t+50 \\ 70-30t \end{bmatrix} - \begin{bmatrix} 40t+10 \\ 100-30t \end{bmatrix} = \begin{bmatrix} 40 \\ -30 \end{bmatrix}.$$

This is the same for all times t; it is constant.

It is fairly obvious that in this example the ship is travelling at *constant velocity*, that is, the ship is travelling in a fixed direction with a constant displacement in each hour. (Remember that velocity is speed in a given direction.) It would be natural to write the velocity, $\mathbf{v}$, of the ship as $\begin{bmatrix} 40 \\ -30 \end{bmatrix}$ where the units are km h⁻¹. The magnitude of the velocity vector, i.e. the *speed*, of the ship is $\sqrt{(40^2+(-30)^2)} = \sqrt{2500} = 50$ km h⁻¹. Its direction in terms of the angle made with due south is $\tan^{-1}\frac{4}{3} \approx 53°$ (see Figure 3). Thus the velocity could be expressed as 50 km h⁻¹ on a bearing of 127° to the nearest degree.

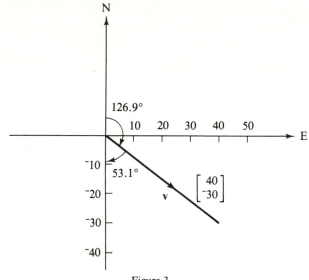

Figure 3

Exercise A

In the following examples, distances are in metres and the axes are east and north. In each case you are given the position vector of a woman t seconds after starting to drive her car. Plot her course, find the displacement vector between t and $t+1$ seconds and state whether her velocity is constant. If the velocity is constant then find its magnitude and bearing. The position vectors are:

1 $\begin{bmatrix} 12t \\ 5t \end{bmatrix}$. **2** $\begin{bmatrix} 7t-5 \\ 24t \end{bmatrix}$. **3** $\begin{bmatrix} t^2 \\ t+1 \end{bmatrix}$. **4** $\begin{bmatrix} 2t-1 \\ 1-2t \end{bmatrix}$.

5 $\begin{bmatrix} 4t-1 \\ 7t+1 \end{bmatrix}$. **6** $\begin{bmatrix} 2-4t \\ 8t-1 \end{bmatrix}$. **7** $\begin{bmatrix} t^2+1 \\ t^2-1 \end{bmatrix}$. **8** $\begin{bmatrix} (t+1)^2 \\ t-1 \end{bmatrix}$.

3. INSTANTANEOUS VELOCITY AND ACCELERATION

It was easy to find the velocity of the ship in the Example 2 because it was constant. How could we proceed if the speed or direction (or both) were changing? We have seen how to deal with motion in one dimension only in Chapters 5 and 6 when we used differentiation and integration to connect distance, velocity and acceleration functions. We shall now extend these ideas to motion in two or three dimensions.

The idea of *average velocity* is very straightforward, as the following example shows.

Example 3

The displacement of a skater on a pond is given by the position vector $\mathbf{p}(t) = \begin{bmatrix} 2t+1 \\ 10-t^2 \end{bmatrix}$ where t is time in seconds and the magnitude is measured in metres. Find the path of the skater and her average velocity between $t = 1$ and $t = 3$.

We have $\mathbf{p}(0) = \begin{bmatrix} 1 \\ 10 \end{bmatrix}$, $\mathbf{p}(1) = \begin{bmatrix} 3 \\ 9 \end{bmatrix}$, etc. Drawing these position vectors allows us to sketch the path of the skater (see Figure 4).

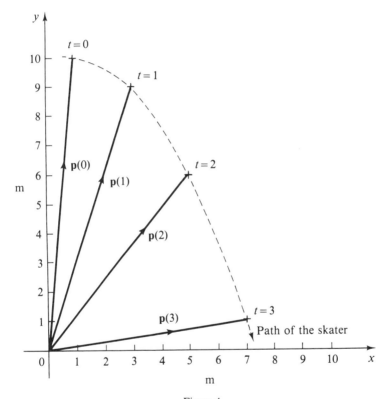

Figure 4

When $t = 3$ her position vector is $\begin{bmatrix} 7 \\ 1 \end{bmatrix}$ m, when $t = 1$ it is $\begin{bmatrix} 3 \\ 9 \end{bmatrix}$ m. Thus her displacement in those 2 seconds is $\begin{bmatrix} 7 \\ 1 \end{bmatrix} - \begin{bmatrix} 3 \\ 9 \end{bmatrix} = \begin{bmatrix} 4 \\ -8 \end{bmatrix}$ m. Since average velocity is displacement per second we must now divide by 2.

Thus her average velocity is $\frac{1}{2}\begin{bmatrix} 4 \\ -8 \end{bmatrix} = \begin{bmatrix} 2 \\ -4 \end{bmatrix}$ ms^{-1}.

Figure 5(a) shows the displacement vector from $t = 1$ to $t = 3$. Figure 5(b) shows the average velocity vector – note that the units are now those of speed.

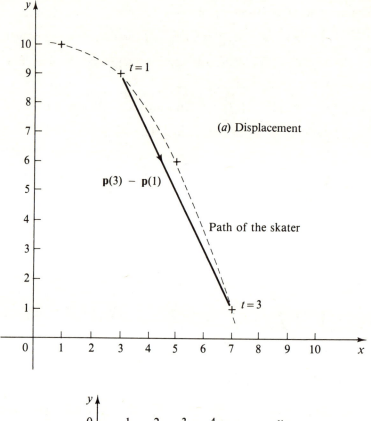

(a) Displacement

$\mathbf{p}(3) - \mathbf{p}(1)$

Path of the skater

$t = 1$

$t = 3$

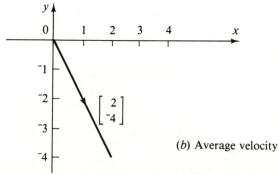

(b) Average velocity

Figure 5

In general, if $\mathbf{p}(t_1)$ and $\mathbf{p}(t_2)$ are the position vectors of an object at times t_1 and t_2,

then the displacement is $\mathbf{p}(t_2) - \mathbf{p}(t_1)$,

and the average velocity is $\dfrac{\mathbf{p}(t_2) - \mathbf{p}(t_1)}{t_2 - t_1}.$

Returning to our problem about the skater, since we know her position vector

$\mathbf{p}(t)$ at all relevant times we can calculate the average velocity from time t to $t+h$ and so deduce the *instantaneous velocity* of the skater at time t.

The skater's displacement from time t to $t+h$ is

$$\begin{bmatrix} 2(t+h)+1 \\ 10-(t+h)^2 \end{bmatrix} - \begin{bmatrix} 2t+1 \\ 10-t^2 \end{bmatrix} = \begin{bmatrix} 2h \\ -2th-h^2 \end{bmatrix},$$

and her average velocity from time t to $t+h$ is

$$\frac{1}{h}\begin{bmatrix} 2h \\ -2th-h^2 \end{bmatrix} = \begin{bmatrix} 2 \\ -2t-h \end{bmatrix}.$$

As h approaches zero, this approaches $\begin{bmatrix} 2 \\ -2t \end{bmatrix}$ and so we deduce that the velocity of the skater at time t is $\begin{bmatrix} 2 \\ -2t \end{bmatrix}$.

Thus when $t = 0, 1, 1\frac{1}{2}$, the instantaneous velocities are $\begin{bmatrix} 2 \\ 0 \end{bmatrix}$ m s^{-1}, $\begin{bmatrix} 2 \\ -2 \end{bmatrix}$ m s^{-1}, $\begin{bmatrix} 2 \\ -3 \end{bmatrix}$ m s^{-1}, respectively.

Comparing the velocity vector $\begin{bmatrix} 2 \\ -2t \end{bmatrix}$ with the position vector $\begin{bmatrix} 2t+1 \\ 10-t^2 \end{bmatrix}$ we see that we could have obtained the velocity vector by separately differentiating the components of the position vector. This is true in general, so that if the position vector of an object is $\mathbf{p}(t) = \begin{bmatrix} r(t) \\ s(t) \end{bmatrix}$ then its velocity vector is $\mathbf{v}(t) = \mathbf{p}'(t) = \begin{bmatrix} r'(t) \\ s'(t) \end{bmatrix}$.

As the instantaneous velocity is a vector, it has magnitude and direction; its magnitude is the instantaneous speed and its direction is the direction of motion at that time. In Figure 6, we can see the path of the skater and *superimposed*

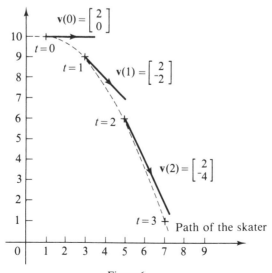

Figure 6

on the same diagram the velocity vectors at some points. (Remember that the units of velocity are not those of position (m s^{-1} not m) and note that the length of the velocity vector is proportional to the speed.) Notice that the velocity is along the tangent to the path at each time, since this is the direction in which the skater is moving.

We can find the average and instantaneous *acceleration* vectors by similar methods to those described above. If the velocity vector at time t is $\mathbf{v}(t)$, then the average acceleration from time t_1 to time t_2 is

$$\frac{\mathbf{v}(t_2) - \mathbf{v}(t_1)}{t_2 - t_1}.$$

If $\mathbf{v}(t) = \begin{bmatrix} u(t) \\ w(t) \end{bmatrix}$ then the *instantaneous* acceleration is

$$\mathbf{a}(t) = \mathbf{v}'(t) = \begin{bmatrix} u'(t) \\ w'(t) \end{bmatrix}.$$

Example 4

A particle moves so that its position vector at time t is $\begin{bmatrix} 2t^3 - 3t^2 \\ t^2 - t \end{bmatrix}$ where the lengths are measured in metres and time is in seconds. Find the velocity and acceleration vectors and determine if the particle is ever at rest. What is the magnitude of the acceleration initially and at time t?

$$\mathbf{v}(t) = \mathbf{p}'(t) = \begin{bmatrix} 6t^2 - 6t \\ 2t - 1 \end{bmatrix}.$$

$$\mathbf{a}(t) = \mathbf{v}'(t) = \begin{bmatrix} 12t - 6 \\ 2 \end{bmatrix}.$$

The particle is at rest when

$$\mathbf{v}(t) = \begin{bmatrix} 0 \\ 0 \end{bmatrix}, \text{ i.e. when } \begin{bmatrix} 6t^2 - 6t \\ 2t - 1 \end{bmatrix} = \begin{bmatrix} 0 \\ 0 \end{bmatrix}.$$

We require $6t^2 - 6t = 0$ and $2t - 1 = 0$. The second equation gives $t = \frac{1}{2}$, but $6(\frac{1}{2})^2 - 6 \times \frac{1}{2} \neq 0$ so the velocity is not $\mathbf{0}$ then. The particle is never at rest.

The magnitude of the acceleration at time t is

$$\sqrt{[(12t - 6)^2 + 4]} = \sqrt{(144t^2 - 144t + 40)} \text{ m s}^{-2}.$$

When $t = 0$ the acceleration has magnitude $\sqrt{40}$ m s^{-2}.

Example 5

The position vector of a particle is given by $\begin{bmatrix} \cos 3t \\ \sin 4t \end{bmatrix}$ where the distance is measured in metres and the time in seconds. Find the velocity and acceleration vectors. What is the magnitude of the acceleration when $t = \frac{1}{4}\pi$?

$$\mathbf{v}(t) = \mathbf{p}'(t) = \begin{bmatrix} -3 \sin 3t \\ 4 \cos 4t \end{bmatrix}$$

$$\mathbf{a}(t) = \mathbf{v}'(t) = \begin{bmatrix} -9\cos 3t \\ -16\sin 4t \end{bmatrix}$$

$$\mathbf{a}(\tfrac{1}{4}\pi) = \begin{bmatrix} -9\cos\tfrac{3}{4}\pi \\ -16\sin\pi \end{bmatrix} = \begin{bmatrix} -9 \times -0.7071 \\ 0 \end{bmatrix}$$

$$= \begin{bmatrix} 6.364 \\ 0 \end{bmatrix}.$$

Hence the magnitude of the acceleration when $t = \tfrac{1}{4}\pi$ is 6.36 m s^{-2} to 3 s.f.

Figure 7 shows the path of the particle from $t = 0$ to $t = \tfrac{1}{4}\pi$, together with $\mathbf{v}(\tfrac{1}{4}\pi)$ and $\mathbf{a}(\tfrac{1}{4}\pi)$.

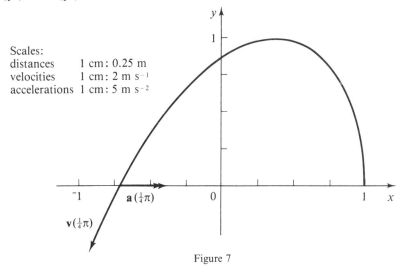

Scales:
distances   1 cm : 0.25 m
velocities   1 cm : 2 m s^{-1}
accelerations 1 cm : 5 m s^{-2}

Figure 7

Exercise B

1   The velocity of a particle is given by $\begin{bmatrix} t+1 \\ t^2 \end{bmatrix}$ where the magnitude is in m s^{-1}. Find the acceleration vector and the magnitude of the acceleration. Determine whether the particle is ever at rest.

2   The position vector of a car is given by $\begin{bmatrix} t^2-1 \\ t+1 \end{bmatrix}$ where the distance is measured in metres and time in seconds. Find the velocity and acceleration vectors. Sketch the path for values of t from 0 to 2, and show the velocity and acceleration vectors at $t = 0, 1, 2$.

3   A ship has position given by the vector $\begin{bmatrix} \tfrac{1}{2}t^2-5t \\ 6t-\tfrac{1}{2}t^2 \end{bmatrix}$ where the axes are east and north. Distance is measured in kilometres, time in hours. At what times is the ship travelling (a) due east, (b) due north? Sketch its path for values of t from 0 to 6.

4   Find the vectors which give the velocity and acceleration of a particle whose position is given by the vector $\begin{bmatrix} 4t^3-2t+1 \\ t^2-4t-6 \end{bmatrix}$. Distance is measured in metres, and time in seconds.

5   A car moves so that its velocity is given by the vector $\begin{bmatrix} t^2 - 3t + 2 \\ t^2 - 1 \end{bmatrix}$ where distance
    is in km and time in hours. When is the car at rest and what is the acceleration vector
    at this time?

6   The position of an ant on a table is given by the vector $\begin{bmatrix} t^2 + 1 \\ 1 - t \end{bmatrix}$ where length is in
    centimetres and time in seconds. Find the velocity vector and the speed of the ant
    at time t. Sketch its path for values of t from 0 to 4, showing the velocity and
    acceleration vectors at $t = 0, 2, 4$.

7   A particle moves in a circle radius 3 m so that its position vector after t seconds is
    given by $\begin{bmatrix} 3 \cos 2t \\ 3 \sin 2t \end{bmatrix}$ referred to the usual axes. Find the velocity and acceleration
    vectors after time t. Find the times of the first two occasions ($t \geqslant 0$) when the particle
    is heading in the y-direction.

8   A particle has position given by the vector $\begin{bmatrix} t^3 \\ 1/t \end{bmatrix}$, ($t > 0$). Find the vectors which
    represent the velocity and acceleration.

9   A particle moves so that its position is given by the vector $\begin{bmatrix} t^4 + 3t^2 \\ t^3 - t^2 \end{bmatrix}$ referred to the
    usual axes where distance is in metres and time in seconds. What is the speed of the
    particle when its acceleration is in the x-direction?

10 A water wheel is rotating steadily and the position vector of one bucket referred
    to horizontal and vertical axes is $\begin{bmatrix} \sin 3t \\ \cos 3t \end{bmatrix}$. Find the velocity and acceleration after
    time t. What is the magnitude of the acceleration when $t = 2$? When is the bucket
    first heading upward at 45°?

4. INTEGRATION OF VECTORS

In Chapter 6 we saw how integration, the inverse process to differentiation,
enables us to deduce the displacement function, $[s(t)]$, from the velocity function
$v(t)$, and the change in the velocity function from the acceleration $a(t)$. It should
be remembered that integration gives the *change* in position and velocity. To
obtain $s(t)$ from $v(t)$ it was necessary to know the position at some time.

    For example, from $v(t) = 3t^2 - 4$, we can deduce that

$$[s(t)] = [t^3 - 4t].$$

If we are then told that $s(1) = 7$, we can write

$$\left[s(t) \right]_1^t = \left[t^3 - 4t \right]_1^t$$
$$s(t) - 7 = t^3 - 4t - (1^3 - 4)$$
$$s(t) = t^3 - 4t + 10.$$

    If we are given no information about the position at any particular time, the
best we can do is to write

$$s(t) = t^3 - 4t + k,$$

where k is an unknown number.

These ideas can readily be extended to motion in two or three dimensions. Since we can obtain $\mathbf{v}(t)$ from $\mathbf{p}(t)$ by differentiating its components, we can write

$$\mathbf{p}(t) = \int \mathbf{v}(t) \, dt$$

and obtain the integral by integrating the components.

Example 6

The velocity of a toy boat at time t is given by the vector $\mathbf{v}(t) = \begin{bmatrix} \sin t \\ 3 \end{bmatrix}$, the units being metres per minute. Find its position $\mathbf{p}(t)$ and acceleration $\mathbf{a}(t)$ at time t given that $\mathbf{p}(0) = \begin{bmatrix} 1 \\ -4 \end{bmatrix}$.

$$\mathbf{p}(t) = \int \mathbf{v}(t) \, dt = \begin{bmatrix} \int \sin t \, dt \\ \int 3 \, dt \end{bmatrix} = \begin{bmatrix} -\cos t + c \\ 3t + d \end{bmatrix},$$

where c and d are constants to be found. Substituting $t = 0$ gives

$$\mathbf{p}(0) = \begin{bmatrix} -1 + c \\ d \end{bmatrix}$$

from which we deduce that $\begin{bmatrix} c \\ d \end{bmatrix} = \begin{bmatrix} 2 \\ -4 \end{bmatrix}$, since $\mathbf{p}(0) = \begin{bmatrix} 1 \\ -4 \end{bmatrix}$.

So

$$\mathbf{p}(t) = \begin{bmatrix} 2 - \cos t \\ 3t - 4 \end{bmatrix}.$$

$\mathbf{a}(t)$ can be found directly from $v(t)$ by differentiation:

$$\mathbf{a}(t) = \begin{bmatrix} \cos t \\ 0 \end{bmatrix}.$$

The path is sketched in Figure 8.

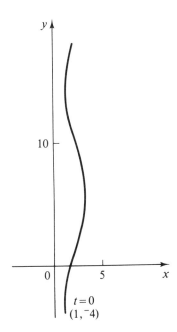

Figure 8

Exercise C

In questions 1–3 and 4–6 find the position vector $\mathbf{p}(t)$ and acceleration $\mathbf{a}(t)$ of a particle whose velocity is given by the vector $\mathbf{v}(t)$. Evaluate $\mathbf{p}(1)$ in each case.

1 $\mathbf{v}(t) = \begin{bmatrix} 1 \\ t \end{bmatrix}$, $\mathbf{p}(0) = \mathbf{0}$.

2 $\mathbf{v}(t) = \begin{bmatrix} t \\ -2t \end{bmatrix}$, $\mathbf{p}(1) = \begin{bmatrix} 3 \\ 2 \end{bmatrix}$.

3 $\mathbf{v}(t) = \begin{bmatrix} \cos 2t \\ 2t \end{bmatrix}$, $\mathbf{p}(0) = \mathbf{0}$.

4 $\mathbf{v}(t) = \begin{bmatrix} 1 \\ 0 \end{bmatrix}$, $\mathbf{p}(0) = \begin{bmatrix} 5 \\ 2 \end{bmatrix}$.

5 $\mathbf{v}(t) = \begin{bmatrix} 1-2t \\ 2t+2 \end{bmatrix}$, $\mathbf{p}(2) = \begin{bmatrix} 1 \\ -1 \end{bmatrix}$.

6 $\mathbf{v}(t) = \begin{bmatrix} \cos 3t \\ \sin 2t \end{bmatrix}$, $\mathbf{p}(0) = \mathbf{0}$.

7 The velocity of a snail in mm s^{-1} at time t seconds is given by the vector function $\mathbf{v}(t) = \begin{bmatrix} 2t+1 \\ 1-t \end{bmatrix}$. Find the position $\mathbf{p}(t)$ and acceleration $\mathbf{a}(t)$, given that $\mathbf{p}(1) = \begin{bmatrix} 1 \\ 1 \end{bmatrix}$.

8 A toy boat on a pond has velocity given by $\begin{bmatrix} 0.25 \\ 0.1 \cos t \end{bmatrix}$ where the axes are horizontal and vertical and the units are m s^{-1}. Find the acceleration and position vectors at time t given that $\mathbf{p}(0) = \mathbf{0}$. What is the direction of motion of the boat when $t = \tfrac{1}{4}\pi$?

9 A particle has velocity given by the vector $\begin{bmatrix} 6t^2 - 2t \\ 2t+3 \end{bmatrix}$ where the units are m s^{-1}. Find the acceleration vector and the position $\mathbf{p}(t)$ if $\mathbf{p}(0) = \begin{bmatrix} 4 \\ -1 \end{bmatrix}$.

10 A stone is dropped and has constant acceleration given by the vector $\begin{bmatrix} 0 \\ -10 \end{bmatrix}$ m s^{-2}. If its initial velocity and position are $\begin{bmatrix} 2 \\ 0 \end{bmatrix}$ m s^{-1} and $\begin{bmatrix} 1 \\ 0 \end{bmatrix}$ m respectively, find the position and velocity vectors after time t. When is the stone at rest?

11 An aphid has acceleration given by the vector $\begin{bmatrix} t \\ t \end{bmatrix}$ where the units are m s^{-2}. Find the position $\mathbf{p}(t)$ and velocity $\mathbf{v}(t)$ given that $\mathbf{p}(0) = \begin{bmatrix} 0 \\ 0 \end{bmatrix}$ and $\mathbf{v}(0) = \begin{bmatrix} 0 \\ 1 \end{bmatrix}$. When does this aphid move instantaneously along one of the coordinate directions?

12 A toy boat has acceleration given by $\begin{bmatrix} 3 \sin t \\ 3 \cos t \end{bmatrix}$ where the units are m s^{-2}. Find the velocity $\mathbf{v}(t)$ and the position vector $\mathbf{p}(t)$, given that $\mathbf{v}(0) = \begin{bmatrix} 2 \\ 1 \end{bmatrix}$ and $\mathbf{p}(0) = \mathbf{0}$.

5. FURTHER APPLICATIONS

We shall often find the techniques of Chapter 8 useful for dealing with conditions which specify the direction of a vector; in particular we often wish to know whether two vectors are perpendicular.

In other cases we may be asked to deal with the magnitude of a vector as a function of t.

Example 7

The velocity of a particle at time t is given by the vector $\begin{bmatrix} t^2-1 \\ 2t \end{bmatrix}$. When is the particle travelling at its least speed? When, if ever, does it have its acceleration in the direction given by $\begin{bmatrix} 1 \\ 1 \end{bmatrix}$?

$$\begin{aligned} \text{The speed is } &\sqrt{((t^2-1)^2+4t^2)} \\ &= \sqrt{(t^4-2t^2+1+4t^2)} \\ &= \sqrt{(t^4+2t^2+1)} \\ &= \sqrt{(t^2+1)^2} \\ &= t^2+1. \end{aligned}$$

It is clear that the least value of t^2+1 is 1 and that this occurs when $t=0$.

$$\mathbf{a}(t) = \mathbf{v}'(t) = \begin{bmatrix} 2t \\ 2 \end{bmatrix}.$$

This is in the direction of $\begin{bmatrix} 1 \\ 1 \end{bmatrix}$ only at $t=1$ when the acceleration is $\begin{bmatrix} 2 \\ 2 \end{bmatrix}$.

Example 8

At time t a passenger on a fairground 'Screamer' has acceleration given by the vector $\begin{bmatrix} 1 \\ 1 \end{bmatrix}$ m s⁻². At time $t=1$ her velocity is $\begin{bmatrix} -9 \\ 1 \end{bmatrix}$ m s⁻¹. It is believed that a particular thrill will be enjoyed if the directions of the velocity and accleration are perpendicular. When should she expect the thrill?

$$\mathbf{a}(t) = \begin{bmatrix} 1 \\ 1 \end{bmatrix} \quad \text{and} \quad \mathbf{v}(t) = \int \mathbf{a}(t)\,dt = \begin{bmatrix} t+C \\ t+D \end{bmatrix}$$

where C and D are constants to be found. Now $\mathbf{v}(1) = \begin{bmatrix} -9 \\ 1 \end{bmatrix}$ gives

$$\begin{bmatrix} -9 \\ 1 \end{bmatrix} = \begin{bmatrix} 1+C \\ 1+D \end{bmatrix} \quad \Rightarrow \quad C = -10 \text{ and } D = 0.$$

Thus
$$\mathbf{v}(t) = \begin{bmatrix} t-10 \\ t \end{bmatrix}.$$

The acceleration and velocity vectors are perpendicular when $\mathbf{v}(t).\mathbf{a}(t) = 0$:

that is, when $\begin{bmatrix} t-10 \\ t \end{bmatrix}.\begin{bmatrix} 1 \\ 1 \end{bmatrix} = 0$;

or when $(t-10) \times 1 + t \times 1 = 0$.

Thus $2t - 10 = 0$ and so $t = 5$.

Hence the special thrill will occur only once, after 5 s.

6. MOTION IN THREE DIMENSIONS

In Section 1 we noted that we could write down the three-dimensional position vector of an object. All the techniques of the preceeding sections may be applied to such vectors.

For instance, if the position is given by $\mathbf{p}(t) = \begin{bmatrix} q(t) \\ r(t) \\ s(t) \end{bmatrix}$ then the velocity is given

by $\begin{bmatrix} q'(t) \\ r'(t) \\ s'(t) \end{bmatrix}$. If the acceleration is given by the vector $\mathbf{a}(t) = \begin{bmatrix} a_1(t) \\ a_2(t) \\ a_3(t) \end{bmatrix}$ then the velocity

is given by the vector $\mathbf{v}(t) = \begin{bmatrix} \int a_1(t)\,dt \\ \int a_2(t)\,dt \\ \int a_3(t)\,dt \end{bmatrix}$, and so on.

Example 9

The velocity of a fly is given by $\begin{bmatrix} t-2 \\ 4-t^2 \\ 2t \end{bmatrix}$ and its vector position after 2 s is $\begin{bmatrix} 6 \\ \frac{16}{3} \\ 3 \end{bmatrix}$ m.

Find the position and acceleration at time t. Does the fly pass through the

point $\begin{bmatrix} 0 \\ 1 \\ 0 \end{bmatrix}$? When, if ever, is the fly travelling straight upwards – that is in the

direction given by $\begin{bmatrix} 0 \\ 0 \\ 1 \end{bmatrix}$ – and what is its speed at this time?

Its position vector

$$\mathbf{p}(t) = \int \mathbf{v}(t)\,dt = \begin{bmatrix} \frac{1}{2}t^2 - 2t + C \\ 4t - \frac{1}{3}t^3 + D \\ t^2 + E \end{bmatrix}$$

where C, D, E are constants to be found. Now $\mathbf{p}(2) = \begin{bmatrix} 6 \\ \frac{16}{3} \\ 3 \end{bmatrix}$; thus $\begin{bmatrix} 6 \\ \frac{16}{3} \\ 3 \end{bmatrix} = \begin{bmatrix} -2+C \\ \frac{16}{3}+D \\ 4+E \end{bmatrix}$,

so $C = 8$, $D = 0$, $E = ^-1$.

Thus $\mathbf{p}(t) = \begin{bmatrix} \frac{1}{2}t^2 - 2t + 8 \\ 4t - \frac{1}{3}t^3 \\ t^2 - 1 \end{bmatrix}$.

Also $\mathbf{a}(t) = \mathbf{v}'(t) = \begin{bmatrix} 1 \\ -2t \\ 2 \end{bmatrix}$.

The fly passes through the point $\begin{bmatrix} 0 \\ 1 \\ 0 \end{bmatrix}$ if, for some t, $\begin{bmatrix} 0 \\ 1 \\ 0 \end{bmatrix} = \begin{bmatrix} \frac{1}{2}t^2 - 2t + 8 \\ 4t - \frac{1}{3}t^3 \\ t^2 - 1 \end{bmatrix}$:

i.e. $\left. \begin{array}{ll} \text{(i)} & \frac{1}{2}t^2 - 2t + 8 = 0 \\ \text{(ii)} & 4t - \frac{1}{3}t^3 = 0 \\ \text{(iii)} & t^2 - 1 = 0 \end{array} \right\}$ simultaneously.

Equation (iii) gives $t = \pm 1$ which does not satisfy (i) or (ii). Thus the fly does not pass through the given point.

$\mathbf{v}(t) = \begin{bmatrix} t - 2 \\ 4 - t^2 \\ 2t \end{bmatrix}$ and so the velocity will be in the direction given by $\begin{bmatrix} 0 \\ 0 \\ 1 \end{bmatrix}$ when

$\begin{bmatrix} t - 2 = 0 \\ 4 - t^2 = 0 \\ 2t \neq 0 \end{bmatrix}$ simultaneously, i.e. when $t = 2$. At this time the velocity is $\begin{bmatrix} 0 \\ 0 \\ 4 \end{bmatrix}$ and

so the speed is $\sqrt{(0 + 0 + 16)} = 4 \text{ ms}^{-1}$.

Example 10

The position vector of a point P on a bolt as it is screwed into a plate is $\begin{bmatrix} 5 \cos t \\ 5 \sin t \\ 2t \end{bmatrix}$

referred to the usual axes where the lengths are in mm and time is in seconds. What are the speed and magnitude of the acceleration of P after 6 s? What angle does the velocity of P make with the positive z-axis after 1 s?

$$\mathbf{v}(t) = \mathbf{p}'(t) = \begin{bmatrix} -5 \sin t \\ 5 \cos t \\ 2 \end{bmatrix}$$

$$\mathbf{a}(t) = \mathbf{v}'(t) = \begin{bmatrix} -5 \cos t \\ -5 \sin t \\ 0 \end{bmatrix}$$

$\mathbf{v}(6) = \begin{bmatrix} -5 \sin 6 \\ 5 \cos 6 \\ 2 \end{bmatrix} = \begin{bmatrix} 1.40 \\ 4.80 \\ 2 \end{bmatrix}$ and speed $= \sqrt{(1.4^2 + 4.8^2 + 2^2)} \text{ mm s}^{-1}$

$$= 5.4 \text{ mm s}^{-1} \text{ to 2 s.f.}$$

$$\mathbf{a}(6) = \begin{bmatrix} -5\cos 6 \\ -5\sin 6 \\ 0 \end{bmatrix} = \begin{bmatrix} -4.80 \\ 1.40 \\ 0 \end{bmatrix};$$

its magnitude is $\sqrt{(4.8^2 + 1.4^2 + 0^2)}$ mm s^{-2} = 5.0 mm s^{-2} to 2 s.f.

$$\mathbf{v}(1) = \begin{bmatrix} -5\sin 1 \\ 5\cos 1 \\ 2 \end{bmatrix} = \begin{bmatrix} -4.21 \\ 2.70 \\ 2 \end{bmatrix}.$$

The positive z-axis has direction given by $\begin{bmatrix} 0 \\ 0 \\ 1 \end{bmatrix}$. Let the required angle be $\theta°$.

Then $\qquad \cos\theta° = \dfrac{\begin{bmatrix} -4.21 \\ 2.70 \\ 2 \end{bmatrix} \cdot \begin{bmatrix} 0 \\ 0 \\ 1 \end{bmatrix}}{\sqrt{(4.21^2 + 2.7^2 + 2^2)} \times 1} = \dfrac{2}{5.39} = 0.371.$

Thus $\theta° = 68°$ to the nearest degree.

Exercise D

1   A bat flies with acceleration given by $\begin{bmatrix} -t \\ 0 \\ 1 \end{bmatrix}$ m s^{-2}. It is known that when $t = 1$ the bat is flying in the positive y-direction with speed 5 m s^{-1}. Find the velocity vector after time t.

2   A particle moves with acceleration $\begin{bmatrix} 1 \\ t \end{bmatrix}$ m s^{-2}. It has speed of $\sqrt{45}$ m s^{-1} when $t = 0$ at which time its velocity is perpendicular to the direction given by $\begin{bmatrix} 1 \\ -2 \end{bmatrix}$. By writing $\mathbf{v}(0) = \begin{bmatrix} v_1 \\ v_2 \end{bmatrix}$ find two equations connecting v_1 and v_2 and then solve them. Deduce the velocity vector after time t.

3   The velocity of a bee is given by $\begin{bmatrix} t-1 \\ t^2 \\ 1-t \end{bmatrix}$ m s^{-1}. Find the acceleration and position vectors at time t given that the position when $t = 3$ is $\begin{bmatrix} 0 \\ 9 \\ \frac{1}{2} \end{bmatrix}$ m. What is the magnitude of the acceleration when the bee is travelling in the y-direction?

4   A bowler delivers an in-swinging full toss. The position of the ball is given by $\begin{bmatrix} 7t \\ t-\frac{1}{4}t^2 \\ 4t-5t^2+2 \end{bmatrix}$ where distances are in metres and time in seconds. The x-, y- and z-axes are taken to be down the pitch, across the pitch and vertically upward respectively. Given that the ball is released when $t = 0$ and strikes the bat 7 m down the pitch, find:
   (a) the velocity at time t;
   (b) the velocity, speed and position of the ball when it strikes the bat;

(c) the angle at impact between the ball and a 'straight', vertical bat which faces straight down the pitch.

5   When Superman flies, his position is given by $\begin{bmatrix} t-3 \\ 2t \\ 20-5t^2 \end{bmatrix}$ where distances are in metres and time in seconds. At what speed will he hit the ground (assuming $t > 0$ and the ground corresponds to the z-component of position being zero)?

Wonderwoman comes to the rescue at the moment of impact with velocity $\begin{bmatrix} 2t \\ t \\ kt \end{bmatrix}$ m s^{-1} which is perpendicular to that of Superman. Find k and determine Wonderwoman's position at time t.

6   A sycamore seed spins as it falls straight to the ground. A point on the wing of the seed has position $\begin{bmatrix} 1.5\cos 2t \\ 1.5\sin 2t \\ 50-2t \end{bmatrix}$ (the z-axis being vertically upwards). Distances are in metres and time in seconds. When does the seed hit the ground? Find the vertical component of the velocity of the seed and the speed of the given point on the wing at this time.

7   The position of the tip of a blade of a rotary snow-plough is $\begin{bmatrix} \cos 20t \\ 0.2t^2 \\ \sin 20t \end{bmatrix}$ where distances are in metres and time in seconds. Find the velocity and acceleration vectors after t seconds. Calculate the speed of the tip of the blade (a) in the plane of its rotation, (b) in space, after 10 s.

7. MOTION WITH CONSTANT ACCELERATION

In many cases of motion we find that the acceleration is constant (or may be treated as constant) and so we shall develop some results applicable to this special case.

It can be shown that such motion must all take place in one plane so our vectors need have only two components if the frame of reference is suitably chosen. The results below are true for both two and three component vectors.

In constant acceleration problems it is customary to use the following notation.

a   is the constant acceleration. We shall use **A** to emphasise that the acceleration is constant.

u   is the velocity at $t = 0$.

v   is the velocity at time t.

r   is the displacement from $t = 0$ to time t.

Example 11

A particle has acceleration $\begin{bmatrix} 2 \\ 1 \end{bmatrix}$ m s^{-2} and initial velocity $\begin{bmatrix} -1 \\ 3 \end{bmatrix}$ m s^{-1}. What is its velocity after 5 seconds and what is its displacement in the first 5 seconds?

The acceleration is constant so $v = \int A \, dt = \begin{bmatrix} 2t \\ t \end{bmatrix} + C$ where C is a constant vector.

$v = \begin{bmatrix} -1 \\ 3 \end{bmatrix}$ when $t = 0$ and so $\begin{bmatrix} -1 \\ 3 \end{bmatrix} = \begin{bmatrix} 0 \\ 0 \end{bmatrix} + C$. Thus $v = \begin{bmatrix} 2t \\ t \end{bmatrix} + \begin{bmatrix} -1 \\ 3 \end{bmatrix} = \begin{bmatrix} 2t-1 \\ t+3 \end{bmatrix}$.

When $t = 5$ we have $v = \begin{bmatrix} 9 \\ 8 \end{bmatrix}$ m s^{-1}.

$p = \int v \, dt$ where p is the position vector of the particle at time t.

$p = \begin{bmatrix} t^2 - t \\ \frac{1}{2}t^2 + 3t \end{bmatrix} + D$ where D is a constant vector.

When $t = 0$ we have $p = D$ and so r, the displacement from time $t = 0$, is

$p - D$,   i.e. $r = \begin{bmatrix} t^2 - t \\ \frac{1}{2}t^2 + 3t \end{bmatrix}$. When $t = 5$ we have $r = \begin{bmatrix} 20 \\ 27\frac{1}{2} \end{bmatrix}$ m.

Alternatively we could argue that since the acceleration is constant

$$r = \text{average velocity} \times \text{time}$$

$$= \tfrac{1}{2}\left(\begin{bmatrix} -1 \\ 3 \end{bmatrix} + \begin{bmatrix} 9 \\ 8 \end{bmatrix} \right) \times 5$$

$$= 2\tfrac{1}{2} \begin{bmatrix} 8 \\ 11 \end{bmatrix} = \begin{bmatrix} 20 \\ 27\frac{1}{2} \end{bmatrix} \text{ m.}$$

If we carry out the integrations of Example 11 algebraically with a constant acceleration A and initial velocity u, we obtain:

$$v = \int A \, dt = tA + C$$

where C is a constant. Now $v = u$ when $t = 0$ and so

$$v = u + tA. \tag{1}$$

$$p = \int v \, dt = tu + \tfrac{1}{2}t^2A + D$$

where D is a constant vector, (using (1)). Now $p = D$ when $t = 0$ and so the displacement is $r = p - D$, giving

$$r = tu + \tfrac{1}{2}t^2A. \tag{2}$$

Rewriting (2) as $2r = t(2u + tA) = t(u + u + tA)$ we have, using (1),

$$2r = t(u + v)$$

and hence $\qquad\qquad r = \tfrac{1}{2}t(u + v). \tag{3}$

A further result may be obtained using the scalar product.

From (1), $\qquad\qquad\qquad v - u = tA.$

From (3), $\qquad\qquad\qquad v + u = \dfrac{2}{t}r.$

The scalar product of the left-hand sides of these equations must be equal to the scalar product of the right-hand sides. Thus

$$(v - u) \cdot (v + u) = tA \cdot \frac{2}{t}r = 2A \cdot r,$$

and so $\qquad\qquad \mathbf{v}.\mathbf{v}+\mathbf{v}.\mathbf{u}-\mathbf{u}.\mathbf{v}-\mathbf{u}.\mathbf{u} = 2\mathbf{A}.\mathbf{r},$

giving $\qquad\qquad\qquad \mathbf{v}.\mathbf{v}-\mathbf{u}.\mathbf{u} = 2\mathbf{A}.\mathbf{r}.$

Now $\mathbf{v}.\mathbf{v} = v^2$, where v is the magnitude of $\mathbf{v}$ (see Chapter 8); that is, v is the speed at time t. Similarly $\mathbf{u}.\mathbf{u} = u^2$ where u is the initial speed.

Thus $\qquad\qquad\qquad\qquad v^2 = u^2+2\mathbf{A}.\mathbf{r}.$ $\qquad\qquad\qquad$ (4)

These formulae may, of course, be re-arranged as required.

Example 12

A particle moves with constant acceleration. Its initial velocity is $\begin{bmatrix}1\\0\end{bmatrix}$ m s^{-1} and

after 4 seconds its displacement is $\begin{bmatrix}6\\5\end{bmatrix}$ m. What is its acceleration and final

velocity?

Since the acceleration is constant and we know values for $\mathbf{u}$, $\mathbf{r}$ and t, using
$\mathbf{r} = t\mathbf{u}+\frac{1}{2}t^2\mathbf{A}$,

$$\begin{bmatrix}6\\5\end{bmatrix} = 4\begin{bmatrix}1\\0\end{bmatrix}+\tfrac{1}{2}\times 4^2\mathbf{A} \quad\Rightarrow\quad \begin{bmatrix}6\\5\end{bmatrix}-\begin{bmatrix}4\\0\end{bmatrix} = 8\mathbf{A}$$

$$\Rightarrow\quad \mathbf{A} = \frac{1}{8}\begin{bmatrix}2\\5\end{bmatrix} = \begin{bmatrix}0.5\\0.675\end{bmatrix}.$$

Using $\mathbf{v} = \mathbf{u}+t\mathbf{A}$,

$$\mathbf{v} = \begin{bmatrix}1\\0\end{bmatrix}+4\begin{bmatrix}0.25\\0.675\end{bmatrix} = \begin{bmatrix}2\\2.5\end{bmatrix}.$$

Its acceleration is $\begin{bmatrix}0.5\\0.675\end{bmatrix}$ m s^{-2} and its final velocity is $\begin{bmatrix}2\\2.5\end{bmatrix}$ m s^{-1}.

Example 13

Find the speed of a fly, initially at rest, after displacement by $\begin{bmatrix}2\\4\end{bmatrix}$ m with

acceleration $\begin{bmatrix}1\\2\end{bmatrix}$ m s^{-2}.

Since the acceleration is constant, using (4) we have

$$v^2 = 0^2+2\begin{bmatrix}1\\2\end{bmatrix}.\begin{bmatrix}2\\4\end{bmatrix} = 2(1\times 2+2\times 4)$$

$$= 2\times 10 = 20,$$

thus $v = 4.47$. So the speed of the fly is 4.47 m s^{-1}.

Exercise E

In this exercise $\mathbf{A}$, $\mathbf{u}$, $\mathbf{v}$, $\mathbf{r}$ and t have meanings described above and the acceleration may always be assumed to be constant.

1    $A = \begin{bmatrix} 4 \\ 1 \end{bmatrix}$, $u = \begin{bmatrix} 1 \\ 2 \end{bmatrix}$, $t = 2$. Find v.

2    $r = \begin{bmatrix} 0 \\ 10 \end{bmatrix}$, $u = \begin{bmatrix} 1 \\ 3 \end{bmatrix}$, $v = \begin{bmatrix} -1 \\ 2 \end{bmatrix}$. Find t.

3    $A = \begin{bmatrix} 1 \\ 2 \end{bmatrix}$, $r = \begin{bmatrix} -1 \\ -4 \end{bmatrix}$, $u = 6$. Find v.

4    $u = \begin{bmatrix} 4 \\ -1 \end{bmatrix}$, $A = \begin{bmatrix} 1 \\ -5 \end{bmatrix}$, $t = 4$. Find r.

5    $A = \begin{bmatrix} 3 \\ -1 \end{bmatrix}$, $r = \begin{bmatrix} 2 \\ -3 \end{bmatrix}$, $v = 5$. Find u.

6    $u = \begin{bmatrix} -1 \\ 3 \end{bmatrix}$, $v = \begin{bmatrix} 4 \\ -1 \end{bmatrix}$, $t = 1\frac{1}{2}$. Find A.

7    $r = \begin{bmatrix} 6 \\ 3 \end{bmatrix}$, $A = \begin{bmatrix} 1 \\ -1 \end{bmatrix}$, $t = 5$. Find u.

8    $u = \begin{bmatrix} 1 \\ 0 \\ 3 \end{bmatrix}$, $v = \begin{bmatrix} -1 \\ 2 \\ 5 \end{bmatrix}$, $t = {}^-2$. Find r.

9   If A and r are perpendicular, what is the connection between u and v? Deduce the value of A. Are you surprised?

10   A particle moves with average velocity $\begin{bmatrix} 4 \\ -7 \end{bmatrix}$ m s^{-1} for 4 seconds. What is its displacement?

11   A particle has acceleration $\begin{bmatrix} -1 \\ 4 \end{bmatrix}$ m s^{-2} for 3 s. By how much does its velocity increase?

12   A particle starting from rest is displaced by $\begin{bmatrix} 2 \\ -3 \end{bmatrix}$ m in 2 s. What is its acceleration?

8. MOTION UNDER GRAVITY: PROJECTILES

An important application of motion with constant acceleration is to motion under gravity.

The acceleration of a particle due to gravity is independent of its mass and referred to axes which are horizontal and vertical (upwards) may be written as

$g = \begin{bmatrix} 0 \\ -g \end{bmatrix}$, where g may be taken to be constant. (See Figure 9.) Note that the

y-component is *negative* as gravity acts *downwards* and the positive y-direction is upwards. The value of g is approximately 9.81. For numerical convenience we often take $g = 9.8$ or 10. (It may be noted that g may be taken to be constant only if the particle does not travel too far in the vertical direction – which is assumed in all examples in this chapter.)

A *projectile* is an object which has been dropped or thrown in the air. If we neglect air resistance the only acceleration is that due to gravity. The following worked examples involve typical problems.

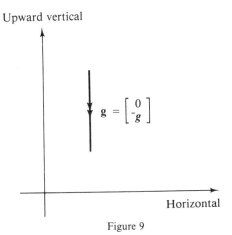

Figure 9

Example 14
A stone is thrown with speed 10 m s^{-1} so as to hit a window 3 m up a building. How fast is it travelling when it hits, if it experiences a constant acceleration of 9.8 m s^{-2} vertically downwards? (See Figure 10.)

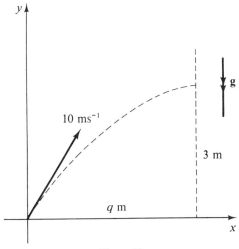

Figure 10

Taking x- and y-axes horizontally and vertically upwards, $\mathbf{A} = \begin{bmatrix} 0 \\ -9.8 \end{bmatrix}$ and $\mathbf{r} = \begin{bmatrix} q \\ 3 \end{bmatrix}$, where q is unknown.

$$v^2 = u^2 + 2\mathbf{A}.\mathbf{r},$$

$$v^2 = 10^2 + 2 \begin{bmatrix} 0 \\ -9.8 \end{bmatrix} . \begin{bmatrix} q \\ 3 \end{bmatrix}.$$

Thus
$$v^2 = 100 + 2(0 - 3 \times 9.8)$$
$$= 100 - 58.8;$$

so
$$v = 6.4.$$

The speed at impact is 6.4 m s^{-1} to 2 s.f.

Note that we did not need to know the horizontal distance of the building. This is typical of the use of the equation $v^2 = u^2 + 2\mathbf{A} \cdot \mathbf{r}$ in problems involving motion under gravity. If $\mathbf{g} = \begin{bmatrix} 0 \\ -g \end{bmatrix}$, $\mathbf{r} = \begin{bmatrix} r_1 \\ r_2 \end{bmatrix}$ then $\mathbf{g} \cdot \mathbf{r} = {}^-gr_2$, independent of r_1.
Note also the importance of a diagram indicating the positive directions.

Example 15
A bullet is fired horizontally at 300 m s^{-1} at the centre of a target 100 m away. How far below the centre does it strike? Take g as 9.8.

Referred to the usual horizontal and vertical axes we have $\mathbf{u} = \begin{bmatrix} 300 \\ 0 \end{bmatrix}$, $\mathbf{a} = \begin{bmatrix} 0 \\ -9.8 \end{bmatrix}$ and $\mathbf{r} = \begin{bmatrix} 100 \\ y \end{bmatrix}$ where y is to be found.
From $\mathbf{r} = t\mathbf{u} + \frac{1}{2}t^2\mathbf{A}$,
$$\begin{bmatrix} 100 \\ y \end{bmatrix} = t \begin{bmatrix} 300 \\ 0 \end{bmatrix} + \frac{1}{2}t^2 \begin{bmatrix} 0 \\ -9.8 \end{bmatrix} = \begin{bmatrix} 300t \\ -4.9t^2 \end{bmatrix}$$

and
$$\left. \begin{array}{l} 100 = 300t \\ y = {}^-4.9t^2. \end{array} \right\} \text{simultaneously.}$$

Thus
$$t = \tfrac{1}{3} \text{ and } y = {}^-4.9 \times \tfrac{1}{9} \approx {}^-0.54.$$

Hence the bullet *drops* about 0.54 m.

Example 16
A football is kicked from the ground at an angle of 40° to the horizontal with a speed of 25 m s^{-1}. What is its maximum height, how far does it go if the ground is horizontal and for how long is it in the air? Take g as 9.8.

Taking axes to be horizontal and vertically upwards then reference to Figure 11 shows that $\mathbf{u} = \begin{bmatrix} 25\cos 40° \\ 25\sin 40° \end{bmatrix} \approx \begin{bmatrix} 19.15 \\ 16.07 \end{bmatrix}$.
$$\mathbf{v} = \mathbf{u} + t\mathbf{A}.$$

So
$$\mathbf{v} = \begin{bmatrix} 19.15 \\ 16.07 \end{bmatrix} + t \begin{bmatrix} 0 \\ -9.8 \end{bmatrix} = \begin{bmatrix} 19.15 \\ 16.07 - 9.8t \end{bmatrix}.$$

Also
$$\mathbf{r} = t\mathbf{u} + \tfrac{1}{2}t^2\mathbf{A}.$$
$$\mathbf{r} = t \begin{bmatrix} 19.15 \\ 16.07 \end{bmatrix} + \tfrac{1}{2}t^2 \begin{bmatrix} 0 \\ -9.8 \end{bmatrix}$$
$$= \begin{bmatrix} 19.15t \\ 16.07t - 4.9t^2 \end{bmatrix}.$$

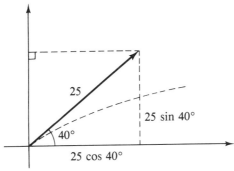

Figure 11

The greatest height is reached when the vertical component of velocity is zero, that is when $16.07 - 9.8t = 0$ or $t = 1.64$. The greatest height is the y-component of $\mathbf{r}$ at this time, which is $16.07 \times 1.64 - 4.9 \times 1.64^2 = 13$ m to 2 s.f.

The horizontal distance travelled (called the *range*) is found by determining when the vertical component of $\mathbf{r}$ is zero.

We require $\qquad 16.07t - 4.9t^2 = 0$

i.e. $\qquad (16.07 - 4.9t)\,t = 0 \quad \Rightarrow \quad t = 0 \quad \text{or} \quad t = 3.28.$

$t = 0$ is clearly the time at which the particle was projected and so we require $t = 3.28$. At this time the horizontal component of position is 19.15×3.28 m $= 63$ m to 2 s.f. Clearly the ball was in the air for 3.28 s.

Note that the horizontal component of velocity is constant and that the greatest height was reached in half the time the ball was in the air.

Example 17
A girl throws a stone at 15 m s^{-1} so as to hit a can standing on a wall 4 m above her hand. How far must she stand from the wall if the stone is thrown at an angle of 53° to the horizontal? Take g as 10.

We shall take the axes to be horizontal and vertical (upward) as usual.

The initial velocity is $\mathbf{u} = \begin{bmatrix} 15\cos 53° \\ 15\sin 53° \end{bmatrix} \approx \begin{bmatrix} 9 \\ 12 \end{bmatrix}$. $\mathbf{A} = \begin{bmatrix} 0 \\ -10 \end{bmatrix}$ and we require

$\mathbf{r} = \begin{bmatrix} x \\ 4 \end{bmatrix}$, where x is the distance from the girl to the wall.

From $\mathbf{r} = t\mathbf{u} + \tfrac{1}{2}t^2\mathbf{A}$ we have

$$\begin{bmatrix} x \\ 4 \end{bmatrix} = t\begin{bmatrix} 9 \\ 12 \end{bmatrix} + \tfrac{1}{2}t^2\begin{bmatrix} 0 \\ -10 \end{bmatrix}$$

and hence $\qquad x = 9t$

$$4 = 12t - 5t^2.$$

The second equation gives $5t^2 - 12t + 4 = 0$

i.e. $\qquad (5t - 2)(t - 2) = 0 \quad \Rightarrow \quad t = 2 \quad \text{or} \quad t = 0.4.$

When $t = 0.4$ we have $x = 3.6$; when $t = 2$ we have $x = 18$.

Note that there are two solutions as the stone could hit the can whilst rising or descending. The girl could be 3.6 m or 18 m from the wall. (See Figure 12.)

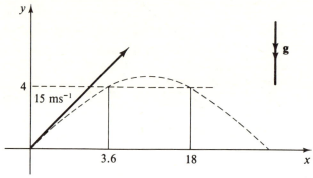

Figure 12

Exercise F

When the value of the acceleration due to gravity is not given assume that it is 9.8 m s^{-2} vertically downwards. Neglect air resistance.

1   A ball is thrown vertically upwards at 10 m s^{-1}. How high does it rise and for how long is it in the air?

2   Water aimed vertically from a fire hose can just reach the top of a 30 m high building. With what velocity does it leave the nozzle?

3   A pellet is fired vertically upwards at 40 m s^{-1}. For how long will its height exceed 50 m?

4   A stone is thrown horizontally at 8 m s^{-1} from the top of a vertical cliff 120 m high. How far out from the base of the cliff will it land?

5   A marble rolls over the square edge of a level table 1 m high and lands 0.5 m from the bottom of the table. How long does it take to fall?

6   A squash ball is hit horizontally so that it drops 0.5 m before hitting the wall 5 m away. What was its initial speed and at what angle does it hit the wall? Take g as 10.

7   How deep is a well if a stone takes 3 seconds to drop to the bottom?

8   A grappling hook is thrown vertically upwards and just reaches a ledge 10 m up a cliff. At what speed was it thrown?

9   A brick is dropped from a 30 m tower at the same instant as another is thrown up from the bottom at 15 m s^{-1}. When and where will the two bricks pass?

10   A dart is thrown horizontally at 16 m s^{-1} at the centre of a dartboard 5 m away. Where and with what speed will it land?

11   A motorcycle is being driven at 60 km h^{-1} along level ground when it goes over a vertical cliff. The remains are found 50 m from the base of the cliff. How high is the cliff and for how long was the motorcycle falling?

12   An object takes $\frac{1}{2}$ s to fall past a 3 m window. How long does it take to pass a similar window 2 m farther down?

13   A stone is thrown at 14 m s^{-1} at 45° to the horizontal. Find where it is after 1 s.

14   A ball is thrown at 20 m s^{-1} at 25° to the horizontal over level ground. Find the time it is in the air, the range and the greatest height attained. Take g as 10.

15   A long-jumper leaps 7.5 m and rises to a height of 0.4 m. What is his velocity as he takes off?

16   Find the range of a shell projected at 800 m s^{-1} at 10° to the horizontal over level ground. For how long is it in the air?

17   A stunt rider wants to leap a line of cars each 1.5 m high and 1.75 m wide. If he takes off at a speed of 20 m s^{-1} at an angle of 30° to the horizontal, how far should the take-off ramp be from the first car and over how many cars can he jump?

18   A particle is projected at a speed v at an angle θ° to the horizontal. Find formulae for the time of flight, the greatest height reached and the range.

9. SUMMARY

If the position vector of a particle at time t is $\mathbf{p}(t)$, then the average velocity between times t_1 and t_2 is

$$\frac{\mathbf{p}(t_2) - \mathbf{p}(t_1)}{t_2 - t_1}.$$

The instantaneous velocity is $\mathbf{v}(t) = \mathbf{p}'(t)$, which can be found by separately differentiating the components of $\mathbf{p}(t)$.

The acceleration $\mathbf{a}(t) = \mathbf{v}'(t)$.

Integrating $\mathbf{a}(t)$ gives the change in the velocity; integrating $\mathbf{v}(t)$ gives the change in the position vector, that is, the displacement.

For motion with constant acceleration $\mathbf{A}$,

$$\mathbf{v} = \mathbf{u} + t\mathbf{A},$$
$$\mathbf{r} = t\mathbf{u} + \tfrac{1}{2}t^2\mathbf{A},$$
$$\mathbf{r} = \tfrac{1}{2}t(\mathbf{u} + \mathbf{v}),$$

and
$$v^2 = u^2 + 2\mathbf{A}\cdot\mathbf{r},$$

where
$$\mathbf{u} = \text{velocity at } t = 0,$$
$$\mathbf{v} = \text{velocity at time } t$$

and
$$\mathbf{r} = \text{the displacement from } t = 0 \text{ to time } t.$$

Miscellaneous exercise

1   A particle moving initially due north with a velocity of 25 cm s^{-1} is subjected to an acceleration of 5 cm s^{-2} in a direction 240° (S 60° W). Find the velocity of the particle 3 seconds later. After what further period of time will the particle be moving instantaneously due west?

2   A cricket ball is thrown with a velocity of 16 m s^{-1} at an angle of 30° above the horizontal and is caught by a fielder at the height at which it was thrown. How long is it in the air, and how far away is the fielder from the thrower?

3   The position vector **r** of a missile is given at any time t by
$$\mathbf{r} = t\mathbf{i}+(8t-t^2)\mathbf{j}$$
where **i** and **j** are unit vectors horizontally and vertically. Calculate:
(a) its position when $t = 2$;
(b) its displacement from $t = 2$ to $t = 3$;
(c) its velocity when $t = 4$;
(d) its greatest height;
(e) when it lands;
(f) the angle at which it hits the ground.   (SMP)

4   A body moving with constant acceleration changes its velocity from $\begin{bmatrix} 3 \\ -2 \end{bmatrix}$ m s⁻¹ to

$\begin{bmatrix} 8 \\ 18 \end{bmatrix}$ m s⁻¹ in five seconds.

(a) Find the acceleration.

(b) Show that the displacement during the five seconds is parallel to $\begin{bmatrix} 11 \\ 16 \end{bmatrix}$.
                                                                             (SMP)

5   A ball is thrown with velocity 20 m s⁻¹ at an angle of 60° to the horizontal. The gravitational attraction of the Earth gives the ball a constant acceleration of 9.8 m s⁻¹ vertically downward. All other forces are negligible. Find by drawing or by calculation:
(a) its speed after 2 seconds;
(b) its displacement after 3 seconds;
(c) the direction in which it is travelling after 4 seconds;
(d) when it is travelling horizontally.   (SMP)

6   A is travelling with velocity $\begin{bmatrix} 2 \\ -4 \end{bmatrix}$ m s⁻¹ and B with velocity $\begin{bmatrix} 7 \\ 5 \end{bmatrix}$ m s⁻¹. Calculate the velocity and the speed with which B is approaching A. When A is at the point $(9, 10)$, B is at the point $(^-3, 1)$. Show that they will never meet.   (SMP)

7   A stone is thrown with velocity $\begin{bmatrix} 10 \\ 20 \end{bmatrix}$ m s⁻¹ from the top of a 300 m cliff. (The x-axis is taken horizontally outwards from the cliff, and the y-axis vertically upwards.) Calculate its velocity and position 3 seconds later, and indicate your results clearly on a sketch. (Take g as 9.8.)   (SMP)

8   A particle moves so that its velocity in m s⁻¹ relative to horizontal and vertical axes, after t seconds, is $\begin{bmatrix} 20 \\ 35-10t \end{bmatrix}$.
(a) Calculate its displacement after t seconds.
(b) Show that the distance from its starting-point after 4 seconds is 100 metres.
(c) Calculate its direction of motion after 2 seconds, giving the angle to the horizontal to the nearest degree.
(d) Plot a graph to show the horizontal and vertical displacements of the particle for the first 7 seconds.
(e) Indicate on your graph the results of parts (b) and (c).   (SMP)

9   The forward speed, v m s⁻¹, of a particle travelling in a straight line is given after t second by $v = 16-3t^2$.
(a) How far does it move in the first second?

    (*b*) After how many seconds does it start to move backwards?

    (*c*) Show that it is back at its starting-point after 4 seconds.

    (*d*) What total distance (forwards and backwards) does it travel in these 4 seconds? (SMP)

10 An UFO is observed to leave a tower from a height of 80 metres and to move so that its displacement *t* seconds later is given by $3t\mathbf{i} + 4t\mathbf{j} + 5t^2\mathbf{k}$, where $\mathbf{i}, \mathbf{j}, \mathbf{k}$ are unit vectors west, north and vertically downwards.

    (*a*) State its velocity after *t* second.

    (*b*) Show that initially it is travelling horizontally and calculate its bearing.

    (*c*) Show that its speed after 2 seconds is $5\sqrt{17}$ m s^{-1}.

    (*d*) Calculate its distance from its starting point after 2 seconds.

    (*e*) Find when it hits the ground.

    (*f*) Calculate its acceleration and hence state, with a brief reason, whether it is powered or not. (SMP)

11 A skier is sliding out of control down a straight, steady slope and accelerating uniformly at 1.5 m s^{-2}.

    (i) While he covers a particular stretch of 25 m his speed is doubled. Find by calculation how fast he was going at the end of 25 m.

    (ii) If he started from rest (with the same acceleration), calculate

        (*a*) the time he would take to cover 25 m,

        (*b*) the speed he would have attained in that time. (SMP)

12 A particle is initially moving with velocity $\begin{bmatrix} -1 \\ 2 \\ -2 \end{bmatrix}$ m s^{-1}.

    (*a*) Calculate its speed.

It is then given a constant acceleration of $\begin{bmatrix} 2 \\ 1 \\ 7 \end{bmatrix}$ m s^{-2}.

    (*b*) Show that after 2 seconds its speed will be 13 m s^{-1}.

    (*c*) Calculate the angle its new direction of travel makes with its initial direction.

    (*d*) Write down its velocity after *t* second.

    (*e*) Hence find when it will be moving at right-angles to its initial direction. (SMP)

13 A missile moves so that its velocity at time *t* after launching is given by $\mathbf{v} = 50\mathbf{i} + (70 - 10t)\mathbf{j}$, where $\mathbf{i}$ and $\mathbf{j}$ are unit vectors horizontally and vertically.

    (*a*) Calculate its speed when $t = 2$.

    (*b*) Find the angle between its direction of motion and the horizontal when $t = 2$.

    (*c*) Calculate its acceleration when $t = 3$.

    (*d*) Write down its displacement from the launching point at time *t*.

    (*e*) State when its height is greatest.

    (*f*) Find its greatest height above the launching point.

    (*g*) State, with reason, whether it is a powered missile. (SMP)

14 A stone was thrown up at 60° from the horizontal at a building 36 metres away and broke a window 3 seconds later.

    (*a*) State the initial horizontal speed and show that the initial vertical speed was approximately 20.8 m s^{-1}.

    (*b*) Calculate the height of the window above the point from which the stone was thrown. (Take *g* to be 10.)

    (*c*) State, with reason, whether the stone was rising or falling when it hit the window. (SMP)

Revision exercise 2

1 Sketch the graph of the function $f: x \to {}^-2x^2 + 3x + 1$ and give approximate values of x for the solution of $f(x) = 0$.
   What is the equation of the line of symmetry?
   Use a suitable formula to find the solution of $f(x) = 0$ correct to 3 significant figures.
   For which values of x is $f(x) > 0$?

2 If t is an integer (not necessarily positive) give, in terms of t, three consecutive integers.
   Form equations in t for which:
   (a) the three consecutive integers add up to 48;
   (b) the sum of twice the first integer plus the third integer plus three times the second integer is 71;
   (c) the sum of the squares of the three integers is 194.
   Solve each of the above equations.

3 A river 1 km wide is flowing west at 3 m s⁻¹. A boat which can travel at 7 m s⁻¹ in still water leaves the south bank and is to travel due north straight across the river. In what direction must the boat be steered and how long will it take to cross the river?

4 In Figure 1, $\underset{\sim}{OA}$, $\underset{\sim}{OB}$ and $\underset{\sim}{OC}$ represent the vectors **a**, **b**, **c**.
   (a) Calculate the lengths of the three vectors (to 2 decimal places).
   (b) On a diagram show the vectors $2\mathbf{a} + 3\mathbf{b}$, $2(\mathbf{c} - 2\mathbf{b})$, $^-2(\mathbf{b} + \mathbf{a})$.
   (c) Calculate the angles between each of the vectors drawn in (b) and the line $y = 0$ (to the nearest degree).

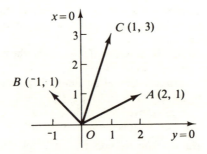

Figure 1

5 Suppose that a particle is moving in a straight line so that, after t seconds, it is $s(t)$ metres away from a fixed point.
   For the time interval $(t, t + m)$: (a) give an expression for the average speed of the particle; (b) evaluate your expression if $s: t \to 4t^3 - 9$.
   What is the derived function (i.e. the limit as m approaches 0)?

190

6   If the cost of operating a snow-plough is $6+3M^2+\dfrac{5}{M}$, where M is the distance ploughed in miles, find the rate of change of the cost with respect to distance.
    Find the average cost per mile if the snow-plough clears (a) 1 mile, (b) 5 miles, (c) 10 miles.
    Comment on your results.

7   The graph of the function $f:x\rightarrow 2x^2-4x+7$ goes through the points $(2,7)$, $(^-1,13)$ and $(1,5)$. Find its gradient at each of these points. Find the value of x when the gradient is (a) 3, (b) $^-3$.

8   A rectangular building site is acquired within the boundaries of a semicircular piece of land of radius m. Since a straight road runs alongside the land one of the boundaries of the site must lie along the diameter. This side of the site is of length $2k$ and the other corners of the site lie on the semicircular boundary of the land. Find an expression for the area of the site in terms of m and k.
    By considering the turning values of the function $f:k\rightarrow k^2(m^2-k^2)$, find the value of k which makes the area of the building site a maximum and hence find the maximum area for the site.

9   On the same diagram sketch the graphs of: (a) $m=\cos\theta$; (b) $m=2\cos\theta$; (c) $m=\frac{1}{2}\cos\theta$; (d) $m=\cos\theta+2$; (e) $m=2\cos\theta+2$; for values of θ between $0°$ and $360°$.
    Give approximate values for θ if $m=0.92$ for as many of the above equations as you can.

10 A pendulum hanging motionless is disturbed so that it swings to a maximum of $9°$ on either side of the vertical. It completes one cycle every 7 seconds. Find
    (a) the angle it makes with the vertical after 2 seconds (to the nearest tenth of a degree);
    (b) the time which elapses between the first two occasions on which it makes an angle of $7°$ with the vertical. Give your answer to 1 decimal place.

11 Sketch the graph of $y=\sin 4x°$ for $0\leqslant x\leqslant 180$.
    Calculate all the values of x in the interval $0\leqslant x\leqslant 180$ for which:
    (a) $\sin 4x°=\frac{\sqrt{3}}{2}$; (b) $\sin 4x°=^-\frac{1}{2}$.
    Write down the complete solution set of $^-\frac{1}{2}\leqslant\sin 4x°\leqslant\frac{\sqrt{3}}{2}$ for $0\leqslant x\leqslant 180$.

12 In the triangle XYZ, $XY=14.2$ cm, angle $Y=47°$ and $XZ=10.8$ cm. Show by scale drawing that there are two different triangles which fit these data. In each case measure YZ and the angles X and Z.
    Now calculate YZ in each case and the sizes of the two angles and complete Table 1.

Table 1

	By drawing	By calculation
YZ		
Angle X		
Angle Z		

Comment on your results.

13 (*a*) Calculate $\underset{\sim}{AB} \cdot \underset{\sim}{AC}$ (Figure 2).

(*b*) Calculate the lengths of $\underset{\sim}{AB}$ and $\underset{\sim}{AC}$ to two decimal places.

(*c*) Calculate the angle CAB to the nearest degree.

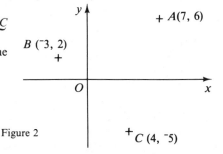

Figure 2

14 Three vectors **p**, **q** and **r** are given in terms of the perpendicular unit vectors **i** and **j** as follows:
$$\mathbf{p} = 3\mathbf{i} - \mathbf{j}, \quad \mathbf{q} = 2\mathbf{i} + 3\mathbf{j}, \quad \mathbf{r} = 4\mathbf{i} + \mathbf{j}.$$

(*a*) Find the value of a if the vector $\mathbf{p} + a\mathbf{q}$ is parallel to **r**.

(*b*) Find the value of b if the vector $\mathbf{p} + b\mathbf{q}$ is perpendicular to **r**.

15 A stone is thrown at a waterfall. It is thrown up at 70° to the horizontal and takes 3 seconds to reach the waterfall 28 m away.

(*a*) Find the initial velocity of the stone, giving its vertical and horizontal components to 2 decimal places.

(*b*) Calculate the height at which the stone reaches the waterfall, assuming that the thrower was standing level with the base of the waterfall and that the stone left her hand when it was at a height of 2 m. Take g to be 10.

(*c*) Was the stone rising or falling when it reached the waterfall?

16 The position vector, **p**, of a missile is given at any time t after its launch by
$$\mathbf{p} = t\mathbf{i} + (6t - t^2)\mathbf{j},$$
where **i** and **j** are horizontal and vertical unit vectors. Calculate:

(*a*) its position when $t = 3$;

(*b*) its displacement from $t = 2$ to $t = 3$;

(*c*) its velocity when $t = 3$;

(*d*) its greatest height;

(*e*) when it lands;

(*f*) the angle at which it hits the ground.

17 Sketch the graph of $f : x \to \sin(2x + \frac{1}{4}\pi)$ for $-\pi \leqslant x \leqslant \pi$. Using the graph you have drawn:

(*a*) state approximate values of x for which $f'(x) = 0$;

(*b*) state approximate values of x for which $f'(x) = 1$;

(*c*) state the values of $f'(\frac{1}{2}\pi)$ and $f'(0)$;

(*d*) estimate the area under the curve for $-\frac{1}{8}\pi \leqslant x \leqslant \frac{1}{8}\pi$.

Calculate answers to (*a*) and (*d*) and comment on your results.

18 After t seconds a particle is s metres from its starting-point where $s = -30t + 18t^2 - 2t^3$. When is its velocity zero? When is the acceleration zero? Find the velocity and acceleration after (*a*) 2 seconds, (*b*) 6 seconds.

19 Find the turning points on the curve $y = x^3 - 5x^2 + 7x - 3$ and distinguish between them. Sketch the curve.

20 Find the 'area under the curve' $y = 3x^2 - 12x - 15$ between the lines $x = {}^-3$ and $x = 2$.

21 Remove the brackets and simplify:
(a) $(2x+1)(4x-3)+(4x+1)(2x-3)$;
(b) $(3x-1)(4x-6)-(3x-2)(2x-1)$;
(c) $(2a+3b)(5c+4d)+(3a-2b)(2c-d)$;
(d) $(4p+2q)(p+3q)-(2p+5q)(2p-3q)$;
(e) $(x+y)\left(\dfrac{1}{x}+\dfrac{1}{y}\right)$; (f) $(a^2+b^2)\left(\dfrac{b^2}{a}+\dfrac{a^2}{b}\right)$;
(g) $(p^3-q^2)\left(\dfrac{p}{q}+\dfrac{q}{p^2}\right)$; (h) $\left(\dfrac{1}{x}-\dfrac{1}{y}\right)\left(\dfrac{x}{2}-\dfrac{xy}{5}+\dfrac{y}{3}\right)$.

22 Factorise the following:
(a) $abc^2+a^2b+ab^2$; (b) $f^4g^2-6f^2g^2+5fg^2$;
(c) x^2+5x+4; (d) $x^2+12x+36$;
(e) $9x^2+6x+1$; (f) $3x^2-5x+2$;
(g) $8x^2-10x+3$.

23 For the following functions:
(i) 'complete the square';
(ii) sketch the graph;
(iii) state the minimum value;
(iv) state the axis of symmetry;
(v) give estimates of the roots of the equation $f(x) = 0$.
(a) $f:x \rightarrow x^2+6x+11$; (b) $f:x \rightarrow x^2+4x+1$;
(c) $f:x \rightarrow x^2-8x+17$; (d) $f:x \rightarrow x^2+5x+3$;
(e) $f:x \rightarrow x^2-x+3$.

24 For the following, sketch the graphs and shade in the region indicated.

Graph	Region
(a) $y = x^2-4$	$y > x^2-4$, $y < 0$
(b) $y = x^2-2$	$y < x^2-2$
(c) $y = x^2$	$y < x^2$,
$x+y = 4$	$x+y > 4$
(d) $y = (x-1)^2$	$y > 4-x^2$
$y = 4-x^2$	$y < (x-1)^2$

25 The speed of a boat in still water is 15 m s^{-1}. A current flowing parallel to the river bank has a speed of 7 m s^{-1}. The boat is steered directly across the river. Find by drawing the actual speed and direction of motion of the boat.

26 On a calm day a small boat wishes to cross the bay to a point due north of the jetty. In still water the boat has a speed of 5 m s^{-1}. There is a strong wind blowing across the bay from the east with velocity 3 m s^{-1}. Find the course the boat must steer if it is to arrive at its destination and find the resultant speed of the boat.

27 A 'mayday' signal is received from a boat 5 nautical miles north of Scarborough. The boat is being blown on a bearing $070°$ at a speed of 15 knots. If the maximum speed of the lifeboat is 20 knots, find the course it must set and the time taken to reach the ship in distress.

28 From an airfield control tower two planes at the same height are sighted. One is

5 km away on a bearing 050° and the other is 6 km away on a bearing of 130°. How far apart are the two planes?

29 The position vectors of two skaters are $\begin{bmatrix} -9 \\ 1 \end{bmatrix}$ and $\begin{bmatrix} -1 \\ -7 \end{bmatrix}$ respectively and their velocity vectors are $\begin{bmatrix} 4 \\ 1 \end{bmatrix} t$ and $\begin{bmatrix} 2 \\ 3 \end{bmatrix} t$, where t is measured in minutes and distances in metres. If the two skaters maintain their present velocities show that they will collide. Give the value of t at the moment of collision and the position vector of the point at which they collide.

30 A particle moves so that its position vector at time t is $\begin{bmatrix} 4t^3 - 24t^2 \\ 2t^2 - t \end{bmatrix}$ where distance is measured in metres and time in seconds.
(a) Find the velocity and acceleration vectors.
(b) Is the particle ever at rest? If so, what is its acceleration at that time?
(c) At what time does the particle have zero velocity in each of the component directions? Find the magnitude of the acceleration at these times.

31 The acceleration of a particle is given by the vector $\begin{bmatrix} 2t \\ 3 \end{bmatrix}$. Find the velocity and position vectors at time t, $\mathbf{v}(t)$ and $\mathbf{p}(t)$, given that

$$\mathbf{v}(1) = \begin{bmatrix} 2 \\ 4 \end{bmatrix} \quad \text{and} \quad \mathbf{p}(1) = \begin{bmatrix} 3 \\ 2 \end{bmatrix}.$$

32 Find the gradient of the tangent to the curve $y = x^2 + 3x + 5$ at the point where $x = 2$. Find also the coordinates of the point where the line through the same point which is perpendicular to the tangent meets the curve again.

PART TWO

11

Probability

1. NOTATION

Suppose a card is selected at random from an ordinary pack of 52 cards. Let A be the event 'the card is an ace', and H the event 'the card is a heart'.

We know that the probability that event A occurs is $\frac{1}{13}$, and so we write

$$p(A) = \tfrac{1}{13}.$$

Similarly $\qquad\qquad\qquad p(H) = \tfrac{1}{4}.$

The probability that the card is *not* an ace is $\frac{12}{13}$. We write this as

$$p(\sim A) = \tfrac{12}{13},$$

where $\sim A$ is a notation for 'not-A'. Hence $p(\sim H) = \frac{3}{4}$.

The probability that the card is *both* an ace *and* a heart is $\frac{1}{52}$. We write

$$p(A \text{ and } H) = \tfrac{1}{52}.$$

The probability that the card is in at least one of the categories 'ace' or 'heart' is $\frac{16}{52}$, since there are 13 hearts and 3 other aces in the pack of 52. We write

$$p(A \text{ or } H) = \tfrac{16}{52}.$$

Note that here 'or' means either one or the other or possibly both. In work on probability we shall always give 'or' this inclusive meaning.

Exercise A

1   A card is selected at random from an ordinary pack of 52 cards.
    Let A be the event 'the card is a red card', B the event 'the card is a picture card', and C the event 'the card is a club'.
    Write down the values of $p(A)$, $p(B)$, $p(C)$, $p(\sim A)$, $p(\sim B)$, $p(\sim C)$, $p(A \text{ and } B)$, $p(B \text{ and } C)$, $p(A \text{ or } C)$, and $p(B \text{ or } \sim C)$.

2   Two ordinary dice, one red and one blue, are thrown together. Let D be the event 'the red die shows a six', E the event 'the blue die shows a three', and F the event 'the total score showing is nine'.
    Write down the values of $p(D)$, $p(\sim E)$, $p(F)$, $p(\sim F)$, $p(D \text{ and } E)$, $p(D \text{ or } E)$, and $p(\sim E \text{ or } F)$. (You may find a diagram helpful in this question).

3   A letter is chosen at random from the letters of the alphabet. V is the event 'the letter is a vowel', and T the event 'the letter is one of the letters of the word TRAPEZIUM'. Write down the values of $p(V)$, $p(\sim T)$, $p(V \text{ and } T)$, $p(V \text{ or } T)$, and $p(\sim V \text{ and } T)$.

4   A class of 30 pupils contains 17 girls. Five of the girls and three of the boys are left-handed. A pupil is chosen at random from the class. G is the event 'the pupil

is a girl', and L the event 'the pupil is left-handed'. Write down the values of $p(\sim G)$, $p(L)$, $p(G \text{ or } L)$, $p(\sim G \text{ and } L)$, and $p(G \text{ and } \sim L)$.

2. CONDITIONAL PROBABILITY

Do you believe that eating an orange a day would reduce your chances of catching a cold? Suppose you wished to test such a claim. You might persuade 100 people of your own age to eat an orange a day from 1 October to 31 March and discover what proportion of them caught a cold during those six months. Suppose 30 caught a cold: what would you conclude? Not much, because you want to compare this with what happens if they don't eat an orange a day. One way of doing that might be to find, say, 50 people who are prepared never to eat an orange for the same six months and who are, as near as you can judge, exposed to the same risk of catching a cold as your original 100. On 1 April you could draw up a table of results, such as Table 1.

Table 1

	Caught a cold	Did not catch a cold	Total
Daily orange-eaters	30	70	100
Abstainers from oranges	40	10	50

With these results it would appear that the orange-eaters have a better chance of not catching a cold compared with the abstainers. In fact if we select one of the 150 people at random, the probability that he or she has had a cold is different depending on whether they are or are not a daily orange-eater: $\frac{30}{100} = 0.3$ or $\frac{40}{50} = 0.8$ respectively. These are *conditional probabilities*, and, using C and D for the events 'caught a cold' and 'eating an orange daily' respectively, we write

$$p(C|D) = 0.3 \quad \text{and} \quad p(C|\sim D) = 0.8.$$

Example 1
Two dice are thrown. If the total score is greater than 8, what is the probability that there is no six showing?

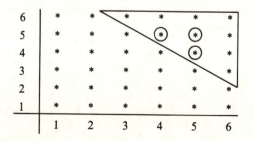

Figure 1

In Figure 1, the ten ways in which a score of more than 8 can be obtained are surrounded by a triangle. Of these ten ways, the three ways in which no six shows are ringed.

Hence p (no six showing | total score > 8) $= \frac{3}{10}$.

Exercise B

1   From Exercise A, question 1, write down the values of $p(A|B)$, $p(B|A)$, and $p(B|C)$.

2   From Exercise A, question 2, write down the values of $p(D|F)$, $p(F|D)$, and $p(F|\sim D)$.

3   From Exercise A, question 3, write down the values of $p(V|T)$, $p(T|V)$, and $p(T|\sim V)$.

4   From Exercise A, question 4, write down the values of $p(G|L)$, $p(L|G)$, and $p(\sim L|\sim G)$.

3. TREE DIAGRAMS

Table 1 of Section 2 can be summarised as in Table 2.

Table 2

	C	$\sim C$	
D	30	70	100
$\sim D$	40	10	50
	70	80	150

The same information is contained in the Venn diagram of Figure 2, where the number in each region is the number of people in the corresponding subset.

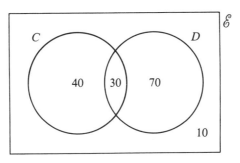

Figure 2

One day you meet one of the 150 people taking part in your trials. She has a cold. What is the probability that she is eating an orange a day?

The answer to this question is $p(D|C) = \frac{30}{70} = \frac{3}{7}$. Notice that this can be obtained not only from the table, but also from the Venn diagram, as

$$\frac{n(D \cap C)}{n(C)}.$$

However, $$\frac{n(D \cap C)}{n(C)} = \frac{n(D \cap C)/n(\mathscr{E})}{n(C)/n(\mathscr{E})} = \frac{p(D \text{ and } C)}{p(C)}$$

so $$p(D|C) = \frac{p(D \text{ and } C)}{p(C)}.$$

In general, for any two events A and B,

$$p(A|B) = \frac{p(A \text{ and } B)}{p(B)}.$$

This form is particularly useful when dealing with tree diagrams.

Note that in the preceding discussion we have used C (and D) in two contexts, first as an event and second as a set. This is permissible providing there is no confusion. Here $n(C)$ means the number of elements in the set C and $p(C)$ means the probability of event C occurring.

Example 2

A bag contains ten coloured beads, seven red and three blue. A bead is selected at random and its colour noted. A second bead is now selected. Draw a tree diagram and calculate the probabilities that:

    (i) both beads are red;

    (ii) the second bead is red;

    (iii) the first was blue if the second was red;

  (a) if the first bead is replaced before the second is drawn;

  (b) if the first bead is *not* replaced before the second is drawn.

Tree diagrams for (a) and (b) are set out side by side for comparison in Figure 3.

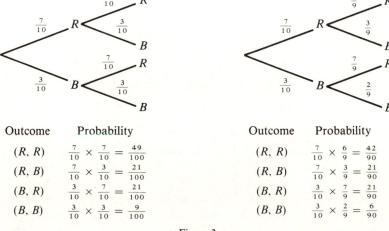

Figure 3

Notice the different results. In fact, in both trees, the probabilities entered on the branches for the *second* bead drawn are conditional probabilities; conditioned by the effect of the first drawing and whether or not the first bead was replaced.
Compare the corresponding numbers in Figure 4.

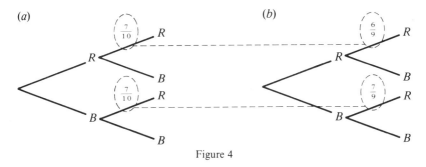

Figure 4

For example, the $\frac{6}{9}$ in (b) reflects the fact that there are now only nine beads of which six are red because the first bead drawn was red. In this case we have

$$p(\text{second red} \mid \text{first red}) = \tfrac{6}{9}.$$

whereas in (a) we have

$$p(\text{second red} \mid \text{first red}) = \tfrac{7}{10}.$$

Note also that the conditional probabilities in (b)

$$p(\text{second red} \mid \text{first blue}) = \tfrac{7}{9}$$

and

$$p(\text{second red} \mid \text{first red}) = \tfrac{6}{9}$$

are not the same.
The answers to the various questions are:

(a) (i) $p(R, R) = \tfrac{49}{100}$.

(ii) $p(\text{second is red})$
$= p(R, R) + p(B, R)$
$= \tfrac{49}{100} + \tfrac{21}{100}$
$= \tfrac{70}{100}$
$= \tfrac{7}{10}$.

(iii) $p(\text{first blue} \mid \text{second red})$
$= \dfrac{p(\text{first blue and second red})}{p(\text{second red})}$
$= \tfrac{21}{100} / \tfrac{70}{100}$
$= \tfrac{21}{70} = \tfrac{3}{10}$.

(b) (i) $p(R, R) = \tfrac{42}{90} = \tfrac{7}{15}$.

(ii) $p(\text{second is red})$
$= p(R, R) + p(B, R)$
$= \tfrac{42}{90} + \tfrac{21}{90}$
$= \tfrac{63}{90}$
$= \tfrac{7}{10}$.

(iii) $p(\text{first blue} \mid \text{second red})$
$= \dfrac{p(\text{first blue and second red})}{p(\text{second red})}$
$= \tfrac{21}{90} / \tfrac{63}{90}$
$= \tfrac{21}{63} = \tfrac{1}{3}$.

Independent events

The answers to part (iii) of Example 2 should be compared. For (a) the probability is just the same as p (first blue); knowing that the second was red does not give us any extra information. We say that the drawings of the first and the second beads are *independent events*. As you can see, this is something like saying that one event does not influence the other. We always assume, for example, that if a coin is tossed twice the result of either toss is not influenced by the result of the other; they are independent. For (*b*), however, knowing that the second bead was red we can say that the probability of the first being blue is higher than we could have said without that extra information.

We say that two events *A* and *B* are independent if $p(A|B) = p(A)$.

One other point about trees is worth noting. Diagrammatically they represent a series of alternative patterns of results, but it is important that at any branching point the alternatives are exclusive; if one happens, others cannot. In Example 2 the first bead was either red or blue; it could not be both. If a card is drawn from the usual pack, trees beginning as in Figure 5 would break this rule because the king of diamonds being drawn would lead us along all branches!

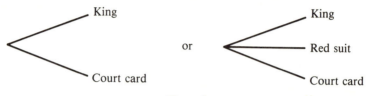

Figure 5

The chains of events that can be traced through the branches of a tree are all mutually exclusive; it is this fact that justifies the *addition* of the probabilities calculated for each appropriate route through a tree.

So far we have been dealing with events which could be subdivided into equally probable outcomes. The ideas of conditional probability can be applied in more general situations.

Example 3

The Smiths and the Browns are neighbours and they always take their summer holidays in either Greece or Spain. If the Smiths go to Greece then the probability that the Browns also go there is $\frac{1}{3}$, but if the Smiths go to Spain then the probability that the Browns go to Greece is $\frac{4}{5}$. Each year the Smiths throw a die to decide where to go for their holidays. As they prefer Greece they decide to go to Greece if the number showing is more than 2, otherwise they go to Spain. They didn't tell me where they were going but I know that the Browns went to Spain. What is the probability that the Smiths were in Greece?

Figure 6 shows the tree diagram for this example.

p(Smiths in Greece | Browns in Spain)

$$= \frac{p(\text{Smiths in Greece and Browns in Spain})}{p(\text{Browns in Spain})}$$

$$= \frac{\frac{2}{3} \times \frac{2}{3}}{\frac{2}{3} \times \frac{2}{3} + \frac{1}{3} \times \frac{1}{5}} = \frac{\frac{4}{9}}{\frac{4}{9} + \frac{1}{15}}$$

$$= \frac{\frac{4}{9}}{\frac{23}{45}} = \frac{20}{23}.$$

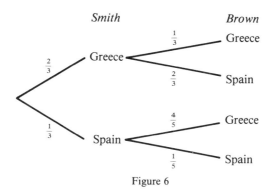

Figure 6

Notice the different answer obtained when the additional information that the Browns were in Spain was known. If I did not know that, then the only information I have is that the Smiths use a die to decide and are therefore in Greece with probability $\frac{2}{3}$. The additional information that the Browns went to Spain has in fact made it now more likely that the Smiths were in Greece.

Exercise C

1   A tin contains nine steel screws and seven brass screws all of the same length. Two screws are selected at random from the tin. Use tree diagrams to find the probabilities that
(i) both are steel;   (ii) the second is steel;
(*a*) if the first is replaced before the second is drawn and (*b*) if the first is not replaced before the second is drawn.
  Two screws are taken from the tin and used. A third screw is then selected at random, which is brass. What is the probability that the first two screws were both steel?

2   A box contains chocolates, five hard-centres and eight soft-centres. In the cinema my companion takes two in succession. What is the probability that one is hard-centred and one soft-centred? I take the third from the box and it is hard. What is the probability that both my companion's were soft?

3   In a class of 17 pupils, 5 are girls. Two pupils are required to collect equipment at the end of a lesson; one for the workcards used and one for the calculators. Find the probabilities that
(i) both are girls, (ii) the calculator collector is a girl,
(*a*) if the same pupil may do both jobs and (*b*) if different pupils do the two jobs.
  In another lesson the teacher requires a pupil to hand out examination papers. He chooses a girl. What is the probability that all three pupils were girls?

4   A box of fireworks holds eight 'fountains' and eleven bangers. If two are selected at random, what is the probability that they are both bangers? If I arrive later and take a third which is a banger, what is the probability that the two already taken were both 'fountains'?

5 I have four dice – three normal ones and one with faces marked 1, 1, 1, 2, 3, 4. If I select one die at random and throw it and it comes down with a 1 showing what is the probability it is one of the normal dice?

6 In a game of cricket a bowler bowls off-breaks with probability $\frac{3}{4}$ and googlies otherwise. The batsman has a probability of $\frac{1}{3}$ of scoring off the off-breaks but only $\frac{1}{5}$ of scoring off the googlies. The batsman was unable to score off the first ball that this bowler bowled. What is the probability that it was a googly?

7 Smith and Green are two lazy pupils who always share their Mathematics homework, one copying the other's work. On any particular evening there is a probability of $\frac{2}{3}$ that Smith does the work; otherwise Green does it. Smith gets it right with probability $\frac{1}{3}$ but Green gets it right with probability $\frac{1}{2}$. Today their homework was wrong. What is the probability that Green did it?

8 A seed merchant sells two types of wallflower seeds – X and Y. 40% of the type X seed produce red flowers whereas 70% of the type Y seed produce red flowers. Unfortunately the label on the packet I buy has been removed. After the seeds have been sown the first flower to open is red. What is the probability I have the type Y seeds? (Assume that I was equally likely to have either type of seed in the first place.)

9 A smallholder has five red hens and seven white ones. The red hens always lay brown eggs but the white hens lay brown eggs with probability 0.6. On a particular morning there is only one egg to collect – a brown one. What is the probability that it came from one of the white hens?

10 A box of chocolates contains eleven soft-centres and five hard-centres. Jane takes the first chocolate and Peter takes the second. If Peter's chocolate is a soft-centre, what is the probability that Jane's was a hard-centre?

11 Two bags A and B contain coloured balls. Bag A has three red and two white ones and bag B has one red and two white. A ball is selected at random from bag A and placed in bag B. Bag B is then shaken and a ball selected. If this ball is white, what is the probability that bag A now contains an equal number of red and white balls?

12 The names of ten children (six boys and four girls) are placed in a hat and three names are selected at random from the hat. If the third name out of the hat is a girl's, what is the probability that the first two names out were both boys'?

13 If $p(A) = \frac{1}{4}$, $p(B|A) = \frac{2}{3}$ and $p(B|\sim A) = \frac{1}{5}$ calculate with the aid of a tree diagram:
(a) $p(A$ and $B)$;   (b) $p(B)$;   (c) $p(A|B)$.

14 If I oversleep on a particular morning I am late for work with probability 0.7 but otherwise I am late for work with probability 0.1. I oversleep with probability 0.02. Today I was late for work. What is the probability that I overslept?

4. BINOMIAL PROBABILITY

Example 4
A die is thrown three times. Draw a tree diagram and calculate the probability of obtaining at least two sixes.

In this example we are interested in sixes or not-sixes and hence the tree diagram would appear as in Figure 7.

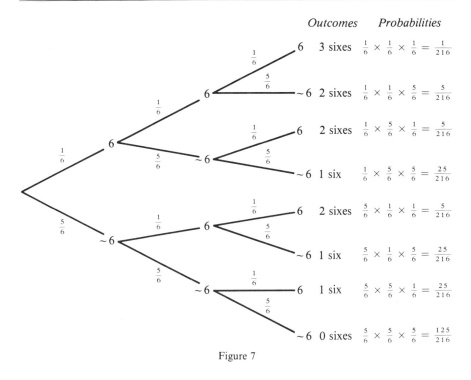

Figure 7

We see that the probability of at least two sixes is

$$\frac{1}{216}+\frac{5}{216}+\frac{5}{216}+\frac{5}{216}=\frac{16}{216}=\frac{2}{27}.$$

Exercise D

*1 When a certain type of drawing-pin is thrown onto a horizontal table there is a probability of 0.7 that it will land point downwards. If two pins are thrown find the probabilities of them landing:
(*a*) both point down; (*b*) one up and one down; (*c*) both point up.

*2 With the same type of pins as in question 1, three pins are thrown onto a table. Find the probabilities of them landing:
(*a*) all point up; (*b*) 2 up and 1 down; (*c*) 1 up and 2 down;
(*d*) all down.

*3 Four drawing-pins of the same type as in question 1 are thrown. Find the probabilities that they land:
(*a*) all point up; (*b*) 3 up and 1 down; (*c*) 2 up and 2 down;
(*d*) 1 up and 3 down; (*e*) all point down.

*4 Repeat question 1 but now suppose that the pins have probability p of falling point down and probability q of falling point up. (In fact $q = 1 - p$, but it is easier to consider the problem with two separate probabilities p and q.)

*5 Repeat question 2 but with the data of question 4.

*6 Repeat question 3 but with the data of question 4.

You should have noticed a pattern developing in your results to questions 4, 5 and 6 of the Exercise D. This can be summarised as in Table 3.

Table 3

		Number landing point down				
		0	1	2	3	4
Number of	2	q^2	$2qp$	p^2		
pins thrown	3	q^3	$3q^2p$	$3qp^2$	p^3	
	4	q^4	$4q^3p$	$6q^2p^2$	$4qp^3$	p^4

Can you predict the results for five pins and six pins?

You probably discovered that the tree diagram for four pins was large and clumsy. A tree diagram for five pins would be even worse. We can, however, considerably simplify the diagram by considering the experiment as the throwing of four pins followed by one extra (Figure 8).

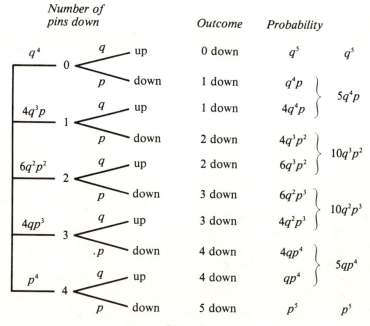

Figure 8

Hence we see that the probabilities for five pins are:

Number of pins down	0	1	2	3	4	5
Probability	q^5	$5q^4p$	$10q^3p^2$	$10q^2p^3$	$5qp^4$	p^5

It should be easy to see the familiar pattern of Pascal's triangle emerging.

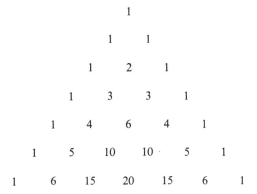

With the aid of Pascal's triangle we can write down the probabilities for six pins without having to draw the tree diagram.

Number of pins landing point down	0	1	2	3	4	5	6
Probability	q^6	$6q^5p$	$15q^4p^2$	$20q^3p^3$	$15q^2p^4$	$6qp^5$	p^6

By continuing Pascal's triangle as far as necessary we could write down the relevant probabilities for any number of pins thrown.

The distribution of probabilities encountered above is called the *Binomial probability distribution* – so called because it consists of a series of repeated trials in each of which one of *two* outcomes can occur with the same probability at each trial. In the example above each pin could either land point up or point down and the probabilities of these two events happening were the same for each pin thrown. In Example 4, for each of the three dice thrown the events we were interested in were 'six' and 'not-six' and for each die the probabilities of these events occurring were the same (i.e. $\frac{1}{6}$ and $\frac{5}{6}$). In each case, the successive 'throws' are independent.

Example 5
The Doghouse missile is so unreliable (due to electrical faults, human error etc.) that its chance of success in any firing is only $\frac{2}{5}$. If four of these missiles are fired find the probabilities of 0 successes, 1 success and of at least 2 successes.

As there are four repeated trials, the relevant line of Pascal's triangle is

$$1 \quad 4 \quad 6 \quad 4 \quad 1$$

and the probabilities of the various numbers of successes are as follows:

Number of successes	0	1	2	3	4
Probability	$(\frac{3}{5})^4$	$4(\frac{3}{5})^3(\frac{2}{5})$	$6(\frac{3}{5})^2(\frac{2}{5})^2$	$4(\frac{3}{5})(\frac{2}{5})^3$	$(\frac{2}{5})^4$

Hence
$$p(0) = (\tfrac{3}{5})^4 = 0.130,$$
$$p(1) = 4(\tfrac{3}{5})^3(\tfrac{2}{5}) = 0.346,$$
$$p(\geqslant 2) = 1 - p(0) - p(1) = 0.524.$$

Notice that it is simpler to calculate $p(\geqslant 2)$ as $1 - p(0) - p(1)$ rather than $p(2) + p(3) + p(4)$.

Example 6
A seedmerchant mixes a large number of white hyacinth bulbs with a large number of blue hyacinth bulbs in the ratio 2:1. If I buy five bulbs and they all produce flowers, calculate the probability that there will be more white flowers than blue flowers.

In this case each bulb has a probability of $\tfrac{2}{3}$ of being blue and $\tfrac{1}{3}$ of being white. In practice this is not quite correct, for suppose the seedmerchant originally mixed 2000 blue bulbs with 1000 white. If the first bulb is blue then the probability that the second is blue is $\tfrac{1999}{2999}$ whereas if the first bulb is white then the probability that the second bulb is blue is $\tfrac{2000}{2999}$. However, both of these fractions are so close to $\tfrac{2}{3}$ that we may assume the probabilities are the same for each bulb chosen. In the question it is the phrase 'a large number' which tells us that we may assume that the removal of a few bulbs will not significantly affect the probabilities that the next bulb is blue or white.

Five bulbs are chosen so the relevant line of Pascal's triangle is

$$1 \quad 5 \quad 10 \quad 10 \quad 5 \quad 1.$$

and the probabilities are:

Number of blue flowers	0	1	2	3	4	5
Probability	$(\tfrac{1}{3})^5$	$5(\tfrac{1}{3})^4(\tfrac{2}{3})$	$10(\tfrac{1}{3})^3(\tfrac{2}{3})^2$	$10(\tfrac{1}{3})^2(\tfrac{2}{3})^3$	$5(\tfrac{1}{3})(\tfrac{2}{3})^4$	$(\tfrac{2}{3})^5$

The probability that I have more white than blue is
$$p(0) + p(1) + p(2) = (\tfrac{1}{3})^5 + 5(\tfrac{1}{3})^4(\tfrac{2}{3}) + 10(\tfrac{1}{3})^2(\tfrac{2}{3})^2$$
$$= 0.210.$$

The following example illustrates how care must be taken in considering whether a problem is a Binomial probability problem or not.

Example 7
A committee of four is to be chosen from a group of pupils consisting of seven boys and three girls. If the committee is chosen at random what is the probability that it contains an equal number of boys and girls?

Here the situation appears Binomial – each selection is either a boy or a girl. However, the probabilities are not the same at each selection – once selected a pupil cannot be selected again. The problem is very similar to Example 6 but whereas in that case the selection of a few more bulbs made little difference to the probabilities because of the 'large number' under consideration, in this case the probabilities will change considerably with each selection.

The problem is best answered with a tree diagram. In Figure 9 only the selections with two boys and two girls are shown.

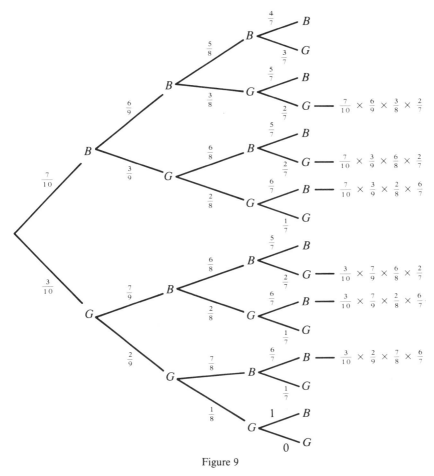

Figure 9

Notice that each of the probabilities is the same, i.e. $\dfrac{7\times6\times3\times2}{10\times9\times8\times7}$, although the figures in the numerator appear in different orders each time. There are six possibilities (note the 6 from the line 1 4 6 4 1 in Pascal's triangle); hence the required probability is

$$6\times\frac{7\times6\times3\times2}{10\times9\times8\times7}=\frac{3}{10}.$$

Exercise E

1 If one child in four is left-handed and I choose a random sample of five children, what is the probability that I choose more than two left-handers?

2 In an examination paper there are six multiple-choice questions where in each case

candidates have to decide between one of four alternatives. If only one answer is correct in each case and a candidate guesses at random, calculate the probabilities of getting:

(*a*) none correct;   (*b*) 1 correct;   (*c*) 2 correct;

(*d*) more than 2 correct.

3   5% of soap powder packets from a certain firm are underweight. If I buy seven packets, what is the probability that I have at least one underweight packet?

4   A coin is tossed five times. Calculate the probability that it lands heads more times than it lands tails.

5   Out of a group of four children what is the probability that at least two have birthdays on a Saturday this year?

6   Three cards are selected without replacement from an ordinary pack of 52 cards. Calculate the probability that at least two diamonds are selected.

7   The probability that a particular marksman hits the bull on a target is 0.4. If he fires seven rounds at the target calculate the probability that he scores

(*a*) fewer than two bulls;   (*b*) at least five bulls.

8   Assuming that 0.52 is the probability that any newborn child is a boy, calculate the probability that in a family of four children there will be more boys than girls.

9   I am collecting plastic soldiers given away, one in each packet, in packets of 'Weetybangs' breakfast cereal. There are twelve in the set and I have eleven of them. If I buy five more packets what is the probability that I will get the missing soldier to complete my set?

10   Eight ordinary dice are thrown together. What is the most probable number of sixes to appear?

11   A certain type of rare flower seed is very difficult to germinate and the probability that one does germinate is 0.2. How many must I buy to be 90% certain that I will get at least one to germinate?

12   A bag contains five red coloured balls and three blue ones. If three balls are chosen at random without replacement, calculate the probability that at least two of them are blue.

5. SUMMARY

$p(A)$ denotes the probability that event A occurs.

$p(\sim A)$ denotes the probability that event A does *not* occur.

$p(A$ and $B)$ denotes the probability of events A and B *both* occurring.

$p(A$ or $B)$ denotes the probability of at least one of events A and B occurring.

The *conditional probability* of A given B is

$$p(A|B) = \frac{p(A \text{ and } B)}{p(B)}.$$

Two events, A and B, are *independent* if $p(A|B) = p(A)$.

Two events are *mutually exclusive* if the occurrence of either makes the other impossible.

In the *Binomial probability* distribution of n trials with probability of 'success' p and of 'failure' $q(= 1 - p)$, the probability of exactly r successes is

$$p(r) = Kp^r q^{n-r}$$

where K is a constant obtained from the line of *Pascal's triangle* that begins $1\ n...$, and is the $(r + 1)$th number in that line.

Miscellaneous exercise

1 The probability that a colony of bees will survive a particularly hard winter is 0.35. I have four colonies and my next door neighbour has eight. What are the probabilities that exactly one colony survives in each case? (SMP)

2 Every year a man puts six nest-boxes in his garden, in the hope that they will be inhabited by nesting birds. The success rate is 70%.
   (a) Assuming that this means that for each box the probability that it will be inhabited is 0.7, show that the probability of exactly three of the boxes being inhabited this year is 0.185.
   (b) Calculate the probability that fewer than half of the boxes will be inhabited. (SMP)

3 It is found by experience that there is a probability of $\frac{1}{5}$ that a toffee is not correctly wrapped. In a bag containing 9 toffees find the probability of (i) 3, (ii) 2 incorrectly wrapped toffees. Determine the most likely number of toffees to be found incorrectly wrapped in the bag. (SMP)

4 Assume that on each day of a week in June the probability that it will rain is 0.2. Show that the probability that it will be dry on exactly 4 of the 5 weekdays is $2^{12} \times 10^{-4}$.
   Obtain similar expressions for the probability that out of the 5 days
   (a) it will be dry on more than three days,
   (b) it will be dry on not more than three days,
   (c) it will rain on not more than three days. (SMP)

5 In one season in a particular habitat, the probability of seeing a redbreast is 0.8 in winter, but only 0.1 in summer; the probability of observing a bluetit is 0.3 in summer but 0.6 in winter. Assume that all these events are independent.
   (i) In summer what are the probabilities of spotting
       (a) a redbreast and a bluetit,
       (b) a bluetit but no redbreast?
   (ii) In winter what are the probabilities of spotting
       (a) a redbreast and a bluetit,
       (b) a redbreast or a bluetit?
   (iii) If both birds are seen, what is the probability that it is summer?
   (iv) If only one species is seen in a summer session, what is the probability that it is a bluetit? (SMP)

6 A certain strain of rabbit produces offspring in equal ratio male to female. One third of the male offspring are black and one half of the female offspring are black.
   In a litter of five offspring, what is the probability that
   (i) exactly two offspring will be male,

(ii) less than four offspring will be female?

In a litter of four offspring, what is the probability that there will be exactly two male offspring which are not black? (MEI)

7 In a routine test for early warning of an unsuspected disease, 8% of those tested react (but do not all have the disease) while 7% in fact have the disease (but do not all react). The probability that a person neither has the disease nor reacts is 0.9. What is the probability that a person with the disease will be detected? (SMP)

12

Statistics

1. MEASURES OF SPREAD

Consider the following sets of marks obtained by the same group of 11 pupils in an English examination and a Mathematics examination.

Pupil	A	B	C	D	E	F	G	H	I	J	K
English	51	54	51	48	52	53	50	53	54	54	52
Mathematics	37	62	81	42	51	52	94	72	11	44	26

It can easily be shown that the mean mark is the same for each subject. However, we could not claim that the performances of the pupils on the two papers were similar. You can see at a glance that the English marks are closely grouped whereas there is a wide variation in the Mathematics marks. In this section we shall be concerned with possible ways of measuring the difference in 'spread' of the two sets of marks.

If we put each set of marks in order we obtain:

English	48	50	51	51	52	52	53	53	54	54	54
Mathematics	11	26	37	42	44	51	52	62	72	81	94

One possible measure of spread might be the *range*. The range of the English marks is $54 - 48 = 6$ marks, and the range for Mathematics is $94 - 11 = 83$ marks. It could, however, be argued that the range might be influenced by odd freak values. For example, the set

$$3 \quad 41 \quad 43 \quad 44 \quad 47 \quad 48 \quad 49 \quad 50$$

has a range of $50 - 3 = 47$ marks, but the one low score of 3 has led to a range which does not give a true idea of the closeness of the rest of the scores. We could overcome this problem by ignoring the extremes – one way would be to find the *inter-quartile range* which is the difference between the first and third quartiles. In this example, with 11 numbers, these two quartiles are the 3rd and 9th in order. Hence the inter-quartile ranges are:

English   54–51 = 3 marks
Mathematics   72–37 = 35 marks.

Another possible way to measure the spread would be to calculate the *average deviation* from the mean. For example, for the English marks which have a mean of 52, we have

Mark	48	50	51	51	52	52	53	53	54	54	54
Deviation from mean	4	2	1	1	0	0	1	1	2	2	2

Total deviation from mean $= 16$

hence average deviation from mean $= \frac{16}{11} = 1.45$ marks.

Notice that we have considered the *absolute* deviation – we have not distinguished between positive and negative deviations. What would happen if we did make this distinction?

The measure of spread we have just discussed is called the *mean absolute deviation*. If x is a typical member of a population of n values with mean m, then the mean absolute deviation is defined as

$$\frac{1}{n}\Sigma|x-m|$$

where Σ means 'the sum of all ...' and $|x-m|$ means the absolute value of the difference between x and m. The calculation is best set out in tabular form. For example, with the Mathematics marks which also have a mean of 52, Table 1 results.

Table 1

x	$|x-52|$
11	41
26	26
37	15
42	10
44	8
51	1
52	0
62	10
72	20
81	29
94	42
	202

Hence the mean absolute deviation (m.a.d.) $= \frac{1}{11}\Sigma|x-52| = \frac{202}{11} = 18.36$ marks.

Exercise A

In questions 1–2 and 3–4, find the mean, range, inter-quartile range and mean absolute deviation.

1 The marks of a group of pupils in a French examination:

43   51   42   48   63   55   37   48   58   50   44

2 The heights in metres of a group of 13-year-old girls:
   1.60 1.50 1.81 1.68 1.52 1.66 1.84 1.62 1.53 1.54 1.63

3 The numbers of letters delivered on a given day to a row of houses:
   3 1 5 0 2 2 4 1 9 0 4 2 1 1 12 3 1

4 The numbers of points scored in 13 rugby matches by Bagford Vipers XV:
   27 12 3 46 15 0 12 8 27 6 58 0 7

2. STANDARD DEVIATION

Although the mean absolute deviation of a sample is fairly easy to calculate, it is rarely used by statisticians because of the mathematical problems of combining expressions within modulus signs. As you may have already realised, the sum of the deviations from the mean is zero if we include their signs:

Mark	48	50	51	51	52	52	53	53	54	54	54
Deviation from mean	−4	−2	−1	−1	0	0	1	1	2	2	2

$$-8 \qquad +8$$

We know that the square of any number is positive, so an alternative to using the modulus of each deviation from the mean is to *square* the deviations and find the average of the total of the squares:

Mark	48	50	51	51	52	52	53	53	54	54	54
Deviation from mean	−4	−2	−1	−1	0	0	1	1	2	2	2
Square of deviation	16	4	1	1	0	0	1	1	4	4	4

Hence the sum of squared deviations from the mean is

$$16+4+1+1+1+1+4+4+4 = 36$$

and the average of the sum of squared deviations from the mean is $\frac{36}{11} = 3.27$ to 3 s.f., where we divide by 11, the total number of marks.

One difficulty remains. Since we squared the deviations, the average (3.27) is 'marks squared'. We overcome this by taking the square root, giving a new measure of spread of $\sqrt{3.27} = 1.81$ marks to 3.s.f.

The mathematical operations involved in calculating this measure of spread do not give rise to the difficulties involved in using the modulus. This measure of spread is called the *standard deviation*. It is calculated in the above way; the steps are:

(1) Calculate the squares of the deviations from the mean and add them up:

$$\Sigma(x-m)^2.$$

(2) Calculate the mean square deviation (the *variance*) by dividing by n, the size of the sample:

$$\frac{1}{n}\Sigma(x-m)^2.$$

(3) Take the square root and obtain the standard deviation:

$$s = \sqrt{\left(\frac{1}{n}(x-m)^2\right)}.$$

The square root is necessary in order to produce a measure of spread with the same dimensions as the original sample. If, for example, the sample was the heights in centimetres of a group of people, then the units of $\frac{1}{n}\Sigma(x-m)^2$ are cm².

Taking the square root gives a measure of spread in the original units – centimetres.

It is usual to give the standard deviation to an accuracy of one more decimal place than the original data.

Example 1

The number of O-levels gained by a group of pupils is as follows:

$$2, 2, 3, 4, 5, 5, 6, 7, 7, 9.$$

Calculate the mean and standard deviation.

Table 2

x	$x-5$	$(x-5)^2$
2	⁻3	9
2	⁻3	9
3	⁻2	4
4	⁻1	1
5	0	0
5	0	0
6	1	1
7	2	4
7	2	4
9	4	16
50		48

From Table 2, $\Sigma x = 50$.

So
$$m = \frac{1}{10}\Sigma x = \frac{50}{10} = 5,$$

and
$$s = \sqrt{\left(\frac{1}{n}\Sigma(x-m)^2\right)}$$
$$= \sqrt{\left(\frac{1}{10}\Sigma(x-5)^2\right)}$$
$$= \sqrt{\left(\frac{48}{10}\right)} = 2.2 \text{ to 2 s.f.}$$

The mean number of O-levels is 5 and the standard deviation is 2.2 O-levels.

Exercise B

Calculate the mean and standard deviation of each of the following sets of data:

1   5, 7, 8, 8, 9, 11, 15.

2   8, 10, 11, 12, 16, 18.

3   4, 5, 8, 8, 10, 11, 14, 15, 19, 22.

4 Two cricketers' scores in eight innings were:

A	21	19	13	30	28	19	34	26
B	64	0	1	33	2	6	81	3

Calculate for each the mean score and the standard deviation.

5 The IQs of a group of girls are 107, 110, 113, 114, 118, 124, 128. Calculate the mean and standard deviation.

3. AN ALTERNATIVE FORMULA

You will have noticed that the arithmetic in some of the examples of the previous exercise became rather cumbersome. This occurred whenever the mean was not an integer. Fortunately this heavy arithmetic can be eliminated by using an alternative version of the formula for standard deviation:

$$s = \sqrt{\left[\frac{1}{n}(\Sigma x^2) - m^2\right]}.$$

A proof that this formula is equivalent to the one used before is given in Section 11.

Example 2
Calculate the mean mark and standard deviation of the following set of marks:

$$7, 9, 10, 12, 12, 13, 14, 15, 19, 20.$$

Using a calculator we obtain $\Sigma x = 131,$
$$\Sigma x^2 = 1869.$$

Hence
$$m = \frac{1}{n}\Sigma x = \frac{131}{10} = 13.1$$

and
$$s = \sqrt{\left[\frac{1}{n}(\Sigma x^2) - m^2\right]}$$
$$= \sqrt{\left[\frac{1869}{10} - (13.1)^2\right]}$$
$$= \sqrt{[186.9 - 171.61]}$$
$$= 3.91 \text{ to 3 s.f.}$$

So the mean and standard deviation are 13.1 and 3.9 marks respectively.

Exercise C

Use the alternative formula $s = \sqrt{\left[\frac{1}{n}(\Sigma x^2) - m^2\right]}$ to find the standard deviation of each of the following sets of data.

*1   2, 3, 4, 6, 7, 7, 11.

*2   3, 4, 5, 7, 8, 8, 12.

*3   20, 30, 40, 60, 70, 70, 110.

*4   13, 18, 23, 33, 38, 38, 58.

*5   Compare your answers for questions 2, 3 and 4 with those for question 1. What do you notice?

4. TRANSFORMATIONS

(i) Translation of the data by addition or subtraction of a constant translates the mean by a corresponding amount but does not alter the standard deviation.

(ii) Enlargement of the data by multiplication or division by a constant causes a corresponding enlargement in both the mean and standard deviation.

For example, the data of question 4 was obtained from that of question 1 by multiplying by 5 and then adding 3. Check that the means are similarly related but the standard deviation of question 4 is just 5 times that of question 1 (the addition of 3 causing no change of standard deviation).

This is, in fact, what we would expect. Translation of the data does not affect the spread, but enlargement of the data should cause a similar enlargement in the spread. The mean is affected by both translation and enlargement.

Exercise D

Calculate the mean and standard deviation of the data of question 1. Hence *write down* the mean and standard deviation of the data of questions 2, 3 and 4.

1   4, 5, 8, 9, 9, 11, 14, 17.

2   6, 7, 10, 11, 11, 13, 16, 19.

3   12, 15, 24, 27, 27, 33, 42, 51.

4   1.04, 1.05, 1.08, 1.09, 1.09, 1.11, 1.14, 1.17.

Calculate the mean and standard deviation in question 5. Hence *write down* the mean and standard deviations in questions 6, 7 and 8.

5   6, 9, 11, 15, 19, 23, 23, 25, 31.

6   13, 16, 18, 22, 26, 30, 30, 32, 38.

7   13, 19, 23, 31, 39, 47, 47, 51, 63.

8   2.6, 2.9, 3.1, 3.5, 3.9, 4.3, 4.3, 4.5, 5.1.

5. FREQUENCY DISTRIBUTIONS

The ideas of the previous sections are readily extended to deal with cases where the data is given in the form of a frequency distribution.

$$n \text{ becomes } \Sigma f,$$
$$\Sigma x \text{ becomes } \Sigma xf,$$
$$\Sigma x^2 \text{ becomes } \Sigma x^2 f.$$

Hence

$$m = \frac{1}{n} \Sigma xf = \frac{\Sigma xf}{\Sigma f}$$

and
$$s = \sqrt{\left[\frac{1}{n}(\Sigma x^2 f) - m^2\right]} = \sqrt{\left(\frac{\Sigma x^2 f}{\Sigma f} - m^2\right)}.$$

Example 3

Four coins are tossed 60 times and the number of heads obtained is noted.

Number of heads	0	1	2	3	4
Frequency		8	11	23	15	3

Calculate the mean number of heads and standard deviation.

Table 3

x	f	xf	x^2f
0	8	0	0
1	11	11	11
2	23	46	92
3	15	45	135
4	3	12	48
	60	114	286

From Table 3, $\qquad \Sigma xf = 114$

and $\qquad \Sigma x^2 f = 286.$

Hence $\qquad m = \dfrac{\Sigma xf}{\Sigma f} = \dfrac{114}{60} = 1.9$

and $\qquad s = \sqrt{\left(\dfrac{\Sigma x^2 f}{\Sigma f} - m^2\right)} = \sqrt{\left(\dfrac{286}{60} - 1.9^2\right)} = 1.08 \text{ to 3 s.f.}$

The mean number of heads is 1.9 and the standard deviation is 1.1.

Example 4

The heights of a group of saplings are given in the following table.

Height in metres	1.3–	1.4–	1.5–	1.6–	1.7–	1.8–	1.9–2.0
Frequency	5	11	24	33	27	17	3

Calculate the mean height and standard deviation.

In this example the data is given in the form of a grouped frequency table. Each group must be represented by the mid-interval value for the purpose of the calculation (Table 4).

A calculator now gives $\Sigma xf = 198.9$ and $\Sigma x^2 f = 332.02$.

Table 4

Height in m	Mid-interval value x	Frequency f
1.3–1.4	1.35	5
1.4–1.5	1.45	11
1.5–1.6	1.55	24
1.6–1.7	1.65	33
1.7–1.8	1.75	27
1.8–1.9	1.85	17
1.9–2.0	1.95	3
		120

Hence
$$m = \frac{\Sigma xf}{\Sigma f} = \frac{198.9}{120} = 1.6575$$

and
$$s = \sqrt{\left(\frac{\Sigma x^2 f}{\Sigma f} - m^2\right)} = \sqrt{\left(\frac{332.02}{120} - 1.6575^2\right)} = 0.1397 \text{ to 4 s.f.}$$

The mean height is 1.66 m and the standard deviation is 0.14 m. (There must be some inaccuracy involved in using the mid-interval values, so it is unreasonable to quote answers to a high degree of accuracy.)

Exercise E

Calculate the mean and standard deviation in each of the following examples.

1 The number of O-levels gained by a certain school in 1980:

Number of O-levels	0	1	2	3	4	5	6	7	8
Number of candidates	3	5	11	16	21	38	14	10	2

2 The distances in km that a class of 30 pupils travelled to school:

Distance (km)	0–	1–	2–	3–	4–	5–6
Number of pupils	3	8	11	4	3	1

3 The IQs of a group of children:

IQ	106–110	111–115	116–120	121–125	126–130	131–135
Number of children	6	15	16	8	3	2

4 The total score obtained when two dice were rolled 50 times:

Score	2	3	4	5	6	7	8	9	10	11	12
Frequency	1	2	4	7	6	10	8	6	3	1	2

5   The times taken by 50 athletes to run 400 m:

Time (seconds)	45–	46–	47–	48–	49–50
Number of athletes	4	10	22	9	6

6   The marks obtained by a group of candidates in a Mathematics examination:

Mark	1–10	11–20	21–30	31–40	41–50	51–60	61–70
Number of candidates	3	11	15	21	29	17	4

6. HISTOGRAMS

You will be familiar with the drawing of *frequency diagrams*. For example a frequency diagram for the data of Exercise E, question 2, is as in Figure 1.

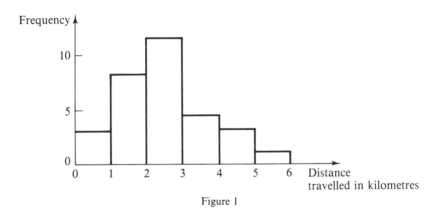

Figure 1

Consider, however, the following data. Two scientists *A* and *B* weigh the same 50 crystals but record their results in different ways as in Table 5.

Table 5

Scientist *A*		Scientist *B*	
Mass (g)	Frequency	Mass (g)	Frequency
0–	2	0–	2
1–	7	1–	7
2–	9	2–	9
3–	12	3–	12
4–	8	4–8	20
5–	7		
6–	4		
7–8	1		

The frequency diagrams corresponding to the two sets of data would be as in Figure 2.

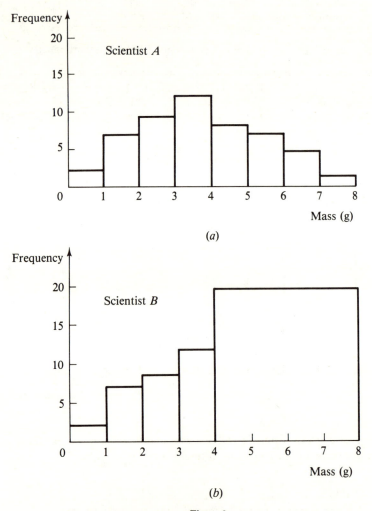

(a)

(b)

Figure 2

Clearly Figure 2(b) is grossly misleading. Why?

Scientist B grouped together the heavier crystals and so his diagram gives the impression that there were far more crystals heavier than 4 g than there were lighter than 4 g.

This difficulty can be overcome by saying that scientist B recorded a frequency of approximately 5 *per unit interval* over the interval 4–8 g. This has the effect of making the *area* instead of the height of the rectangle correspond to the frequency. The vertical axis now measures *frequency density* (in this case it measures 'frequency per 1 g interval' or 'number of crystals per 1 g interval'). The two diagrams now appear as in Figure 3.

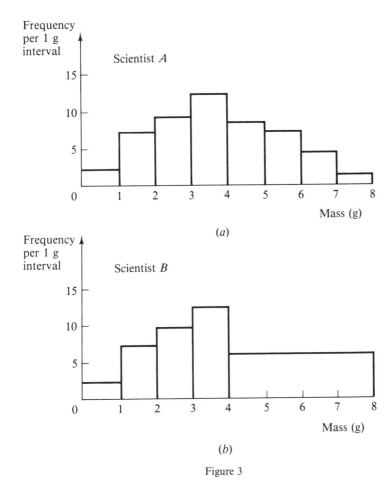

Figure 3

Notice that the diagram for A's data appears exactly the same as before. This is because her class intervals are all the same length, and consequently area is proportional to height on her diagram. The impression given by Figure 3(b) is now much more reasonable and we would interpret it as an average of 5 crystals per 1 g interval for each of the intervals 4–5, 5–6, 6–7 and 7–8.

Such a frequency diagram, where *area* corresponds to frequency, is called a *histogram*.

Example 5

The number of letters delivered at 30 houses in a road one morning is as follows:

Number of letters	2–3	4	5	6	7–9
Frequency	4	5	7	8	6

Draw a histogram to illustrate this.

In this case the data is discrete. It is impossible to have 3.47 letters delivered, whereas it is quite possible for a crystal to have its mass measured as 3.47 g. Mass is an example of a continuous variable.

To represent this situation on a histogram we draw, for example, the column for the value 4 to 3.5 to 4.5 and that for the values 7–9 from 6.5 to 9.5 as shown in Table 6 and Figure 4.

Table 6

Number of letters	Frequency	Interval on histogram	Width of interval	Height of rectangle (frequency density)
2–3	4	1.5–3.5	2	2
4	5	3.5–4.5	1	5
5	7	4.5–5.5	1	7
6	8	5.5–6.5	1	8
7–9	6	6.5–9.5	3	2

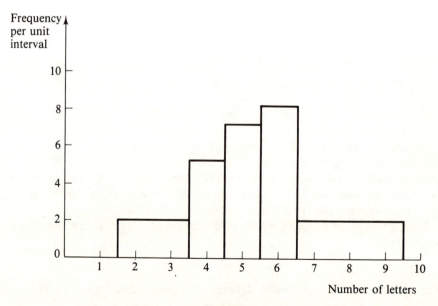

Figure 4

Exercise F

Draw histograms to represent the following data.

1 The durations of telephone calls made through an office switchboard during a given day:

Duration of call in minutes	0–1	1–2	2–3	3–6	6–12
Number of calls	14	19	31	30	24

2 The heights of a group of children:

Height in centimetres	120–	125–	130–	135–	145–160
Number of children	7	11	18	13	6

3 The score obtained when a die was rolled 50 times:

Score	1	2	3	4	5	6
Frequency	8	9	5	12	9	7

4 The number of eggs laid by 45 hens in a given week:

Number of eggs	0–3	4	5	6–7
Frequency	12	15	13	5

7. HISTOGRAMS AND STANDARD DEVIATION

Standard deviation is a rather indirect measure of spread, and it is useful to relate it to the 'spread' of a histogram. This idea is explored in the next example and the following exercise.

Example 6
The number of customer complaints received per day by a large manufacturing firm was recorded for 200 working days, with the following results:

Number of complaints	0–3	4–5	6–7	8–9	10–11	12–15
Frequency	22	40	61	42	21	14

Calculate the mean number of complaints per day (m) and the standard deviation (s). Draw a histogram and mark on it lines representing $m-2s$, $m-s$, $m+s$ and $m+2s$. Estimate the percentage of days where the number of complaints lies: (*a*) within 2 standard deviations of the mean; (*b*) within 1 standard deviation of the mean.

Table 7

No. of complaints	Mid-value x	Frequency f
0–3	1.5	22
4–5	4.5	40
6–7	6.5	61
8–9	8.5	42
10–11	10.5	21
12–15	13.5	14
		200

From Table 7, a calculator gives

$$\Sigma xf = 1376$$

and
$$\Sigma x^2 f = 11338.5.$$

Hence
$$m = \frac{1376}{200} = 6.88$$

and
$$s = \sqrt{\left(\frac{11338.5}{200} - 6.88^2\right)} = 3.06 \text{ to 3 s.f.}$$

Hence, to 1 decimal place, $m = 6.9,$
$$s = 3.1.$$

The mean number of complaints per day is 6.9; the standard deviation is 3.1 complaints per day. With these values,

$$m - 2s = 0.7, \; m - s = 3.8, \; m + s = 10.0, \; m + 2s = 13.1.$$

When drawing the histogram (Figure 5), note that the intervals are

0–3	represented by ⁻0.5 to 3.5,
4–5	represented by 3.5 to 5.5,
6–7	represented by 5.5 to 7.5.
8–9	represented by 7.5 to 9.5.
10–11	represented by 9.5 to 11.5
12–15	represented by 11.5 to 15.5.

Notice also that the two outer intervals are twice the width of the others, hence the heights must be reduced by half on the histogram.

Remembering that the area gives the frequency we seen that between $m - s$ and $m + s$ there are approximately:

$$(1.7 \times 20) + 61 + 42 + (0.5 \times 10.5) \approx 142 \text{ days,}$$

i.e. 71% of days.

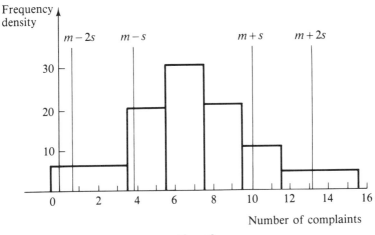

Figure 5

Between $m-2s$ and $m+2s$ there are approximately
$$15.4+40+61+42+21+5.6 = 185 \text{ days}$$
i.e. approximately 93% of days.

Exercise G

1   A case of 250 apples was opened and each apple was weighed. The masses were distributed according to the following table:

Mass in grams	80–	100–	120–	140–	160–180
Number of apples	24	52	90	64	20

Calculate the mean mass (m g) and standard deviation (s g). Draw a histogram and by marking lines representing $m-2s$ and $m+2s$, estimate the percentage of apples whose masses lie within 2 standard deviations of the mean.

*2   Whilst testing police radar equipment, 120 cars were checked at a certain point and their speeds recorded as follows:

Speed in m.p.h.	less than 10	10–	20–	30–	35–	40–	50–60
Number of cars	3	7	18	26	35	22	9

Calculate the mean speed (m m.p.h.) and the standard deviation (s m.p.h.). Draw a histogram and mark on it lines representing $m-s$ and $m+s$. Estimate the percentage of cars whose speeds lie within 1 standard deviation of the mean.

*3   In order to establish a case for the building of a by-pass, a town council arranged for 500 lorries passing through the town on a given day to be weighed at a vehicle weighbridge. The following results were obtained.

Weight in tonnes	less than 10	10–	15–	20–	25–	30–	35–	40–50
Number of lorries	30	72	125	107	75	48	28	1

Calculate the mean weight (m tonne) and standard deviation (s tonne). Draw a histogram and mark lines representing m and $m + 1.5s$. Hence estimate the percentage of lorries whose weights lie between the mean and 1.5 standard deviations above the mean.

8. STATISTICAL DATA AND PROBABILITY DISTRIBUTIONS

If we look back at Example 6 and the answers to Exercise G, we can see considerable similarities between the different situations.

(i) The histograms all have a 'humped' shape, with the modal class near the centre, and are roughly symmetrical.

(ii) About two-thirds of the values lie within a band which is 1 standard deviation wide on either side of the mean.

(iii) About 95% of the values lie within a band which is 2 standard deviations wide on either side of the mean.

Another way of looking at the statement in (iii) is to say that, in question 1, for example, there is a probability of about 0.95 that an apple chosen at random from this case will have a mass within the $m \pm 2s$ band, that is between about 87 g and 174 g. Similar probabilistic statements could be made about the other situations.

The common pattern underlying these different situations occurs frequently; so frequently that it is useful to have a mathematical model of it. Such a model must follow the common pattern, but it must be independent of the particular quality or variable under consideration and applicable whatever the size of the sample or the scale of any measurement.

We have already used *area* to represent frequency. However, if we had two consignments of apples, one of which was of 100 apples, while the other consisted of 350 apples, we could still not compare the two histograms relating to their masses directly. If we use area to represent *relative* frequency this difficulty is overcome; you will also remember that relative frequency can help us to estimate probability, so we have moved towards the probabilistic statements in (ii) and (iii) above. What effect does the change to relative frequency for area have?

It is easy to see that the *shape* of the histogram is unchanged, for we simply divide each height by the total frequency (250 in question 1). But now the *total* area must be 1 for all histograms; we recall that 1 is also the total of probabilities of the outcomes of any experiment. We are led to the idea that, to set up a model to use with many different situations we should work in terms of probabilities and set up a *probability distribution*. For any observed pattern of results we need a mathematical function, the shape of whose graph is close to the shape of the histogram and with the scales adjusted so that the 'area under the graph' is 1.

The Normal probability distribution

One of the commonest patterns is known as the *Normal probability distribution.* The equation of the curve is complicated and beyond our scope at the moment,

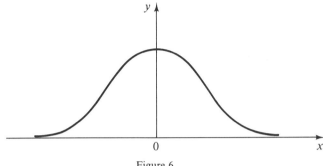

Figure 6

but the symmetrical 'bell-shaped' curve shown in Figure 6, centred on the *y*-axis, with area under it of 1, is so useful that areas under it are tabulated for us. For this Normal model the mean is zero; it is easy to apply the model to situations where the mean is not zero by a simple translation, while an enlargement, using the standard deviation of the data, can be used to match the spreads.

For this theoretical model or distribution, 68.2% of the population lies within one standard deviation of the mean and 95.4% within two standard deviations of the mean (Figure 7). (Compare these figures with those quoted in (ii) and (iii)

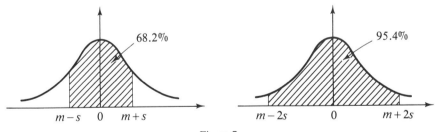

Figure 7

above.) If we believe that a given population would be well-modelled by this theoretical model we can use our knowledge of the model to make predictions about the underlying situation. As 68.2% is approximately two-thirds we can see that, in a Normal distribution, about two-thirds of the population lies in an interval two standard deviations wide situated symmetrically on either side of the mean. The middle two-thirds of a population is known as the *inter-sextile range*; hence half the inter-sextile range is approximately equal to one standard deviation. This is sometimes a useful rough method of calculating the standard deviation.

Approximately 95% of the population lies within two standard deviations of the mean. This forms the basis of a useful quick check to see that the calculated value of a standard deviation is reasonable. For example, forgetting to take the square root when calculating the standard deviation would show up easily on a quick 'two standard deviations' check.

The percentage of the Normal distribution lying below a given value, measured in multiples of the standard deviation above the mean, is given in Table 8.

Table 8

	0.0	0.1	0.2	0.3	0.4	0.5	0.6	0.7	0.8	0.9
0	50.0	54.0	57.9	61.8	65.5	69.1	72.6	75.8	78.8	81.6
1	84.1	86.4	88.5	90.3	91.9	93.3	94.5	95.5	96.4	97.1
2	97.7	98.2	98.6	98.9	99.2	99.4	99.5	99.7	99.7	99.8
3	99.87	99.90	99.93	99.95	99.97	99.98	99.98	99.99	99.99	100

Thus 93.3% of the population lies below $m + 1.5\,s$ (Figure 8). As the graph is symmetrical, 93.3% will lie above $m - 1.5s$.

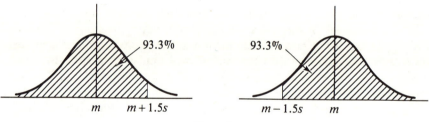

Figure 8

We could also say that $93.3 - 50 = 43.3\%$ lies between m and $m + 1.5s$, and $43.3 + 43.3 = 86.6\%$ lies between $m - 1.5s$ and $m + 1.5s$ (Figure 9).

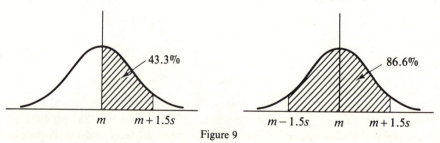

Figure 9

Example 7
Calculate the percentage of a Normal population lying between 1.2 standard deviations below the mean and 0.7 standard deviations above.

From Table 8, 88.5% lies below $m + 1.2s$, so 38.5% lies between m and $m + 1.2s$.
By symmetry 38.5% lies between $m - 1.2s$ and m (Figure 10).
From Table 8, 75.8% lies below $m + 0.7s$, so 25.8% lies between m and $m + 0.7s$.
Hence $38.5 + 25.8 = 64.3\%$ lies between $m - 1.2s$ and $m + 0.7s$ (Figure 11).

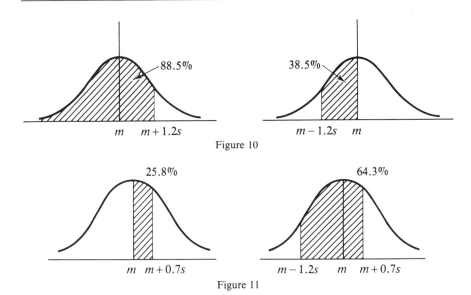

Figure 10

Figure 11

Exercise H

Use Table 8 to calculate the percentages of a Normal population lying:

1 between $m - 1.8s$ and m;

2 below $m - 1.3s$;

3 between $m + s$ and $m + 2s$;

4 above $m - 1.7s$;

5 between $m - 0.8s$ and $m + 1.3s$;

6 between $m - 2.3s$ and $m - 1.3s$.

9. USING THE NORMAL MODEL

Example 8

An intelligence test produces a mark which has Normal distribution with mean 100 and standard deviation 15. If 500 children take the test, estimate how many will (*a*) score more than 130, (*b*) score less than 82.

(*a*) From Table 8, 97.7% lies below $m + 2s$; hence 2.3% lies to the right of $m + 2s$ (Figure 12). 2.3% of 500 is 11.5. Hence about 12 score more than 130.

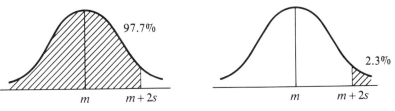

Figure 12

(*b*) 82 is 18 marks below the mean, i.e. 1.2 standard deviations below the mean. From Table 8, 88.5% lies below $m + 1.2s$. Hence 11.5% lies to the right of $m + 1.2s$, and 11.5% lies to the left of $m - 1.2s$ by symmetry (Figure 13). 11.5% of 500 is 57.5, and so about 58 of the children score less than 82.

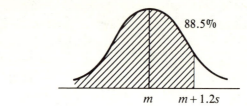

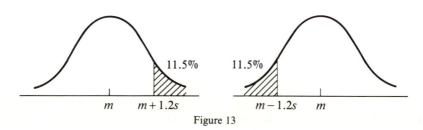

Figure 13

Example 9

The distribution of weights of Ruritanian army recruits is Normal. If 5.5% weigh more than 180 kg and 2.3% weigh less than 126 kg, estimate the mean weight and standard deviation.

If 5.5% weigh more than 180 kg then 94.5% weigh less than 180.
From Table 8, 180 kg is therefore 1.6 standard deviations above the mean,
i.e. $m + 1.6s = 180.$ (i)
If 2.3% weigh less than 126 kg then 126 must be 2 standard deviations below the mean,
i.e. $m - 2s = 126.$ (ii)

From (i) and (ii) we have $3.6s = 54$, hence $s = 15$ kg and $m = 156$ kg. The mean is 156 kg and the standard deviation is 15 kg.

Exercise I

In this exercise assume all populations are Normal.

1   The heights of 100 13-year-olds at a certain school have mean 160 cm and standard deviation 10 cm. Estimate how many are (*a*) taller than 175 cm, (*b*) shorter than 150 cm.

2   Morberry Jams fill their $\frac{1}{2}$ kg jam jars with a machine that delivers 505.4 g on average with a standard deviation 2.2 g. If 3000 jars are filled in a day, estimate how many will be underweight.

3   A commuter takes 43 minutes on average for her journey to work, with a standard deviation of 3 minutes. How long must she allow to be sure of being punctual 99% of the time?

4   Mr Smith catches the same train to work every morning. He timed its arrival on 100 successive days and found that it arrived after 8:43 on 1 day out of 100 and before 8:32 on 9 days out of 100. Estimate to the nearest minute the mean and standard deviation of the arrival times.

5   The number of words on a page of a certain book has mean 230 and standard deviation 10. If the book has 340 pages, estimate how many pages (a) have more than 248 words, (b) have fewer than 215 words.

6   It is found that the time taken by candidates to complete an intelligence test has mean 45 minutes and standard deviation 4.2 minutes. How long should be allowed for the test if we require 95% of the candidates to be able to finish the test?

7   In an examination, the marks have mean 48 and standard deviation 11. If the top 10% gain a distinction and the bottom 20% fail, estimate the minimum marks required for (a) a distinction, (b) a pass.

8   A machine filling cereal packets produces a standard deviation of 4.5 g in its process. If only 1 out of 200 packets filled weighs more than 263 g, estimate the mean weight of packets filled.

10. SUMMARY

If x is a typical member of a population occurring with frequency f,

$$\text{the mean } m = \frac{\Sigma xf}{\Sigma f}$$

and the standard deviation $s = \sqrt{\left(\frac{\Sigma(x-m)^2 f}{\Sigma f}\right)} = \sqrt{\left(\frac{\Sigma x^2 f}{\Sigma f} - m^2\right)}$.

The standard deviation is the positive square root of the mean of the squared deviations from the mean.

A histogram shows frequency densities for each group in the population:

$$\text{frequency density} = \frac{\text{frequency}}{\text{width of class interval}}.$$

Many populations can be modelled by the Normal probability distribution; in this model about $\frac{2}{3}$ of the population is within 1 standard deviation of the mean and about 95% of the population is within 2 standard deviations of the mean.

11. ALTERNATIVE FORMULAE FOR STANDARD DEVIATION

We now prove that the two formulae for standard deviation given in Section 3 are equivalent.

By definition $s = \sqrt{\left[\frac{1}{n}\Sigma(x-m)^2\right]}$.

So
$$s^2 = \frac{1}{n}\Sigma(x-m)^2$$

$$= \frac{1}{n}\Sigma(x^2 - 2mx + m^2)$$

$$= \frac{1}{n}(\Sigma x^2) - \frac{1}{n}(\Sigma 2mx) + \frac{1}{n}(\Sigma m^2).$$

But m is a constant, hence

$$s^2 = \frac{1}{n}(\Sigma x^2) - 2m\frac{1}{n}(\Sigma x) + \frac{1}{n}nm^2,$$

and
$$m = \frac{1}{n}\Sigma x,$$

so
$$s^2 = \frac{1}{n}(\Sigma x^2) - 2m^2 + m^2$$

$$= \frac{1}{n}(\Sigma x^2) - m^2$$

and
$$s = \sqrt{\left[\frac{1}{n}(\Sigma x^2) - m^2\right]}.$$

Miscellaneous exercise

1   A survey of the weekly earnings of a sample of 1000 young people who had found
    employment after leaving school gave the following results:

Earnings (£)	5–10	10–20	20–30	30–40	40–45
Frequency	80	360	380	120	60

(i) Illustrate the data by means of a histogram.
(ii) From your histogram, or otherwise, calculate an estimate of the probability
that a member of the sample chosen at random had a weekly income between
£14 and £24.
(iii) Calculate an estimate of the mean of the sample, and the standard
deviation.   (MEI)

2   In an agricultural experiment, 320 plants were grown on a plot, and the lengths of
    the stems were measured ten weeks after planting. The lengths were found to be
    distributed as in Table 9.

Table 9

Length (cm)	Number of plants
20–32	30
32–38	80
38–44	90
44–50	60
50–68	60

(a) On squared paper, draw a histogram to illustrate the data.

(b) Using your histogram, or otherwise, estimate the probability that the length of stem of a plant chosen at random from the plot exceeds 48 cm.

(c) From the table, calculate an estimate of the mean length of stem of a plant from this experiment.

(d) Calculate also the standard deviation. (MEI)

3 The life (in months) of eight car batteries of a certain make were measured and the following numbers obtained:

$$18, 19, 20, 20, 21, 23, 23, 24.$$

Calculate the mean life of this sample and the standard deviation.

Assuming that the sample is drawn from a population which is Normally distributed with the same mean and standard deviation, estimate:

(a) the probability of a battery of this make lasting for more than 26 months;

(b) the percentage of batteries which will last between 19 and 21 months;

(c) the time after which all but 1% of the batteries will have failed. (MEI)

4 The heights, correct to 2 significant figures, of thirty-two students are as shown:

Height (m)	1.5	1.6	1.7	1.8	1.9	2.0
Frequency	1	4	9	15	2	1

Calculate the mean and verify that the standard deviation is 0.1.

Assuming that the statistics you have calculated are representative of the entire student population and that it is Normally distributed, estimate the proportion whose height is between 1.60 and 1.80 metres, indicating your method clearly.

(SMP)

5 The times, to the nearest second, that a group of twenty students took to run an obstacle race are as shown:

Time	26	28	29	30	31	32	33
Frequency	1	5	4	2	1	4	3

Verify that their mean time is half a minute.

Plot the cumulative frequency curve and read off the inter-sextile range, that is, the range of the middle two-thirds of the group. This should give a good approximation to twice the standard deviation, if the distribution is Normal.

Calculate their actual standard deviation. Comment, in one sentence only, on your results. (SMP)

6 The times taken by fifty children to get to school one morning are shown below:

Time (min)	0–10	10–20	20–30	30–40	40–50
Frequency	15	24	8	2	1

Show that, to the nearest minute, the mean is 15 and the standard deviation is 9.

Draw a frequency diagram, with time along the horizontal axis. On it show, by drawing vertical lines, the proportion taking up to one, two, three standard deviations more, or less, than the mean time. Hence complete a table like the one shown below:

Estimated number of children taking less than the time shown, measured in units of standard deviation from the mean:

Time	3	2	1	0	$^-1$	$^-2$	$^-3$
Number	49	48					0

What would these numbers have been if it were a Normal population?   (SMP)

13

Correlation

1. SCATTER DIAGRAMS

Fifteen pupils took two examination papers in the same subject, and obtained the following marks shown in Table 1.

Table 1

Name	Paper 1	Paper 2
Anne	44	37
Brian	65	54
Catherine	24	23
David	30	36
Elizabeth	48	46
Frances	80	57
Geoff	74	52
Helen	45	42
Ian	37	28
John	56	49
Keith	25	30
Len	59	45
Margaret	10	24
Nicola	57	44
Pat	38	41

Oliver scored 40 on paper 1, but was ill for paper 2. What mark should he be given for that paper?

Careful study of the table of marks does show that, on the whole, those who obtained higher marks on paper 1 also scored higher marks on paper 2, but it is quite difficult to see this. The relationship is easier to see if we plot the scores on a graph, using the paper 1 mark as the x-coordinate and the paper 2 mark as the y-coordinate. (See Figure 1.) Such a graph is called a *scatter diagram*. Now it is clearer that, generally, pupils with higher marks in paper 1 also have higher marks in paper 2: we say that there is some degree of *positive correlation* between the marks. Figure 2 shows possible scatter diagrams for negative correlation and no correlation.

In order to decide what might be a fair mark for Oliver, we could try to draw a line to fit the points as closely as possible. In this case it is fairly straightforward to draw such a line by eye, but often it is useful to have a method for calculating

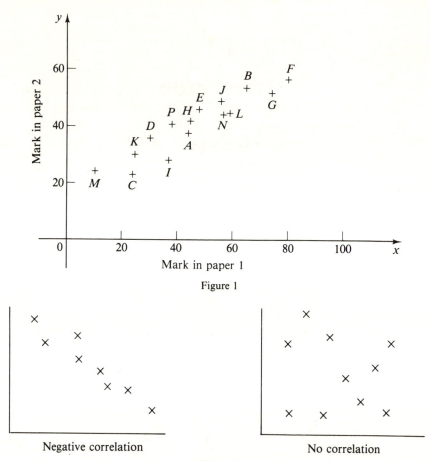

Figure 1

Figure 2

the equation of such a line. There are various methods, depending on how we decide what constitutes the best fit. The method described here is one of the quickest and simplest to use.

First we place the candidates in order of their marks on paper 1, and divide them into three roughly equal groups – in this case, this can be done exactly. (See Table 2.)

We now find the median marks in each paper for the bottom group and for the top group (ringed in Table 2). We take these as giving the coordinates of two points – (25, 28) and (65, 52) – and draw the line through these two points. (See Figure 3.) Since the line passes through (40, 37) we would, on this basis, consider 37 to be a fair mark for Oliver's paper 2. If necessary, the equation of the line can be calculated from the two points: in this case the gradient is $\dfrac{52-28}{65-25} = 0.6$, and the equation is $y = 0.6x + 13$.

Table 2

Name	Paper 1	Paper 2
Margaret	10	24
Catherine	24	23
Keith	(25)	30
David	30	36
Ian	37	(28)
Pat	38	41
Anne	44	37
Helen	45	42
Elizabeth	48	46
John	56	49
Nicola	57	44
Len	59	45
Brian	(65)	54
Geoff	74	(52)
Frances	80	57

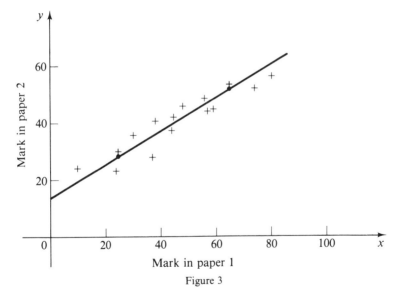

Figure 3

Exercise A

Draw scatter diagrams for each of the following sets of data, state whether they suggest positive correlation, negative correlation or no correlation, and draw a 'best-fit' line where appropriate.

1 The weights and heights of a group of pupils:

Weight in kilograms	46	43	51	48	45	39	42	40	50
Height in metres	155	154	150	160	152	145	149	152	155

2   Some data collected by the Mudwich Chamber of Trade over a period of 12 months:

Number of refrigerators sold	83	91	98	100	104	110	114	116	106	94	86	92
Number of juvenile court cases	78	52	58	74	74	66	70	48	54	64	75	72

3   The size of shoes and the number of O-levels gained by a group of fifth form boys:

Shoe size	7	6	$10\frac{1}{2}$	9	8	$9\frac{1}{2}$	8
Number of O-levels	4	5	7	5	0	3	6

4   The IQ and the time taken to complete a puzzle:

IQ	115	118	110	103	120	104	124
Time taken (seconds)	14	15	21	27	11	25	9

2. KENDALL'S RANK CORRELATION COEFFICIENT

It is often useful to have a measure of correlation. Various measures have been devised, giving 1 for perfect positive correlation and $^-1$ for perfect negative correlation. Some measures use the actual quantitites being measured, others just use the orders (or *ranks*) obtained from these measurements. We shall consider one of these latter measures.

If you look back at Table 2 it is fairly apparent that the marks in the two papers are positively correlated because the orders appear to be almost the same in the two papers. Table 3 shows the two orders.

Table 3

	F	G	B	L	N	J	E	H	A	P	I	D	K	C	M
Rank in paper 1	1	2	3	4	5	6	7	8	9	10	11	12	13	14	15
Rank in paper 2	1	3	2	6	7	4	5	8	10	9	13	11	12	15	14

The orders are nearly the same, but how nearly? We will consider a simpler situation first before returning to this question.

Suppose there were only five pupils, V, W, X, Y and Z, taking the two papers, and that the ranks were as in Table 4.

Table 4

Pupil	V	W	X	Y	Z
Rank in paper 1	1	2	3	4	5
Rank in paper 2	1	4	2	5	3

In this case we see that

V scored higher marks than W, X, Y, Z in both papers;

W scored higher marks than Y in both papers, but scored higher marks than X and Z in one paper and lower marks in the other paper;

X scored higher marks than Y and Z in both papers;

Y scored a higher mark than Z in one paper but not the other.

We have now compared all possible pairs of pupils:

for the comparisons made with V, we obtained 4 agreements about the ranks;

for the comparisons made with W, we obtained 1 agreement and 2 disagreements;

for the comparisons made with X, we obtained 2 agreements;

for the comparison made with Y, we obtained 1 disagreement.

These are most conveniently set out in a table (see Table 5).

Table 5

Pupil	V	W	X	Y	Z	
Rank in paper 1	1	2	3	4	5	
Rank in paper 2	1	4	2	5	3	
Agreements	4	1	2	0		7
Disagreements	0	2	0	1		3

Notice that, with the pupils placed in order according to paper 1, it is easy to construct the table by working from left to right and considering only the ranks in paper 2. Thus, for example, we compare W and X, Y and Z, and, since 4 (the rank of W) comes before 5 (the rank of Y) but after 2 and 3 (the ranks of X and Z), there are 1 agreement and 2 disagreements. We do not compare W with V at this stage because that comparison has already been made when comparing V with the other four pupils.

If the orders for the two papers had been exactly the same, there would have been $4+3+2+1 = 10$ agreements. Similarly, if one order had been the reverse of the other there would have been 10 disagreements. It follows that the fraction

$$\frac{\text{number of agreements} - \text{number of disagreements}}{\text{total number of pairs}}$$

must be between 1 (perfect agreement) and $^-1$ (total disagreement). This fraction is called *Kendall's rank correlation coefficient* (τ). It was devised by Kendall in 1948.

In this case, the coefficient is $\dfrac{7-3}{10} = 0.4$, indicating a fair degree of agreement.

Example 1

Calculate Kendall's rank correlation coefficient between the ranks obtained by 15 pupils in two papers, as in Table 6. (This is the data of Tables 1 and 3.)

Table 6

Pupil	A	B	C	D	E	F	G	H	I	J	K	L	M	N	P
Rank in paper 1	9	3	14	12	7	1	2	8	11	6	13	4	15	5	10
Rank in paper 2	10	2	15	11	5	1	3	8	13	4	12	6	14	7	9

Before making the comparisons we place the pupils in order of their rank in paper 1 (Table 7).

Table 7

Pupil	F	G	B	L	N	J	E	H	A	P	I	D	K	C	M	
Rank in paper 1	1	2	3	4	5	6	7	8	9	10	11	12	13	14	15	
Rank in paper 2	1	3	2	6	7	4	5	8	10	9	13	11	12	15	14	
Agreements	14	12	12	9	8	9	8	7	5	5	2	3	2	0		96
Disagreements	0	1	0	2	2	0	0	0	1	0	2	0	0	1		9

Total number of pairs $= 14 + 13 + 12 + 11 + \ldots + 3 + 2 + 1$

$$\tau = \frac{96-9}{105} = 0.83 \text{ to 2 s.f.}$$

Notice two points of detail:

(i) It can be shown that for n ranks the total number of pairs is $\frac{1}{2}n(n-1)$.

(ii) If two (or more) ranks are equal, we cannot say that there is either an agreement or a disagreement and so we omit the corresponding comparison from consideration. Table 8 shows an example, in which the comparisons (W, X) and (W, Z) cannot be made.

Table 8

	V	W	X	Y	Z
Rank 1	1	2⁻	2⁻	4	5
Rank 2	1	3⁻	2	5	3⁻
Agreements	4	1	2	0	7
Disagreements	0	0	0	1	1

$$\tau = \frac{7-1}{\frac{1}{2} \times 5 \times 4} = 0.6.$$

Exercise B

Calculate Kendall's rank correlation coefficient in each of the following questions and keep your answers for Exercise C.

1 Two housewives put seven soap powders in order of preference:

Powder	A	B	C	D	E	F	G
Housewife X	1	2	3	4	5	6	7
Housewife Y	3	4	1	7	6	2	5

2 The marks of six pupils in a Mathematics and in a Physics examination:

Mathematics	63	41	68	53	57	70
Physics	69	52	74	61	58	66

3 The distance walked and the amount per mile earned when eleven people took part in a sponsored walk:

Distance (miles)	25	23	22	21	19	18	12	10	9	5	3
Rate per mile (p)	14	16	12	13	11	7	9	6	10	4	8

4 Two panels of judges rank the songs in a Eurovision Song Contest:

Song	UK	Ireland	France	Germany	Norway
Panel A	4	2	1	3	5
Panel B	3	5	4	2	1

5 The nicotine yield and tar yield of six brands of cigarette:

Tar yield (g)	20	14	21	17	18	19
Nicotine yield (g)	1.4	0.9	1.3	1.0	1.4	1.2

6   The number of goals scored by the top ten teams in the Football League at home and away:

Home	40	43	34	44	36	47	30	37	46	19
Away	32	31	25	18	20	26	21	22	24	20

3. THE SIGNIFICANCE OF KENDALL'S CORRELATION COEFFICIENT

Can you distinguish whiter shades of white? In a consumer test Simon was asked to place six paint samples in order of whiteness. His order, and that of the manufacturer, are shown in Table 9.

Table 9

Manufacturer's order	1	2	3	4	5	6	
Simon's order	3	1	4	2	6	5	
Agreements	3	4	2	2	0		11
Disagreements	2	0	1	0	1		4

$$\tau = \frac{11-4}{\frac{1}{2} \times 6 \times 5} = \frac{7}{15}.$$

We see that Simon agrees reasonably well with the manufacturer. But might he have done as well as this by pure chance? If he had just guessed, he would have obtained one of the 720 possible orders of the numbers 1, 2, 3, 4, 5, 6 – and each of these orders would be equally probable. On this basis we can find the probability of obtaining each of the possible values of Kendall's rank correlation coefficient.

Clearly just one order (123456) gives $\tau = 1$.

If we interchange one adjacent pair of numbers, 2 and 3 for example, we have introduced one disagreement and so reduced τ to $\frac{14-1}{15} = \frac{13}{15}$. (See Table 10.)

Table 10

	1	2	3	4	5	6	
	1	3	2	4	5	6	
Agreements	5	3	3	2	1		14
Disagreements	0	1	0	0	0		1

There are five such adjacent pairs, so there are 5 out of the 720 possible orders which yield a coefficient of $\frac{13}{15}$ (213456, 132456, 124356, 123546, 123465).

To obtain two disagreements we need to interchange two pairs or make one cyclic change of three as in Tables 11 and 12.

Table 11

	1	2	3	4	5	6	
	2	1	3	5	4	6	
Agreements	4	4	3	1	1		13
Disagreements	1			1			2

Table 12

	1	2	3	4	5	6	
	3	1	2	4	5	6	
Agreements	3	4	3	2	1		13
Disagreements	2						2

There are 6 orders which change two pairs and 8 orders which have one cyclic change of three, hence 14 orders out of 720 with a coefficient of $\tau = \frac{11}{15} = 0.73$. (See if you can list all 14 of these orders.)

The distribution of the frequencies of all the possible coefficients when six objects are rearranged is approximately Normal and they are as in Table 13.

Table 13

τ	$\frac{15}{15}$	$\frac{13}{15}$	$\frac{11}{15}$	$\frac{9}{15}$	$\frac{7}{15}$	$\frac{5}{15}$	$\frac{3}{15}$	$\frac{1}{15}$	$\frac{-1}{15}$	$\frac{-3}{15}$	$\frac{-5}{15}$	$\frac{-7}{15}$	$\frac{-9}{15}$	$\frac{-11}{15}$	$\frac{-13}{15}$	$\frac{-15}{15}$
Frequency	1	5	14	29	49	71	90	101	101	90	71	49	29	14	5	1

If Simon had been guessing, so that each of the 720 orders had been equally probable, a score of $\frac{7}{15}$ or higher would occur with probability $\frac{98}{720} \approx 0.14$, which is by no means improbable.

What would be considered improbable? Statisticians tend to take notice when the probability is 0.05 or lower. If a particular value of τ (or higher) could occur by chance with probability less than 0.05 (or 5%), then that value of τ is said to be 'significant at the 5% level'. The significant values of τ for different values of n can be shown in a table or displayed on a graph as shown in Figure 4. From this diagram we see that, for example, if $n = 15$, τ has to be at least 0.35 to be significant at the 5% level.

In Example 1 we obtained $\tau = 0.83$ with $n = 15$. Reference to Figure 4 shows that this value of τ is significant at the 0.1% level, so such a value of τ, or higher, would occur by chance with a probability of less than 0.001. This is extremely improbable, and so we have strong evidence of correlation between the two orders.

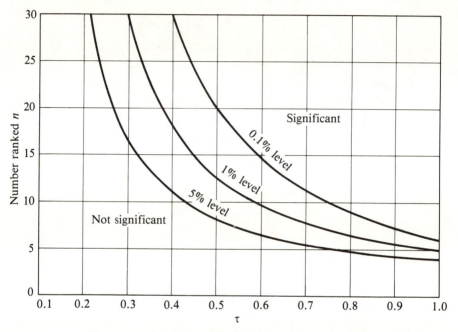

Significance levels of Kendall's rank correlation coefficient

Figure 4

Exercise C

1–6 Investigate the significance of the correlation coefficients found in Exercise B.

7 A, B, C, D, E was the order of merit for the five marrows entered in a local vegetable show. One gardener put them in the order B, A, C, D, E; show that she got 0.8 as a Kendall rank correlation coefficient on comparing her order with the official result. List all the other possible orders which would give the same coefficient when compared with A, B, C, D, E. How many different orders could there be for the five marrows? How likely is the gardener to obtain a value of τ as high or higher than she did, purely by chance?

8 List all the rearrangements of the figures 1234567 which will give a Kendall's rank correlation coefficient of $\frac{19}{21}$ and all the rearrangements which will give a value of $\frac{17}{21}$. What is the probability of getting a value of τ of at least $\frac{17}{21}$ when $n = 7$, by chance?

4. SUMMARY

A scatter diagram can be used to illustrate positive, negative or no correlation between two sets of data.

Kendall's rank correlation coefficient measures the amount of agreement between two orders. It is defined as

$$\tau = \frac{\text{number of agreements} - \text{number of disagreements}}{\text{total number of pairs}}.$$

τ lies between 1 (perfect agreement) and $^-1$ (total disagreement).

A value of τ is said to be significant at the 5% level if it, or a higher value, could occur by chance with probability less than 0.05.

Miscellaneous exercise

1   In the ten days of a holiday an angler kept a record of the number of fish which he caught and the amount of sunshine on each day. The results are as follows:

Hours of sunshine	8	13	9	$10\frac{1}{2}$	12	14	10	$9\frac{1}{2}$	11	$12\frac{1}{2}$
Number of fish caught	5	4	7	7	6	3	8	9	2	1

Show that the KRCC for these two sets of figures is $\frac{-4}{9}$, and give the level of significance for this correlation. What does this suggest about fishing?   (SMP)

2   In two sailing races the eight boats are placed in order as shown:
Order in the first race: *A   B   C   D   E   F   G   H*
Order in the second race: *C   A   D   B   H   F   E   G*
   (*a*) Calculate Kendall's rank correlation coefficient.
   (*b*) If the order of the first two boats in the second race had been reversed, calculate what the KRCC would have been.
   (*c*) Give the levels of significance of the correlations in (*a*) and (*b*).   (SMP)

3   In a competition housewives are asked to put in order of importance the following items of household equipment: oven (*O*), freezer (*F*), refrigerator (*R*), extractor (*E*), toaster (*T*), mixer (*M*), waste disposal unit (*W*). The judges have decided on the order *R, F, M, W, O, E, T*.

Mrs *A* chooses *M, F, W, R, T, E, O*. Show that the KRCC between her order and the judges' is $\frac{1}{3}$.

Mrs *B* chooses *F, R, M, O, W, E, T*. Find the KRCC between her order and the judges'.

In the event of nobody choosing their order the judges will award prizes to those obtaining a KRCC significant at the 1% level. What is the least value of the KRCC which would qualify for a prize?

Would Mrs *A* win a prize? Would Mrs *B* win a prize? Give a reason in each case.
   (SMP)

4   Two sixth-formers, a boy and a girl, are applying for entry to university. Each has to choose five university courses and put them in order of preference. They choose

the same five courses, but put them in different orders, the boy listing A, B, C, D, E and the girl listing B, E, A, D, C. Calculate the KRCC.

If the boy now brackets his last three choices together, so that C, D and E are equal third, calculate the new KRCC.

What is meant by saying that this correlation is not significant?   (SMP)

Revision exercise 3

1 For 7 consecutive weeks I decide to have a night out chosen at random but never a Sunday, once in each (6 day) week.
  (i) What is the probability
     (a) that my first night out is on a Wednesday,
     (b) that my second night out is not on a Saturday,
     (c) that four of the seven nights will be Wednesdays?
  (ii) What are the answers to each of the above questions if there is the additional restriction that I never go out on the same night two weeks in succession?
(SMP)

2 John washes up after lunch three times a week and his sister Mary washes up the other four times. John's days are chosen at random each week. The probability that he will break one or more dishes during washing-up is 0.1 and the probability that Mary will is 0.05. One day after lunch Dad, hearing a dish crash said 'Apparently this is John's day for doing the washing up.' What is the probability that he was right? (SMP)

3 When I oversleep there is a probability of 0.8 that I shall miss breakfast. There is a probability of 0.3 that I shall miss breakfast even when I do not oversleep. What is the probability that I am in time for breakfast on a day that I oversleep?

I find that I have a probability of 0.4 of oversleeping. Draw a tree diagram to show the probabilities of my oversleeping or not, with the consequent probabilities of my having breakfast or not.

On a day chosen at random what is the probability:
(a) that I oversleep and miss breakfast;
(b) that I do not oversleep but miss breakfast;
(c) that I do not miss breakfast?
If I am observed to have missed breakfast, what is the probability that I overslept?
(SMP)

4 (a) In a box of counters for playing the game of Ludo there are 5 green, 4 red, 5 blue and 3 yellow counters. I take out from the box three counters at random. What is the probability that I pick out:
    (i) 3 red counters;
    (ii) 1 green, 1 yellow, 1 blue in that order;
    (iii) 1 green, 1 yellow, 1 blue in any order?
(b) In a street of 17 houses, 4 have no washing machine, 3 have a Rondomat and the rest are equally divided between Duotub and Spinall washing machines. A market researcher wants to find one housewife who uses each type of machine. If she calls at the first three houses in the street, what does (a) suggest as her probability of success? Can you give a reason why (a) might not be a good model of the situation? (SMP)

5 A bag contains ten identical marbles, except that six are red and four are green. On five successive occasions one marble is drawn at random from the bag.

(i) Each marble is replaced before the next draw is made. Calculate the probability that:

(a) no green marble is drawn;

(b) at least one green marble is drawn;

(c) exactly two green marbles are drawn;

(d) all the marbles drawn are green.

(ii) If none of the marbles is replaced after it has been drawn, calculate what each of these probabilities would be.   (SMP)

6 On average I get up early three days out of ten and get up late one day out of ten. I forget something on two out of every five days on which I am late and on one-third of the days on which I am early. I do not forget anything on the days when I get up on time. Draw a tree diagram showing all of this information.

(i) State the probability that I am:

(a) early and forgetful;

(b) late and forgetful;

(c) punctual and forgetful.

(ii) Hence calculate:

(a) the probability of my not forgetting anything;

(b) the probability that, if I have forgotten something, I got up late.

(iii) Nevertheless confirm that for 80% of the time I am neither forgetful nor late.
                                                                      (SMP)

7 In a large batch of manufactured drinking glasses, 25% are considered to be substandard and are labelled 'seconds'. Find the probability that a random sample of four glasses drawn from the batch contains

(a) not more than one 'second',

(b) exactly two 'seconds'.

A wholesaler will accept the batch if either (i) a sample of four glasses contains not more than one 'second', or (ii) a first sample contains exactly two 'seconds', but a further sample of four glasses contains no 'seconds'. Calculate the probability that the wholesaler will accept the batch.   (MEI)

8 A group of children takes examinations in Mathematics and English. The proportions passing are respectively 80% and 70%, while 10% fail both examinations. Find the probabilities that

(a) a child passes both examinations,

(b) a child who passes in Mathematics also passes in English.   (MEI)

9 Large numbers of bulbs producing a red flower and bulbs producing a yellow flower are mixed in a box in the ratio 1:2 respectively. If four bulbs are selected at random and planted in a bowl, what is the probability that just one red flower will be produced?

If at the same time three more bulbs are planted in a second bowl, what is the probability that

(a) both bowls will produce just one red flower,

(b) there will be one red flower in one bowl but no red flowers in the other?
                                                                      (MEI)

10 Two guns fire at a target. For each shot by gun A the probability of hitting is $\frac{2}{5}$. For each shot by gun B the probability of hitting is $\frac{1}{2}$. Gun A fires three shots at the target. Find the probability that there is at least one hit.

This series of shots is in fact unsuccessful. In a second series of shots, gun A fires 3 shots and gun B fires 2 shots. Find the probabilities of (i) no hits, (ii) at least one hit, (iii) exactly one hit.   (MEI)

11 A canvasser knocks on the door of each house in a row of six. For each house the probability of there being a reply to his knock is $\frac{2}{3}$. Find the probabilities that
(i) he receives exactly five replies,
(ii) he receives at least five replies.
Given that he receives a reply, the probability that it is satisfactory is $\frac{4}{5}$. Find, to 3 significant figures, the probability that he receives exactly five satisfactory replies.   (MEI)

12 A number is selected from the integers 1 to 1000 inclusive. Each number has an equal chance of being selected. Evaluate exactly
(i) the probability that the number is divisible by 3,
(ii) the probability that the number is divisible by 7,
(iii) the probability that the number is divisible by both 3 and 7.
Using your answers, or otherwise, evaluate
(iv) exactly, the probability that the number is divisible by 3 or 7 or both,
(v) to 3 decimal places, the probability that the number is divisible by 7, given that it is divisible by 3.   (MEI)

13 Calculate the mean and standard deviation of the numbers of letters in the words of this sentence.   (SMP)

14 A shoe factory producing a new style of shoe tested six pairs to see how many months they would last before needing repair. The results were as follows:

$$17, 21, 20, 24, 14, 18.$$

For these numbers calculate (a) the mean, (b) the standard deviation.
(For the remainder of this question, assume that the 'lives' of all pairs of shoes in this style are Normally distributed with the mean and standard deviation given by the sample.)
(i) The factory proposes to replace free any pair which needs repair in less than one year. What percentage of shoes sold will they have to replace?
(ii) 10% of pairs of this style of shoe should last at least x months without repair. Calculate x.   (MEI)

15 A street trader selling mechanical toys outside a department store records that the number he sells on each of six consecutive days is as follows:

$$40, 45, 85, 60, 70, 60.$$

(i) Calculate the mean and the standard deviation of these numbers.
(ii) Assuming that this is the true mean and true standard deviation of the daily demand for these toys, calculate:
(a) the probability that, if he decides to bring six dozen toys with him each day, he will sell them all on any particular day;
(b) the number of toys he should bring with him each day if he wishes to reduce the probability of his being sold out at the end of the day to 2%.   (MEI)

16 A machine fills packets of sugar which are marked as containing 1 kg. Assuming that the masses vary according to the Normal distribution, with mean 1.03 kg and standard deviation 0.02 kg, calculate the percentage of packets which you expect to be underweight.
If the machine can be adjusted so that the mean is changed, but the standard

deviation remains the same, calculate the lowest value at which to set the mean so that we may expect no more than 2% of the packets to be underweight.   (SMP)

17   A machine produces ball-bearings with a mean diameter of 3 mm, and it is found that 6.3% of the production is being rejected as below the lower tolerance limit of 2.9 mm, and a further 6.3% is being rejected as above the upper tolerance limit of 3.1 mm. Assuming that the diameters are Normally distributed, calculate the standard deviation of the distribution.

The setting of the machine now 'wanders' so that the standard deviation remains the same, but the mean changes to 3.05 mm. Calculate the total percentage of the production which will now fall outside the given tolerance limits.   (MEI)

18   After all the History exam scripts have been marked, those near the pass mark are looked at again, primarily to try to find some reason to award the candidate a pass. The alteration in marks for the first 100 scripts looked at was:

Mark change	$^-2$	$^-1$	0	1	2	3	4
Frequency		5	10	20	30	20	10	5

State the percentage of the candidates whose score is raised by (a) exactly one mark, (b) at least one mark. Assuming the distribution above to be typical of the whole population of scripts and that 40 candidates scored any given mark in the vicinity of the pass mark, (c) estimate how many of the 40 candidates who scored one below the pass mark will have their total raised sufficiently to get a pass, and (d) verify that 48 candidates altogether will be expected to be raised to the pass mark or above.

Calculate (e) the mean change, and (f) the standard deviation. On the assumption that the distribution of mark changes is Normal, (g) estimate the percentage of candidates who would be expected to have their score raised by more than 4.

(SMP)

19   A firm produces packets of flour which are labelled as containing 500 g. In fact the masses of the packets are found to be Normally distributed with a mean mass of 505 g and a standard deviation of 3 g. Calculate the percentage of packets which are underweight.

The packing machinery now develops a fault, and it is found that 10% of packets are underweight.

(a) If in fact the mean is unchanged, calculate the new standard deviation.

(b) If, alternatively, the standard deviation is unchanged, calculate the new mean.

(MEI)

20   The marks within different subjects in an examination are all Normally distributed, but with different means and standard deviations. In order to simplify the comparison of each individual candidate's performances in different subjects, the examining body asks the examiners to convert their 'raw' marks to a standardised scale.

On the standardised scale, the pass-mark will be 40, and 70% of candidates will pass; the distinction mark will be 76, and 15% of candidates will be awarded a distinction. Denoting the mean of the standardised scale by m and its standard deviation by s, write down two equations of the form

$$m - ps = 40$$
$$m + qs = 76$$

where p and q are approximate decimal fractions to be found. Solve this pair of simultaneous equations to find the values of m and s to the nearest whole number.

The 'raw' marks for Geography have a mean of 56 and a standard deviation of 16. A particular candidate scores 73 marks on the paper. How many standard deviations above the mean is this? To what number should this mark of 73 be converted on the standardised scale? (MEI)

21 A certain examination has a mean mark of 100, and a standard deviation of 15. The marks can be assumed to be Normally distributed.
(a) What is the least mark needed to be in the top 35% of pupils taking this examination?
(b) Between which two marks will the middle 90% of the pupils lie?
(c) 150 pupils take this examination. Calculate the number of pupils likely to score 110 or over. (MEI)

22 Two judges in a singing competition award the following marks to the nine vocalists:

$$A (89, 90), B (94, 81), C (96, 75), D (81, 86), E (75, 69),$$
$$F (73, 54), G (58, 67), H (62, 58), I (67, 52).$$

For example F (73, 54) indicates that vocalist F was awarded 73 marks by the first judge and 54 marks by the second.
(a) Give the ranked order for each judge.
(b) State the vocalist upon whose position they are agreed.
(c) Show that their overall agreement, as measured by Kendall's rank correlation coefficient, is $\frac{4}{9}$.
(d) Write down, from the graph on p. 246, the probability of getting this measure of agreement purely by chance.
(e) Calculate the number of ranked pairs over which the judges would have to agree to obtain a measure of agreement twice as great.
(f) Estimate the probability of getting a measure of agreement twice, and also three times, as great. (SMP)

23 The batting and bowling averages of seven of the members of a cricket team are put in order as follows:

Player:	A	B	C	D	E	F	G
Batting order:	1	2	3	4	5	6	7
Bowling order:	7	4	6	5	3	1	2

(a) Calculate Kendall's rank correlation coefficient between these two orders.
(b) Is this significant at the 1% level?
(c) If the positions of A and C in the bowling order were reversed, show that Kendall's rank correlation coefficient would be significant at the 5% level. (SMP)

24 State the number of possible orders in which four different items can be arranged. Hence show that there is about 4% probability of getting a Kendall's rank correlation coefficient of 1 by chance.
   Show that, compared with numerical order, the arrangement 1, 3, 2, 4 gives $\text{KRCC} = \frac{2}{3}$ and state the other two arrangements which give the same KRCC as this. Hence show that you can expect to get a KRCC of $\frac{2}{3}$ or more just over 16% of the time.
   A forecaster in a horse race with 10 runners got the first five horses in the correct order but the last five in their reverse order. Verify that the $\text{KRCC} = 0.56$ approximately. Estimate how many of the 3.6 million possible forecasts will give a KRCC greater than 0.56. Is this a particularly significant level of correlation? Would you think this was a good measure of the forecaster's ability? (SMP)

25 The finalists in the shapely shin contest were placed by the two judges in the same order, except that the second and fourth were interchanged and the third and sixth were interchanged.

(i) If there were ten contestants, find Kendall's rank correlation coefficient.

(ii) With the above data the KRCC reduces to the formula

$$1 - \frac{24}{n(n-1)}$$

where n is the number of finalists. Show that, if there were only seven shapely shins in the final, then this correlation might have occurred by chance even at the 5% level of significance.

(iii) How many shapely shins would be required for the correlation to be significant at the 1% level? Give the correlation coefficient in this case.

(SMP)

14

Dynamics of a particle

1. FORCES

In Chapter 10 we investigated some problems in *kinematics*, that is the motion of particles and bodies. In this chapter on *dynamics* we shall find out how to relate the motion of a body to the forces acting on it.

Before discussing the effects they produce we shall look more closely at forces themselves.

You are familiar with forces as pushes or pulls; a force acts when we *push* a door, *pull* a rope or *press* a button. We know that we can vary both the size and the direction of the force, which means that forces are most satisfactorily represented as vectors.

In a diagram we represent a force as a directed line-segment, e.g. $\xrightarrow{\quad F \quad}$, where the length of the line is proportional to the magnitude $|\mathbf{F}|$ of the vector and where the direction of the force is taken to be that of the line (with sense shown by the arrow). We may also represent forces in column-matrix form where the components refer to known directions. Referred to the usual axes,

$\mathbf{F} = \begin{bmatrix} 1 \\ 2 \\ -3 \end{bmatrix}$ would indicate that the force had components 1, 2, 3 in the positive x, positive y and negative z directions respectively.

What forces are to be found in everyday situations? Apart from forces which we apply such as when we push a car, pull the curtains or throw a ball, there are others which may affect an object. The most important of these is the *weight* of the object. We make the definition:

The *weight* of an object is the force acting on it due to the gravitational attraction of the Earth.

We take the direction of the weight to be towards the centre of the Earth.

You know that the effect of the weight of, say, a book is to cause it to fall when released from a hand. But what happens if the book is resting on a table when it is released? Why does it no longer fall? Obviously the table is supporting the book by means of a force due to the contact between the book and the table (See Figure 1.)

Note that $\mathbf{S}$ has been assumed to be at 90° to the table.

Note also that the book has been drawn not in contact with the table. This is the usual practice so that forces acting on each part of a complex system can be shown without confusion.

What would happen if we tried to push the book along the table with a force

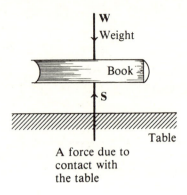

A force due to
contact with
the table

Figure 1

parallel to the table? It depends on how hard we push. A gentle push will not move the book but a harder one will. This suggests thst there is a resistance to sliding. It would seem that now the contact force **S** must no longer be at right-angles to the table so that there is a component of it resisting the push as well as a component opposing the weight (See Figure 2.)

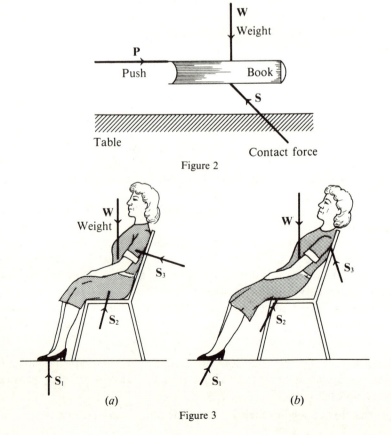

Figure 2

Figure 3

Consider now the forces acting on a woman sitting on a chair. In addition to her weight we would expect contact forces at each area of contact, say on her feet, her seat and her back (see Figure 3(*a*)). Other forces would be added for other contacts (e.g. if she rested her arm on the chair). What would happen if she slid down in the chair? To support her in the position shown in Figure 3(*b*) the contact forces will have changed their directions (and perhaps magnitudes) to prevent her sliding further.

In Figures 1, 2 and 3 we have attempted only to indicate the existence and directions of the forces acting. We cannot assume that the forces act at a particular point. Also, for our convenience, when there has been a considerable area of contact (as, for example, between the book and the table or between the woman and the chair) we have represented the total contact force as a single force, not by a collection of forces, one for each of the millions of points of contact.

Frequently in the following examples and exercise we shall wish to indicate forces, velocities and accelerations in the same diagram. To avoid confusion we shall use double arrows for acceleration and block arrows for velocity (Figure 4).

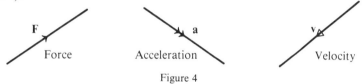

Figure 4

Example 1
Represent the forces acting on a block which is sliding down a plane inclined at 30° to the horizontal:
   (*a*) assuming no resistance to sliding;
   (*b*) assuming resistance to sliding.

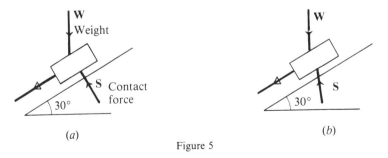

Figure 5

See Figure 5. In case (*a*) we would expect the contact force to act at 90° to the plane, simply preventing the block from sinking into the plane. In case (*b*) we would expect the contact force to have a component acting up the plane as there is resistance to sliding; so it will be inclined at a different angle to the plane (and have a different magnitude).

Example 2

Represent the forces acting on a rounders ball in flight.

We assume that there is resistance due to the motion through the air and that this resistance is in the opposite direction to the motion at any instant and may be regarded as a contact force (Figure 6).

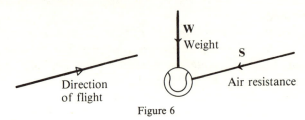

Figure 6

Exercise A

For each question, represent the weight and any contact forces you think might act, attempting to give approximately correct directions.

1   A sledge sliding down a slope with negligible resistance to motion.

2   (*a*) A car travelling along a horizontal road.
     (*b*) A car climbing a steep hill.

3   A ladder sliding down a wall with no resistance to motion from the wall but with sliding resistance from the floor.

4   A three-legged stool on a horizontal floor.

5   A conker hanging on a string.

6   An aeroplane in level flight.

7   A girl sliding on ice.

8   A boy sliding on a polished floor.

9   A child sliding down a banister.

10   A man sinking into quicksand.

11   A woman going up in a lift.

2. SLIDING

Our attempts to represent weight and contact forces are a first step towards producing a *mathematical model* of the real world. That is, we represent the real world in a way which makes our ideas clearer and calculations simpler. Sometimes our model will describe the real world exactly, but usually the description will be only approximate. We use our (relatively simple) model to make predictions about the real world that are at least approximately true. For greater accuracy we require more sophisticated models which are usually more difficult to understand and use.

One simplification we have already incorporated into our model is that some surfaces may be regarded as *smooth*, that is they offer negligible resistance to sliding. Another simplification that is useful is that some bodies are *light*, that is they have negligible weight in the context of the situation being studied. (For instance, when weighing an elephant we would ignore some fleas on its back.)

A further simplification we shall use is in the way that we represent the contact force between surfaces. In Section 1 we considered the forces acting on a book resting on and being pushed along a table. We expected that the contact force would be different in the two cases. However, it would seem reasonable to suppose that, when sliding, the component of the contact force at 90° to the table (that is, opposing the weight of the book) is the same as when the book is at rest.

The total contact force has changed because of the resistance to sliding, an effect which can be represented by a force parallel to the table. Thus the contact force could be regarded as having two components, one, which is constant, at 90° to the table and another parallel to the table, which varies according to the resistance to sliding.

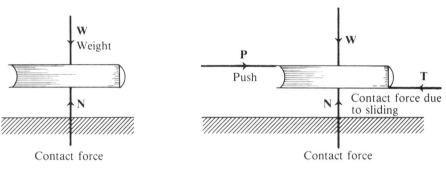

(a) Book at rest

(b) Book sliding, showing contact force as two components

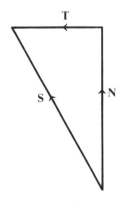

(c) Adding vectors **T** and **N** gives resultant **S**

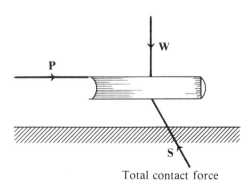

(d) Book sliding, showing contact force as single force

Figure 7

Experience tells us that if we push the book gently it will not move but pushing harder will cause movement. This suggests that the resistance to sliding has a maximum size.

We shall incorporate these ideas into our model and assume the following:

(1) Surfaces in contact can always exert a force at 90° to the surfaces at the point of contact. This is called the *normal contact force* or *normal reaction*.

(2) Surfaces in contact may exert a resistance to sliding which is parallel to their common surfaces and which may not exceed a certain value. The direction of this resistive force is opposite to that in which the body is sliding or would slide.

Example 3

Represent the forces acting on a block sliding down a plane inclined at 45° to the horizontal with resistance to sliding.

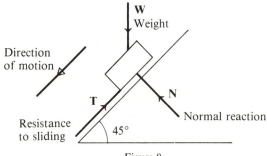

Figure 8

We have already noted that in diagrams we have indicated only the existence and directions of the forces acting. We do not know at which particular point or points the forces actually act. Although we might expect the resistance to sliding, for example, to act *along* the surface of contact, it is not at all clear where the weight or normal reaction would act. We shall discover the answer to this problem in the next section.

It was Isaac Newton (1642–1727) who first stated the principles which seem to describe the effect that forces have on the motion of particles and so they are known as *Newton's laws of motion*.

These laws cannot be proved but they are accepted because they give results which can mostly be confirmed by experiment and observation. It has been shown during the last hundred years that these laws are not *exactly* true; however, the errors are only significant when we attempt to solve some problems on a very large scale (as in astronomy) or on a very small scale (as in nuclear physics). For most practical purposes the results we obtain by applying Newton's laws are more accurate than we require and so we can incorporate them into our model.

The first principle we shall apply solves the problem about where the forces on an object should be considered to act. It may be shown that if we adopt a model of the real situation in which the object is taken to be a particle (occupying a single point) and all the forces are represented as acting at that point, then this will not affect the calculations about the *linear* motion of the object, although we shall have ignored any turning effect that the forces might have. For example, in the case of a book being pushed across a table the situation shown in Figure 9(*a*) can be modelled as shown in Figure 9(*b*). (Note that it is customary to represent forces as acting away from the particle.) In this case the model describes the motion across the table but ignores the possible turning effect due to **P** being applied at different points (see Figure 9(*c*) and (*d*)).

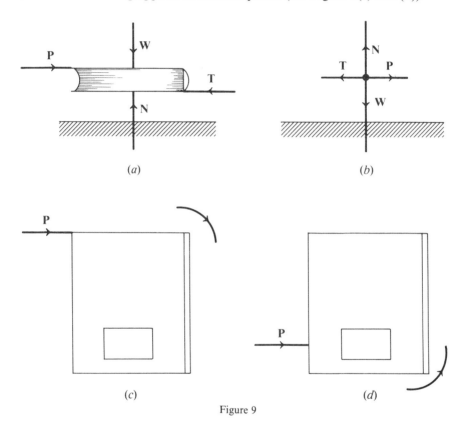

Figure 9

In this chapter we shall usually give both a picture and a model of the forces acting at a point.

Example 4
Repeat Example 3, representing the object as a particle.

See Figure 10.

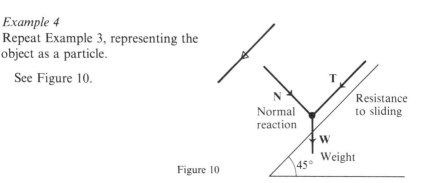

Figure 10

Exercise B

Repeat the questions of Exercise A drawing diagrams showing the forces as if they acted on a particle. Show the contact forces as normal contact forces and as components due to the resistance to sliding (i.e. as in Example 4).

3. FORCES AND ACCELERATION

Often it seems as if forces can act on objects without moving them or that objects need a force acting on them to keep them moving but, as you may have begun to suspect, this is not accurate. If we put in *all* the forces acting we find that:

Any object continues with constant velocity (which may be zero) unless a resultant force acts on it.

This is *Newton's first law (N1L)*. It is a vector relationship and so is valid in any component direction.

Referring to the examples in Section 2 (Figure 7); in (*a*) we can see that since there is zero acceleration perpendicular to the plane the resultant force in that direction is zero, i.e. $\mathbf{N} + \mathbf{W} = \mathbf{0}$, or $\mathbf{N} = {}^-\mathbf{W}$. These forces have equal size but are in opposite senses.

In (*b*) $\mathbf{N} = {}^-\mathbf{W}$ again and the book moves at constant velocity (including, perhaps, being at rest) if $\mathbf{T} + \mathbf{P} = \mathbf{0}$, but accelerates if $\mathbf{P} + \mathbf{T} \neq \mathbf{0}$.

We shall now find the amount of acceleration caused by the action of a resultant force.

You know that a given force does not have the same effect on all objects. Leaving aside all resistance, some objects have more matter in them and so have a greater 'inertia' to acceleration. This property of an object due to the matter in it is called its *mass*.

Note carefully that the mass of an object is an *intrinsic* property: it depends only on the matter in it, whereas the weight depends on the Earth's attraction for the matter. An object has a fixed mass under *all* circumstances but its weight may easily change; on the moon the gravitational attraction is much less. The earth certainly has mass but what could we mean by the question 'What is the weight of the Earth?'.

Consider a boy pushing a pram. If resistances are negligible, we know that the pram would change its velocity whilst being pushed (*N1L*). If the pram had twice the mass, or if the boy pushed three times as hard, what would be the effects on the acceleration? The answer is given by *Newton's second law (N2L)*:

$$\mathbf{F} \propto m\mathbf{a}.$$

The resultant applied force is proportional to the mass of the body multiplied by its acceleration.

The following points should be noted.

(1) This is a vector relationship. The result holds in any component direction.

(2) The acceleration depends on both the force and the mass.

(3) Newton's second law links force and acceleration, *not* force and velocity.

The constant of proportionality may be taken as 1 if the appropriate units are used, giving

$$\mathbf{F} = m\mathbf{a}$$

where m is in kg, $\mathbf{a}$ in m s^{-2} and $\mathbf{F}$ in newton (N). The kg is the mass of the standard kilogram, kept at Sèvres, near Paris, and we can see that 1 N is the force required to give a mass of 1 kg the acceleration of 1 m s^{-2}.

The effect of Newton's second law is to relate the system of forces acting on an object to the kinematics of the object. Once we know the acceleration we can use the methods of Chapter 10 to find the velocity, distance travelled, etc.

Example 5

A mass of 12 kg has acceleration $\begin{bmatrix} 2 \\ 3 \end{bmatrix}$ m s^{-2}. What force is acting?

Using *N2L*, if the force acting is $\mathbf{F}$, then

$$\mathbf{F} = 12\begin{bmatrix} 2 \\ 3 \end{bmatrix} = \begin{bmatrix} 24 \\ 36 \end{bmatrix}.$$

Thus a force of $\begin{bmatrix} 24 \\ 36 \end{bmatrix}$ N is acting.

Example 6

A particle of mass 12 kg is acted on by a force $\begin{bmatrix} 24 \\ 6 \\ 12 \end{bmatrix}$ N.

What is its acceleration? What is its speed 3 s after starting from rest?

By *N2L*, $\begin{bmatrix} 24 \\ 6 \\ 12 \end{bmatrix} = 12\mathbf{a}$. Thus $\mathbf{a} = \begin{bmatrix} 2 \\ 0.5 \\ 1 \end{bmatrix}$ and the acceleration is $\begin{bmatrix} 2 \\ 0.5 \\ 1 \end{bmatrix}$ m s^{-2}.

Since $\mathbf{a}$ is constant we may use $\mathbf{v} = \mathbf{u} + t\mathbf{a}$ ((1) from Chapter 10) giving

$$\mathbf{v} = \begin{bmatrix} 0 \\ 0 \\ 0 \end{bmatrix} + \begin{bmatrix} 2 \\ 0.5 \\ 1 \end{bmatrix} \times 3 = \begin{bmatrix} 6 \\ 1.5 \\ 3 \end{bmatrix}.$$

Thus the speed is $\sqrt{(6^2 + 1.5^2 + 3^2)} \approx 6.87$ m s^{-1}.

Exercise C

1 A force of $\begin{bmatrix} -1 \\ 2 \end{bmatrix}$ N is applied to a particle of mass 3 kg. What is its acceleration?

2 A particle of mass 4 kg has acceleration $\begin{bmatrix} 2 \\ 0 \\ -3 \end{bmatrix}$ m s^{-2}. What force is being applied?

3 A car of mass 1 tonne has acceleration $\begin{bmatrix} 4 \\ 3 \end{bmatrix}$ m s^{-2}. What resultant force is being applied?

4   A particle of mass 8 kg is acted on by a force of $\begin{bmatrix} 2 \\ 1 \\ 3 \end{bmatrix}$ N. What is its acceleration and change in velocity over 4 seconds?

5   A constant force acts for 6 s on a particle of mass 3 kg which has initial velocity $\begin{bmatrix} 2 \\ 3 \end{bmatrix}$ m s^{-1} and final velocity $\begin{bmatrix} 5 \\ 8 \end{bmatrix}$ m s^{-1}. Find the force.

6   A force of $\begin{bmatrix} 2 \\ 6 \\ 1 \end{bmatrix}$ N acts on a particle of mass 4 kg. What is its acceleration?

7   A ship of mass 4000 tonne has acceleration $\begin{bmatrix} 0.1 \\ 0.15 \end{bmatrix}$ m s^{-2}. What resultant force is being applied?

8   A force of $\begin{bmatrix} 2 \\ 4 \\ k \end{bmatrix}$ N gives a particle the acceleration $\begin{bmatrix} 1 \\ 2 \\ 4 \end{bmatrix}$ m s^{-2}. Find the mass of the particle, the value of k, and the magnitude of the force.

9   A force of $\begin{bmatrix} 20 \\ 15 \end{bmatrix}$ N acts on a particle of mass 5 kg. Find the acceleration and velocity after 8 seconds if the particle was initially at rest.

10   A force $\begin{bmatrix} 2t \\ t \\ 3t \end{bmatrix}$ N acts on a particle of mass 2 kg at time t. What is the acceleration at time t? If the particle is initially at rest find the velocity after 3 seconds.

4. APPLICATIONS

The following examples are chiefly concerned with practical situations in which it is necessary to consider carefully how to apply our techniques.

Example 7
A woman slides a packing case of mass 52 kg across a floor with acceleration 0.25 m s^{-2}. What force is required if the resistance to sliding is 15 N?

Let us suppose that the packing case slides in the $\begin{bmatrix} 1 \\ 0 \\ 0 \end{bmatrix}$ direction. Then from

Figure 11 we have **G**, the resistance, is $\begin{bmatrix} -15 \\ 0 \\ 0 \end{bmatrix}$ and $\mathbf{a} = \begin{bmatrix} 0.25 \\ 0 \\ 0 \end{bmatrix}$.

By *N2L*, $\mathbf{P} + \begin{bmatrix} -15 \\ 0 \\ 0 \end{bmatrix} = 52 \begin{bmatrix} 0.25 \\ 0 \\ 0 \end{bmatrix}$, so $\mathbf{P} = \begin{bmatrix} 15 \\ 0 \\ 0 \end{bmatrix} + \begin{bmatrix} 13 \\ 0 \\ 0 \end{bmatrix} = \begin{bmatrix} 28 \\ 0 \\ 0 \end{bmatrix}$, and

the required force is $\begin{bmatrix} 28 \\ 0 \\ 0 \end{bmatrix}$ N or 28 N in the direction in which the case is sliding.

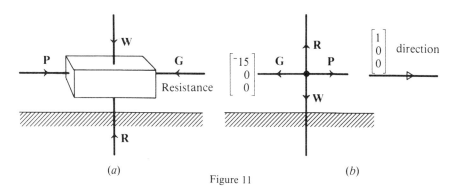

(a)   (b)

Figure 11

In this example it was not necessary to use three-component vectors to represent all the forces. Vectors of the form $\begin{bmatrix} x \\ y \end{bmatrix}$ would have been quite adequate. Indeed, in problems of this type, where *N2L* is applied in a single direction, the calculation can be done without vector notation.

Taking $\mathbf{P} = \begin{bmatrix} p \\ 0 \\ 0 \end{bmatrix}$ and applying *N2L* in the $\begin{bmatrix} 1 \\ 0 \\ 0 \end{bmatrix}$ direction we have, for the component in this direction,

$$p - 15 = 52 \times 0.25,$$

giving
$$p = 15 + 13 = 28.$$

Remember that each of the terms is a vector component, not a scalar. We shall adopt this practice where possible. Notice that such a simplification would not have been possible in Example 6.

Example 8
A bird of mass 0.5 kg is gliding due east at constant height at 4 m s^{-1} when it is blown by a wind with constant force. After 5 s its velocity has become 5 m s^{-1} in a direction E 37° S. What is the force of the wind?

In this case we require two-component vectors to express the velocities and forces referred to axes north and east. If the initial velocity of the bird is $\begin{bmatrix} 4 \\ 0 \end{bmatrix}$ m s^{-1},

then from Figure 12 the final velocity of the bird is

$$\begin{bmatrix} 5 \cos 37° \\ 5 \sin 37° \end{bmatrix} = \begin{bmatrix} 4 \\ 3 \end{bmatrix} \text{ m s}^{-1}$$

approximately.

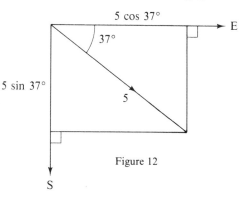

Figure 12

The change in velocity is $\begin{bmatrix} 4 \\ 3 \end{bmatrix} - \begin{bmatrix} 4 \\ 0 \end{bmatrix}$ m s^{-1} and the acceleration is

$\frac{1}{5}\left(\begin{bmatrix} 4 \\ 3 \end{bmatrix} - \begin{bmatrix} 4 \\ 0 \end{bmatrix}\right) = \begin{bmatrix} 0 \\ 0.6 \end{bmatrix}$ m s^{-2}.

By *N2L*, $\mathbf{P} = 0.5 \begin{bmatrix} 0 \\ 0.6 \end{bmatrix} = \begin{bmatrix} 0 \\ 0.3 \end{bmatrix}$; i.e. a force of 0.3 N due south.

Example 9
One man can push (horizontally) a piano of mass 350 kg with an acceleration of magnitude 0.15 m s^{-2}. With help from another man pushing just as hard (i.e. with the same force) the piano accelerates at 0.4 m s^{-2}. How hard is each pushing and what is the resistance to sliding?

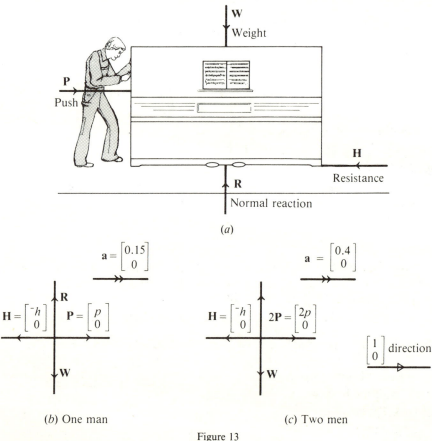

(a)

(b) One man   (c) Two men

Figure 13

See Figure 13. Two-component vectors are required to represent all of the forces but we need to apply *N2L* only in one direction. Take the push of one man to be $\mathbf{P} = \begin{bmatrix} p \\ 0 \end{bmatrix}$; then the push of two men is $\begin{bmatrix} 2p \\ 0 \end{bmatrix}$. Take the resistance to sliding as

$\mathbf{H} = \begin{bmatrix} -h \\ 0 \end{bmatrix}$, (note the sign of the component). An acceleration of magnitude

0.15 m s^{-2} in the $\begin{bmatrix} 1 \\ 0 \end{bmatrix}$ direction may be written $\begin{bmatrix} 0.15 \\ 0 \end{bmatrix}$.

Writing down only the component in the $\begin{bmatrix} 1 \\ 0 \end{bmatrix}$ direction:

With one man, by N2L, $p - h = 350 \times 0.15$, so $p - h = 52.5$;
with two men, by N2L, $2p - h = 350 \times 0.4$ so $2p - h = 140$.
Solving these equations simultaneously we have

$$p = 87.5 \quad \text{and} \quad h = 35.$$

Thus each man pushes with a force of 87.5 N and the resistance is 35 N.

Note that we could have written $\mathbf{H} = \begin{bmatrix} h \\ 0 \end{bmatrix}$ in which case we should have found

$h = {}^{-}35$.

Exercise D

1 What force is needed to give a 60 kg mass an acceleration of magnitude 2 m s^{-2}?

2 What force would give an α-particle, of mass 6×10^{-21} kg, an acceleration of magnitude 7×10^{10} m s^{-2}? How long would the force need to be applied for the particle to reach, from rest, a velocity of 3×10^{-6} m s^{-1}?

3 Four aero-engines, each developing a thrust of magnitude 5000 N, give a plane of mass 72000 kg an acceleration of magnitude 0.25 m s^{-2}. What is the magnitude of the air resistance?

4 An object of mass 1 kg is travelling at 20 m s^{-1} when a constant force of 5 N is applied at 150° to the direction of motion. What will be the direction of motion after 3 seconds?

5 What force is necessary to accelerate a train of mass 200 tonnes at 0.2 m s^{-2} against a resistance of 20000 N? What will be the acceleration if the train free-wheels against the same resistance?

6 A stone of mass 20 g is accelerated to 15 m s^{-1} horizontally in 0.5 s by means of a catapult. Find the average force acting.

7 A car of mass 750 kg is travelling at 50 km h^{-1} and is brought to rest in 10 seconds. What is the constant force required to achieve this?

8 A bag of cement, mass 50 kg, is dragged along the ground by a horizontal force of magnitude 80 N with an acceleration of magnitude 0.2 m s^{-2}. What is the magnitude of the resistance?

9 A bullet of mass 50 g is acted on by a force of 90 N. What is its resulting acceleration? Will it break the sound barrier if the force acts for $\frac{1}{2}$ s? (The speed of sound is 330 m s^{-1}.)

10 A force of magnitude 12 N gives an object an acceleration of magnitude 6 m s^{-2}. What is the mass of the object?

11 A satellite of mass 150 kg has a speed of 100 m s^{-1} which becomes one of 75 m s^{-1} at right-angles to its original direction of motion when a constant force is applied for 10 seconds. What is the magnitude and direction of this force?

12 A car of mass 1 tonne is travelling at constant speed of 50 m s⁻¹ when a driving force of magnitude 1000 N is acting. What is the resistance to motion? If the resistance remains constant how long will it take for the car to come to rest if the driving force is removed?

13 A lorry of mass 25 tonnes is travelling at 50 km h⁻¹ when its brakes are applied. What constant force is required to bring it to rest in 10 seconds?

14 A twin-engined aeroplane has acceleration of magnitude 2 m s⁻² with both engines working, but of only 0.75 m s⁻² if one engine has cut out. Assuming that the mass of the plane is 4000 kg, that the engines produce equal thrust and that the resistances to motion remain constant, find the thrust of each engine and the resistance to motion.

15 A force of magnitude 150 N accelerates a mass of 12 kg at 4 m s⁻². What is the magnitude of the resistance to motion?

5. WEIGHT

We shall now find the weight of an object whose mass is known – that is, the force with which the Earth attracts the object. It was Newton, again, who solved this problem with his *universal law of gravitation*. This states that any two particles of matter attract each other with force of size

$$\frac{GMm}{d^2}$$

where M and m are the masses of the particles, d is the distance between them and G is a universal constant. The direction of the force of attraction is along the line joining the particles (Figure 14).

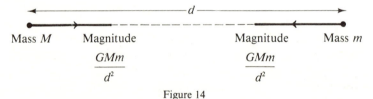

Figure 14

This law also applies to objects of finite size in which case the attraction is as if the whole mass of each object were concentrated in a single point. In simple cases this point is the 'centre' of the object.

G is very small, approximately 6.7×10^{-11} in SI units, and so the force of attraction between everyday-sized masses is slight. For instance, the attraction between two 1 kg masses 1 m apart is, as can be seen by substituting in the expression above, 6.7×10^{-11} N. We usually ignore the gravitational attraction of small objects for each other.

We defined the weight of an object to be the attraction of the Earth on it and so an object of mass m has weight

$$\mathbf{w} = \frac{GMm}{d^2}\mathbf{k}$$

where M is the mass of the earth, d is the distance to the centre of the Earth and **k** is a unit vector directed towards the centre of the Earth.

We may deduce (see Section 9) that

$$\mathbf{w} = m\mathbf{g}$$

where **g** is approximately constant. We referred to **g** in Chapter 10. It has a magnitude of about 9.81 m s^{-1} towards the centre of the Earth. It is often convenient and sufficiently accurate to take **g** to have magnitude 9.8 m s^{-2} or even 10 m s^{-2}.

Notice that the *weight* of a 1 kilogram mass is approximately 9.8 N and that 1 N is the weight of an average-sized apple!

Example 10
Calculate the force of attraction between two particles 0.1 m apart, each of mass 10 kg.

Taking G as 6.7×10^{-11}, the force of attraction is

$$\frac{6.7 \times 10^{-11} \times 10 \times 10}{(0.1)^2} \, \mathrm{N} = 6.7 \times 10^{-7} \, \mathrm{N}$$

directed along the line joining the particles.

Example 11
What is the weight of a 70 kg man?

His weight **w** is $m\mathbf{g}$, i.e. $\mathbf{w} = 70\,\mathbf{g}$.
Taking g, the magnitude of **g**, as 9.8, his weight is $70 \times 9.8 = 690$ N to 2 s.f. (vertically downwards).

Exercise E

Take g as 9.8 and G as 6.7×10^{-11}.

1   Calculate the force of attraction between two people, each of mass 50 kg, standing 5 m apart.

2   Two particles 10 m apart have a mutual force of attraction of 1 N. If the particles have the same mass what is it?

3   What is the weight of a 1 tonne car?

4   If the weight of an object is 1 N, what is its mass?

5   Find the weight of a 75 kg man on the moon. (The moon has a mass of 7×10^{22} kg and radius 1.6×10^6 m.)

6   Two objects are at the same height. Explain why, in the absence of significant air resistance:
    (a) they will land at the same time if dropped;
    (b) they may land at different times if thrown with the same downward force.

7   What is the weight of a 15 tonne truck?

8   What is the mass of a bicycle of weight 245 N?

6. VERTICAL MOTION

In this section we shall consider examples involving the weights of objects which are moving vertically.

Example 12
A brick of mass m is dropped over a cliff. Ignoring resistance to motion find its speed after 1 s.

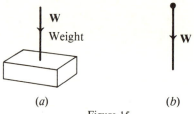

(a)   (b)

Figure 15

The only force acting is the weight, **W** (Figure 15). Since $\mathbf{W} = m\mathbf{g}$ we have $\mathbf{W} = m\begin{bmatrix} 0 \\ g \end{bmatrix}$, taking axes horizontally and vertically. Applying *N2L*, we have

$\mathbf{W} = m\begin{bmatrix} 0 \\ a \end{bmatrix}$, where $\begin{bmatrix} 0 \\ a \end{bmatrix}$ is the acceleration, or $m\begin{bmatrix} 0 \\ g \end{bmatrix} = m\begin{bmatrix} 0 \\ a \end{bmatrix}$ giving $a = g$.

Since a is constant we may use $\mathbf{v} = \mathbf{u} + t\mathbf{a}$, giving $\mathbf{v} = \begin{bmatrix} 0 \\ 0 \end{bmatrix} + 1\begin{bmatrix} 0 \\ g \end{bmatrix}$ after 1 s, i.e.

$\mathbf{v} = \begin{bmatrix} 0 \\ g \end{bmatrix}$ and the speed is g m s^{-1}.

Of course, you solved problems of this type in Chapter 10 assuming that the acceleration was $\begin{bmatrix} 0 \\ g \end{bmatrix}$. It is usual to make this assumption when there is no resistance to motion. However, in any situation other than free fall, we must always consider the forces acting.

Example 13
A stone of mass 2 kg is dropped. What is its acceleration if it experiences a constant retarding force of 5 N?

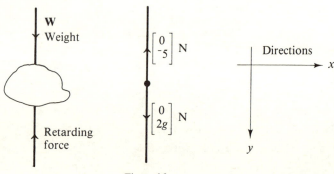

Figure 16

Taking axes horizontally and vertically downwards (Figure 16), the weight of the stone is $m\mathbf{g} = 2\begin{bmatrix} 0 \\ g \end{bmatrix} = \begin{bmatrix} 0 \\ 2g \end{bmatrix}$.

We may write the retarding force as $\begin{bmatrix} 0 \\ -5 \end{bmatrix}$ N and the acceleration as $\begin{bmatrix} 0 \\ a \end{bmatrix}$.

Writing down only the components in the $\begin{bmatrix} 0 \\ 1 \end{bmatrix}$ direction, using *N2L*,

$$2g - 5 = 2a.$$

Taking g as 9.8, $a = \dfrac{14.6}{2} = 7.3$. Thus the acceleration is 7.3 m s^{-2} downwards.

Example 14
What is the normal reaction between a 60 kg woman and the floor of a lift if the lift is accelerating downwards at a rate of 0.4 m s^{-2}?

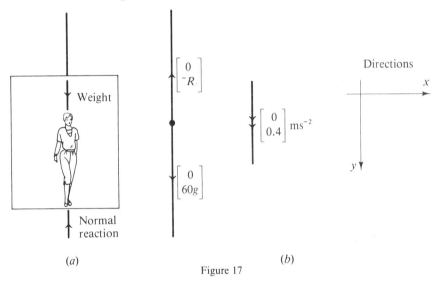

(a)   (b)

Figure 17

See Figure 17. The weight of the woman is $60\,\mathbf{g} = 60\begin{bmatrix} 0 \\ g \end{bmatrix} = \begin{bmatrix} 0 \\ 60g \end{bmatrix}$.

We may write the normal reaction as $\begin{bmatrix} 0 \\ -R \end{bmatrix}$ and the acceleration as $\begin{bmatrix} 0 \\ 0.4 \end{bmatrix}$ m s^{-2}.

Taking components in the $\begin{bmatrix} 0 \\ 1 \end{bmatrix}$ direction, using *N2L*,

$$60g - R = 60 \times 0.4 = 24.$$

Thus $R = 60g - 24 = 564$ (taking g as 9.8).

The normal reaction on the woman is 560 N to 2 s.f.

   Now a woman, mass M, not accelerating, experiences a normal reaction of 9.8 M. Thus the woman in the lift experiences the same force as a woman, not accelerating, of mass 58 kg to 2 s.f.

Exercise F

Take g as 10.

1 A coin is dropped from the Eiffel Tower, 300 m above the ground. Ignoring air resistance, at what speed does it hit the ground?

2 Calculate the normal contact force between a woman of mass 45 kg and the floor of a lift when accelerating upwards at 2 m s^{-2}.

3 A boy of mass 30 kg falls down a 20 m deep well. By bracing himself against the sides he manages to achieve a constant retarding force of 80 N upwards. At what speed does he hit the water?

4 A girl of mass 35 kg is accelerated from rest to a vertical speed of 3 m s^{-1} in 2 s by a trampoline. What is the magnitude of the average force acting over this time?

5 An object falls with constant acceleration of magnitude 5 m s^{-2} whilst subject to a retarding force of magnitude 80 N. What is its mass?

6 A car of mass 1 tonne falls vertically off a cliff with acceleration of magnitude 3 m s^{-2}. What is the magnitude of the air resistance?

7 A ball is thrown vertically upwards with speed 4 m s^{-1}. How high will it rise? (Ignore air resistance.)

8 A woman of mass 55 kg is suspended by a rope from a helicopter. What must the pull on the rope be if the woman is accelerating upwards at 0.5 m s^{-2} whilst experiencing a constant downward air resistance of 15 N?

9 A dropped brick of mass 0.8 kg accelerates at 5 m s^{-2}. What is the magnitude of the air resistance? How long will it take to drop 25 m from rest?

10 What is the apparent weight of a 50 kg woman whilst in a lift which is accelerating at 0.4 g (*a*) upwards; (*b*) downwards?

7. MOTION AND WEIGHT

We shall now consider some problems which involve the weight of bodies but where the motion is not simply vertical.

Example 15

A particle of mass 3 kg slides down a smooth slope inclined at 60° to the horizontal. What is its acceleration?

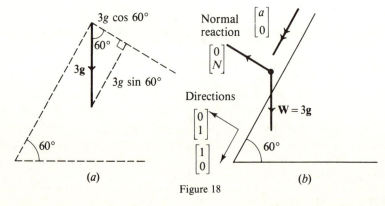

Figure 18

Since we are concerned with motion parallel to the plane it is sensible to choose our reference directions parallel and perpendicular to the plane. Using Figure 18(a) we can express the weight in terms of components in these directions. Thus

$$3\,\mathbf{g} = \begin{bmatrix} 3g \sin 60° \\ -3g \cos 60° \end{bmatrix}.$$

Using the component in the $\begin{bmatrix} 1 \\ 0 \end{bmatrix}$ direction and writing the acceleration as $\begin{bmatrix} a \\ 0 \end{bmatrix}$ we have by *N2L*

$$3g \sin 60° = 3a \Rightarrow a = g \sin 60° = 8.49 \quad \text{(taking g as 9.8).}$$

Thus the acceleration is 8.49 m s^{-2} down the plane.

Note that, by choosing the reference directions suitably we did not need to calculate the normal reaction. Should be wish to find the normal reaction we could do so by observing that there is zero acceleration perpendicular to the plane.

Using the components in the $\begin{bmatrix} 0 \\ 1 \end{bmatrix}$ direction, we have by *N2L*

$$N - 3g \cos 60° = 0 \Rightarrow N = 3g \cos 60° = 14.7.$$

Thus the normal reaction has magnitude 14.7 N.

Example 16
A book of mass 1.5 kg slides down a desk top inclined at 30° to the horizontal. There is a constant resistance to sliding of 4 N. What is acceleration down the plane and how long does it take the book to accelerate from 2 m s^{-1} to 5 m s^{-1}?

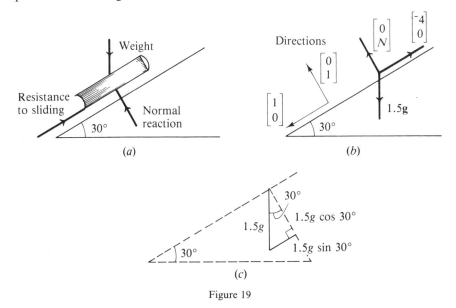

Figure 19

The weight of the book is $1.5\,g$ N. We are again concerned with motion parallel to the plane and so we take our reference directions parallel and perpendicular to the plane. Using Figure 19(a) we may express the weight in terms of components in these directions: $1.5\,\mathbf{g} = \begin{bmatrix} 1.5g\sin 30° \\ -1.5g\cos 30° \end{bmatrix}$. The resistance and acceleration may be written as $\begin{bmatrix} -4 \\ 0 \end{bmatrix}$ N and $\begin{bmatrix} a \\ 0 \end{bmatrix}$ respectively. Using components in the $\begin{bmatrix} 1 \\ 0 \end{bmatrix}$ direction we have, by N2L,

$$1.5g\sin 30° - 4 = 1.5a$$
$$\Rightarrow \qquad 7.35 - 4 = 1.5a$$
$$\Rightarrow \qquad a \approx 2.33.$$

Since **a** is constant, we may use $\mathbf{v} = \mathbf{u} + t\mathbf{a}$ and so

$$5 = 2 + 2.33t$$
$$t \approx 1.29.$$

The time taken is 1.3 s to 2 s.f.

Exercise G

Take *g* as 10.

1 A stone of mass 0.5 kg slides down a smooth roof at an angle of 45° to the horizontal. What is its acceleration?

2 A ship of mass 1000 tonnes is to be launched down a slipway of angle 6° to the horizontal. Assuming a constant resistance to sliding of 8×10^5 N what is the acceleration of the ship? How long does it take to slide from rest down 400 m of slipway?

3 A child of mass 30 kg goes down a slide of length 2.5 m inclined to the horizontal. If she reaches a speed of 5 m s^{-1} at the bottom, what is the angle of inclination of the slide? (Ignore resistances to sliding.)

4 An eccentric inventor claims to have produced an entirely smooth material. To attempt to substantiate his claim he has erected a slide of length 20 m made of the material, inclined at 30° to the horizontal. If he slides a block of the material down the slope what should be its final speed? How long should it take to slide down?

5 A ski-jumper of mass 80 kg descends a ski slope against a constant resistance of 50 N in order to take off at the bottom at a speed of 30 m s^{-1}. If the slope is 150 m long what is its inclination to the horizontal?

6 A particle slides down a smooth slope inclined at 45° to the horizontal. What is its acceleration?

7 A particle slides down a slope inclined at 30° to the horizontal against a constant resistance of magnitude 100 N with acceleration of magnitude 2 m s^{-2}. What is the mass of the particle?

8 A book of mass 1.5 kg is sliding *up* a plane which is inclined at an angle of 30° to the horizontal with an acceleration of magnitude 0.75 m s^{-2}. If it experiences a constant resistance to sliding of magnitude 8 N what force must be applied parallel to the plane to cause this acceleration?

9 A 60 kg woman skis down a slope which makes an angle of 60° with the horizontal and has acceleration of magnitude 8 m s^{-2}. What is the size of the resistive force?

10 A book of mass 2 kg lies on a smooth desk which slopes at 20° to the horizontal. What magnitude must a *horizontal* force applied to the book have if it is to accelerate *up* the desk at 1.25 m s⁻²?

8. SUMMARY

Types of forces

There are contact forces between two surfaces touching each other. Such a contact force may be regarded as the sum of two forces:
  (i) a normal reaction perpendicular to the surfaces;
  (ii) a resistance to sliding which is parallel to the surfaces and in a direction opposite to that in which the body is sliding or would slide.
The weight of an object is the force acting on it due to the gravitational attraction of the Earth. It is directed towards the centre of the Earth and is of magnitude $m\mathbf{g}$, where m is the mass of the object, and $\mathbf{g}$ the acceleration due to gravity.

Laws of motion

Any object continues with constant velocity (which may be zero) unless a resultant force acts on it.   (Newton's first law)
  The resultant force on an object is proportional to the mass of the object multiplied by its acceleration. The acceleration is in the same direction as the resultant force.   (Newton's second law)
  1 newton is the force required to give a mass of 1 kg an acceleration of 1 m s⁻².

9. WEIGHT AND MASS

From Newton's law of gravitation, the weight $\mathbf{w}$ of a body of mass m is given by

$$\mathbf{w} = \frac{GMm}{d^2}\mathbf{k},$$

where M is the mass of the Earth, d is the distance to the centre of the Earth and $\mathbf{k}$ is a unit vector towards the centre of the Earth.
  Now M is constant, G is constant and for ordinary small movements not exceeding, say, 10 km, d is essentially constant.

Thus $\dfrac{GM}{d^2}$

is constant.

  Now, by *N2L*, $\mathbf{w} = m\mathbf{a}$

where $\mathbf{a}$ is the acceleration of the particle, so

$$m\mathbf{a} = \frac{GM}{d^2}m\mathbf{k},$$

Thus
$$\mathbf{a} = \frac{GM}{d^2}\mathbf{k},$$

i.e. the acceleration is towards the centre of the Earth and, since $\dfrac{GM}{d^2}$ is constant,

it is constant.

We write this acceleration as $\mathbf{g}$ and so have

$$\mathbf{w} = m\mathbf{g}.$$

Miscellaneous exercise

1 The mechanism of a clockwork toy lorry can be assumed to exert a constant propulsive force under all conditions. Starting from rest on a level table, the lorry will attain a speed of 1.2 m s^{-1} in 3 seconds. The mass of the lorry is 0.15 kg.
   (a) Find the propulsive force in newtons.
   (b) If the lorry is loaded with blocks of total mass 0.3 kg find how long it will take to reach a speed of 1.2 m s^{-1} from rest.
   (c) If the unloaded lorry is placed on a plank sloping at an angle whose sine is $\frac{1}{50}$, find how long it will take to reach the same speed.
   (d) Show that the loaded lorry cannot climb this plank.   (SMP)

2 A mass of 3 kg is acted on by a constant force of $\begin{bmatrix} 6 \\ 3 \end{bmatrix}$ newton. Find the acceleration.

   At time $t = 0$ it is travelling with velocity $\begin{bmatrix} -3 \\ 2 \end{bmatrix}$ m s^{-1}. Find the displacement from
   $t = 0$ to $t = 4$ (seconds) and the velocity at the end of that time.   (SMP)

3 A mass of 10 kg is to be accelerated at 5×10^{-2} m s^{-2} in a direction 060° (N 60° E). If it is known that a constant force of 5 newton acts in a direction due south, determine the forces acting in the directions due east, and due north, which will produce the required acceleration.   (MEI)

4 A crane can lift a load of 1000 kg by means of a single vertical cable which can withstand a maximum tension of 10^4 newton. What is the greatest possible upward vertical acceleration that can be given to the load?

   To speed the process, two such cranes are used, but in this case the two cables lie each at 30° on either side of the vertical. If the tensions in the cables are now not to exceed 7500 newton for safety reasons, what is the maximum upward vertical acceleration that can be given to the load? (Give your answer correct to 2 significant figures.)   (MEI)

5 A ship of mass 500 tonne is towed by two tugs exerting forces of 4000 and 7000 newton. The hawsers are horizontal and the angle between them is 60°. Using the cosine rule, or otherwise, find the magnitude of the resulting force exerted by the hawsers.

   The resistance to motion is 300 newton. Show that the acceleration of the ship is about 0.013 m s^{-2}. Also, using the sine rule or otherwise, calculate the direction of motion of the ship, to the nearest degree, in relation to the direction of the hawser which carries a force of 4000 N.   (SMP)

6 A particle of mass 3.5 kg initially moving with velocity $\begin{bmatrix} 3 \\ 6 \end{bmatrix}$ m s^{-1} is acted on by a constant force of $\begin{bmatrix} 3.5 \\ -7 \end{bmatrix}$ newton.

(a) Write down the acceleration after 2 seconds.

(b) Show that the speed after 4 seconds is about 7.3 m s⁻¹.

(c) Calculate the displacement after 6 seconds.

(d) Determine when it is moving in the direction perpendicular to its initial velocity.   (SMP)

7   A skier of mass 70 kg is at the top of a run which is inclined at 20° to the horizontal. He is already moving at 4 m s⁻¹ in the direction straight down the slope. The resistances amount to a force of 95 N up the slope.

Show that his acceleration will be about 2 m s⁻². (Take $g = 9.8$.)

If the slope is 77 m long and he skis straight down it, calculate how long he will take to get to the bottom and how fast he will then be travelling.   (SMP)

8   A particle moves so that its velocity, in m s⁻¹ relative to horizontal and vertical axes after t seconds is $\begin{bmatrix} 20 \\ 35 - 10t \end{bmatrix}$.

(a) Calculate its displacement after t seconds.

(b) Show that its distance from its starting-point after 4 seconds is 100 metres.

(c) Calculate its direction of motion after 2 seconds, giving the angle to the horizontal to the nearest degree.

(d) Prove that the force on it is constant and state its direction.

(e) Plot a graph to show the horizontal and vertical displacements of the particle for the first 7 seconds.

(f) Indicate on your graph the results of parts (b) and (c).   (SMP)

9   A mass of 3 kg has position vector $\mathbf{r} = (3t^2 + 2t)\mathbf{i} - 4t^2\mathbf{j}$ where $\mathbf{r}$ is measured in metres and t is the time in seconds.

Find its velocity and acceleration vectors in terms of t.

Show that it is subject to a force of magnitude 30 N, and find the direction of the force.   (SMP)

10   A mass of $\frac{1}{2}$ kg at the origin is acted on by forces of magnitude 6 N, 7 N and 3 N

in the directions of $\begin{bmatrix} 1 \\ -2 \\ 2 \end{bmatrix}$, $\begin{bmatrix} 2 \\ 3 \\ 6 \end{bmatrix}$ and $\begin{bmatrix} 2 \\ 2 \\ 1 \end{bmatrix}$ respectively.

(a) Show that the resultant acceleration is $\begin{bmatrix} 12 \\ 2 \\ 22 \end{bmatrix}$ m s⁻².

(b) Two of the forces are at right-angles. Say which these are and justify your answer.   (SMP)

15

Interaction

1. ROPES AND RODS

We shall now consider problems which involve two bodies, either in contact or connected with rod or string.

Consider a light rope which is being pulled from each end (we say it is in *tension*). See Figure 1.

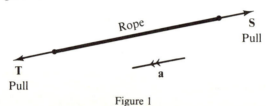

Figure 1

Applying *N2L* to the whole rope:

$$\mathbf{T} + \mathbf{S} = 0 \times \mathbf{a}$$

where **a** is the acceleration, since the mass of the rope is negligible.

Thus $$\mathbf{T} + \mathbf{S} = \mathbf{0},$$

that is, $$\mathbf{T} = {}^{-}\mathbf{S}.$$

We see that the pull on each end must be equal in magnitude but opposite in sense and this is independent of the acceleration of the rope; **a** does not have to be **0**.

A similar result may be obtained for a rod of negligible mass. In this case we can also argue in a similar way if the rod is pushed at each end (i.e. is in *compression*). Ropes cannot be put into compression.

We shall assume that strings (ropes) and rods are *inextensible*, that is they do not stretch when in tension (or contract under compression in the case of a rod).

Suppose that particles, or bodies, A and B are connected by a light rod and have common acceleration **a** caused by an applied force **P** parallel to the rod. Let the particles A and B have masses m_A, m_B respectively.

The effect on each particle due to the connection between them through the rod is completely described by the force transmitted by the rod; so long as this force is remembered and shown in diagrams the particles can be considered separately (Figure 2).

S and **T** are the forces transmitted by the rod to each particle. We know that $\mathbf{S} + \mathbf{T} = \mathbf{0}$.

278

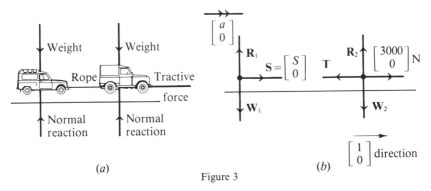

Figure 2

For body A, by N2L, $\mathbf{S} = m_A\,\mathbf{a}$ (1)

For body B, by N2L, $\mathbf{P} + \mathbf{T} = m_B\,\mathbf{a}$ (2)

Adding (1) and (2) gives

$$\mathbf{P} + (\mathbf{S} + \mathbf{T}) = (m_A + m_B)\,\mathbf{a}.$$

Thus $\mathbf{P} = (m_A + m_B)\,\mathbf{a}$ and the two bodies behave as if they were one body of mass $m_A + m_B$, acted on by $\mathbf{P}$. The pair of forces $\mathbf{S}$ and $\mathbf{T}$ due to the tension or compression of the connection between the bodies are said to be *internal* when the system is treated as a single particle and in that case may be omitted; when the system is treated as two separate particles these forces are no longer internal and must be included.

Example 1

A car of mass 1 tonne tows another car of mass 0.75 tonne with a light tow-rope. If the towing car exerts a tractive force of magnitude 300 N and resistances to motion may be neglected, find the acceleration of the two cars and the tension in the rope, when they are travelling on a horizontal road.

Figure 3

See Figure 3. Since $\mathbf{S}$ and $\mathbf{T}$ are the tension transmitted by the rope we have $\mathbf{S} + \mathbf{T} = \mathbf{0}$. Choosing the reference direction $\begin{bmatrix} 1 \\ 0 \end{bmatrix}$ to be in the direction of motion of the cars we may represent the acceleration as $\begin{bmatrix} a \\ 0 \end{bmatrix}$, $\mathbf{S}$ as $\begin{bmatrix} S \\ 0 \end{bmatrix}$ and $\mathbf{T}$ as $\begin{bmatrix} -S \\ 0 \end{bmatrix}$.

Writing out the components in the $\begin{bmatrix} 1 \\ 0 \end{bmatrix}$ direction we have:

(a) For both cars, by N2L,

$$3000 = (750 + 1000)\,a.$$

Thus
$$a = \frac{3000}{1750} = \frac{12}{7} \approx 1.71.$$

(*b*) For the car being towed, using *N2L*,
$$S = 750a$$
$$= 750 \times 1.71 \approx 1286.$$

Thus the acceleration is 1.7 m s^{-2} to 2 s.f. in the direction of motion and the tension in the coupling has magnitude 1300 N to 2 s.f.

Note that in (*b*) we could have applied *N2L* to the driving car giving $3000 - S = 1000a$ where $a \approx 1.71$. We usually choose the easier calculation.

You will notice that in Figure 3 we have not given $\mathbf{R}_1, \mathbf{R}_2, \mathbf{W}_1, \mathbf{W}_2$ in component form. Although the forces should be represented there is little point in specifying components which will not be used.

Example 2

A car of mass 1 tonne has a tractive force of magnitude 4000 N and is subject to resistances of magnitude 500 N. On a level road, the car pulls a caravan of mass 800 kg, also subject to resistance, with acceleration of magnitude 1.5 m s^{-2}. Find the resistance to motion of the caravan and the tension in the coupling.

See Figure 4. Since $\mathbf{S}$ and $\mathbf{T}$ are the forces transmitted by the coupling we have $\mathbf{S} + \mathbf{T} = \mathbf{0}$. Choosing the reference direction $\begin{bmatrix} 1 \\ 0 \end{bmatrix}$ to be in the direction of motion we may represent the acceleration as $\begin{bmatrix} 1.5 \\ 0 \end{bmatrix}$ m s^{-2}, the resistance to motion of the car as $\begin{bmatrix} -500 \\ 0 \end{bmatrix}$ N, the tractive force as $\begin{bmatrix} 4000 \\ 0 \end{bmatrix}$ N, the resistance of the caravan as $\begin{bmatrix} -D \\ 0 \end{bmatrix}$ N, $\mathbf{S}$ as $\begin{bmatrix} S \\ 0 \end{bmatrix}$ and $\mathbf{T}$ as $\begin{bmatrix} -S \\ 0 \end{bmatrix}$.

We shall use only the components in the $\begin{bmatrix} 1 \\ 0 \end{bmatrix}$ direction.

For car and caravan together, using *N2L*,
$$4000 - D - 500 = 1800 \times 1.5 \tag{1}$$
Thus
$$D = 3500 - 2700 = 800.$$
For the car only, using *N2L*,
$$4000 - 500 - S = 1000 \times 1.5.$$
Thus
$$S = 3500 - 1500 = 2000.$$

Hence the resistance of the caravan has magnitude 800 N and the tension in the coupling has magnitude 2000 N.

Note that, in (1) above, as in the previous example, we have not included the internal forces $\mathbf{S}$ and $\mathbf{T}$ when considering the combined system.

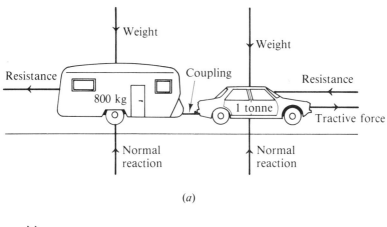

(a)

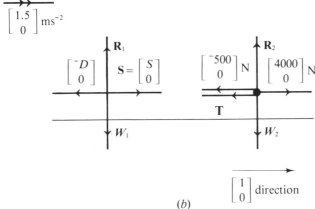

(b)

Figure 4

Exercise A

1   A car and caravan, each of mass 1 tonne, have acceleration of magnitude 3 m s⁻². Ignoring resistances to motion find the tractive force of the car and the tension in the coupling.

2   A car of mass 0.75 tonne exerts a tractive force of magnitude 3500 N. The car pulls a trailer of mass 0.5 tonne. The resistance to motion of the trailer is of magnitude 500 N. Ignoring resistances to motion on the car find the acceleration of the car and trailer and the tension in the coupling.

3   A locomotive of mass 20 tonne pulls a truck of mass 10 tonne with acceleration of magnitude 1.5 m s⁻². If the truck has resistance to motion of magnitude 5000 N what is the tension in the coupling? If the coupling were to break what would be the acceleration of the locomotive if its tractive force and resistance to motion are unchanged?

4   A lorry of mass 10 tonne pulls a trailer of mass 5 tonne with acceleration of magnitude 2 m s⁻² when exerting a tractive force of magnitude 40 000 N. If the trailer

has resistance to motion of magnitude 750 N what is the tension in the coupling? What is the resistance to motion of the lorry?

5   A car and caravan have acceleration of magnitude 4 m s⁻². Find the tension in the coupling if the caravan has mass 0.75 tonne and resistance to motion of magnitude 1000 N.

6   A model car and trailer have masses 1 kg and 0.75 kg respectively. Ignoring resistances to motion, find the tractive force of the car and the tension in the coupling if the acceleration is 2.5 m s⁻².

7   A 1 tonne car pulls a trailer of mass 0.5 tonne with acceleration of magnitude 1 m s⁻². If the resistance to motion of the trailer has magnitude 1000 N, find the tension in the coupling. If the tractive force of the car has magnitude 3000 N, find the resistance to motion of the car.

8   A man pulls a sledge of mass 55 kg by means of a rope along a horizontal path against constant resistance of magnitude 500 N, with acceleration of magnitude 0.5 m s⁻². If the rope is inclined at 30° to the horizontal find the magnitude of the tension in it.

2. NEWTON'S THIRD LAW

We have noted already that there are always normal contact forces between surfaces in contact, but so far we have only been interested in the reaction of the table on a book, for example. May we assume that the book pushes down on the table with an equal force? In other words, do we have a pair of forces, as we found to be the case with gravitational attraction and tension in a string? The answer to that question is 'Yes' and this is *Newton's third law* (*N3L*) which states:

Whenever object *A* exerts a force on object *B* then object *B* exerts a force of the same magnitude but opposite sense on object *A*.

It should be noted carefully that this law is true for *all* forces and remains true even if the objects are in motion.

This last point sometimes causes difficulty. Consider a man sinking into a swamp. He sinks not because the force of the swamp on his feet is less than the force of his feet on the swamp. These forces are equal in size. He sinks because the maximum value this contact force can take is less than his weight. Suppose the man's weight is 600 N but the maximum upward force the swamp can exert on his feet is 500 N. His feet exert a force of 500 N downward onto the swamp. The net force on the man is thus 100 N downwards and so he accelerates.

We do not necessarily put the 'other half' of each pair of forces on our diagrams. If they do not act on the object whose motion we are investigating then they are not relevant.

In Figure 5, which shows a book sliding on a table, we would not normally include the 'other half' of the normal contact force (R_2) or of the resistance (S_2). Notice also that we have still not shown the attraction of the book for the Earth nor the reaction of the table on the floor, the floor on the supporting joists, the joists on the walls, etc.

When two objects are in contact their motion may be determined in the same

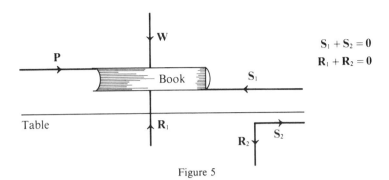

Figure 5

way as discussed in Section 1 when we considered two objects joined by a rod or rope. That is, they may be regarded as a single object with the contact forces *internal* (and usually omitted from the discussion), *or* as two separate objects, each considered to have a contact force acting upon it. The following examples introduce further extensions to our model of the real world, allowing us to analyse some fairly complex situations.

Example 3
A locomotive of mass 55 tonnes pulls three trucks, each of mass 15 tonnes. The resistance to motion of the locomotive has magnitude 3000 N and that of each truck is 1500 N. If the tractive force exerted by the locomotive has magnitude 30 000 N, find the acceleration of the train and the tension in each coupling.

See Figure 6. Choosing the reference direction $\begin{bmatrix} 1 \\ 0 \end{bmatrix}$ to be in the direction of motion of the train we may represent the acceleration as $\begin{bmatrix} a \\ 0 \end{bmatrix}$ and the tensions in the couplings as

$$\mathbf{T}_6 = \begin{bmatrix} T_6 \\ 0 \end{bmatrix}, \mathbf{T}_5 = \begin{bmatrix} T_5 \\ 0 \end{bmatrix}, \mathbf{T}_4 = \begin{bmatrix} T_4 \\ 0 \end{bmatrix}, \mathbf{T}_3 = \begin{bmatrix} T_3 \\ 0 \end{bmatrix}, \mathbf{T}_2 = \begin{bmatrix} T_2 \\ 0 \end{bmatrix}, \mathbf{T}_1 = \begin{bmatrix} T_1 \\ 0 \end{bmatrix}.$$

Considering the forces transmitted by each coupling, we have, as usual, $\mathbf{T}_1 + \mathbf{T}_2 = 0, \mathbf{T}_3 + \mathbf{T}_4 = 0, \mathbf{T}_5 + \mathbf{T}_6 = 0$. We shall consider only components in the $\begin{bmatrix} 1 \\ 0 \end{bmatrix}$ direction.

If we consider the whole train as a single system all the tensions in the couplings are internal forces and so the only forces of which we need take account are the tractive force of 30 000 N and the total resistance of 7500 N; they act on a mass of $55 + 3 \times 15 = 100$ tonnes.

By *N2L*, $$30\,000 - 7500 = 100\,000a.$$

Thus $$a = \frac{22\,500}{100\,000} = 0.225.$$

The acceleration is 0.23 m s^{-2} to 2 s.f. in the direction of motion of the train.
In order to find the tensions in the couplings we first notice that in the case

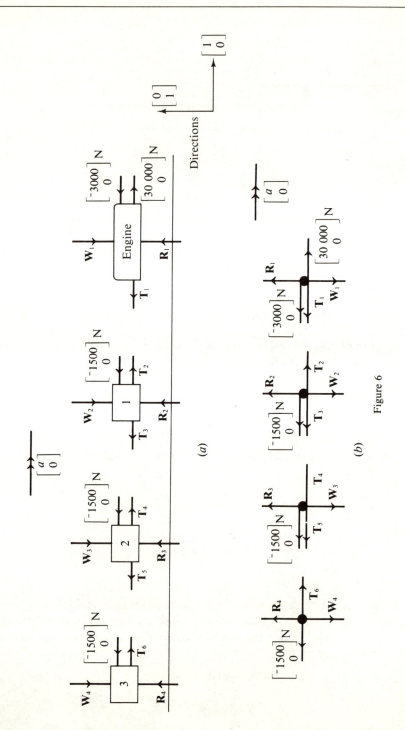

Figure 6

of the locomotive and also for the last truck (3) there is only one unknown force acting, whereas for the other two trucks there are two unknown forces. Consider the last truck (3). It has the acceleration of the whole train and so, by N2L,

$$T_6 - 1500 = 15000 \times 0.225.$$

Thus
$$T_6 = 1500 + 3375 = 4875,$$

and the tension in this coupling is of magnitude 4900 N to 2 s.f. Since $\mathbf{T}_5 + \mathbf{T}_6 = \mathbf{0}$ we have $\mathbf{T}_5 = \begin{bmatrix} -4875 \\ 0 \end{bmatrix}$ and now the only unknown force for the next truck in line (2) is $\mathbf{T}_4$.

By N2L applied to this truck (2),
$$T_4 - 4875 - 1500 = 15000 \times 0.225.$$

Thus
$$T_4 = 4875 + 1500 + 3375 = 9750,$$

and the tension in this coupling is of magnitude 9800 N to 2 s.f.

Note that, since the trucks all have the same mass and resistance to motion, we could have deduced that the tension in this coupling was twice that of the first one considered (i.e. $9750 = 2 \times 4875$).

To find the tension in the final coupling (between the locomotive and truck (1)) we can either

(i) argue that it is $3 \times 4875 = 14625$ N;

(ii) deduce that, since $\mathbf{T}_3 + \mathbf{T}_4 = \mathbf{0}$, Then $\mathbf{T}_3 = \begin{bmatrix} -9750 \\ 0 \end{bmatrix}$ N and apply N2L to the next truck (1);

(iii) apply N2L to the locomotive.

We shall apply N2L to the locomotive:
$$30000 - 3000 - T_1 = 55000 \times 0.225.$$

Thus
$$T_1 = 27000 - 12375 = 14625,$$

and the tension in this coupling is of magnitude 15000 N to 2 s.f.

Note the importance of planning the order of the calculations and the possibility of doing them in a different order.

Example 4
Masses of 10 kg and 5 kg hang by an inextensible string over a smooth pulley and are released from rest. What is the acceleration of the system and how long does it take for the masses to reach a speed of 10 m s^{-1}?

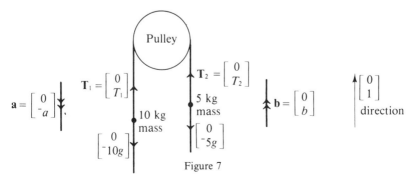

Figure 7

See Figure 7. This is the first time that we have considered the effect of a pulley. In our model of the situation we shall assume that a 'smooth pulley' requires no effort to turn it. Thus there is no change in the magnitude of the tension in the string as it passes over the pulley; the effect of the pulley is to change the *direction* of the string and hence of the tension.

In this example the masses move vertically and so we take the reference direction $\begin{bmatrix} 0 \\ 1 \end{bmatrix}$ vertically upwards. The forces acting on the masses due to the tension in the string may be represented as $\begin{bmatrix} 0 \\ T_1 \end{bmatrix}, \begin{bmatrix} 0 \\ T_2 \end{bmatrix}$. Since the pulley is smooth, the magnitude of the tension is constant and so $T_1 = T_2$.

Note that, since the two masses are moving in different directions it is convenient to assign to them different acceleration vectors $\begin{bmatrix} 0 \\ -a \end{bmatrix}, \begin{bmatrix} 0 \\ b \end{bmatrix}$; the signs are opposite to reflect the fact that as one particle rises the other falls. Since the string is inextensible $a = b$, i.e. one rises with exactly the same acceleration as the other falls.

The weights of the 10 kg and 5 kg masses are $\begin{bmatrix} 0 \\ -10g \end{bmatrix}, \begin{bmatrix} 0 \\ -5g \end{bmatrix}$ respectively (see Chapter 14). Writing down only components in the $\begin{bmatrix} 0 \\ 1 \end{bmatrix}$ direction:

applying *N2L* to the 10 kg mass
$$T_1 - 10g = {}^-10a;$$
applying *N2L* to the 5 kg mass
$$T_2 - 5g = 5b.$$
Now using the information that $T_1 = T_2$ and $a = b$ these equations may be rewritten:
$$T_1 - 10g = {}^-10a,$$
$$T_1 - 5g = 5a;$$
giving $15a = 5g$ and thus $a = \frac{1}{3}g$.

Since a is constant we may use $v = u + ta$, and so the time taken to reach a speed of 10 m s^{-1} is given by
$$10 = 0 + t \times \tfrac{1}{3}g \quad \Rightarrow \quad t = \frac{30}{g}.$$

The acceleration is 3.3 m s^{-1} to 2 s.f. The time taken to reach a speed of 10 m s^{-1} is 3.1 s to 2 s.f.

Note the significance of giving accelerations **a** and **b** opposite senses. Had **a** been represented as $\begin{bmatrix} 0 \\ a \end{bmatrix}$ then we should have had $a = {}^-b$ as the condition that the string was inextensible.

Example 5
A block of mass 5 kg slides on a horizontal table against a resistance of 15 N. It is connected by a horizontal, light inextensible string over a pulley to a mass

of 8 kg which hangs freely. Assuming that the pulley is smooth, and that the system is released from rest, what is the tension in the string and the acceleration of the system?

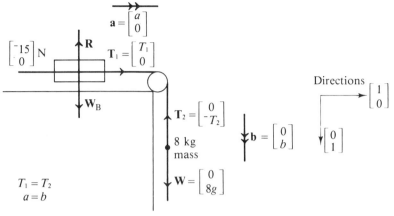

Figure 8

See Figure 8. In this problem the smooth pulley has the effect of turning the string through 90° whilst leaving the tension unchanged in magnitude. Again, since the string is inextensible the acceleration vectors also have the same magnitude. A sensible choice of reference directions simplifies the representation and calculations.

For the block, which is moving only in the $\begin{bmatrix} 1 \\ 0 \end{bmatrix}$ direction, we apply *N2L* in this direction, giving

$$T_1 - 15 = 5a.$$

For the hanging mass, which is moving only in the $\begin{bmatrix} 0 \\ 1 \end{bmatrix}$ direction, we apply *N2L* in this direction:

$$8g - T_2 = 8b.$$

Now $T_1 = T_2$ since the magnitude of the tension is the same at each end of the string, and $a = b$ since block and hanging mass have accelerations with the same magnitude. Thus the equations become

$$T_1 - 15 = 5a,$$
$$8g - T_1 = 8a,$$

giving $\qquad 8g - 15 = 13a \quad \Rightarrow \quad a = \dfrac{8g - 15}{13}$

$$= 4.88 \text{ to 3 s.f.} \quad \text{(taking } g \text{ as 9.8).}$$

Since $\qquad T_1 - 15 = 5a, \ T_1 = 5a + 15 = 39.4 \text{ to 3 s.f.}$

The acceleration is 4.9 m s⁻² to 2 s.f. and the tension in the string is 39 N to 2 s.f.

Exercise B

Unless instructed otherwise take g as 9.81.

1 Two particles with masses of 20 kg and 30 kg hang by means of a light inextensible string of length 4 m over a small smooth pulley. When released from rest the particles are at the same height. Find the acceleration of the system, the tension in the string and the time taken before the lighter particle reaches the pulley. Take g as 10.

2 A block of mass 10 kg resting on level ground is pulled by means of a light rope inclined at 60° to the horizontal. The tension in the rope has magnitude 50 N. What is the normal contact force on the block?

3 A toy train has a locomotive with 5 trucks each of mass 0.2 kg and is moving with acceleration 2 m s^{-2}. Ignoring resistances to motion find the tension in each coupling.

4 A block of mass 4 kg slides on a horizontal table with resistance of 10 N. It is connected by a horizontal, light, inextensible string passing over a smooth pulley, to a mass of 12 kg which hangs freely; the situation is similar to that in Example 5. Find the tension in the string when the system is released from rest and the time taken for the block to move 1 m.

5 A 70 kg workman, standing on the ground, tries to lower a 78 kg bucket of rubble from a height of 12 m by means of a smooth pulley, but is pulled off his feet! Find his acceleration and the speed with which he hits the bucket (halfway up). What is the speed of the bucket relative to the workman?

6 Two masses, 8 kg and 10 kg, are suspended by a light, inextensible string over a smooth pulley. Find the tension in the string and the acceleration of the system.

7 If, in question 6, the smaller mass slides up a wall, which has the effect of halving the acceleration, what is the sliding resistance and the new tension in the string?

8 A light, inextensible rope hangs over a smooth pulley. On one side of the pulley the rope is attached to a block resting on a slope inclined at 30° to the horizontal. The block is of mass 100 kg and the rope on this side is parallel to the plane. On the other side of the pulley the rope is attached to a mass of 75 kg which hangs freely. Find the normal contact force on the block and the minimum resistance to sliding if the block does not move. Take g as 10.

9 A 50-tonne locomotive has a 20-tonne tender, a 15-tonne truck and a 12-tonne brake-van (at the end). The locomotive, tender, truck and brake-van have resistances to motion of 3000 N, 1000 N, 1500 N and 2500 N respectively. If the tension in the coupling to the brake-van is 8000 N find the acceleration of the train and the tensions in the other couplings. What is the tractive force of the locomotive?

10 A block of mass 10 kg slides on a horizontal table with resistance of 15 N. It is connected by a light, inextensible string over a smooth pulley to a mass of 20 kg which hangs freely. Find the tension in the string and the acceleration when the system is released.

3. IMPULSE AND MOMENTUM

In Chapter 14 we learnt how to calculate the acceleration of an object subject to a force acting on it; we can then calculate the change in velocity over a given time. We now investigate the possibility of finding this change directly without calculating the acceleration.

Example 6

A force of $\begin{bmatrix} 2 \\ 1 \\ 3 \end{bmatrix}$ N acts on an object of mass 2 kg for 3 seconds. What is the change in the velocity of the mass?

By *N2L* the force causes acceleration **a** given by

$$\begin{bmatrix} 2 \\ 1 \\ 3 \end{bmatrix} = 2\mathbf{a} \quad \Rightarrow \quad \mathbf{a} = \begin{bmatrix} 1 \\ 0.5 \\ 1.5 \end{bmatrix}.$$

Since this acceleration is constant we may use the result $\mathbf{v} = \mathbf{u} + t\mathbf{a}$. The change in velocity in 3 s is

$$\mathbf{v} - \mathbf{u} = t\mathbf{a} = \begin{bmatrix} 1 \\ 0.5 \\ 1.5 \end{bmatrix} \times 3$$

$$= \begin{bmatrix} 3 \\ 1.5 \\ 4.5 \end{bmatrix} \text{ m s}^{-1}.$$

In general, suppose a constant force **F** acts on a mass m for a time t causing the velocity to change from **u** to **v**.

By *N2L* $\qquad\qquad\qquad \mathbf{F} = m\mathbf{a}.$

But also $\qquad\qquad\qquad \mathbf{v} - \mathbf{u} = t\mathbf{a}.$

Thus $\qquad\qquad\qquad \mathbf{F} = m(\mathbf{v} - \mathbf{u}) \times \dfrac{1}{t}$

or $\qquad\qquad\qquad t\mathbf{F} = m\mathbf{v} - m\mathbf{u}$ $\qquad\qquad$ (1)

We call the quantity $m\mathbf{v}$ the *momentum* of the object of mass m moving with velocity **v**.

The right-hand side of equation (1) is often written

final momentum – initial momentum.

or the change in momentum of the object. The terms initial and final refer to the times between which the force acted.

The left-hand side of equation (1) is called the *impulse* of the force acting for a time t. Thus equation (1) could be written

Impulse = final momentum – initial momentum. $\qquad$ (2)

The units of impulse and momentum are obviously the same and are newton seconds (N s) where the force is in newtons and the time in seconds. (It is interesting to note that the unit N s follows naturally from the definition of impulse as $t\mathbf{F}$. It does, however, seem strange as a unit for momentum which we might expect to have units kg m s^{-1} from $m\mathbf{v}$. Of course, kg m s^{-1} = kg m s$^{-2} \times$ s = N s.)

The following examples show how we can:

(i) calculate the momentum (and hence velocity) change of an object given the impulse of the force acting;

(ii) deduce the impulse of a force acting on an object from the momentum change it causes. If we also know the time for which the force acts we can deduce the constant force required.

Example 7

A force of $\begin{bmatrix} 3 \\ -2 \end{bmatrix}$ N acts for 10 s on an object of mass 4 kg which was originally

moving with velocity $\begin{bmatrix} -2 \\ 15 \end{bmatrix}$ m s^{-1}. What is its final velocity?

Let the final velocity be v m s^{-1}. Then by equation (1)

$$10\begin{bmatrix} 3 \\ -2 \end{bmatrix} = 4v - 4\begin{bmatrix} -2 \\ 15 \end{bmatrix}$$

$$\Rightarrow \quad 4v = \begin{bmatrix} 30 \\ -20 \end{bmatrix} + \begin{bmatrix} -8 \\ 60 \end{bmatrix} = \begin{bmatrix} 22 \\ 40 \end{bmatrix}.$$

Thus $v = \begin{bmatrix} 5.5 \\ 10 \end{bmatrix}$ and so the final velocity is $\begin{bmatrix} 5.5 \\ 10 \end{bmatrix}$ m s^{-1}.

Example 8

A constant force acts for 5 s on an object of mass 20 kg causing its velocity to change from $\begin{bmatrix} -1 \\ 4 \end{bmatrix}$ m s^{-1} to $\begin{bmatrix} 3 \\ 6 \end{bmatrix}$ m s^{-1}. Find the impulse of the force and the force acting.

The initial momentum is $20\begin{bmatrix} -1 \\ 4 \end{bmatrix} = \begin{bmatrix} -20 \\ 80 \end{bmatrix}$ N s; the final momentum is

$20\begin{bmatrix} 3 \\ 6 \end{bmatrix} = \begin{bmatrix} 60 \\ 120 \end{bmatrix}$ N s. So the impulse is $\begin{bmatrix} 60 \\ 120 \end{bmatrix} - \begin{bmatrix} -20 \\ 80 \end{bmatrix} = \begin{bmatrix} 80 \\ 40 \end{bmatrix}$ N s. Since the

force **F** which acts is constant, the impulse is t**F**. Thus $5\mathbf{F} = \begin{bmatrix} 80 \\ 40 \end{bmatrix}$ and $\mathbf{F} = \begin{bmatrix} 16 \\ 8 \end{bmatrix}$,

so the constant force is $\begin{bmatrix} 16 \\ 8 \end{bmatrix}$ N.

Exercise C

1   A force of $\begin{bmatrix} 2 \\ 7 \end{bmatrix}$ N acts on an object of mass 4 kg for 6 s. Find the momentum change and the velocity change of the object.

2   An object of mass 5 kg has its velocity changed from $\begin{bmatrix} 2 \\ 4 \end{bmatrix}$ m s^{-1} to $\begin{bmatrix} 15 \\ 6 \end{bmatrix}$ m s^{-1} by the action of a constant force over 10 s. What is the impulse of the force?

3   A force of $\begin{bmatrix} 3 \\ -4 \end{bmatrix}$ N acts for 2 s on an object of mass 1.5 kg with initial velocity of

$\begin{bmatrix} 2 \\ 7 \end{bmatrix}$ m s^{-1}. What is the final velocity?

4   A particle of mass 4 kg is given an impulse of 30 N s. What is the change in its velocity?

5   A car of mass 1 tonne has initial and final velocities of $\begin{bmatrix} 9 \\ 10 \end{bmatrix}$ m s^{-1} and $\begin{bmatrix} 12 \\ 8 \end{bmatrix}$ m s^{-1} respectively. What is the magnitude of the impulse of the force which produced the change?

6   An object of mass 2 kg has its velocity increased from $\begin{bmatrix} 2 \\ -1 \end{bmatrix}$ m s^{-1} to $\begin{bmatrix} 3 \\ 5 \end{bmatrix}$ m s^{-1} in 4 s. Find the impulse of the force acting and the magnitude of the constant force required.

7   A force of $\begin{bmatrix} 4 \\ -5 \end{bmatrix}$ N acts on an object of mass 3 kg for 5 s. What is the momentum change?

8   A force of $\begin{bmatrix} 1 \\ -1 \end{bmatrix}$ N acts on a particle of mass 0.5 kg. Find the velocity of the particle 6 s after it has velocity $\begin{bmatrix} 2 \\ 1 \end{bmatrix}$ m s^{-1}.

9   An object with momentum of $\begin{bmatrix} 7 \\ 6 \end{bmatrix}$ N s is struck by a bat giving an impulse of $\begin{bmatrix} -1 \\ 5 \end{bmatrix}$ N s. What is the new momentum of the object?

10   A particle of mass 10 kg is given an impulse of 50 N s. What is the change in velocity?

4. APPLICATIONS

We shall now apply the results concerning impulse and momentum to some practical situations.

Example 9
A man whose mass is 70 kg, initially at rest, is parachuting to the ground. If the average air resistance to the man and his parachute is 600 N, what is his velocity after 14 s? (Take g as 10.)

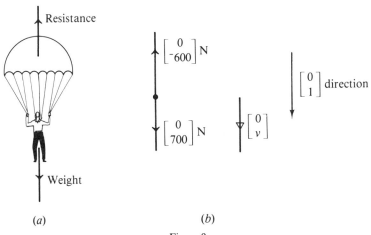

Figure 9

Since the motion is vertically downwards we shall choose this as our reference direction. The man has weight $\begin{bmatrix} 0 \\ 70g \end{bmatrix} = \begin{bmatrix} 0 \\ 700 \end{bmatrix}$ N and the resistance is $\begin{bmatrix} 0 \\ -600 \end{bmatrix}$ N.

Taking components in the $\begin{bmatrix} 0 \\ 1 \end{bmatrix}$ direction, the force acting is $700 - 600 = 100$ N.

Let the final velocity of the man be $\begin{bmatrix} 0 \\ v \end{bmatrix}$ m s^{-1}, then, using

$$\text{Impulse} = \text{change in momentum,}$$
$$14 \times 100 = 70v - 70 \times 0,$$
so $$v = 20.$$

His velocity after 14 s is 20 m s^{-1} downwards.

Note from this example that when more than one force is acting it is the *resultant* which is used to calculate the impulse. You may also notice that, as before, although **F**, **v**, **u** are vectors we may only need to write down the components in one direction.

Example 10

A car travelling at 15 m s^{-1} runs into a brick wall. What is the average force on a 60 kg passenger if:

   (*a*) she is brought to rest in 0.01 s;

   (*b*) she is brought to rest in 0.5 s because she is wearing a safety belt which takes time to stretch?

   (15 m s^{-1} is about 30 m.p.h.)

Suppose the car is moving in the $\begin{bmatrix} 1 \\ 0 \end{bmatrix}$ direction. The initial velocity of the passanger is $\begin{bmatrix} 15 \\ 0 \end{bmatrix}$; her final velocity is $\begin{bmatrix} 0 \\ 0 \end{bmatrix}$. Let the force acting be $\begin{bmatrix} P \\ 0 \end{bmatrix}$.

Taking components of the impulse–momentum equation in the $\begin{bmatrix} 1 \\ 0 \end{bmatrix}$ direction, we have

   (*a*) $$0.01P = 60 \times 0 - 60 \times 15$$
$$P = {}^{-}90\,000.$$

The force is 90 000 N in the opposite direction to the original velocity (the opposite direction being indicated by the negative sign). This is a huge force – enough to lift about 9 tonnes (say 12 cars).

   (*b*) $$0.5P = 60 \times 0 - 60 \times 15$$
$$P = {}^{-}1800$$

The force in this case is 1800 N in the opposite direction to the original velocity – the equivalent of being sat on by three people.

Example 11

In a game of rounders the ball, of mass 0.1 kg, travelling at 12 m s^{-1} immediately before impact, leaves the bat with speed 10 m s^{-1}. Find the impulse acting if the direction of the ball after the impact is (a) straight back, (b) at 60° to the initial direction.

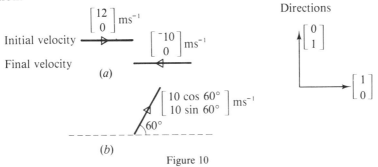

Figure 10

See Figure 10. Take the initial velocity as $\begin{bmatrix} 12 \\ 0 \end{bmatrix}$ m s^{-1} and the impulse as **I**.

(a) The final velocity is $\begin{bmatrix} -10 \\ 0 \end{bmatrix}$ m s^{-1}. (Note the negative sign.)

$$\mathbf{I} = 0.1 \times \begin{bmatrix} -10 \\ 0 \end{bmatrix} - 0.1 \times \begin{bmatrix} 12 \\ 0 \end{bmatrix} = \begin{bmatrix} -2.2 \\ 0 \end{bmatrix}.$$

Thus the impulse is $\begin{bmatrix} -2.2 \\ 0 \end{bmatrix}$ N s, i.e. of magnitude 2.2 N s in the direction opposite to that of the initial velocity.

(b) The final velocity is $\begin{bmatrix} 10\cos 60° \\ 10\sin 60° \end{bmatrix} = \begin{bmatrix} 5 \\ 8.66 \end{bmatrix}$ m s^{-1}.

$$\mathbf{I} = 0.1 \times \begin{bmatrix} 5 \\ 8.66 \end{bmatrix} - 0.1 \times \begin{bmatrix} 12 \\ 0 \end{bmatrix} = \begin{bmatrix} -0.7 \\ 0.87 \end{bmatrix}.$$

This impulse of $\begin{bmatrix} -0.7 \\ 0.87 \end{bmatrix}$ N s has magnitude 1.12 N s in a direction which makes an angle of 128.8° with the direction of the initial velocity.

You may notice that a larger impulse was required in (a) than in (b) because the ball had been deflected further from its original direction in (a). Clearly, an impulse of $\begin{bmatrix} -1.2 \\ 0 \end{bmatrix}$ N s would be required to stop the ball 'dead' and a further component of impulse in the $\begin{bmatrix} -1 \\ 0 \end{bmatrix}$ direction would be required if the ball is to rebound.

Note that it is vital that the velocity vectors be given the correct sense; clear indication of the positive vector component directions in a diagram is helpful, even when the choice of such directions is arbitrary, as in this example.

In Section 2 we saw that forces act in opposing pairs (N3L). That is, to each force there is a reaction of equal magnitude but opposite sense. Consider a force **F**; to it there will be a reaction $\mathbf{R} = {}^-\mathbf{F}$. If the force **F** acts for a time t causing an impulse $t\mathbf{F}$, the reaction **R** must act for time t causing a reactive impulse $t\mathbf{R} = {}^{-}t\mathbf{F}$. Thus in Example 11 the impulse given to the ball is associated with a reactive impulse given to the bat. In (a) the impulse on the bat is $\begin{bmatrix} 2.2 \\ 0 \end{bmatrix}$ N s

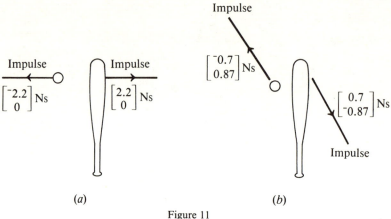

(a)   (b)

Figure 11

and in (b) the impulse on the bat is $\begin{bmatrix} 0.7 \\ -0.87 \end{bmatrix}$ N s. See Figure 11. Such a reactive impulse is eventually transmitted through the handle of the bat to the hand. You may be familiar with such reactions as a jarring sensation.

Exercise D

1   A 1-tonne car and a bicycle, mass 75 kg, are both travelling at 10 m s⁻¹ when their brakes are applied so as to produce, in each case, a resistance to motion of 150 N. How long will each take to stop?

2   A 7 kg shot is put with a force of 50 N. What velocity does this give to the shot if the force is applied for (a) 0.1 s, (b) 1 s?

3   A jet of water emerges at 10 m s⁻¹ from a nozzle of radius 2 cm. What mass of water emerges in 1 s if the density of water is taken to be 1000 kg m⁻³? If this jet is stopped by a wall what impulse is given to the water in 1 s? What impulse is given to the wall in 1 s? What is the force on the wall?

4   A bird of mass 0.2 kg is flying at 5 m s⁻¹. At a time 0.25 s later it is flying at 7 m s⁻¹ at 12° to its original direction of motion. What is its change in momentum? What is the magnitude and direction of the constant force required to effect this change?

5   What is the momentum of a mass of 0.4 kg, initially at rest, after a force of $\begin{bmatrix} 0.1 \\ 0.2 \end{bmatrix}$ N has acted on it for 5 s? What is its final speed?

6   A brick of mass 0.75 kg is dropped from the roof of a tower block. What is its velocity after 3 s? (Take g as 10.)

7   A diver of mass 60 kg hits the surface of the water vertically with speed 8 m s⁻¹. If her speed is halved on impact what is the impulse she receives from the surface? What impulse does she transmit to the surface?

8 A 800 kg car is brought to rest from 50 km h^{-1} in 2.5 s. What constant force must be produced by the brakes?

9 A skier, mass 75 kg, slides from rest down a slope inclined at 45° to the horizontal against a constant resistance of 250 N. How long does it take him to reach a speed of 6 m s^{-1}?

10 A spaceship is travelling at 200 m s^{-1} when its retro-rockets give it an impulse of 3×10^6 N s at 90° to the direction of motion. What is the resulting velocity if the mass of the spaceship is 50 000 kg?

11 A hockey ball travelling at 8 m s^{-1} is deflected through 35° by an impulse applied at 90° to its direction of motion. Find the resulting speed.

12 A beam of electrons is deflected through 20° by the deflector plate of a cathode ray tube, its speed being unaltered. Find the magnitude of the change in momentum as a percentage of the magnitude of the momentum before deflection.

5. CONSERVATION OF MOMENTUM

We have assumed that the force causing an impulse is constant. In real situations this is seldom the case and our model of constant force leads us to calculate the *average* force acting.

In some of the examples above we have calculated the impulse of a force acting over several seconds. Of course, a force may act for any length of time and it will have an associated impulse. However, the word 'impulse' suggests a blow or sudden happening and many of our examples do involve impulses in which a very large force acts for a very short period of time. Large forces do not necessarily lead to large impulses. For example, an impulse of magnitude 10 N s could be caused by:

a force of magnitude 10 N acting for 1 s;

a force of magnitude 100 N acting for 0.1 s;

a force of magnitude 1000 N acting for 0.01 s;

a force of magnitude 10^6 N acting for 10^{-5} s, etc.

Striking a ball with a bat or club involves huge forces of very short duration as slow motion films of the deformations caused clearly show. In cases like these it is usual to describe the blow in terms of the impulse given, not in terms of the force and its duration.

It follows from the impulse–momentum equation that if, over a certain period of time, no resultant force acts on an object, then

$$t0 = m\mathbf{v} - m\mathbf{u} \quad \text{or} \quad m\mathbf{v} = m\mathbf{u};$$

clearly the momentum has not changed.

Of course it is not true to say that because there is no *resultant* force then no forces act; consider a book at rest on a table on which both the weight and the normal contact force act, but with equal magnitude and opposite senses. We also know from *N3L* that there will be forces of equal magnitude but opposite sense between the particles making up the book.

We have mentioned these forces of interaction between parts of an object or parts of a system before; we referred to them as *internal* forces. They have zero resultant and so zero resultant impulse. It is only *external* forces (i.e. forces from outside an object or a system) which, if not balanced by other external forces, will give a resultant impulse to the object or system. This fact has useful consequences as it may be applied in a general way. We shall adopt the following in our model:

Any object or collection of objects on which no external forces act has constant total momentum.

This is called the *principle of conservation of (linear) momentum* and may be applied in any component direction.

As we shall see in the following examples, we may apply this principle to the firing of guns and to the collision of objects since in neither case does an *external* force act. You will notice that when we have several particles or objects forming a system, any forces between them are regarded as *internal*; also we do not claim that each one has constant momentum, only that the *total* momentum is constant in the absence of a resultant external force.

Example 12

Two particles, one of mass 2 kg and moving with velocity $\begin{bmatrix} 3 \\ 2 \end{bmatrix}$ m s^{-1} and the other of mass 3 kg with velocity $\begin{bmatrix} -3 \\ 1 \end{bmatrix}$ m s^{-1}, collide and coalesce (that is, stick together). What is the velocity of the combined particle?

We consider the system as consisting of both particles, so no external forces act. Let the velocity of the combined particle after the impact be **v**. The total momentum of the system before impact is

$$2\begin{bmatrix} 3 \\ 2 \end{bmatrix} + 3\begin{bmatrix} -3 \\ 1 \end{bmatrix} = \begin{bmatrix} -3 \\ 7 \end{bmatrix} \text{ N s.}$$

The total momentum of the system after impact is

$$2\mathbf{v} + 3\mathbf{v} = 5\mathbf{v}.$$

Applying the principle of conservation of momentum

$$5\mathbf{v} = \begin{bmatrix} -3 \\ 7 \end{bmatrix} \quad \Rightarrow \quad \mathbf{v} = \begin{bmatrix} -0.6 \\ 1.4 \end{bmatrix},$$

and so the combined velocity is $\begin{bmatrix} -0.6 \\ 1.4 \end{bmatrix}$ m s^{-1}.

Notice that if we had considered the system as containing only the particle of mass 2 kg then the collision would have involved an external force and the momentum of that particle would not have been conserved. This makes sense – the external force is that which provides the impulse in the collision. See Figure 12.

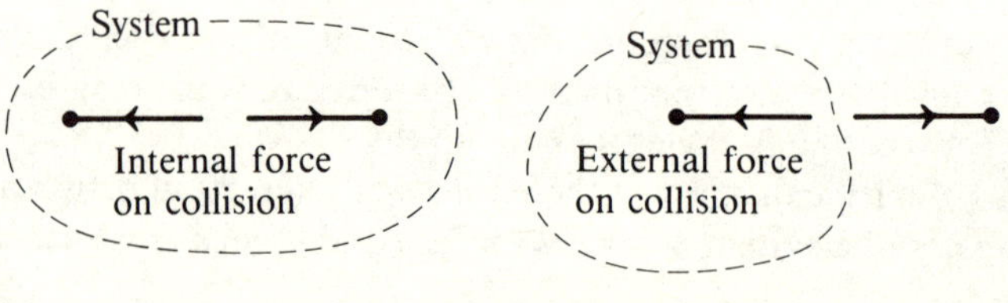

Internal force on collision   External force on collision

Figure 12

We could find the impulse due to the collision as follows. The momentum of the 2 kg particle after the collision is $2\begin{bmatrix} -0.6 \\ 1.4 \end{bmatrix} = \begin{bmatrix} -1.2 \\ 2.8 \end{bmatrix}$ N s, and before collision was $2\begin{bmatrix} 3 \\ 2 \end{bmatrix} = \begin{bmatrix} 6 \\ 4 \end{bmatrix}$ N s.

Let the impulse be $\mathbf{I}$ N s, then

$$\mathbf{I} = \begin{bmatrix} -1.2 \\ 2.8 \end{bmatrix} - \begin{bmatrix} 6 \\ 4 \end{bmatrix} = \begin{bmatrix} -7.2 \\ -1.2 \end{bmatrix}.$$

The impulse on the 3 kg particle is, of course, $\begin{bmatrix} 7.2 \\ 1.2 \end{bmatrix}$ N s.

Example 13
A field gun of mass 16 tonnes fires a shell of mass 20 kg horizontally with speed 500 m s^{-1} relative to the ground. What is the initial speed of recoil of the gun and what constant force must act to bring the gun to rest in 2 s?

Figure 13

See Figure 13. Consider the system containing both gun and shell. At the moment of firing, no external force acts. The explosion is internal, the force on the shell being balanced by the reaction on the gun, and so the total momentum is conserved.

Let the initial speed of recoil be $\mathbf{v}$. The total momentum before the shell is fired is $\mathbf{0}$. After firing the momentum of the shell is $20\begin{bmatrix} 500 \\ 0 \end{bmatrix}$ N s $= \begin{bmatrix} 10000 \\ 0 \end{bmatrix}$ N s and of the gun $16000\mathbf{v}$.

Applying the principle of conservation of momentum

$$\mathbf{0} = 16000\mathbf{v} + \begin{bmatrix} 10000 \\ 0 \end{bmatrix},$$

thus $\qquad 16000\mathbf{v} = -\begin{bmatrix} 10000 \\ 0 \end{bmatrix}$ and $\mathbf{v} = \begin{bmatrix} -0.625 \\ 0 \end{bmatrix}$.

Hence the velocity of recoil is 0.625 m s^{-1} in the opposite direction to the motion of the shell (since the component is negative).

The momentum of the gun is $\begin{bmatrix} -10000 \\ 0 \end{bmatrix}$ N s. To bring it to rest in 2 s we require a force $\mathbf{F}$ so that, using the impulse–momentum equation,

$$\mathbf{F} \times 2 = \mathbf{0} - \begin{bmatrix} -10000 \\ 0 \end{bmatrix}, \quad \text{and} \quad \mathbf{F} = \begin{bmatrix} 5000 \\ 0 \end{bmatrix},$$

i.e. a force of 5000 N in the direction of the shell's motion.

Problems involving particles ejected from rockets may be dealt with in a similar fashion. Rocket propulsion depends on a small mass of gas being ejected at high speed in order to give the rocket a small extra velocity in the opposite direction. The increase in momentum of the rocket has the same magnitude as the change in momentum of the gases ejected.

You may find the following result useful. If no external forces act on a system of two particles, masses m_1 and m_2, whose initial velocities are $\mathbf{u}_1$ and $\mathbf{u}_2$ and whose final velocities (after a collision) are $\mathbf{v}_1$ and $\mathbf{v}_2$, then the principle of conservation of momentum gives

$$m_1\mathbf{v}_1 + m_2\mathbf{v}_2 = m_1\mathbf{u}_1 + m_2\mathbf{u}_2.$$

Exercise E

1   A railway truck of mass 2 tonnes travelling at 4 m s^{-1} collides with a stationary truck of mass 1.75 tonnes. If the two trucks move on together what is their speed immediately after the collision?

2   A 10 g bullet is fired at 300 m s^{-1} from a 3 kg rifle. Find the velocity of recoil of the rifle.

3   A 1 tonne car travelling at 50 km h^{-1} runs into a stationary 200 kg cow. What is their common velocity after the collision?

4   A drop of water of mass 2 g moving at 0.1 m s^{-1} runs into a stationary drop of mass 1.5 g. With what speed does the combined drop move off?

5   A girl of mass 40 kg is sitting on a stationary sledge of mass 10 kg which is resting on horizontal ice. She throws a snowball of mass 0.5 kg horizontally with speed 12 m s^{-1}. With what speed does the sledge move?

6   The empty fuel tanks of a rocket are jettisoned at a relative velocity of 20 m s^{-1}. If they account for three-quarters of the rocket's mass, what additional velocity does this give to the remainder?

7   A particle of mass 2 kg and velocity $\begin{bmatrix} 3 \\ 4 \end{bmatrix}$ m s^{-1} collides with a second particle of mass 3 kg and velocity $\begin{bmatrix} 1 \\ -1 \end{bmatrix}$ m s^{-1} and the two particles coalesce and move on together. Find their common velocity. This combined particle then meets a third particle of mass 5 kg, coalesces with it and moves off with velocity $\begin{bmatrix} 3 \\ 0 \end{bmatrix}$ m s^{-1}. Find the velocity of the third particle before impact.

8   Two buckets, each of mass 4 kg, hang by a light inextensible string over a smooth pulley. A hammer of mass 2 kg is dropped from a height of 3 m into one bucket. With what velocity does the system begin to move? (Take g as 10.)

9   A gun of mass 1000 kg is mounted on a smooth horizontal surface and has a horizontal barrel. It can fire a shell of mass 40 kg at a speed of 300 m s^{-1}. What is the speed of recoil of the gun and what force must act to bring it to rest in 2.5 s?

10   Two ice skaters of masses 44 kg and 48 kg have speeds of 6 m s^{-1} and 8 m s^{-1} respectively at an angle of 45° to each other before they collide. On impact they cling to each other. Find their velocity and speed immediately after the collision.

11   A 60 kg man dives, at 4 m s⁻¹, from the back of a stationary boat of mass 300 kg. Find the horizontal velocity with which the boat moves off if he dives (*a*) horizontally, (*b*) upwards at 20° to the horizontal.

12   A stationary nucleus disintegrates, a third of it flying off at 6×10^4 m s⁻¹. What will be the speed of the other portion?

13   A 6 g lump of putty moving at 3 m s⁻¹ coalesces with a 4.5 g lump moving at 5 m s⁻¹. Find their common velocity immediately afterwards if they collide (*a*) 'head-on', (*b*) at right-angles.

14   A 90 g ball, travelling at 4 m s⁻¹, is hit by a 700 g block travelling with a velocity of 6 m s⁻¹, and this brings the block to rest. How fast will the ball then travel if it was hit: (*a*) straight back in the direction it came from, (*b*) at 75° to the direction it came from?

15   A 60-kilogram load of coal falls into a small wagon of mass 40 kg moving with speed of 5 m s⁻¹. What is the new speed of the wagon?

16   By considering components of total momentum prove that: (*a*) When an object disintegrates into two portions they will move off in the same line. (*b*) When an object disintegrates into three portions they will move off in the same plane.

6. SUMMARY

Newton's third law: Whenever an object *A* exerts a force on an object *B* then object *B* exerts a force of the same magnitude but of opposite sense on object *A*.

When two objects are in contact (or joined by a rope or rod of negligible mass) their motion may be investigated either by regarding them as a single object, with the contact forces (or tensions or thrusts in the ropes or rods) considered as internal, or as two separate objects. In this latter case, the contact forces, etc., must be taken into consideration.

A constant force **F** acting for a time *t* produces an impulse *t***F**.

The momentum of an object moving with velocity **v** is *m***v**.

The impulse exerted on an object is equal to the change in momentum of the object.

Any object or collection of objects on which no external forces act has constant total momentum.

Miscellaneous exercise

1   A batsman drives a ball of mass 0.1 kg which is coming to him at 12 m s⁻¹ straight back to the bowler at 6 m s⁻¹. Find the impulse given to the ball by the bat and compare it with that needed to deflect the ball through 60° without change of speed. Ignore vertical motion.   (SMP)

2   An object of mass 10 kg is at the point with coordinates (5, 8), taking metres as units. Its velocity is $\begin{bmatrix} 2 \\ 6 \end{bmatrix}$ m s⁻¹. Two seconds later it is at (11, 16) and has velocity $\begin{bmatrix} 4 \\ 2 \end{bmatrix}$ m s⁻¹.

Find the change in momentum.

If this is the result of a constant force acting during two seconds, find the force. State your units carefully throughout. (SMP)

3   A ball of mass 0.4 kg travels at speed 13 m s⁻¹ towards a boy, who kicks it straight back with an impulse of 12 newton seconds. Find the speed at which it leaves him.

   Calculate and show clearly on a sketch the direction in which he should have kicked it (with the same magnitude of impulse) in order to deflect it through a right-angle. (SMP)

4   A one-tonne (1000 kg) car tows a half-tonne caravan along a straight level road. The resistance to motion may be taken to be 400 newtons per tonne for the car and another 400 newtons per tonne for the caravan. What is the tension in the coupling:

   (a) when accelerating at 2 m s⁻²;

   (b) at a steady speed of 60 m s⁻¹? (SMP)

5   A mass of 2 kg moving with velocity $\begin{bmatrix} 3 \\ 5 \end{bmatrix}$ m s⁻¹ collides and combines with a mass

of 3 kg moving with velocity $\begin{bmatrix} 1 \\ -2 \end{bmatrix}$ m s⁻¹. Calculate the velocity of the single mass so

formed. Find also the impulse which each mass has received. (SMP)

6   A breakdown lorry is to tow a car along a horizontal road by means of an inextensible horizontal bar. The mass of the car is 1000 kg and that of the lorry and its driver is 2000 kg. Assuming that the resistances to motion can be taken as equivalent to 500 newtons acting horizontally on the lorry, and 100 newtons acting horizontally on the car, and that the horizontal tractive force exerted by the lorry engine is 2100 newtons, calculate the acceleration of the two vehicles. Find also the tension in the bar during the period of acceleration. (MEI)

7   A car of mass 1 tonne, travelling at 100 km h⁻¹, strikes at right-angles a lorry of mass 3 tonnes, travelling at 60 km h⁻¹ and becomes embedded in the lorry. What is their common velocity immediately after impact? (SMP)

8   A mass of 10 kg moves along a straight line in such a way that its velocity in metres per second, t seconds after passing through a point A, is given by the equation $v = 64 - t^3$. If the mass subsequently comes to instantaneous rest at a point B, determine the distance AB.

   State the force (in newtons) which is acting on the mass at the instant when it is at B, and which way this force is acting. (MEI)

9   A force **P**, magnitude 5 N, direction 090°, acts on a mass of 2 kg initially at rest. What is the momentum of the mass after 6 seconds? Hence determine the velocity of the mass at that instant.

   An additional force **Q**, magnitude 30 N, direction 180°, acts on the mass during the next 4 seconds. What is the magnitude of the velocity of the mass at the end of the 10-second period? (MEI)

10 A mass of 3 kg moving with velocity $\begin{bmatrix} 2 \\ 3 \\ -2 \end{bmatrix}$ m s⁻¹ collides with a mass of 5 kg moving

with velocity $\begin{bmatrix} -1 \\ 2 \\ 6 \end{bmatrix}$ m s⁻¹. After the collision they each have the same velocity.

Find this velocity and also the impulse which each receives as a result of the collision. (SMP)

Revision exercise 4

1 A crane is lifting two concrete blocks, each of mass 800 kg, joined together by a cable which will break if the tension, T_1, reaches 1.04×10^4 N. (See Figure 1.)
    (i) Show that the acceleration must be less than 3.2 m s^{-2}.
    (ii) If the acceleration is half this value, calculate:
        (a) T_1, the tension in the cable connecting the two blocks;
        (b) T_2, the force applied by the crane.
Make your method clear.
(SMP)

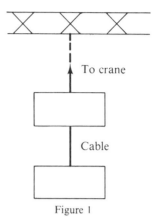

Figure 1

2 A Fizzler, of mass 2 kg and velocity $\begin{bmatrix} 3 \\ 2 \\ 6 \end{bmatrix}$ m s^{-1}, collides with a Slurp, of mass 8 kg and velocity $\begin{bmatrix} 1 \\ 2 \\ 3 \end{bmatrix}$ m s^{-1}, and has its velocity reduced to $\begin{bmatrix} 2 \\ 2 \\ 1 \end{bmatrix}$ m s^{-1}. Calculate:

    (a) the final velocity of the Slurp;
    (b) the impulse on the Fizzler;
    (c) the angle of deflection of the Fizzler.
(SMP)

3 A mass of 5 kg is observed moving instantaneously on a bearing of 180° with a velocity of 6 m s^{-1}. After a certain interval of time the mass is observed moving instantaneously on a bearing of 270° with a velocity of 8 m s^{-1}. What is the impulse that has been given to the mass?

   If an equal impulse is applied during a further period of time equal to the first interval, find the magnitude of the final velocity of the mass.
(MEI)

4 A motorboat tows a water-skier by means of a horizontal, inextensible rope. When the boat and the skier are moving with a constant velocity the engine of the boat is producing a forward force of 60 N. The mass of the motorboat and its occupant is 500 kg, and that of the skier 80 kg. What is the total resistance to motion on the boat and the skier?

   If these resistances are assumed to remain constant, and are divided between the boat and the skier in the ratio 5:1 respectively, calculate the acceleration of the boat and the skier and the tension in the rope when the forward force provided by the engine of the boat is increased to 1800 N.

   If the rope suddenly snaps, by how much will the acceleration of the boat increase?
(MEI)

5 A charged particle, of mass 5×10^{-6} kg, is moving across a rectangular grid of centimetre squares under a constant force of $\begin{bmatrix} 10^{-5} \\ -10^{-5} \end{bmatrix}$ newtons maintained by deflector plates. It is observed at the point (3, 3) to be travelling with velocity $\begin{bmatrix} -1 \\ 2 \end{bmatrix}$ m s^{-1}. What is its velocity t seconds later?

Show that it passes through the point (9, 0) and state the values of t for which this occurs.   (SMP)

6 A boat of mass 5 tonnes is being driven before the wind at a constant speed of 1 m s^{-1}. It is dragging a sea anchor, and the tension in the cable by which it is attached is 6000 N in the direction horizontally backwards. If the cable suddenly breaks, show that the boat starts to accelerate at 1.2 m s^{-2}.

If the acceleration now decreases at a constant rate and is reduced to zero in 3 seconds, sketch the time–acceleration graph.

Find the speed at which the boat will be travelling at the end of the 3 seconds.   (SMP)

7 An engine pulls a 50-tonne train with a force of 8 kilonewtons. Assuming that the resistance to motion is negligible,
   (a) calculate its acceleration on the level;
   (b) show that a gradient of $\frac{1}{200}$ will reduce its acceleration by about 30%. (Take $g = 10$.)
   (c) The train consists of a 30-tonne truck with a 20-tonne truck behind. Calculate the force transmitted by the coupling between the two trucks in the situation given in part (a) and also in part (b).   (SMP)

8   (i) A ball of mass 0.2 kg travelling at 15 m s^{-1} is hit so as to give it a speed of 10 m s^{-1} in the opposite direction. Calculate the magnitude and direction of the impulse received.
   (ii) If the ball had been deflected through 90° by an impulse of the same magnitude, show that its resultant speed would have been 20 m s^{-1} and state the direction of the impulse.
   (iii) If the deflection is only to be 60°, find by how much the magnitude of the impulse may be reduced, yet still give the same change in speed as for part (ii).   (SMP)

9 A shunting engine of mass 20 tonnes pushes two trucks, each of mass 2 tonnes, along a horizontal track. The resistance to the engine is 300 N and to each truck 150 N. If the tractive force produced by the engine is 15 kN, calculate the acceleration of the engine and the trucks. Calculate also the thrust between the buffers of the engine and the first truck.

Subsequently, the two trucks on their own run down a slope whose angle of inclination to the horizontal is α, where $\sin \alpha = 0.1$. What must the resistance now be to each truck if they are to descend at a constant speed?   (MEI)

10 The spaceship *Apleno* of mass 3 with velocity $\begin{bmatrix} 4 \\ 5 \end{bmatrix}$ docks with the spaceship *Sayes* of mass 2 and velocity $\begin{bmatrix} 3 \\ 4 \end{bmatrix}$.

   (a) Show that the common velocity afterwards is $\begin{bmatrix} 3.6 \\ 4.6 \end{bmatrix}$.

   (b) Calculate the impulse of *Apleno* on *Sayes*.

(c) Calculate whether *Sayes* has been deflected in the positive or negative sense.
(SMP)

11 A toboggan of mass 15 kg is placed on an icy slope which is inclined to the horizontal at an angle of 18°. It is held in position by applying a horizontal force **R**. (Friction and other resistances may be neglected.)
   (a) Draw a diagram to show the forces acting on the toboggan.
   (b) Show by calculation that the magnitude of **R**, correct to 2 significant figures, is 48 N. (Take $g = 9.8$.)
   (c) Find the acceleration of the toboggan if the force **R** is now removed. (SMP)

12 A worker of mass 70 kg, standing on the ground, lowers a light bucket containing 20 bricks each of mass 4 kg from a height of 24 metres, by means of a rope over a light pulley.
   (a) Show that the worker is carried off the ground and accelerates upwards at about $\frac{2}{3}$ m s^{-2},
   (b) Calculate the tension in the rope.
   (c) After how long with the bucket hit the ground?   (SMP)

13 A mass of 80 kg hanging on the end of a string, is held by a horizontal force **F** so that the string is inclined at 70° to the vertical.
   (i) Calculate **F** and the tension, **T**, in the string.
   (ii) If the force **F** is suddenly removed, in which direction will the mass accelerate immediately? Draw a new diagram of forces, and calculate the new tension in the string and the acceleration of the mass.   (SMP)

14 Two particles of masses 2 kg and 3 kg, with speeds of 6 m s^{-1} and 8 m s^{-1} respectively are travelling directly towards each other. On collision, the lighter of the two gets deflected through 90° and the heavier through 45°. Draw a diagram showing all this information.
   (a) Prove that the speed of the heavier in the direction in which it was originally travelling is halved.
   (b) State the speed of the heavier afterwards perpendicular to its original direction.
   (c) Hence show that the speed of the lighter is unchanged.
   (d) Calculate the magnitude and direction of the impulse on the lighter.
   (e) State the final speed of the heavier particle.   (SMP)

15 A toy truck of mass 2 kg is pulled along the level floor by a string inclined at 30° to the horizontal. When it just moves, explain why the tension in the string is not more than 40 N. Hence prove that, if it does move, the resistance to motion must certainly be less than 35 N.   (SMP)

16 A ball of mass 0.4 kg travelling at 20 m s^{-1} is struck with a impulse of 6 Ns. Calculate the speed with which it continues:
   (a) if it is struck in the direction opposite to its velocity;
   (b) if it is struck in a direction at right-angles to its velocity;
   (c) if it is struck so as to deflect it as much as possible.   (SMP)

17 A buoy of mass 3 kg is held 5 m below the surface of the water by a vertical cable. There is an upward buoyancy force of 42 N acting on the buoy. Write down the tension in the cable (take $g = 10$).
   Suddenly the cable breaks. Show that while in the water the buoy will accelerate at 4 m s^{-2}.
   The buoy maintains this constant acceleration while it is in the water. Calculate

the time taken to reach the surface and show that the velocity on emergence will be 6.3 m s⁻¹ approx.

Ignoring air resistance, calculate the height above the surface which the buoy will reach.   (SMP)

18 A helicopter with mass 0.6 tonne experiences two constant forces in addition to its weight. These are $(2000\mathbf{i} + 5000\mathbf{j})$ N and $(^{-}1000\mathbf{i} + 2000\mathbf{j})$ N, where $\mathbf{i}$ and $\mathbf{j}$ are unit vectors horizontally and vertically. What extra force is required for it to hover (i.e. to remain stationary)?

If a 'thrust' of 2000 N is applied, find, by drawing or calculation, the direction in which it should be applied for level flight. Give your answer as an angle from the horizontal.   (SMP)

19 An object of mass 2 kg is on smooth level ground. A force of 8 N applied at 30° to the vertical causes it to accelerate at 1.5 m s⁻². Show that the resistance to motion is 1 N.

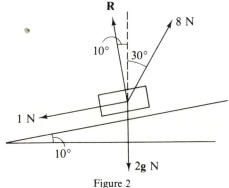

Figure 2

If instead the ground slopes at 10° to the horizontal, so that the forces are as shown in Figure 2, verify that the force, R, of interaction between the object and the ground is about 13.6 N. Also calculate the acceleration of the object up the slope. (Take $g = 10$.)   (SMP)

20 A mass of 2 kg is connected by a light string to a mass of 3 kg and projected with speed 5 m s⁻¹ due north across a smooth horizontal floor. After a short while the string is observed to be taut and aligned north–south.

Calculate their common component of velocity due north.

The heavier mass is then travelling on a bearing of 060°. Prove that its speed is 4 m s⁻¹.

Hence calculate the speed of each mass eastwards.

Verify that the lighter mass will then be moving on a bearing of about 291°.

16

Plane transformations

1. ISOMETRIES

An isometry is a transformation which does not alter lengths. Isometries are either direct or opposite. In the plane there are just four possible isometries – translation, rotation, reflection and glide – as the following considerations of the possibilities shows.

(a) Direct isometries

If AB is mapped to $A'B'$ by a direct isometry, then

either AB is parallel to $A'B'$, in which case $AA'B'B$ is a parallelogram, and the isometry is the translation given by the vector $\underset{\sim}{AA'}$ (see Figure 1),

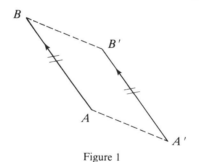

Figure 1

or AB is not parallel to $A'B'$. In this latter case, there are two possibilities:

(i) AA' is parallel to BB', so that $AA'B'B$ is an isosceles trapezium and the isometry is a rotation about X, the point of intersection of AB and $A'B'$ (see Figure 2);

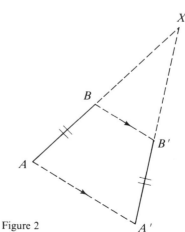

Figure 2

305

(ii) AA' is not parallel to BB', so that the perpendicular bisectors of AA' and BB' meet at a point X, and this is the centre of the rotation (see Figure 3).

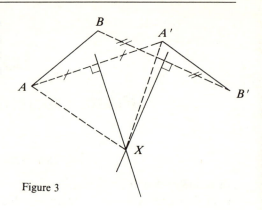

Figure 3

(b) Indirect isometries

If AB is mapped to $A'B'$ by an indirect isometry, then let L and M be the midpoints of AA' and BB' respectively and l be the line LM. If L and M coincide, take l as the line through L perpendicular to AB. The isometry is a glide with axis l, which can be thought of as reflection in the line l (to obtain $A''B''$) followed by the translation $\underset{\sim}{A''A'}$ parallel to l. If it so happens that A'' and A' coincide, then the isometry is a simple reflection. The vector $\underset{\sim}{A''A'}$ is the *throw* of the glide; a reflection may be considered to be a glide with zero throw. (See Figure 4.)

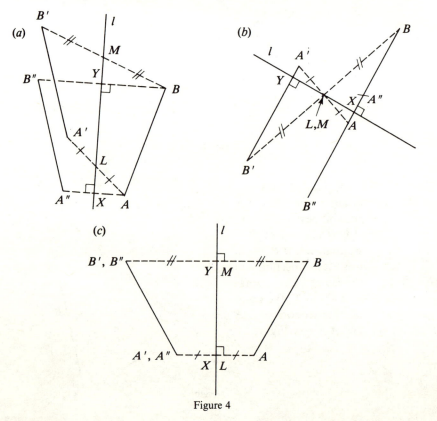

Figure 4

Invariance

All the isometries of a plane can be conveniently shown in the following table:

Isometries (preserving lengths)	Direct	Opposite
With a fixed point	Rotation	Reflection
Without a fixed point	Translation	Glide

The other geometric transformations do not preserve lengths but some have valuable properties. You will be familiar with the useful ones:

a *shear* preserves area and has an invariant line;

a *stretch* preserves ratio and has an invariant line;

an *enlargement* preserves angle and has an invariant point.

Exercise A

1 *ABCDE* is a regular pentagon. What is (i) the direct isometry, (ii) the opposite isometry which maps *AB* onto:
(a) *AE*; (b) *CD*; (c) *DC*?
State the centres and angles for rotations and axes for reflections.

2 *A'B'* is the image of *AB* under rotation centre *X*. Show that *XY* bisects the angle *AYB'*, where *Y* is the intersection of *AB* and *A'B'*.

3 *A* and *B* are equidistant from *O*. *AB* is enlarged with centre *O* and rotated about *O* to *XY*. Show that *AX* = *BY* and state an angle equal to the angle that *AX* makes with *BY*.

4 If *ABC* is rotated about a point *X* to *A'B'C'*, say whether the following must be true or may be false, giving reasons:
(a) *CX* = *C'X*; (b) angle *ABC* = angle *A'B'C'*;
(c) *BC'* = *B'C*; (d) angle between *BC* and *B'C'* = angle *AXA'*.

5 In Figure 4(a) say whether the following must be true or may be false, giving reasons:
(a) *BB''* = *BB'*; (b) *B''B'* = 2*XL*; (c) *XY* = *LM*;
(d) if a point C is mapped to *C'*, then the midpoint of *CC'* lies on *LM*.

6 A glide sends *ABC* to *A'B'C'*. The axis of the glide meets *AA'* at *R* and *BB'* at *S*. A half-turn about *R* followed by a reflection will send *ABC* to *A'B'C'*. Draw the axis for this reflection. Similarly a half-turn about *S* followed by a reflection will also send *ABC* to *A'B'C'*. Draw the axis for this reflection. Comment, with reasons if possible, on the three axes you have drawn.

7 Given a triangle *ABC* it is required to construct a square *P'Q'R'S'* touching the triangle with *P'S'* along *BC*, *Q'* on *AB* and *R'* on AC. Show that an enlargement from *B* of *PQRS* will give the required square and make the necessary construction. (See Figure 5.)

Figure 5

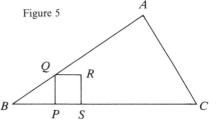

8 *PQRSTU* is a regular hexagon. What is (i) the direct isometry, (ii) the opposite isometry which maps *PQ* onto:
   (a) *PU*;   (b) *RS*;   (c) *SR*?

9 *ABC* is any triangle. Equilateral triangles *BXC*, *CYA*, *AZB* are drawn on each side of triangle *ABC*. Show that *AX* = *BY* = *CZ*.

10 If *ABC* is rotated about a point *X* to *A'B'C'*, say whether the following must be true or may be false, giving reasons:
   (a) *AA'* = *CC'*;   (b) angle *BAC* = angle *B'A'C'*;
   (c) angle *CXC'* = angle *BXB'*;
   (d) angle between *BC* and *B'C* = angle *AXA'*.

11 In Figure 4(a) say whether the following must be true or may be false, giving reasons:
   (a) *A'B''* = *AB*;   (b) *AA'* = *AA''*;   (c) *XL* = *YM*;
   (d) *XY* bisects *DD'*, where *D* is the midpoint of *AB* and *D'* is the midpoint of *A'B'*.

12 A glide sends *ABC* to *A'B'C'*. Translation of *A* to *A'* followed by reflection will also send *ABC* to *A'B'C'*. Similarly translation of *B* to *B'* or *C* to *C'* followed by reflections will send *ABC* to *A'B'C'*. Draw these three reflection axes and the glide axis. Comment, with reasons if possible, on any properties of these axes.

13 *P'Q'R'S'* is the image of the line *PQRS* under an isometry. *K*, *L*, *M*, *N* are the midpoints of *PP'*, *QQ'*, *RR'*, *SS'*. Show (Hjelmslev's theorem) that *K*, *L*, *M*, *N* are either collinear or coincident.

2. MATRIX TRANSFORMATIONS

Base vectors

You will know that one way to write down a formula for a transformation is to consider what happens to the general point (x, y) and find its image (x', y').

Example 1
Find the matrices for:
   (a) enlargement, centre the origin, with scale factor 2;
   (b) the quarter-turn about the origin.

   (a) For example, $(3, 4) \rightarrow (6, 8)$ and, in general, $(x, y) \rightarrow (2x, 2y)$.

So
$$x' = 2x$$
$$y' = 2y$$

and hence
$$\begin{bmatrix} x' \\ y' \end{bmatrix} = \begin{bmatrix} 2 & 0 \\ 0 & 2 \end{bmatrix} \begin{bmatrix} x \\ y \end{bmatrix}.$$

The matrix is $\begin{bmatrix} 2 & 0 \\ 0 & 2 \end{bmatrix}$.

   (b) For example $(3, 4) \rightarrow (^{-}4, 3)$ (see Figure 6) and, in general, $(x, y) \rightarrow (^{-}y, x)$.

So
$$x' = {}^{-}y$$
$$y' = x$$

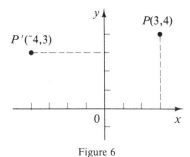

Figure 6

and hence
$$\begin{bmatrix} x' \\ y' \end{bmatrix} = \begin{bmatrix} 0 & -1 \\ 1 & 0 \end{bmatrix}\begin{bmatrix} x \\ y \end{bmatrix}.$$

The matrix is $\begin{bmatrix} 0 & -1 \\ 1 & 0 \end{bmatrix}$.

However, it is not always easy to see just what happens to the general point (x, y), for reflection in $y = 2x$ or rotation through 30°, for example. The simpler approach, as you may know, is to consider what happens to the specific base points $I(1, 0)$ and $J(0, 1)$. The images of their associated base vectors $\mathbf{i} = \begin{bmatrix} 1 \\ 0 \end{bmatrix}$ and $\mathbf{j} = \begin{bmatrix} 0 \\ 1 \end{bmatrix}$ give the columns of the required matrix.

Example 1 (second method)
  (a) $(1, 0) \rightarrow (2, 0); (0, 1) \rightarrow (0, 2)$
$$\begin{bmatrix} 1 \\ 0 \end{bmatrix}, \begin{bmatrix} 0 \\ 1 \end{bmatrix} \rightarrow \begin{bmatrix} 2 \\ 0 \end{bmatrix}, \begin{bmatrix} 0 \\ 2 \end{bmatrix}$$
giving matrix $\begin{bmatrix} 2 & 0 \\ 0 & 2 \end{bmatrix}$.

  (b) $\begin{bmatrix} 1 \\ 0 \end{bmatrix} \rightarrow \begin{bmatrix} 0 \\ 1 \end{bmatrix}; \begin{bmatrix} 0 \\ 1 \end{bmatrix} \rightarrow \begin{bmatrix} -1 \\ 0 \end{bmatrix}$
giving matrix $\begin{bmatrix} 0 & -1 \\ 1 & 0 \end{bmatrix}$.

We can try this approach for a less convenient rotation.

Example 2
Find the matrix for rotation through 30° with O as centre. See Figure 7.

  $I(1, 0) \rightarrow I'$ (cos 30, sin 30);
  $J(0, 1) \rightarrow J'$ (⁻sin 30, cos 30).

Hence $\begin{bmatrix} 1 \\ 0 \end{bmatrix}, \begin{bmatrix} 0 \\ 1 \end{bmatrix} \rightarrow \begin{bmatrix} 0.87 \\ 0.50 \end{bmatrix}, \begin{bmatrix} -0.50 \\ 0.87 \end{bmatrix}$

and thus the matrix
  $\mathbf{R}_{30°} \approx \begin{bmatrix} 0.87 & -0.50 \\ 0.50 & 0.87 \end{bmatrix}$ to 2.s.f.

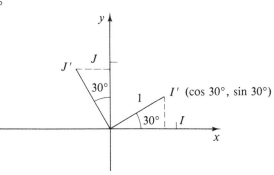

Figure 7

Similarly for rotation through any angle α about the origin:

$$\underset{\mathbf{i}}{\begin{bmatrix}1\\0\end{bmatrix}}\rightarrow\underset{\mathbf{i'}}{\begin{bmatrix}\cos\alpha\\\sin\alpha\end{bmatrix}},\quad\underset{\mathbf{j}}{\begin{bmatrix}0\\1\end{bmatrix}}\rightarrow\underset{\mathbf{j'}}{\begin{bmatrix}^-\sin\alpha\\\cos\alpha\end{bmatrix}},$$

and so the matrix $\mathbf{R}_\alpha=\begin{bmatrix}\cos\alpha & ^-\sin\alpha\\\sin\alpha & \cos\alpha\end{bmatrix}$.

Drawing the images of $\mathbf{i}$ and $\mathbf{j}$ can give a good approximation for any matrix, without calculation, as in the following example.

Example 3
Find to one decimal place accuracy the matrix for reflection in the line $y = 2x$.

Folding the graph paper along the line $y = 2x$ and pricking through gives $I'(^-0.6, 0.8)$, $J'(0.8, 0.6)$. (See Figure 8.)

so the matrix $\mathbf{M}=\begin{bmatrix}^-0.6 & 0.8\\0.8 & 0.6\end{bmatrix}$

to one decimal place accuracy.

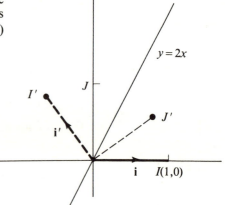

Figure 8

Linearity

But in what circumstances does the method of obtaining a transformation matrix from base vectors work? Consider a typical point $P(3, 4)$ and its image P' under a transformation represented by the matrix $\mathbf{M}=\begin{bmatrix}a & c\\b & d\end{bmatrix}$.

$$\mathbf{M}\begin{bmatrix}3\\4\end{bmatrix}=\begin{bmatrix}a & c\\b & d\end{bmatrix}\begin{bmatrix}3\\4\end{bmatrix}=\begin{bmatrix}3a+4c\\3b+4d\end{bmatrix}=3\begin{bmatrix}a\\b\end{bmatrix}+4\begin{bmatrix}c\\d\end{bmatrix},$$

$$\mathbf{i'}=\mathbf{Mi}=\begin{bmatrix}a & c\\b & d\end{bmatrix}\begin{bmatrix}1\\0\end{bmatrix}=\begin{bmatrix}a\\b\end{bmatrix},\quad \mathbf{j'}=\mathbf{Mj}=\begin{bmatrix}a & c\\b & d\end{bmatrix}\begin{bmatrix}0\\1\end{bmatrix}=\begin{bmatrix}c\\d\end{bmatrix}.$$

So $$\begin{bmatrix}3\\4\end{bmatrix}=3\mathbf{i}+4\mathbf{j}\quad\text{and}\quad\mathbf{M}\begin{bmatrix}3\\4\end{bmatrix}=3\mathbf{i'}+4\mathbf{j'}.$$

So the image P' is related to $\mathbf{i'}$, $\mathbf{j'}$ in exactly the same way as P is related to $\mathbf{i}$, $\mathbf{j}$. Similarly for every point. It is in effect like having new axes with base vectors $\mathbf{i'}$, $\mathbf{j'}$ in place of $\mathbf{i}$, $\mathbf{j}$. We can write old $(3, 4) \rightarrow$ new $(3, 4)$. Similarly old $(x, y) \rightarrow$ new (x, y). (See Figure 9.)

So a matrix can only describe a transformation which behaves in this

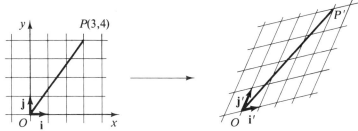

Figure 9

way – such a transformation is called *linear*. Linear transformations map straight lines to straight lines, the origin to itself, and any grid of parallelograms to another grid of parallelograms.

Exercise B

1 Write down the matrix for rotation through $60°$ about the origin. Hence find the position of P (234, 567) after such a rotation.

2 Write down the matrix **R** for rotation of $240°$ about the origin. Calculate **R²** and state what transformation it represents.
   Calculate **R³** and comment on your result.

3 Find, correct to one place of decimals, the matrix **M** for reflection in $y = 4x$. Calculate **M²** and comment on your result.

4 Find to one place of decimals the matrices **A, B, C, D**, for:
   (*a*) reflection in $y = 3x$; (*b*) reflection in $y = {}^-3x$;
   (*c*) reflection in $3y = x$; (*d*) reflection in $3y = {}^-x$.
   Calculate the products **AD** and **BC** and comment.

5 Show that $M(xi+yj) = xMi+yMi$, where

   (*a*) $M = \begin{bmatrix} 2 & 4 \\ 3 & 5 \end{bmatrix}$, $x = 7$, $y = 9$; (*b*) $M = \begin{bmatrix} a & c \\ b & d \end{bmatrix}$.

6 Find to one place of decimals the matrix for a shear with invariant line $y = 2x$ and shearing constant 3.

7 Describe fully the single transformation represented by
   (*a*) $\begin{bmatrix} 0.96 & {}^-0.28 \\ 0.28 & 0.96 \end{bmatrix}$, (*b*) $\begin{bmatrix} 0.96 & 0.28 \\ 0.28 & {}^-0.96 \end{bmatrix}$,
   (*c*) $\begin{bmatrix} 0.96 & 0.28 \\ {}^-0.28 & 0.96 \end{bmatrix}$, (*d*) $\begin{bmatrix} {}^-0.96 & 0.28 \\ 0.28 & 0.96 \end{bmatrix}$.

8 Write down the matrix for rotation through $30°$ about the origin. Hence find the position of P (654, 321) after such a rotation.

9 Write down the matrix **R** for rotation of $120°$ about the origin. Calculate **R²** and state what transformation it represents. Calculate **R³** and comment on your result.

10 Find, correct to one place of decimals, the matrix **M** for reflection in $y = 5x$. Calculate **M²** and comment on your result.

11   Find to one place of decimals the matrices **A, B, C, D** for:
     (a) reflection in $y = 2x$;   (b) reflection in $x = 2y$;
     (c) reflection in $y = -2x$;   (d) reflection in $x = -2y$.
     Calculate the products **AD** and **BC** and comment.

12   Show that $\mathbf{M}(a\mathbf{i} - b\mathbf{j}) = a\mathbf{Mi} - b\mathbf{Mj}$, where

     (a) $\mathbf{M} = \begin{bmatrix} 4 & -5 \\ 6 & 7 \end{bmatrix}$, $a = 3, b = 2$;   (b) $\mathbf{M} = \begin{bmatrix} p & r \\ q & s \end{bmatrix}$.

13   Find to one place of decimals the matrix for a shear with invariant line $y = 3x$ and
     scale factor 2.

14   Describe fully the single transformations represented by

     (a) $\begin{bmatrix} 0.352 & 0.936 \\ 0.936 & -0.352 \end{bmatrix}$,   (b) $\begin{bmatrix} 0.936 & -0.352 \\ 0.352 & 0.936 \end{bmatrix}$,

     (c) $\begin{bmatrix} -0.936 & 0.352 \\ -0.352 & -0.936 \end{bmatrix}$,   (d) $\begin{bmatrix} 0.352 & 0.936 \\ -0.936 & 0.352 \end{bmatrix}$.

3. ROTATION MATRICES

We have already seen in Section 2 that the matrix for the rotation of α about

the origin is $\begin{bmatrix} \cos\alpha & -\sin\alpha \\ \sin\alpha & \cos\alpha \end{bmatrix}$. We can use this in a variety of ways, as the next

two examples show.

Example 4
Find the matrix for the rotation through a (positive) acute angle which maps
the x-axis onto the line $y = 3x$.

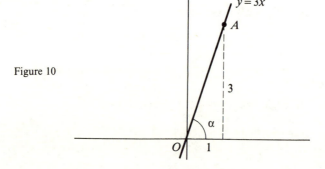

Figure 10

The angle required is shown in Figure 10. By Pythagoras' theorem,
$OA = \sqrt{(1^2 + 3^2)} = \sqrt{10}$, so $\cos\alpha = \dfrac{1}{\sqrt{10}}$ and $\sin\alpha = \dfrac{3}{\sqrt{10}}$. Hence the
matrix is

$$\begin{bmatrix} \dfrac{1}{\sqrt{10}} & \dfrac{-3}{\sqrt{10}} \\ \dfrac{3}{\sqrt{10}} & \dfrac{1}{\sqrt{10}} \end{bmatrix}.$$

Notice that the matrix $\begin{bmatrix} 1 & -3 \\ 3 & 1 \end{bmatrix}$, which is the matrix above multiplied by $\sqrt{10}$, rotates the x-axis onto the line $y = 3x$ but also enlarges by a factor of $\sqrt{10}$. A transformation like this, a combination of a rotation and an enlargement (with the same centres), is called a *spiral similarity*. Spiral similarities will be studied in more detail in Chapter 20.

Example 5

Find the matrix for the positive rotation which maps the direction of $\begin{bmatrix} 3 \\ 2 \end{bmatrix}$ to the direction of $\begin{bmatrix} -1 \\ 8 \end{bmatrix}$.

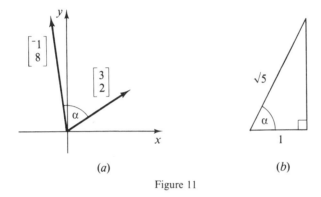

(a)   (b)

Figure 11

In order to write down the matrix, we need to find the angle α between the two vectors. This can be obtained by using their scalar product.

If
$$\mathbf{p} = \begin{bmatrix} 3 \\ 2 \end{bmatrix} \quad \text{and} \quad \mathbf{q} = \begin{bmatrix} -1 \\ 8 \end{bmatrix} \quad \text{then}$$
$$\mathbf{p}.\mathbf{q} = pq \cos \alpha$$
$$^-3 + 16 = \sqrt{13} \sqrt{65} \cos \alpha$$

and so
$$\cos \alpha = \frac{13}{\sqrt{(13 \times 65)}} = \frac{1}{\sqrt{5}}.$$

Using Pythagoras' theorem (see Figure 11(b)), $\sin \alpha = \dfrac{2}{\sqrt{5}}$, so the matrix is

$$\begin{bmatrix} \dfrac{1}{\sqrt{5}} & \dfrac{-2}{\sqrt{5}} \\ \dfrac{2}{\sqrt{5}} & \dfrac{1}{\sqrt{5}} \end{bmatrix}.$$

Exercise C

1 Find, without using decimals, the rotation matrix from the x-axis to the line $y = 4x$.

2 Write down the matrix of a spiral enlargement from the x-axis to the line $y = 2x$.

3   Write down, without using decimals, the matrix **M** for rotation from the x-axis to the line $y = x$. Calculate **M²** and comment.

4   Find, without decimals, the matrix for rotation from the y-axis to the line $y = 3x$.

5   Find the matrix for rotation from the x-axis to the line $y = {}^-3x$, without using decimals.

6   Find, using scalar product, the matrix for rotation from the line $y = 2x$ to $y = 5x$.

7   Write down the matrices, without decimals, for rotation from the x-axis to each of the lines $y = 2x$ and $y = 5x$. Hence obtain the matrix in question 6.

---

8   Find, without using decimals, the rotation matrix from the x-axis to the line $y = 7x$.

9   Write down the matrix of a spiral enlargement from the x-axis to the line $y = 3x$.

10 Write, without using decimals, the matrix **M** for rotation from the x-axis to the line $x + y = 0$. Calculate **M²** and comment.

11 Find, without decimals, the matrix for rotation from the y-axis to the line $y = 2x$.

12 Find the matrix for rotation from the x-axis to the line $3y = 2x$, without using decimals.

13 Find, using scalar product, the matrix for rotation from the line $y = 3x$ to $y = 4x$.

14 Write down the matrices, without decimals, for rotation from the x-axis to each of the lines $y = 3x$ and $y = 4x$. Hence obtain the matrix in question 13.

4. REFLECTION

Although we can calculate the matrix for any rotation about the origin, for most reflections we have had to draw the base vectors and estimate the matrix. However, it is possible to find the matrix for any reflection using a combination of simple isometries.

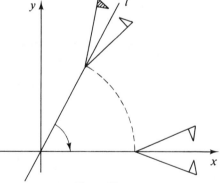

Figure 12

Suppose we wish to find the matrix which will reflect a flag in the line l. Instead we can rotate to the x-axis, reflect in the x-axis, then rotate back again. (See Figure 12.)

Example 7
Find the matrix **M** for reflection in $l : y = 3x$.

We rotate the line $y = 3x$ to the x-axis (matrix **A**), reflect in the x-axis (matrix **X**), and then rotate back (matrix **B**). It follows that **M = BXA** (note the order).

A and **B** will, of course, be inverses of each other and it is easier to find **B** first as it is rotation from the x-axis to the direction $\begin{bmatrix} 1 \\ 3 \end{bmatrix}$.

We take $\qquad\qquad \mathbf{B} = \dfrac{1}{\sqrt{10}}\begin{bmatrix} 1 & -3 \\ 3 & 1 \end{bmatrix}$, as in Example 4.

So $\qquad\qquad \mathbf{A} = \dfrac{1}{\sqrt{10}}\begin{bmatrix} 1 & 3 \\ -3 & 1 \end{bmatrix}$, the inverse of $\mathbf{B}$.

Also $\qquad\qquad \mathbf{X} = \begin{bmatrix} 1 & 0 \\ 0 & -1 \end{bmatrix}$,

$$\begin{aligned}\mathbf{M} &= \frac{1}{10}\begin{bmatrix} 1 & -3 \\ 3 & 1 \end{bmatrix}\begin{bmatrix} 1 & 0 \\ 0 & -1 \end{bmatrix}\begin{bmatrix} 1 & 3 \\ -3 & 1 \end{bmatrix}\\ &= \frac{1}{10}\begin{bmatrix} 1 & -3 \\ 3 & 1 \end{bmatrix}\begin{bmatrix} 1 & 3 \\ 3 & -1 \end{bmatrix}\\ &= \frac{1}{10}\begin{bmatrix} -8 & 6 \\ 6 & 8 \end{bmatrix}\\ &= \begin{bmatrix} -0.8 & 0.6 \\ 0.6 & 0.8 \end{bmatrix}.\end{aligned}$$

Similarly the matrix for a shear or stretch with invariant line l can be obtained by rotating l to the x-axis first.

Exercise D

1 Find the matrix $\mathbf{M}$ for reflection in $y = 2x$. Calculate $\mathbf{M}^2$ and comment.

2 Find the matrix $\mathbf{N}$ for reflection in $x = {}^-2y$. Calculate $\mathbf{N}^2$ and comment.

3 For the matrices $\mathbf{M}$ and $\mathbf{N}$ of the first two questions calculate $\mathbf{MN}$ and $\mathbf{NM}$ and comment.

4 Find the matrix for a shear with shearing constant 2 and invariant line $y = 3x$.

5 Find the matrix $\mathbf{P}$ for a stretch with scale factor 2 and the line $y = 7x$ invariant, and the matrix $\mathbf{Q}$ for a stretch with scale factor 2 and the line $x = {}^-7y$ invariant. Calculate $\mathbf{PQ}$ and $\mathbf{QP}$ and comment.

6 Find the matrix $\mathbf{M}$ for reflection in $y = \frac{1}{2}x$. Calculate $\mathbf{M}^2$ and comment.

7 Find the matrix $\mathbf{N}$ for reflection in $y = {}^-2x$. Calculate $\mathbf{N}^2$ and comment.

8 For the matrices $\mathbf{M}$ and $\mathbf{N}$ of questions 6 and 7, calculate $\mathbf{MN}$ and $\mathbf{NM}$ and comment.

9 Find the matrix for a shear with shearing constant 3 and invariant line $y = 2x$.

10 Find the matrix $\mathbf{P}$ for a stretch with scale factor 3 and with $y = 5x$ invariant, and the matrix $\mathbf{Q}$ for a stretch with scale factor 3 and with $x = {}^-5y$ invariant. Calculate $\mathbf{PQ}$ and $\mathbf{QP}$ and comment.

5. INVARIANCE

In Section 1 we used invariant points to help to classify isometries. In general, the invariant features of a transformation can give considerable information about it.

Example 8

Find the invariant points for the transformation given by the matrix $\frac{1}{5}\begin{bmatrix} 3 & 4 \\ 4 & -3 \end{bmatrix}$.

$$\mathbf{P}\,(a,\,b) \text{ is invariant} \quad \Leftrightarrow \quad \frac{1}{5}\begin{bmatrix} 3 & 4 \\ 4 & -3 \end{bmatrix}\begin{bmatrix} a \\ b \end{bmatrix} = \begin{bmatrix} a \\ b \end{bmatrix}$$

$$\Leftrightarrow \quad \begin{bmatrix} 3 & 4 \\ 4 & -3 \end{bmatrix}\begin{bmatrix} a \\ b \end{bmatrix} = 5\begin{bmatrix} a \\ b \end{bmatrix}$$

$$\Leftrightarrow \quad \left. \begin{matrix} 3a+4b = 5a \\ 4a-3b = 5b \end{matrix} \right\} \quad \Leftrightarrow \quad \left. \begin{matrix} 4b = 2a \\ 4a = 8b \end{matrix} \right\} \quad \Leftrightarrow \quad \left. \begin{matrix} a = 2b \\ a = 2b \end{matrix} \right\} .$$

So every point for which $a = 2b$, such as (6, 3), is invariant. This gives a whole set of invariant points all lying on the line $x = 2y$.

Notice that the new base vectors $\begin{bmatrix} 3 \\ 4 \end{bmatrix}$, $\begin{bmatrix} 4 \\ -3 \end{bmatrix}$ are perpendicular; also that the determinant $(0.6 \times {}^-0.8 - 0.8 \times 0.8) = {}^-1$, so this is an opposite transformation with area invariant. It is in fact the reflection in $x = 2y$.

Example 9

Find the invariant points for $\begin{bmatrix} 5 & -6 \\ 6 & 5 \end{bmatrix}$.

$$(a,\,b) \text{ invariant} \quad \Leftrightarrow \quad \begin{bmatrix} 5 & -6 \\ 6 & 5 \end{bmatrix}\begin{bmatrix} a \\ b \end{bmatrix} = \begin{bmatrix} a \\ b \end{bmatrix}$$

$$\Leftrightarrow \quad \left. \begin{matrix} 5a-6b = a \\ 6a+5b = b \end{matrix} \right\} \quad \Leftrightarrow \quad \left. \begin{matrix} 2a = 3b \\ 3a = {}^-2b \end{matrix} \right\} \quad \Leftrightarrow \quad \left. \begin{matrix} a = 0 \\ b = 0 \end{matrix} \right\} .$$

So the origin is the only invariant point.

Notice that the new base vectors $\begin{bmatrix} 5 \\ 6 \end{bmatrix}$ and $\begin{bmatrix} -6 \\ 5 \end{bmatrix}$ are perpendicular and equal in length, so there is enlargement. This is, in fact, a spiral similarity.

Exercise E

1   Find the invariant points for the isometries given by
(a) $\begin{bmatrix} 0.8 & -0.6 \\ 0.6 & 0.8 \end{bmatrix}$, (b) $\begin{bmatrix} -0.8 & 0.6 \\ 0.6 & 0.8 \end{bmatrix}$, (c) $\begin{bmatrix} 0.8 & 0.6 \\ 0.6 & -0.8 \end{bmatrix}$.
Hence describe each transformation fully.

2   Find whether area or any points are invariant under the transformation $\begin{bmatrix} 0.2 & 0.4 \\ -1.6 & 1.8 \end{bmatrix}$.
Hence describe the transformation fully.

3   Find the invariant line for $\begin{bmatrix} 2.8 & -0.6 \\ -0.6 & 1.2 \end{bmatrix}$. Find the image of a vector perpendicular to this line. Hence describe the transformation fully.

4   Describe the transformations given by
(a) $\begin{bmatrix} 1.08 & -0.56 \\ -0.56 & 4.92 \end{bmatrix}$, (b) $\begin{bmatrix} 0.3 & 4.9 \\ -0.1 & 1.7 \end{bmatrix}$, (c) $\begin{bmatrix} 0.96 & 0.28 \\ 0.28 & -0.96 \end{bmatrix}$.

5 Find the invariant points for the isometries given by

(a) $\begin{bmatrix} 0.28 & 0.96 \\ -0.96 & 0.28 \end{bmatrix}$, (b) $\begin{bmatrix} -0.28 & 0.96 \\ 0.96 & 0.28 \end{bmatrix}$, (c) $\begin{bmatrix} 0.28 & 0.96 \\ 0.96 & -0.28 \end{bmatrix}$.

Hence describe each transformation fully.

6 Find whether area or any points are invariant under the transformation $\begin{bmatrix} 0.58 & 0.06 \\ -2.94 & 1.42 \end{bmatrix}$. Hence describe the transformation fully.

7 Find the invariant line for $\begin{bmatrix} -\frac{40}{13} & -\frac{18}{13} \\ -\frac{18}{13} & \frac{25}{13} \end{bmatrix}$. Find the image of a vector perpendicular to this line. Hence describe the transformation fully.

8 Describe fully the transformations given by

(a) $\begin{bmatrix} -0.2 & -0.6 \\ 2.4 & 2.2 \end{bmatrix}$, (b) $\begin{bmatrix} \frac{61}{53} & \frac{28}{53} \\ \frac{28}{53} & \frac{151}{53} \end{bmatrix}$, (c) $\begin{bmatrix} 0.96 & -0.28 \\ 0.28 & 0.96 \end{bmatrix}$.

6. NON-CENTRAL TRANSFORMATIONS

Rotation and reflection

All the matrix transformations we have considered so far have had the origin as an invariant point. But, by using translations, we can deal with rotations about centres other than the origin. The method is similar to that in Section 4.

Example 10
Find an algebraic rule for the quarter-turn about the point (2, 3).

First we translate the point to the origin, then make the quarter-turn, then translate back.

$$\begin{bmatrix} x \\ y \end{bmatrix} \rightarrow \begin{bmatrix} x \\ y \end{bmatrix} + \begin{bmatrix} -2 \\ -3 \end{bmatrix} \rightarrow \begin{bmatrix} 0 & -1 \\ 1 & 0 \end{bmatrix}\begin{bmatrix} x-2 \\ y-3 \end{bmatrix} \rightarrow \begin{bmatrix} 0 & -1 \\ 1 & 0 \end{bmatrix}\begin{bmatrix} x-2 \\ y-3 \end{bmatrix} + \begin{bmatrix} 2 \\ 3 \end{bmatrix}$$

$$\Rightarrow \begin{bmatrix} x \\ y \end{bmatrix} \rightarrow \begin{bmatrix} -y+3 \\ x-2 \end{bmatrix} + \begin{bmatrix} 2 \\ 3 \end{bmatrix} = \begin{bmatrix} -y+5 \\ x+1 \end{bmatrix}.$$

Check for invariant points:

(x, y) is invariant $\Leftrightarrow \left.\begin{array}{r} -y+5 = x \\ x+1 = y \end{array}\right\} \Leftrightarrow x = 2, y = 3$ as expected.

Example 11
Express algebraically the transformation of reflection in $y = 3x+4$.

Translate any point on the line, e.g. (0, 4), to the origin, reflect in the line $y = 3x$, then translate back.

$$\begin{bmatrix} x \\ y \end{bmatrix} \rightarrow \begin{bmatrix} x \\ y-4 \end{bmatrix} \rightarrow \begin{bmatrix} -0.8 & 0.6 \\ 0.6 & 0.8 \end{bmatrix}\begin{bmatrix} x \\ y-4 \end{bmatrix} \rightarrow \begin{bmatrix} -0.8 & 0.6 \\ 0.6 & 0.8 \end{bmatrix}\begin{bmatrix} x \\ y-4 \end{bmatrix} + \begin{bmatrix} 0 \\ 4 \end{bmatrix}$$

$$\Rightarrow \begin{bmatrix} x \\ y \end{bmatrix} \rightarrow \begin{bmatrix} -0.8x+0.6y-2.4 \\ 0.6x+0.8y +0.8 \end{bmatrix}.$$

Note that we have used the reflection matrix found in Example 7.

Check for invariant points:

$$\left.\begin{array}{r}^{-}0.8x+0.6y-2.4=x\\0.6x+0.8y+0.8=y\end{array}\right\} \Leftrightarrow \left.\begin{array}{r}^{-}1.8x+0.6y-2.4=0\\0.6x-0.2y+0.8=0\end{array}\right\} \Leftrightarrow \left.\begin{array}{r}y=3x+4\\y=3x+4\end{array}\right\}.$$

So all the points for which $y=3x+4$ are invariant, as expected.

Exercise F

1   Express algebraically the three-quarter-turn about the point (5, 6) and check for invariant points.

2   Express algebraically the rotation through 60° about (4, 3) and check for invariant points.

3   Express algebraically the reflection in the line $y=3x+5$ and check for invariant points.

4   Express algebraically the reflection in the line $4x+5y=6$ and check for invariant points.

5   Express algebraically (*a*) the stretch with scale factor 5;   (*b*) the shear with shearing constant 3, each with the line $y=7x-4$ invariant.
    Find the images of P (3, 1) and Q (1, 3) in each case.

6   Find the image of (19, 84) after each of the following transformations:
    (*a*) reflection in $y=5x+9$;
    (*b*) half-turn about (98, 76);
    (*c*) stretch with scale factor 4 and the line $y=6x-7$ invariant.

7   Express algebraically the enlargement with scale factor 9 and centre (5, $^{-}$7). Check for invariant points.

8   Express algebraically the three-quarter-turn about the point (4, 5) and check for invariant points.

9   Express algebraically the rotation through 60° about (3, 2) and check for invariant points.

10   Express algebraically the reflection in the line $y=4x-3$ and check for invariant points.

11   Express algebraically the reflection in the line $2x+3y=4$ and check for invariant points.

12   Express algebraically the enlargement with scale factor 4 and centre (6, $^{-}$5). Check for invariant points.

13   Express algebraically:
    (*a*) the reflection in the line $y=3x+1$;
    (*b*) the stretch with scale factor 4 and with the line $y=5x-3$ invariant;
    (*c*) the shear with shearing constant 3 and the line $y=6x-5$ invariant.
    State the images of P (7, 2) and Q (2, 7) in each case.

7. GLIDE

It is easy to express a glide algebraically if we are told the axis and the throw. First we find the reflection; then we combine it with a suitable translation.

Example 12

Find the image of P (6, 7) under a glide with axis $y = 3x + 4$ and throw 5 in the direction $\begin{bmatrix} 1 \\ 3 \end{bmatrix}$.

A displacement of $\begin{bmatrix} 1 \\ 3 \end{bmatrix}$ has throw $\sqrt{10}$, so a displacement of $\dfrac{5}{\sqrt{10}}\begin{bmatrix} 1 \\ 3 \end{bmatrix}$ has throw 5.

From Example 11, we know that reflection in $y = 3x + 4$ is given by

$$\begin{bmatrix} x \\ y \end{bmatrix} \rightarrow \begin{bmatrix} -0.8x + 0.6y - 2.4 \\ 0.6x + 0.8y + 0.8 \end{bmatrix},$$

so the glide is given by

$$\begin{bmatrix} x \\ y \end{bmatrix} \rightarrow \begin{bmatrix} -0.8x + 0.6y - 2.4 + \frac{5}{\sqrt{10}} \\ 0.6x + 0.8y + 0.8 + \frac{15}{\sqrt{10}} \end{bmatrix} \approx \begin{bmatrix} -0.8x + 0.6y - 0.82 \\ 0.6x + 0.8y + 5.54 \end{bmatrix}.$$

From this, we find that (6, 7) is mapped to ($^{-}$1.42, 14.74) to 2 decimal places.

The method of Example 12 is not practicable when we are given the positions of the object and image rather than the axis and throw. But the transformation can be found as a combination of simpler ones.

Example 13

Express algebraically the glide which transforms A (2, 3), B (3, 7) and C (4, 4) to A' (9, 4), B' (5, 3) and C' (8, 2).

There are various ways in which this can be split into a sequence of other isometries. One possibility is as follows:

(1) Translate A to the origin to give triangle $OB_1 C_1$.
(2) Reflect in the x-axis to give triangle $OB_2 C_2$.
(3) Rotate direction OB_2 to direction $A'B'$ to give triangle $OB_3 C_3$.
(4) Translate triangle $OB_3 C_3$ to triangle $A'B'C'$.

The four stages are illustrated in Figure 13.

The mappings for each stage are:

(1) $\begin{bmatrix} x \\ y \end{bmatrix} \rightarrow \begin{bmatrix} x \\ y \end{bmatrix} + \begin{bmatrix} -2 \\ -3 \end{bmatrix}$.

(2) $\begin{bmatrix} x \\ y \end{bmatrix} \rightarrow \begin{bmatrix} 1 & 0 \\ 0 & -1 \end{bmatrix}\begin{bmatrix} x \\ y \end{bmatrix}$.

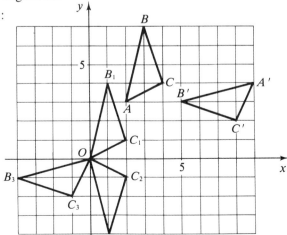

Figure 13

(3) B has coordinates $(1, 4)$, so $\underset{\sim}{QB_2} = \begin{bmatrix} 1 \\ -4 \end{bmatrix}$. This has to be rotated to the

direction of $\underset{\sim}{A'B'}$, which is $\begin{bmatrix} -4 \\ -1 \end{bmatrix}$. This is a three-quarter-turn about O, so

the mapping is $\begin{bmatrix} x \\ y \end{bmatrix} \to \begin{bmatrix} 0 & 1 \\ -1 & 0 \end{bmatrix} \begin{bmatrix} x \\ y \end{bmatrix}$.

(4) $\begin{bmatrix} x \\ y \end{bmatrix} \to \begin{bmatrix} x \\ y \end{bmatrix} + \begin{bmatrix} 9 \\ 4 \end{bmatrix}$.

Combining these four mappings, we have

$$\begin{bmatrix} x \\ y \end{bmatrix} \to \begin{bmatrix} 0 & 1 \\ -1 & 0 \end{bmatrix}\begin{bmatrix} 1 & 0 \\ 0 & -1 \end{bmatrix}\begin{bmatrix} x-2 \\ y-3 \end{bmatrix} + \begin{bmatrix} 9 \\ 4 \end{bmatrix}$$

$$= \begin{bmatrix} 0 & -1 \\ -1 & 0 \end{bmatrix}\begin{bmatrix} x-2 \\ y-3 \end{bmatrix} + \begin{bmatrix} 9 \\ 4 \end{bmatrix}$$

$$= \begin{bmatrix} -y+3 \\ -x+2 \end{bmatrix} + \begin{bmatrix} 9 \\ 4 \end{bmatrix}$$

$$= \begin{bmatrix} -y+12 \\ -x+6 \end{bmatrix}.$$

The glide is given by

$$\begin{bmatrix} x \\ y \end{bmatrix} \to \begin{bmatrix} -y+12 \\ -x+6 \end{bmatrix}.$$

(Check that this is consistent with the original information about the images of A, B and C.)

Note that the invariant points are given by

$$\left. \begin{matrix} -y+12 = x \\ -x+ 6 = y \end{matrix} \right\} \Rightarrow \left. \begin{matrix} x+y = 12 \\ x+y = 6 \end{matrix} \right\}$$

which are contradictory, so there are no invariant points, as we would expect.

It is, however, possible to find the equation of the axis of the glide. Suppose that the axis is $y = kx+c$. If (x, y) is on this line, than so is its image $(-y+12, -x+6)$.

So $y = kx+c \Rightarrow -x+6 = k(-y+12)+c$

$$= -ky+12k+c$$

$$\Rightarrow \quad ky = x+12k+c-6,$$

$$y = \frac{1}{k}x+12+\frac{c}{k}-\frac{6}{k}.$$

So we require $k = \dfrac{1}{k}$ and $c = 12+\dfrac{c}{k}-\dfrac{6}{k}$.

$k = \dfrac{1}{k} \Rightarrow k^2 = 1 \Rightarrow k = \pm 1$.

$k = 1 \Rightarrow c = 12+c-6 = c+6$, which is impossible.

$$k = {}^-1 \Rightarrow c = 12 - c + 6 = 18 - c$$
$$\Rightarrow 2c = 18$$
$$\Rightarrow c = 9.$$

The axis of the glide is $y = {}^-x + 9$.

As a check, we notice that the midpoint of AA', $(5\frac{1}{2}, 3\frac{1}{2})$, and the midpoint of BB', $(4, 5)$, lie on this line.

Finally, the transformation may be written to show that it is reflection in the axis with a translation parallel to it.

Reflection in the line $y = {}^-x + 9$ can be achieved by translating the line so that it passes through the origin – translation vector $\begin{bmatrix} 0 \\ -9 \end{bmatrix}$ – reflecting in the parallel line $y = {}^-x$, and then translating back. This gives

$$\begin{bmatrix} x \\ y \end{bmatrix} \rightarrow \begin{bmatrix} 0 & {}^-1 \\ {}^-1 & 0 \end{bmatrix}\begin{bmatrix} x \\ y-9 \end{bmatrix} + \begin{bmatrix} 0 \\ 9 \end{bmatrix} = \begin{bmatrix} {}^-y+9 \\ {}^-x+9 \end{bmatrix}.$$

Comparing this with the glide $\begin{bmatrix} x \\ y \end{bmatrix} \rightarrow \begin{bmatrix} {}^-y+12 \\ {}^-x+6 \end{bmatrix}$ shows that the glide is reflection in the line $y = {}^-x + 9$ followed by the translation $\begin{bmatrix} 3 \\ -3 \end{bmatrix}$, which is parallel to the line.

Exercise G

1 Express algebraically the glide with axis $y = 2x + 3$ and translation $\begin{bmatrix} 3 \\ 6 \end{bmatrix}$.

2 Find the image of $P\,(17, 18)$ after a glide with axis $3y = 4x + 5$ and throw 6 in the direction $\begin{bmatrix} 3 \\ 4 \end{bmatrix}$.

3 Find the axis and throw for the glide
$$\begin{bmatrix} x \\ y \end{bmatrix} \rightarrow \begin{bmatrix} {}^-0.96x + 0.28y + 0.6 \\ 0.28x + 0.96y + 14.2 \end{bmatrix}.$$

4 Express algebraically the glide for which the images of $A\,(1, 3)$, $B\,(2, 6)$, $C\,(4, 5)$ are $A'\,(2, 6)$, $B'\,(5, 7)$, $C'\,(4, 9)$. Also find the glide axis and throw.

5 Find the image of $P\,(7, {}^-4)$ under the glide in which the images of $A\,(3, 4)$, $B\,({}^-2, {}^-1)$, $C\,(8, {}^-6)$ are $A'\,({}^-9.8, 3.6)$, $({}^-6.4, 9.8)$, $C'\,(1.2, 1.6)$.

6 Express algebraically the glide with axis $y = 3x + 5$ and translation $\begin{bmatrix} 2 \\ 6 \end{bmatrix}$.

7 Find the image of $P\,(11, 17)$ after a glide with axis $4y = 5x - 6$ and throw 7 in the direction $\begin{bmatrix} 4 \\ 5 \end{bmatrix}$.

8 Find the axis and throw for the glide
$$\begin{bmatrix} x \\ y \end{bmatrix} \rightarrow \begin{bmatrix} {}^-0.28x + 0.96y + 6.84 \\ 0.96x + 0.28y + 1.12 \end{bmatrix}.$$

9 Express algebraically the glide for which the images of $A(2, 5)$, $B(3, 8)$, $C(4, 1)$ are $A'(6, 7)$, $B'(7, 10)$, $C'(2, 5)$. Also find the glide axis and throw.

8. SUMMARY

(1)

Isometries (preserving lengths)	Direct	Opposite
With a fixed point	Rotation	Reflection
Without a fixed point	Translation	Glide

The centre of a rotation lies on the mediator of the line joining each point and its image.

The axis of a glide goes through the midpoint of the line joining each point and its image.

(2) A matrix gives a linear transformation.

$$\mathbf{M}(x\mathbf{i} + y\mathbf{j}) = x\mathbf{M}(\mathbf{i}) + y\mathbf{M}(\mathbf{j})$$

The image of a straight line is a straight line.

The images of the base vectors $\begin{bmatrix} 1 \\ 0 \end{bmatrix}$ and $\begin{bmatrix} 0 \\ 1 \end{bmatrix}$ give the columns of the matrix for the transformation.

(3) Reflection $\mathbf{L}$ in a line l through the origin is given by $\mathbf{L} = \mathbf{RXR}^{-1}$ where $\mathbf{X}$ is reflection in the x-axis and $\mathbf{R}$ is rotation from the x-axis to l. The invariant points give the axis for a reflection.

(4) Rotation about a general point P is obtained by translating P to the origin, rotating about the origin, then translating from the origin back to P.

Reflection in a general line is obtained by translating the line to the origin, reflecting, then translating back again.

Miscellaneous exercise

1 Under a shear, which leaves the origin invariant, $(1, 2)$ is mapped to $(1, 4)$. Find the matrix $\mathbf{A}$ representing this shear.

Show that the transformation which is given by the matrix product

$$\mathbf{A}\begin{bmatrix} 1 & ^-1 \\ ^-2 & 3 \end{bmatrix}$$

is a shear, and describe it fully.   (SMP)

2 Illustrate the transformations of the unit square, with vertices $(0, 0)$, $(1, 0)$, $(0, 1)$ and $(1, 1)$, when the following matrices are applied to the appropriate column-vectors:

(a) $\mathbf{A} = \begin{bmatrix} 2 & 0 \\ 0 & 3 \end{bmatrix}$;   (b) $\mathbf{B} = \begin{bmatrix} 0 & ^-3 \\ ^-3 & 0 \end{bmatrix}$;   (c) $\mathbf{AB}$.

Draw a fresh diagram for each answer.   (SMP)

3 Each of the following matrices, when used to premultiply column position vectors, gives a transformation in the plane. In each case *either* describe the transformation geometrically *or* illustrate it by a sketch showing the image of the unit square $OPQR$, where O is $(0, 0)$, P is $(1, 0)$, Q is $(1, 1)$ and R is $(0, 1)$.

(i) $\begin{bmatrix} 3 & 0 \\ 0 & 1 \end{bmatrix}$; (ii) $\begin{bmatrix} 0 & 0 \\ 0 & 1 \end{bmatrix}$; (iii) $\begin{bmatrix} -1 & 0 \\ 0 & 1 \end{bmatrix}$;

(iv) $\begin{bmatrix} 0 & -1 \\ -1 & 0 \end{bmatrix}$; (v) $\begin{bmatrix} 1 & 0 \\ 0 & 2 \end{bmatrix}$; (vi) $\begin{bmatrix} 1 & 0 \\ 0 & 0 \end{bmatrix}$.

Choose matrices **A**, **B** from these six satisfying:
(*a*) **AB** = **BA** $\neq$ **0**;   (*b*) **AB** $\neq$ **BA**;   (*c*) **AB** = **0**.   (SMP)

4   Plot on squared paper the points A $(2, 0)$ and B $(0, 3)$. Show on your diagram, stating the coordinates, their images A' and B' under the transformation whose matrix is

$$T = \begin{bmatrix} 0.8 & -0.6 \\ 0.6 & 0.8 \end{bmatrix}.$$

Hence find the equation of the image of the line $3x + 2y = 6$ under this transformation and show that it meets the x-axis at C' $(5, 0)$. Describe the geometric effect of the transformation **T** and write down, with a brief reason, the area of the triangle $OA'B'$, where O is the origin. Find the coordinates of C (whose image is C') and verify that it lies on AB.   (SMP)

5   (*a*) The position vectors of the vertices of the triangle O $(0, 0)$, A $(1, 0)$, B $(0, 1)$ are multiplied on the left by the matrix

$$P = \begin{bmatrix} 1 & 3 \\ 0 & 1 \end{bmatrix}.$$

Sketch the position of the transformed triangle, labelling its vertices $O_1 A_1 B_1$.
(*b*) The new triangle $O_1 A_1 B_1$ is now transformed by using in the same way the matrix

$$Q = \begin{bmatrix} 1 & 0 \\ 2 & 1 \end{bmatrix}.$$

Sketch the position of the final triangle $O_2 A_2 B_2$. Explain why the area of this final triangle is the same as that of the triangle OAB.   (SMP)

6   **X** is the matrix for reflection in the line $y = 0$, **Z** is the matrix for reflection in the line $y = x$, **R** is the matrix for rotation about the origin through 60° and **M** is the matrix for reflection in the line through the origin at 30° to the x-axis. Write down the matrices **X**, **Z** and **R**.

'The product of two reflections is equivalent to a rotation through twice the angle between their axes.' Verify that the product **ZX** satisfies this statement and write down the relation between **X**, **R** and **M** which also satisfies this statement. Hence prove that **M** = **RX** and so calculate the matrix **M**.   (SMP)

7   A reflection **M** in the x-axis is followed by a rotation **R** about the origin through a positive (counter-clockwise) angle 2θ.
(*a*) Show that the line $y = x \tan \theta$ is unchanged.
(*b*) Describe the single transformation which has the same effect as **RM** (i.e. **M** followed by **R**).
(*c*) Find the matrices for **M**, **R** and **RM**.   (SMP)

8   The line l passes through the origin O and makes an angle θ with the x-axis. **X** is the reflection in the x-axis and **L** is the reflection in l (Figure 14).
(*a*) If **LX** = **R**, what transformation is **R**? Describe it fully.

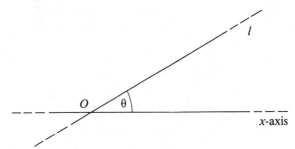

Figure 14

(b) Given that the matrix $\begin{bmatrix} \cos \alpha & -\sin \alpha \\ \sin \alpha & \cos \alpha \end{bmatrix}$ represents a positive rotation through an angle α about the origin, write down the matrix representing **R**.

(c) Write down: (i) the matrix representing **X**; (ii) the inverse of this matrix.

(d) Hence solve equation **LX** = **R** to find the matrix representing **L**.   (SMP)

9 Express algebraically the two isometries in which $P(5, {}^-6) \rightarrow P'(2, {}^-3)$, and $Q(8, 5) \rightarrow Q'\ (9, 6)$.

10 Draw the image of the unit square under the transformations represented by each of the following matrices:

(a) $\begin{bmatrix} 1 & 2 \\ 0 & 1 \end{bmatrix}$; (b) $\begin{bmatrix} 1 & 0 \\ -1 & 1 \end{bmatrix}$; (c) the product $\begin{bmatrix} 1 & 0 \\ -1 & 1 \end{bmatrix}\begin{bmatrix} 1 & 2 \\ 0 & 1 \end{bmatrix}$.

Find the area of the image in (c).   (SMP)

17

Isometries

1. REFLECTIONS

Reflections are particularly important because every transformation can be expressed as a combination of a number of reflections.

Two reflections

Since reflection is an opposite isometry, two reflections will give a direct isometry. This is a rotation, unless the mirror lines are parallel, in which case it is a translation.

The flag in Figure 1 is reflected in the line l and then in the line m. The point B goes to B' then B''; so that

$$\mathbf{L}(OB) = OB' \quad \text{and} \quad \mathbf{ML}(OB) = OB''.$$
$$OB'' = OB' = OB.$$

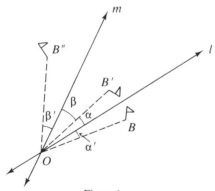

Figure 1

Also angle $BOB'' = \alpha' + \alpha + \beta + \beta'$
$$= 2\alpha + 2\beta \qquad \text{since } \alpha' = \alpha \text{ and } \beta' = \beta,$$
$$= 2\theta \qquad \text{where } \theta \text{ is the angle from } l \text{ to } m.$$

So a rotation centre O through 2θ will map B on B''. Similarly it will map every point P onto its image P''. We may write this

$$\mathbf{ML} = \mathbf{R}_{2\theta}.$$

Reflection in l followed by reflection in m is equivalent to rotation through twice the angle from l to m about their point of intersection.

The useful thing about this result is that rotation α about C can be replaced by reflections in a pair of lines l and m through C where the angle from l to m is $\frac{1}{2}\alpha$. For example, the quarter-turn about the origin is the same as reflection in $y = x$ followed by reflection in the y-axis.

When, as in Figure 2, the lines l and m are parallel there is no rotation.

$$\mathbf{BB}'' = \mathbf{a}' + \mathbf{a} + \mathbf{b} + \mathbf{b}'$$
$$= 2\mathbf{a} + 2\mathbf{b} \quad \text{since } \mathbf{a}' = \mathbf{a} \text{ and } \mathbf{b}' = \mathbf{b},$$
$$= 2\mathbf{d} \quad \text{where } \mathbf{d} \text{ is the displacement from } l \text{ to } m, \text{ and perpendicular to } l$$
and m.

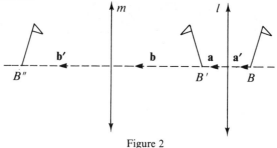

Figure 2

So a translation of $2\mathbf{d}$ will map B onto B''. Similarly it will map every point P onto its image P''. We may write this

$$\mathbf{ML} = \mathbf{T}_{2\mathbf{d}}.$$

Reflection in parallel lines l, m is equivalent to translation through twice the displacement from l to m, and perpendicular to the lines.

So a translation $\mathbf{t}$ can be replaced by reflection in any two parallel axes so long as the displacement between them is $\frac{1}{2}\mathbf{t}$. Thus translation $\begin{bmatrix} 6 \\ 0 \end{bmatrix}$ is the same as reflection in $x = 2$ followed by reflection in $x = 5$, for example.

Three reflections

A glide is equivalent to reflection in the axis followed by translation parallel to the axis. Since the translation can be replaced by two reflections, a glide is

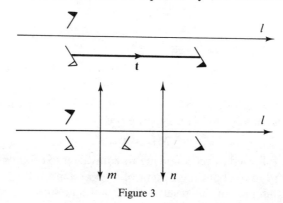

Figure 3

equivalent to three reflections, two of the axes being parallel, and perpendicular to the third. (See Figure 3). But three reflections, although they must give an opposite isometry, do not always give a glide.

Case (*i*). The three mirror lines are all parallel. (See Figure 4.)

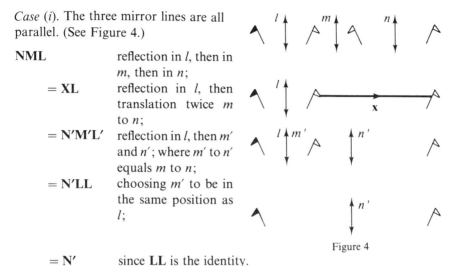

NML	reflection in *l*, then in *m*, then in *n*;
= **XL**	reflection in *l*, then translation twice *m* to *n*;
= **N'M'L'**	reflection in *l*, then *m'* and *n'*; where *m'* to *n'* equals *m* to *n*;
= **N'LL**	choosing *m'* to be in the same position as *l*;
= **N'**	since **LL** is the identity.

Figure 4

So in this case the three reflections reduce to a single reflection.

Case (*ii*). The three lines meet at one point. (See Figure 5.)

In this case also the three reflections reduce to a single reflection. The method is shown for a particular case in the following example:

Example 1

X, Y, Z denote reflection in the *x*-axis, *y*-axis and line $x + y = 0$ respectively. Find the single transformation equivalent to **ZYX**.

ZYX = **Z'Y'X**	where *y'*, *z'* are any lines through *O* with angle 45° from *y'* to *z'*;
= **Z'XX**	taking *y'* as the *x*-axis, and thus *z'* as the line $y = x$;
= **Z'**	as **XX** is the identity.

So **ZYX** is equivalent to reflection in the line $y = x$.

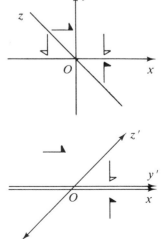

In all other cases three reflections give a glide.   Figure 5

Case (iii). The second and third lines are parallel. (See Figure 6.)

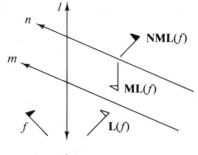

NML	reflection in *l*, then in *m*, then in *n*;

= **TL**	reflection in *l*, then translation **t**, twice *m* to *n*;
= **YXL**	replacing the translation by its components **x**, **y** perpendicular to and parallel to *l*;

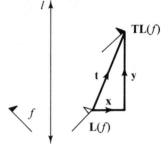

= **YN′**	reducing **XL** to **N′** as in Figure 4.

A similar technique can be used if the first and second lines are parallel.

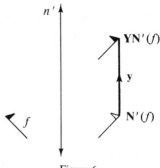

Example 2
Find the axis and throw for the glide equivalent to quarter-turn about the origin followed by reflection in the line $x = 6$.

Figure 6

NQ	quarter-turn then reflection in $x = 6$;
= **NML**	reflection in $y = x$ and in $x = 0$ then in $x = 6$, choosing *m* to be parallel to *n*, thus as case (iii) (see Figure 7(a));
= **TL**	reflection in $y = x$ then translation $\begin{bmatrix} 12 \\ 0 \end{bmatrix}$:
= **YXL**	reflection in $y = x$ then translation $\begin{bmatrix} 6 \\ -6 \end{bmatrix}$ and $\begin{bmatrix} 6 \\ 6 \end{bmatrix}$; perpendicular to and along $y = x$ to get glide (see Figure 7(b));
= **YPLL**	where translation $\begin{bmatrix} 6 \\ -6 \end{bmatrix}$ is replaced by reflection in lines *l* and *p* whose displacement is $\begin{bmatrix} 3 \\ -3 \end{bmatrix}$;
= **YP**	reflection in $x - y = 6$ then translation $\begin{bmatrix} 6 \\ 6 \end{bmatrix}$ (see Figure 7(c)).

This is a glide with axis $x - y = 6$ and throw $6\sqrt{2}$.

(a)

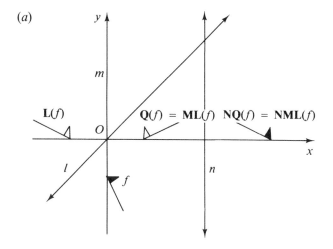

(b)

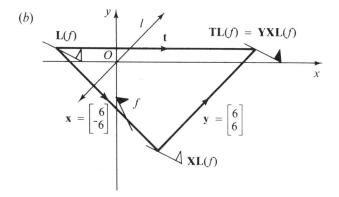

(c)

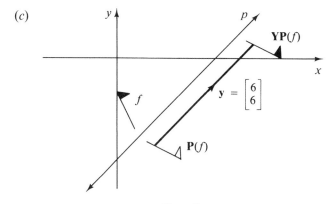

Figure 7

Case (*iv*). The three lines form a triangle. (See Figure 8.)

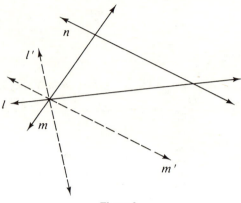

Figure 8

Replace *l*, *m* by *l′*, *m′* meeting at the same point and with the angle from *l′* to *m′* the same as the angle from *l* to *m*, choosing *m′* to be parallel to *n*.
This is now the same as case (iii).

Case (*v*). The first and third lines are parallel. (See Figure 9.)

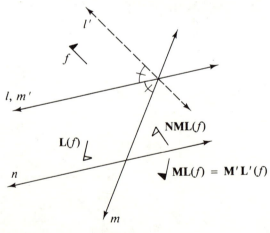

Figure 9

Replace *l*, *m* by *l′*, *m′* meeting at the same point and with the angle from *l′* to *m′* the same as the angle from *l* to *m*, choosing *m′* to be parallel to *n* (i.e. *m′* is coincident with *l*).

This is now the same as case (iii).

So three reflections give a glide, unless the three mirror lines are concurrent or parallel, in which cases the equivalent transformation is a reflection.

Exercise A

1 $OA'' = OA$ and $OB'' = OB$. Draw a diagram to show that this is not enough to ensure that AB can be rotated onto $A''B''$. What is the extra property in Figure 1 which does ensure rotation?

2 **ML**, reflection in l followed by reflection in m, is equivalent to rotation through 2θ, where θ is the angle from l to m. To what rotation is **LM** equivalent?
   Find the rotation equivalent to **ML** followed by **LM**.

3 **L** is reflection in $y = 0$, **M** is reflection in $y = x$, **N** is reflection in $y = {}^-x$. Find the single reflection equivalent to (*a*) **LMN**, (*b*) **NML**, (*c*) **MLN**.

4 Find the axis and throw for the glide equivalent to reflection in $x = 8$ followed by (*a*) half-turn, (*b*) quarter-turn about the origin.

5 Find the single transformation equivalent to (*a*) **MT**, (*b*) **TR**, where **M** is reflection in $y = 6$, **T** is translation $\begin{bmatrix} 3 \\ 4 \end{bmatrix}$, **R** is rotation through 90° centre (2, 0).

6 **L**, **M**, **N** denote reflections in the lines l, m, n. What property do the lines l, m, n have if (*a*) **LM** = **ML**, (*b*) **LMN** = **NML**?

7 $AA'' = BB''$. Draw a diagram to show that this is not enough to ensure that AB can be translated onto $A''B''$. What is the extra property in Figure 2 which does ensure translation?

8 The displacement from l to the parallel line m is **d**; then **ML**, reflection in l followed by reflection in m, is equivalent to translation 2**d**. What translation is equivalent to **LM**? Find the translation equivalent to **LM** followed by **ML**.

9 **L** is reflection in $x = 0$, **M** is reflection in $x = 2$, **N** is reflection in $x = 6$. Find the single reflection equivalent to (*a*) **NML**, (*b*) **MNL**, (*c*) **LMN**.

10 Find the axis and throw for the glide equivalent to reflection in $y = 10$ followed by (*a*) half-turn, (*b*) quarter-turn about the origin.

11 Find the single transformation equivalent to (*a*) **MT**, (*b*) **RT**, where **M** is reflection in $x = 8$, **T** is translation $\begin{bmatrix} 4 \\ -3 \end{bmatrix}$, **R** is rotation through 90° centre (0, 6).

12 **M** is a reflection, **R** is a rotation, **T** is a translation. What geometric properties must they have if (*a*) **RT** = **TR**, (*b*) **RM** = **MR**?

2. COMBINATIONS OF ISOMETRIES

Every isometry is equivalent to three or fewer reflections. So every combination of isometries can be conveniently investigated as a combination of a number of reflections. We shall look in particular at two rotations and two glides.

Two rotations

It is easy to see that rotation about the same centre, of α then β, is equivalent to rotation $\alpha + \beta$ about that centre. But what happens when the centres of rotation are different? What is equivalent to a rotation through α, centre A, followed by rotation through β, centre B?

Replace rotation α by **ML**, reflection in l then in m, where the angle from l to m is $\frac{1}{2}\alpha$, and m is the line AB. Replace rotation β by **NM**, reflection in m then in n, where the angle from m to n is $\frac{1}{2}\beta$. (See Figure 10.)

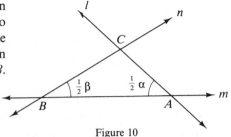

Figure 10

The combined rotations $=$ **NMML** $=$ **NL**, which is equivalent to rotation through twice the angle from l to n about C, the point where l meets n.

The angle from l to n is angle $ACB = (180° - \frac{1}{2}\alpha - \frac{1}{2}\beta)$ clockwise,

$\Rightarrow$ the angle of equivalent rotation $= 360° - \alpha - \beta$ clockwise,
$= \alpha + \beta$ anticlockwise.

Thus rotation α followed by rotation β is equivalent to rotation $\alpha + \beta$; which is hardly surprising. But that the resulting centre of rotation is found by halving these angles of rotation is unexpected.

Notice that, as always, the order in which the reflections are carried out is important. **ML**, reflection in l then m, gives an anticlockwise positive rotation through angle α. **LM** gives a clockwise negative rotation of α.

Example 3
XYZ is an equilateral triangle labelled anticlockwise.
Find the single transformation equivalent to:
(a) rotation through 120° about X then rotation through 120° about Y;
(b) rotation through 120° about Y then rotation through 240° about X.

(a) Let C be the centre of the resultant rotation of 240° (Figure 11). Then from CX to YX is 60° anticlockwise and from XY to CY is 60° anticlockwise. So C is the reflection of Z in XY.

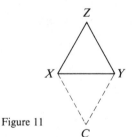

Figure 11

(b) Let D be the centre of the resultant rotation of 360°. Then from DY to XY is 60° anticlock- wise, so D lies on YZ, and from YX to DY is 120° anticlockwise, so D lies on CX. So there is no centre of rotation; as CX is parallel to YZ they do not meet. The resulting transformation is equivalent to reflection in YZ then in CX; which is translation through twice the displacement from YZ to CX, that is through twice the height of the triangle.

Two glides

Two opposite isometries are equivalent to a direct isometry. So two glides are equivalent to a translation when their axes are parallel, or to a rotation through twice the angle between their axes. But where is the centre of such a rotation?

If glide $\mathbf{G}_1$ is reflection in m_1, with translation $\mathbf{t}_1$ along m_1,

then
$$\mathbf{G}_1 = \mathbf{M}_1 \mathbf{T}_1$$
$$= \mathbf{M}_1 \mathbf{B}_1 \mathbf{A}_1,$$

where $\mathbf{A}_1, \mathbf{B}_1$ are reflections in parallel lines a_1, b_1 with displacement $\frac{1}{2}\mathbf{t}_1$ from a_1 to b_1.

If, similarly,
$$\mathbf{G}_2 = \mathbf{B}_2 \mathbf{A}_2 \mathbf{M}_2$$

then
$$\mathbf{G}_2 \mathbf{G}_1 = \mathbf{B}_2 \mathbf{A}_2 \mathbf{M}_2 \mathbf{M}_1 \mathbf{B}_1 \mathbf{A}_1$$

where (Figure 12) b_1 and a_2 are chosen to go through O, the point where m_1 and m_2 meet at an angle α.

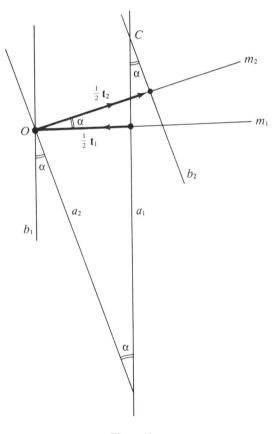

Figure 12

Now $\mathbf{M_2\,M_1}$ can be replaced by $\mathbf{A_2\,B_1}$, since both are equivalent to rotation 2α about O. Thus

$$\mathbf{G_2\,G_1} = \mathbf{B_2\,A_2\,A_2\,B_1\,B_1\,A_1}$$
$$= \mathbf{B_2\,A_1}$$

which is rotation through 2α about C, the point of intersection of a_1 and b_2. So the centre C is on the line a_1, perpendicular to m_1 and displaced $-\tfrac{1}{2}t_1$ from O, and on the line b_2, perpendicular to m_2 and displaced $\tfrac{1}{2}t_2$ from O.

Example 4
Find the coordinates of the centre of the rotation equivalent to a glide of 6 along the x-axis followed by a glide of 8 along the y-axis.

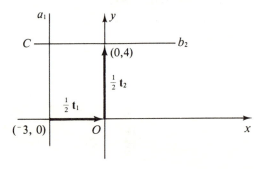

Figure 13

See Figure 13. The centre C lies on the line $x = ^-3$ and on the line $y = 4$. So the resulting transformation is a half-turn about $(^-3, 4)$.

Exercise B

1   XYZ is an equilateral triangle. What combination of rotations about X and Y is equivalent to rotation about Z of (a) 240°, (b) 120°, (c) 60°?

2   Calculate the coordinates of the centre of the rotation equivalent to rotation of 180° about the origin followed by rotation about the point $(6, 0)$ of (a) 60°, (b) 240°.

3   Show that reflection in the four sides of a square taken in anticlockwise order is equivalent to translation through twice one diagonal.

4   The lines a, b, c, d form the sides of a parallelogram with a parallel to c and b parallel to d. **A, B, C, D** denote reflection in a, b, c, d respectively. Find the single transformation equivalent to (a) **ABCD**, (b) **ACBD**.

5   Find the single transformation equivalent to glide 4 along the y-axis followed by glide 10 along (a) the x-axis, (b) the line $x = 6$.

6   Find the single transformation equivalent to glide $\begin{bmatrix} 2 \\ 6 \end{bmatrix}$ along $y = 3x$ followed by glide 5 along the y-axis.

7   XYZ is an equilateral triangle. Find the combination of rotations about X and Z which is equivalent to rotation about Y of (a) 240°, (b) 120°, (c) 60°.

8   Calculate the coordinates of the centre of the rotation equivalent to rotation of 180° about the origin followed by rotation about the point $(0, 4)$ of (a) 120°, (b) 50°, (c) 180°.

9   **A**, **C** denote reflection in opposite sides a, c of a square and **L**, **M** denote reflection in the diagonals l, m. Show that **ALMC** is equivalent to half-turn about the centre of the square. Also show that **LACM** = **MCAL**.

10   **A**, **B**, **C**, **D** denote reflection in a, b, c, d respectively where a is parallel to c and b is parallel to d. Find the single transformation equivalent to (a) **ABDC**, (b) **ACDB**.

11   Find the single transformation equivalent to glide $^-8$ along the x-axis followed by glide $^-6$ along (a) the y-axis, (b) the line $y = 10$.

12   Find the single transformation equivalent to a glide 5 along the x-axis followed by glide $\begin{bmatrix} 2 \\ 4 \end{bmatrix}$ along $y = 2x$.

3. STRIP PATTERNS

Pattern generators

M, **N** denote reflection in axes m, n with angle α between them and meeting at O (see Figure 14). We know that **NM** gives rotation 2α and that **MN** gives rotation $^-2\alpha$ about O. Now we extend our ideas to consider every possible combination of reflections in m and n.

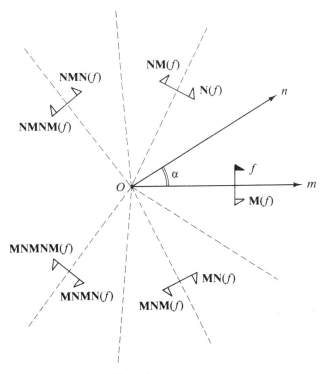

Figure 14

Figure 14 shows the images of the shaded flag f under some of the possible combinations of reflections. The collection of all possible images is called a *pattern*. In this case it is fairly clear that the pattern is a double-flag rotated through $\pm 2\alpha$ any number of times.

Whether or not the motif is a flag, the pattern generated by **M, N** will have this property of doubling and rotation. If $\alpha = 30°$ we shall only get one more double-flag image before the images start to coincide with one of those we have already. This gives a finite pattern with the motif occurring twelve times, as six double-flags. If $\alpha = 40°$ it requires two circuits before images coincide, giving a complete pattern with the motif occurring 18 times, as nine double-flags. How many times does the motif occur in the complete pattern when $\alpha = 50°$?

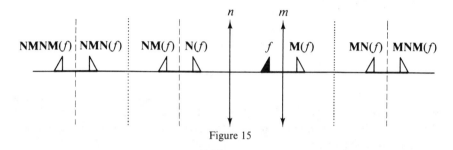

Figure 15

When the axes m, n are parallel Figure 14 turns into Figure 15 and **M, N** generate an infinite pattern. It is in the form of a long narrow strip called a *frieze pattern*, and does not spread out all over the plane. m and n are the only axes for reflection and the complete pattern has symmetry about m and n. However, there are many other axes, shown dotted and dashed, like m and n, about which the complete pattern also has symmetry and indeed which could have been used instead of m, n to generate the pattern. The pattern has the same symmetry as ...bdbdbd..., for which the motif is b or d instead of a flag. Similarly the pattern ...UUU... has the same symmetry. What is the motif in this case? These patterns are all classified as being of the same type because they can be generated by reflections in two parallel axes; this type is called **mm**.

Frieze patterns

There are only a limited number of transformations which will generate a pattern which does not spread all over the plane but is confined to a strip.

One possibility is obviously a single translation **T**, though we should have to apply it in the negative direction as well as in the positive direction (Figure 16), to get a complete strip.

A translation **T** can also be combined with half-turn **H** about a fixed point to generate a frieze pattern as in Figure 17. Notice that **HTH** is equivalent to T^{-1}. This type of pattern is called **t2**, because it is generated by a translation and a rotation of order two.

The complete pattern has many other centres of symmetry of order 2 and is

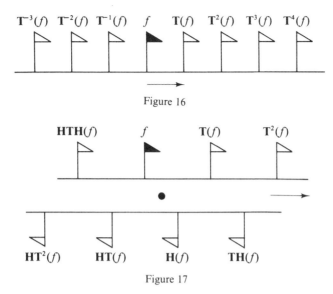

Figure 16

Figure 17

of the same type as ...bqbq... with motif b or q, and as ...NNN... with motif
Ⱶ or Ⱶ.

What happens if we combine translation with quarter-turn **Q**? We shall get
everything in Figure 16; but we shall also have **QT**, **QT²**, **QT³**, which give a set
of images up the page. These, combined with further translation, as **TQT**, **T²QT**,
etc., will spread over the whole plane. Clearly for a frieze pattern rotation must
be confined to half-turn.

In fact there are only seven possible types of frieze pattern, three of which
we have already considered (Figures 15, 16, 17).

t:	... bbbbbb ...	or	... JJJJ ...
t2:	... bqbqbq ...		... NNNN ...
g:	... bpbpbp ...		... LⱵLⱵ ...
mm:	... bdbdbd ...		... UUUU ...
m2:	... bdpqbd ...		... U∩U∩ ...
tm:	... bbbbbb pppppp ...		... DDDD ...
tmm:	... bdbdbd pqpqpq ...		... HHHH ...

(In this coding, **t** denotes a translation, **2** a half-turn, **g** a glide, and **m** a reflection.
Note that to get a complete pattern may require using the inverses for
translations and glides, the other isometries involved being self-inverse.)

Example 5
Classify the pattern ...SSSS...

This has no axis of symmetry, but it has half-turn symmetry and translational symmetry. So it is of the type **t2**.

Example 6
Classify the pattern generated by reflection in a line m and translation **T** perpendicular to m.

Although this is generated by translation it is seen, Figure 18, to be the same type as ...bdbd... and so is type **mm**. This is because the same pattern could be generated by replacing the translation by reflection in an axis $\frac{1}{2}$**T** from m.

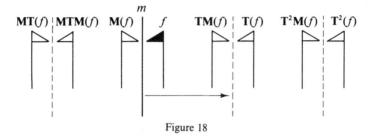

Figure 18

Exercise C

1   Classify the following frieze patterns:
    (*a*) ...AAAA...;   (*b*) ...CCCC...;   (*c*) ...FFFF...;
    (*d*) ...IIII...;   (*e*) ...XXXX...;   (*f*) ...ZZZZ....

2   Take a motif with its base at $(1,0)$. Show the complete patterns generated by **G** and **H** between $x = \pm 5$ when **G** is a glide of 2 along the x-axis, and **H** is half-turn about (*a*) $(0,0)$, (*b*) $(0,1)$. Classify the patterns.

3   Take a motif with its base at $(1,0)$. Show the complete patterns generated by **Y** and **T** between $x = \pm 5$ and $y = \pm 5$ when **Y** is reflection in the y-axis, and **T** is translation

    (*a*) $\begin{bmatrix} 0 \\ 2 \end{bmatrix}$,   (*b*) $\begin{bmatrix} 2 \\ 0 \end{bmatrix}$,   (*c*) $\begin{bmatrix} 2 \\ 2 \end{bmatrix}$.

    Classify the patterns.

4   **X** is reflection in the x-axis, **Y** is reflection in the y-axis, and **Z** is reflection in $x = 3$. Take a motif with its base at $(1,0)$. Show the complete pattern generated by **X**, **Y** and **Z** between $x = \pm 7$ and $y = \pm 7$ and classify the pattern.

5   To what type must a frieze pattern belong if it has symmetry (*a*) of **t2** and of **mm**, (*b*) of **g** and of **tm**?

6   In Figure 14, find the number of times the motif occurs when α is
    (*a*) 6°,   (*b*) 7°,   (*c*) $(p/q)°$ where p, q are integers.

7   Classify the following frieze patterns:
    (*a*) ...EEEE...;   (*b*) ...LLLL...;   (*c*) ...MMMM...;
    (*d*) ...OOOO...;   (*e*) ...PPPP...;   (*f*) ...TTTT....

8   Take a motif with its base at $(1,0)$. Show the complete patterns generated by **G** and **H** between $x = 7$ and $x = {}^-7$ when **G** is a glide of 3 along $y = 1$, and **H** is half-turn about (*a*) $(0,1)$, (*b*) the origin. Classify the patterns.

9   Take a motif with its base at $(1,0)$. Show the complete patterns generated by **X** and **T** between $x = \pm 5$ and $y = \pm 5$ when **X** is reflection in the x-axis, and **T** is translation

(a) $\begin{bmatrix} 0 \\ 2 \end{bmatrix}$,   (b) $\begin{bmatrix} 2 \\ 0 \end{bmatrix}$,   (c) $\begin{bmatrix} 2 \\ 2 \end{bmatrix}$.

Classify the patterns.

10   **X** is reflection in the x-axis, **Y** is reflection in the y-axis, and **Z** is reflection in $y = {}^{-}3$. Take a motif with its base at $(1,0)$. Show the complete pattern between $x = \pm 7$ and $y = \pm 7$ and classify it.

11   To what type must a strip pattern belong if it has symmetry (a) of **t2** and of **m2**, (b) of **g** and of **mm**?

12   In Figure 14, find the number of times the motif occurs when α is (a) $60°$, (b) $70°$. For what value of α is the pattern not discrete?

4. PLANE PATTERNS

Direct isometries

Consider the pattern generated by rotations **R**, **S**, where **R** is half-turn about A and **S** is rotation of $120°$ about B. Part of the pattern is shown in Figure 19, with motif the shaded flag, f.

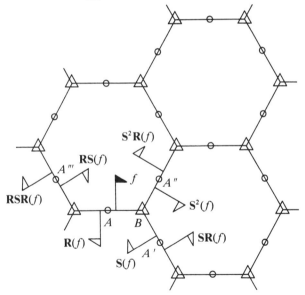

Figure 19

R gives rotational symmetry of order 2 about A. Applying **S** transforms A to A' and any pair of motifs symmetrical about A will become a pair symmetrical about A', such as **S**(f) and **SR**(f). So A' as well as A will be a centre of symmetry of order 2 for the complete pattern. Similarly **S** transforms A' to A'' with its motifs **S²**(f), **S²R**(f), while **R** transforms A' to A''' with its motifs **RS**(f), **RSR**(f), giving two further centres of symmetry of order 2 for the complete pattern. Similarly all images of *A*, some of which are indicated by a circle in Figure 19, will be centres of symmetry of order 2.

In the same way *B* is a centre of symmetry of order three and the points to which it is transformed by any combinations of **R** and **S** will also be centres of order three for the complete pattern. Some of them are shown by triangles in Figure 19.

The complete pattern will be spread over the whole plane, the centres of symmetry of order 2 and 3 being regularly spaced out. We also find centres of order 6 as well. As with friezes, it is the symmetry which fixes the nature of the pattern; the particular motif chosen merely affects its design. So we shall now concentrate on the centres of symmetry only.

What sort of pattern is generated by rotations of order 2 and 5?

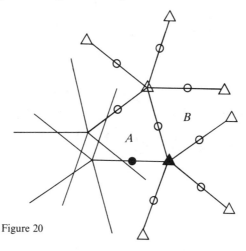

Figure 20

Figure 20 shows a few of the centres of symmetry generated by rotations of order 2 about *A* and order 5 about *B*. As we continue to develop it we find the centres becoming more and more densely packed. Any motif would have innumerable images overlapping in a hopeless jumble.

In fact the possible rotations which give clear, discrete patterns are surprisingly limited, as Barlow's theorem shows.

Barlow's theorem

If there is more than one centre of rotation, the only possible orders of rotation for a discrete pattern are 2, 3, 4 or 6.

Let *C* be one centre of rotation of order *p*. Then there will be many other centres of rotational symmetry of order *p*. Among them, since the pattern is discrete, must be one which is closest to *C*; or one of several equally near. Take this to be *D* (Figure 21). Then rotation

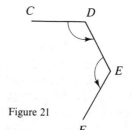

Figure 21

through $360°/p$ about D takes C to a new centre E, and the same rotation about E takes D to a further centre F.

If F is the same as C, then the triangle CDE is equilateral; all the angles are $60°$ and $p = 6$.

If F is not the same as C, CF can not be less than CD, since D is a centre nearest to C.

Either $CF = CD$, so that $CDEF$ is a square and $p = 4$, or $CF > CD$, in which case the angles must be obtuse and $p = 3$ or 2.

Note that for $p = 5$, as in Figure 20, we get $CF < CD$. That is, for any centre D near C there is always another centre F even closer. So the centres are not spaced out and the pattern is not discrete.

So the only possible angles of rotation are $180°$, $120°$, $90°$, $60°$. You might think that even these could be combined together in a large number of ways but this is not the case. There are, in fact, just five types of pattern involving only direct isometries, rotations or translations (**pi**). They are named by the order of the highest rotation they contain **p1**, **p2**, **p3**, **p4**, **p6**. Barlow's theorem limits us to these orders and it so happens that there is only one pattern for each order of rotation; they are illustrated in Figure 22.

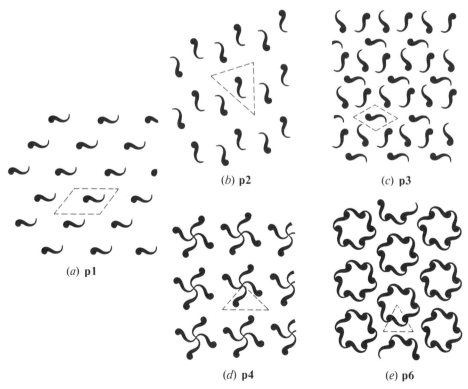

(a) p1

(b) p2

(c) p3

(d) p4

(e) p6

Figure 22

Example 7
Classify the pattern in Figure 19.

Although it is generated by rotations of order 2 and 3, the resulting pattern contains centres of symmetry of order 6 (the centres of the hexagons). So this is an example of **p6**.

Patterns involving reflections

Here there are several types of patterns for each order of rotation. We can obtain them from the five rotation patterns by introducing reflection or glide. There are twelve of them in all. They are shown below with their international symbols and illustrated in Figure 23.

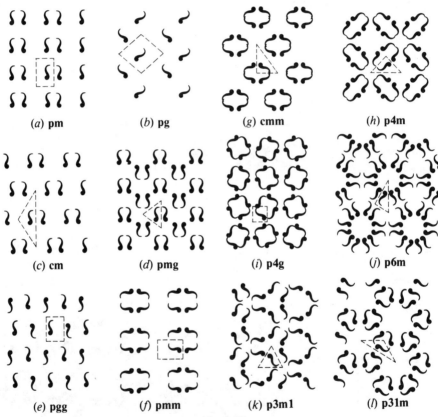

Figure 23

p1 becomes, taking a rectangular lattice of motifs,

	pm	by adding a reflection,
	pg	by adding a glide,
	cm	by adding both.
p2 becomes	**pmg**	by adding a reflection,
	pgg	by adding a glide,
	pmm⎱	⎰by adding two reflections in perpendicular
	cmm⎰	⎱lines in two different ways.

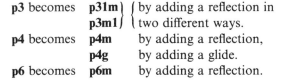

p3 becomes **p31m** } by adding a reflection in
             p3m1 } two different ways.
p4 becomes **p4m** by adding a reflection,
             p4g by adding a glide.
p6 becomes **p6m** by adding a reflection.

Exercise D

1   If to each of the strip-groups in turn we add a second translation perpendicular to the strip, which of the 17 plane pattern groups is produced in each case?

2   Classify the patterns shown in Figure 24.

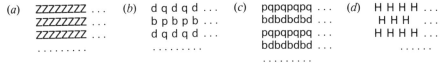

(a) ZZZZZZZZ ... (b) d q d q d ... (c) pqpqpqpq ... (d) H H H H ...
    ZZZZZZZZ ...   b p b p b ...   bdbdbdbd ...   H H H ...
    ZZZZZZZZ ...   d q d q d ...   pqpqpqpq ...   H H H H ...
    . . . .   bdbdbdbd ...   .
                                       . . . .

Figure 24

3   Classify the symmetry of an infinite sheet of:
     (a) squared paper;
     (b) hexagonal wire netting;
     (c) isometric paper (ruled in equilateral triangles).

4   Discover the eight ways of covering a plane with regular polygons of more than one kind (the so-called *semi-regular* or *Archimedean* tessellations). Allocate them to their different symmetry types.
     (*Hint.* There is one of each of the types **p4m**, **p4g**, **cmm**, **p6**, and four of the type **p6m**.)

5   Find the types of symmetry groups of the patterns of block and bricks shown in Figure 25.

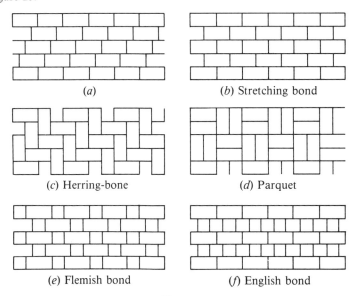

               (a)   (b) Stretching bond

            (c) Herring-bone   (d) Parquet

            (e) Flemish bond   (f) English bond

Figure 25

6   Try making up your own patterns in the various symmetry types. Figure 26 shows
    an example, which might be used for wallpaper.

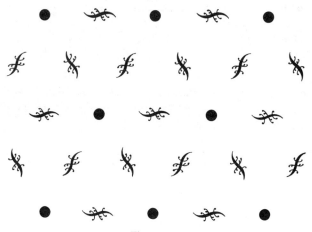

Figure 26

7   Cut out several of each of the shapes in Figure 27 from coloured paper and try to
    fit sets of them into patterns. The first will make a pattern in **p6**; the second one
    in **cm** and one in **pmm**; the third in **p3** or **p6**, depending on the shading; while the
    fourth will work in **cm**, **pmm**, and **p4m**.

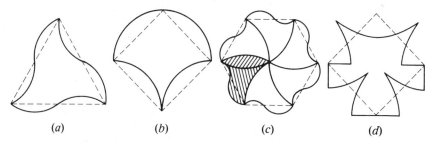

(a)   (b)   (c)   (d)

Figure 27

8   In each of the patterns shown in Figure 23 there appears a triangle or parallelogram
    outlined in pecked lines. This figure is called a *fundamental region* of the pattern;
    it contains just one 'tadpole' motif, and its images under the rotations, translations,
    reflections, or glide reflections of the pattern-group will form a tiling of the plane
    which shows the pattern concerned.
        Draw each fundamental region, and mark on it:
        (a) any axes of reflection (which will have to be along edges of the regions);
        (b) any centres of half-turns (which will have to be at vertices of the region, or
             at midpoints of its sides);
        (c) any centre of rotations of order 3, 4 or 6 (which will have to be at vertices);
        (d) any glide-reflection axes which are not mirrors (mark these⟍).
        Must a vertex where two mirrors intersect be a centre of rotation?

9   Classify the patterns in Figure 28 and in Figure 29: (i) if each motif is to be
    transformed onto one of the same colour; (ii) if each motif is to be transformed onto
    one of the same shape regardless of colour.

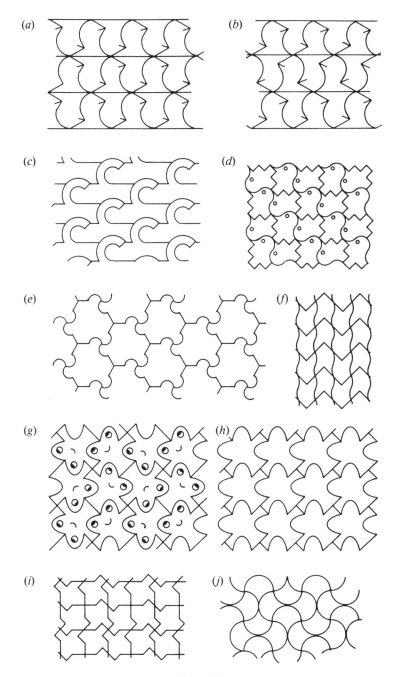

Figure 28

(a)   (b)   (c)   (d)

Figure 29

5. SUMMARY

Every transformation can be expressed as a combination of reflections. Reflection in two parallel axes gives translation through twice the displacement between the axes. Reflection in two intersecting axes gives rotation about the point of intersection through twice the intersection angle. Three reflections give a glide or a single reflection.

Rotation through α about A followed by rotation through β about B is equivalent to rotation through $\alpha + \beta$ about C, where the rotation from AB to AC is $\frac{1}{2}\alpha$ and from BA to BC is $\frac{1}{2}\beta$.

Two glides are equivalent to a rotation or translation.

There are seven possible frieze patterns.

t:	... F F F F ...
t2:	... Z Z Z Z ...
g:	... L Γ L Γ ...
mm:	... W W W W ...
m2:	... v ∧ v ∧ ...
tm:	... E E E E ...
tmm:	... X X X X ...

There are five possible direct plane patterns: **p1, p2, p3, p4, p6**. There are twelve possible plane patterns involving reflections.

6. CRYSTALS

It is when we come to consider three-dimensional patterns that the subject of this chapter becomes important. It is fascinating to be able to classify wallpaper and tiling patterns, no doubt; but we live in a three-dimensional world in which many regular patterns can be found. If the atoms of a chemical compound are packed in regular pattern then it is a crystal and the structure of the pattern is of considerable interest for many different purposes. Classifying such patterns is one of the main tasks of the science of crystallography. It is a task much more elaborate and formidable than the plane classification we have considered in this chapter, but the same methods have to be used as here. In particular Barlow's theorem tells us that we cannot have pentagonal symmetry. No natural crystal therefore can occur in the form of a regular dodecahedron or icosahedron. Crystals in the form of the other regular solids are possible, and, indeed, all are actually found.

There are only 14 possible lattices, localised patterns of points. These are the Bravais lattices shown in Figure 30. They can be combined with translations to give a total of 230 possible patterns in space. Some of these are shown in Figure 31.

CUBIC
$$\alpha = \beta = \gamma = 90°$$
$$a = b = c$$

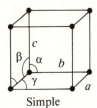

Simple   Face-centred   Body-centred

TETRAGONAL
$$\alpha = \beta = \gamma = 90°$$
$$a = b \neq c$$

MONOCLINIC
$$\alpha = \gamma = 90° \; \beta \neq 90° \text{ obtuse}$$
$$a \neq b \neq c$$

 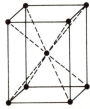 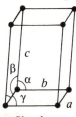 

Simple   Body-centred   Simple   End face-centred

ORTHORHOMBIC
$$\alpha = \beta = \gamma = 90°$$
$$a \neq b \neq c$$

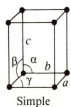

Simple   Body-centred   End face-centred   Face-centred

TRICLINIC
$$\alpha \neq \beta \neq \gamma \neq 90°$$
$$a \neq b \neq c$$

TRIGONAL
$$\alpha = \beta = \gamma \neq 90°$$
$$a = b = c$$

HEXAGONAL
$$\alpha = \beta = 90°; \gamma = 120°$$
$$a = b = \neq c$$

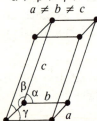 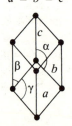 

Bravais lattices

α is the angle between positive b and positive c. β is the angle between positive c and positive a. γ is the angle between positive a and positive b.

Figure 30

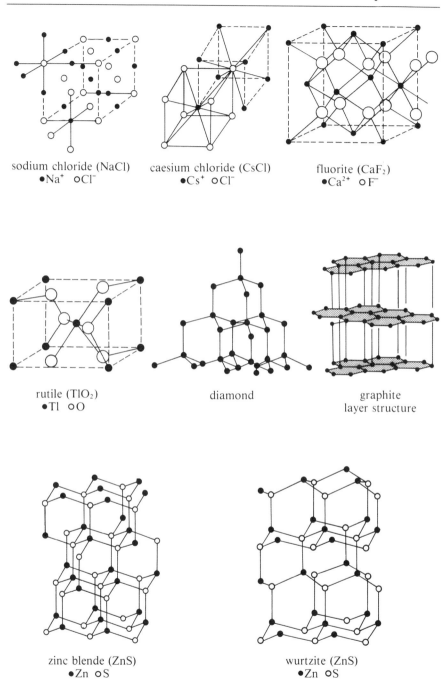

sodium chloride (NaCl)
•Na⁺ ○Cl⁻

caesium chloride (CsCl)
•Cs⁺ ○Cl⁻

fluorite (CaF₂)
•Ca²⁺ ○F⁻

rutile (TlO₂)
•Tl ○O

diamond

graphite
layer structure

zinc blende (ZnS)
•Zn ○S

wurtzite (ZnS)
•Zn ○S

Figure 31

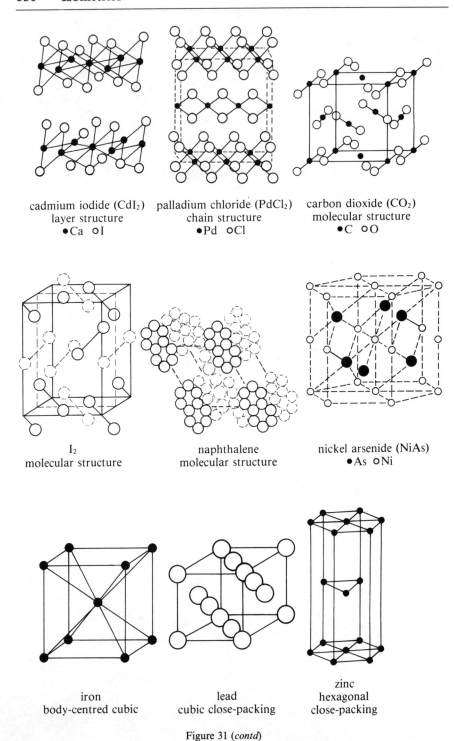

cadmium iodide (CdI₂)
layer structure
●Ca ○I

palladium chloride (PdCl₂)
chain structure
●Pd ○Cl

carbon dioxide (CO₂)
molecular structure
●C ○O

I₂
molecular structure

naphthalene
molecular structure

nickel arsenide (NiAs)
●As ○Ni

iron
body-centred cubic

lead
cubic close-packing

zinc
hexagonal
close-packing

Figure 31 (*contd*)

Miscellaneous exercise

1   The reflection in the line through the origin at 25° to the *x*-axis is denoted by **A**, and **B** denotes the reflection in the *x*-axis. Give a complete description of the geometrical transformations
    (i) **AB**, (ii) **BA**, (iii) **BABA**.
    Explain why the inverse of **AB** is **BA** and find the inverse of **BABA** in terms of **A** and **B** only. (SMP)

2   The operations **L**, **M** are reflections in parallel mirrors *l*, *m*. What single operation is the combination **ML** (first **L**, then **M**)? If **ML** = **LK**, what is the operation **K**? Show **LML** is a reflection in the line *m**, where *m** is the image of *m* after reflection in *l*. (A simple reasoned argument is required, not merely a sketch.)
    Is this last statement still true when *m* and *l* are not parallel? Give reasons or a counter-example. (SMP)

3   A cardboard triangle *ABC* rests on a sheet of graph paper with its corner *A* on the point $(0,0)$, *B* on $(1,0)$ and *C* on $(0,1)$. It is given a positive (counter-clockwise) quarter-turn **R** about $(0,1)$ and then a positive quarter-turn **S** about $(1,0)$. We call the combination of these operations in this order **SR**.
    Draw a sketch to show the final positions of *A*, *B* and *C*, labelling them *A**, *B** and *C**. What point, if any, is unchanged by the transformation **SR**, and what single transformation would have the same effect?
    Show that **RS** leaves a point fixed. Give the coordinates of this point, and describe completely the single transformation equivalent to **RS**. (SMP)

4   Figure 32 shows part of a pattern which has a centre of symmetry of order 4 and an axis of symmetry, *m*.
    Which of the points *A*, *B*, *C* is the centre of symmetry?
    Draw sufficient of the pattern to show these symmetries and mark on it the axis *m*.
    What other symmetries must the pattern have? (SMP)

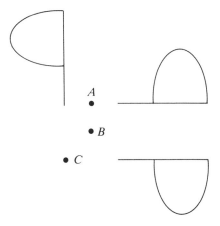

Figure 32

5   Use graph paper and a scale of 1 cm to one unit.
    (i) Copy Figure 33 and draw the flag motif together with its images under
        (*a*) a half-turn about the origin *O*,
        (*b*) a half-turn about the centre *A* $(3,2)$.
    (ii) Apply to the three flags now on your diagram the same half-turns (*a*) and (*b*) above.
    Continued applications of these operations successively would produce an infinite pattern. Give a translation under which the pattern is invariant. (SMP)

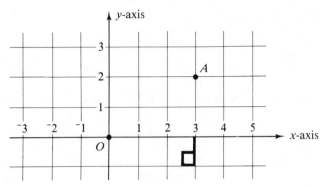

Figure 33

6   **Q** denotes a quarter-turn clockwise about X. **T** denotes a translation two units to the right. [$AB = 2$ units.] Copy Figure 34 onto squared paper and illustrate on your diagram the effects of **Q**, **Q²**, **Q³** on the flag motif. Which is the inverse of **Q**?

On another diagram illustrate **Q²**, **TQ²**, **Q²T**, **TQ²T**, **Q²TQ²**. Which is the inverse of **T**? Hence find a combination of the transformations **Q** and **T** which gives a quarter-turn clockwise about the base A of the flag.   (SMP)

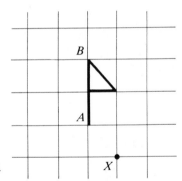

Figure 34

7   The tile shown in Figure 35 can be used to form a tessellation with centres of sixfold rotational symmetry. (The tiles may not be turned over.)
   (*a*) Sketch a part of the pattern.
   (*b*) On a drawing of the tile, mark:
        (i) a centre of sixfold rotational symmetry (mark it S);
        (ii) a centre of rotational symmetry of order 2 only (mark it H);
        (iii) a third centre of rotational symmetry of order different from S and H, stating the order.   (SMP)

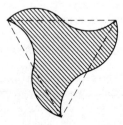

Figure 35

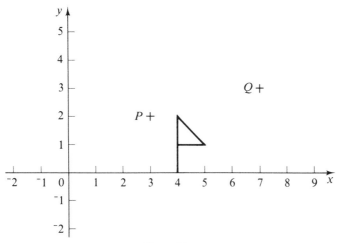

Figure 36

8   Copy Figure 36 showing points P (3, 2) and Q (7, 3) and the flag motif. (Use graph paper and a scale of 1 cm to one unit.)

T is the translation $\begin{bmatrix} 4 \\ 1 \end{bmatrix}$ which maps P onto Q, and **R** is a rotation of 90° about P.

(i) On your figure show and label the image of the motif under the transformations

(a) **T**⁻¹,   (b) **RT**⁻¹,   (c) **TRT**⁻¹.

(ii) State whether **TRT**⁻¹ is a direct or an opposite isometry, and state the angle through which it rotates lines.

(iii) State the images of the point Q under (a) **T**⁻¹, (b) **RT**⁻¹, (c) **TRT**⁻¹.

(iv) Describe **TRT**⁻¹ completely as a single isometry.   (SMP)

9   **M** denotes reflection in $x = 0$. **T** denotes the translation $\begin{bmatrix} 2 \\ 1 \end{bmatrix}$. The combined transformation **TM** is called a glide and is denoted by **A**. It may be thought of as reflection in a line (the axis) together with translation (throw) along the axis.

(a) By considering the unit square, or otherwise, show that the axis is $x = 1$ and state the throw.

(b) What single transformation is equivalent to **A**²?

The combined transformation **MT** is also a glide and is denoted by **B**.

(c) Show that **B** has throw of 1 and state its axis.

(d) Give the single transformations equivalent to **B**², **BA**.

(e) Hence state in terms of **M** and **T** the translation $\begin{bmatrix} -8 \\ 6 \end{bmatrix}$.   (SMP)

10   Draw, on graph paper, two perpendicular axes and label the x-axis from ⁻5 to 10 at 1 cm intervals. Draw any simple flag motif, height 2 units, width 1 unit, with base at the origin.

The glide **G** consists of reflection in the x-axis together with translation 3 units along this axis. **M** denotes reflection in the line $x = 2$.

On your diagram show the images, clearly labelled, of the flag under each of the transformations **M**, **G**, **MG**, **GM**, **MGM**, **GMG**. (*Question continues overleaf.*)

(a) State which of these transformations are self-inverse and which is the inverse of **G**.

(b) Hence write down and simplify the inverse of **G²**.

(c) State the magnitude of the translation given by **G²**.

(d) Hence give combinations of **G** and **M** which will map the base of the original flag to (i) (10, 0), (ii) (61, 0), (iii) (‾21, 0).   (SMP)

11 A strip pattern for British Rail carpeting is based on the motif BR shown in Figure 37. (The point and the lines are not part of the pattern.) The total pattern is

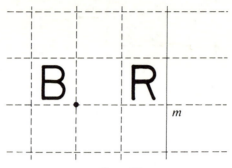

Figure 37

formed by any combination of (i) half-turns **H** about the point, (ii) reflections **M** in the line *m*, and (iii) translations **T** of twice the displacement of the point from the line.

Illustrate on graph paper the effect of (a) **H**, (b) **M**, (c) **T**, and also of (d) **HM**, (e) **MH**, (f) **HMH**, all clearly labelled on separate diagrams.

On a separate diagram sketch enough to indicate the whole pattern and describe two other isometries which leave the whole pattern invariant.   (SMP)

12 Figure 38 represents a square with vertices 1, 2, 3, 4 and centre *O*. Let

$$a = \begin{pmatrix} 1 & 2 & 3 & 4 \\ 3 & 2 & 1 & 4 \end{pmatrix} \quad \text{and} \quad b = \begin{pmatrix} 1 & 2 & 3 & 4 \\ 3 & 4 & 1 & 2 \end{pmatrix}$$

be symmetry transformations of the square (in which each vertex in the top rows is sent to the one directly beneath it in the corresponding bottom row). Interpret *a* and *b* geometrically in terms of rotations about axes through *O*.

Express, in terms of *a* and *b*, the transformation which is a rotation of 180° about the axis 13.

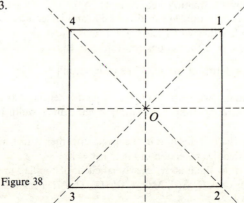

Figure 38

18

Equations and matrices

1. TWO DIMENSIONS

We begin with a review of some familiar ideas, in a way which can readily be extended to three or more dimensions.

Lines and normals

In two-dimensional space, in which a general point P is represented by the coordinates (x, y), a linear equation $ax + by = c$ restricts P to lie on a line.

For example, $3x + 4y = 24$ is a line through $(8, 0)$, $(0, 6)$ and $(^-4, 9)$ as these are some of the possible pairs of values of x and y which satisfy the restriction $3x + 4y = 24$. (See Figure 1.)

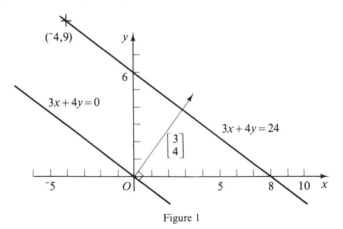

Figure 1

The line $3x + 4y = 0$ clearly passes through the origin $(0, 0)$. It does not meet the line $3x + 4y = 24$ since there can be no pair (x, y) for which $3x + 4y = 0$ and $3x + 4y = 24$. So these two lines are parallel.

The equation $3x + 4y = 0$ can also be written as

$$\begin{bmatrix} 3 \\ 4 \end{bmatrix} \cdot \begin{bmatrix} x \\ y \end{bmatrix} = 0 \quad \text{or} \quad \begin{bmatrix} 3 \\ 4 \end{bmatrix} \cdot \underset{\sim}{OP} = 0$$

where $P(x, y)$ is any point on the line. This shows that OP is perpendicular to $\begin{bmatrix} 3 \\ 4 \end{bmatrix}$. So $3x + 4y = 0$ is a line perpendicular to $\begin{bmatrix} 3 \\ 4 \end{bmatrix}$. Since $3x + 4y = 24$ is a parallel line, it is also perpendicular to $\begin{bmatrix} 3 \\ 4 \end{bmatrix}$. We say that $\begin{bmatrix} 3 \\ 4 \end{bmatrix}$ is *normal* to these lines.

Example 1

Find the equation of the line through $(2, 3)$ and with normal $\begin{bmatrix} 4 \\ 5 \end{bmatrix}$.

Since the normal is $\begin{bmatrix} 4 \\ 5 \end{bmatrix}$, the equation of the line must be $4x + 5y = k$ for some k.

But it passes through $(2, 3)$ so $4 \times 2 + 5 \times 3 = k$, and hence $k = 23$.
The equation of the line is $4x + 5y = 23$.

The angle between two lines

The angle between two lines can conveniently be found by using normals. Since the normals are perpendicular to their lines, the angle between the lines is the same as the angle between their normals.

Example 2

Find the angle between the lines $2x + 3y = 4$ and $4x - 5y = 6$.

The normals to these lines are

$$\mathbf{p} = \begin{bmatrix} 2 \\ 3 \end{bmatrix} \quad \text{and} \quad \mathbf{q} = \begin{bmatrix} 4 \\ -5 \end{bmatrix}.$$

$$p = \sqrt{(2^2 + 3^2)} = \sqrt{13}; \quad q = \sqrt{(4^2 + 5^2)} = \sqrt{41}; \quad \mathbf{p}.\mathbf{q} = pq \cos \theta°.$$

So
$$8 + {}^-15 = \sqrt{13} \times \sqrt{41} \cos \theta°,$$

$$\cos \theta° = \frac{-7}{\sqrt{13}\,\sqrt{41}},$$

$$\theta° = 108° \text{ to the nearest degree.}$$

The acute angle between the lines is $72°$ to the nearest degree.

Exercise A

1   Find the equations of the following lines:

    (*a*) through $(1, 2)$ with normal $\begin{bmatrix} 3 \\ 4 \end{bmatrix}$;

    (*b*) through $(5, {}^-4)$ with normal $\begin{bmatrix} 2 \\ 5 \end{bmatrix}$;

    (*c*) through $({}^-3, 7)$ with normal $\begin{bmatrix} -5 \\ 2 \end{bmatrix}$.

2   Write down the normal vectors to the following lines and hence state which pairs are parallel and which perpendicular.
    (*a*) $2x + y = 3$;   (*b*) $3x - 6y = 7$;   (*c*) $4x + 2y = 5$;
    (*d*) $5x - 10y = 0$;   (*e*) $6x + 3y = {}^-4$.

3   Find the angle between the lines:
    (*a*) $3x - 4y = 5$ and $6x + 7y = 9$;
    (*b*) $2x - 3y = 7$ and $4x - y = 8$.

4 Find all the angles of the triangle formed by the lines
$$9x + 8y = 7, \quad 6x - 5y = 4, \quad 3x + 2y = 1.$$

5 Find the equations of the following lines:

    (a) through $(7, 5)$ with normal $\begin{bmatrix} 3 \\ 6 \end{bmatrix}$;

    (b) through $(^-4, 6)$ with normal $\begin{bmatrix} 5 \\ -2 \end{bmatrix}$;

    (c) through $(^-1, ^-3)$ with normal $\begin{bmatrix} -4 \\ -7 \end{bmatrix}$.

6 Write down the normal vectors to the following lines and hence state which pairs are parallel and which perpendicular.

    (a) $x + 3y = 3$; (b) $3x - 9y = 8$; (c) $2x + 6y = 7$;
    (d) $6x - 2y = 5$; (e) $6x + 2y = 1$.

7 Find the angle between the lines:

    (a) $7x - 5y = 3$ and $2x + 4y = 9$;   (b) $3x + 8y = 5$ and $5x - 3y = 8$.

8 Find all the angles of the triangle formed by the lines
$$2x - 7y = 6, \quad 8x + y = 5, \quad 2x - 9y = 4.$$

2. ELIMINATION

$3x + 4y = 24$ restricts $P(x, y)$ to lie on a line perpendicular to $\begin{bmatrix} 3 \\ 4 \end{bmatrix}$. $5x + 6y = 37$ restricts $P(x, y)$ to lie on a line perpendicular to $\begin{bmatrix} 5 \\ 6 \end{bmatrix}$. These two equations taken simultaneously will restrict P to be at the intersection of these two lines.

    The point of intersection may be calculated by elimination.

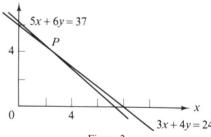

Figure 2

Equation one:	E_1:	$3x + 4y = 24,$
Equation two:	E_2:	$5x + 6y = 37.$

New equation one = 5 times old
    equation one: $E_1' = 5E_1$: $15x + 20y = 120,$
New equation two = 3 times old
    equation two: $E_2' = 3E_2$: $15x + 18y = 111.$

Subtraction will now eliminate x
from equation two: $E_1' = E_1$: $15x + 20y = 120,$
         $E_2' = {}^-E_1 + E_2$: $^-2y = {}^-9.$

We can now eliminate y from
equation one: $E_1' = E_1 + 10E_2$: $15x \qquad = 30,$
         $E_2' = E_2$: $^-2y = {}^-9.$

So $x = 2, \quad y = 4.5.$

Note that it may sometimes be more convenient to eliminate y first, rather than x.

Example 3

Solve
$$\left.\begin{array}{r} 4x - 3y = 2, \\ 7x + y = 5. \end{array}\right\}$$

$$\begin{array}{ll} E'_1 = E_1: & \left.\begin{array}{r} 4x - 3y = 2, \\ 21x + 3y = 15. \end{array}\right\} \\ E'_2 = 3E_2: & \\ E'_1 = E_1: & \left.\begin{array}{r} 4x - 3y = 2, \\ 25x = 17. \end{array}\right\} \\ E'_2 = E_1 + E_2: & \\ E'_2 = \frac{1}{25}E_2: & x = 0.68. \\ \text{Substitute in } E_1: & 4(0.68) - 3y = 2. \\ & y = 0.24. \end{array}$$

The solution is $x = 0.68$, $y = 0.24$.

Note that there is no solution when the lines are parallel, unless they coincide, in which case there is a whole line of solutions.

Exercise B

1   Use elimination to solve the following simultaneous equations:
(a) $3x + 2y = 1$, $4x + 3y = 2$;
(b) $5x + 2y = 4$, $6x - 4y = 5$;
(c) $7x - 6y = 3$, $8x - 9y = {}^-6$.

2   Find the equations of the lines through the points:
(a) $(3, 5)$ and $(7, 4)$; (b) $(2, {}^-6)$ and $({}^-5, 8)$;
by assuming that the equations of the line is $ax + by = 1$ and solving simultaneous equations to get the values of a and b.

3   Find values of k for which the equations
$$x + ky = 1, \quad kx + y = 2 - k$$
have (a) no solution, (b) one solution, (c) many solutions.

4   Use elimination to solve the following simultaneous equations:
(a) $4x + 5y = 3$, $5x + 6y = 4$;
(b) $3x - 2y = 1$, $4x + 8y = 7$;
(c) $5x - 4y = 2$, $7x - 3y = 8$.

5   Find the equations of the lines through the points:
(a) $(4, 7)$ and $(3, 2)$; (b) $(5, {}^-1)$ and $({}^-9, 6)$;
by assuming that the equation of the line is $ax + by = 1$ and solving simultaneous equations to get the values of a and b.

6   Find values of k for which the equations
$$kx - 2y = 16, \quad 8x - ky = 2k^2$$
have (a) no solution, (b) one solution, (c) many solutions.

3. MATRICES

The elimination process of the previous section may be expressed in matrix notation.

$$3x+4y = 24, \\ 5x+6y = 37. \qquad \begin{bmatrix} 3 & 4 \\ 5 & 6 \end{bmatrix} \begin{bmatrix} x \\ y \end{bmatrix} = \begin{bmatrix} 24 \\ 37 \end{bmatrix}.$$

Multiplying both sides by

$$E_1' = 5E_1 + 0E_2: \quad 15x + 20y = 120, \\ E_2' = 0E_1 + 3E_2: \quad 15x + 18y = 111. \qquad \begin{bmatrix} 5 & 0 \\ 0 & 3 \end{bmatrix} \text{ gives } \begin{bmatrix} 15 & 20 \\ 15 & 18 \end{bmatrix} \begin{bmatrix} x \\ y \end{bmatrix} = \begin{bmatrix} 120 \\ 111 \end{bmatrix}.$$

Multiplying both sides by

$$E_1' = 1E_1 + 0E_2: \quad 15x + 20y = 120, \\ E_2' = {}^-1E_1 + 1E_2: \quad {}^-2y = {}^-9. \qquad \begin{bmatrix} 1 & 0 \\ {}^-1 & 1 \end{bmatrix} \text{ gives } \begin{bmatrix} 15 & 20 \\ 0 & {}^-2 \end{bmatrix} \begin{bmatrix} x \\ y \end{bmatrix} = \begin{bmatrix} 120 \\ {}^-9 \end{bmatrix}.$$

Multiplying both sides by

$$E_1' = E_1 + 10E_2: \quad 15x = 30, \\ E_2' = E_2: \quad {}^-2y = {}^-9. \qquad \begin{bmatrix} 1 & 10 \\ 0 & 1 \end{bmatrix} \text{ gives } \begin{bmatrix} 15 & 0 \\ 0 & {}^-2 \end{bmatrix} \begin{bmatrix} x \\ y \end{bmatrix} = \begin{bmatrix} 30 \\ {}^-9 \end{bmatrix}.$$

Multiplying both sides by

$$E_1' = \tfrac{1}{15}E_1: \quad x = 2, \\ E_2' = {}^-\tfrac{1}{2}E_2: \quad y = 4.5. \qquad \begin{bmatrix} \tfrac{1}{15} & 0 \\ 0 & {}^-\tfrac{1}{2} \end{bmatrix} \text{ gives } \begin{bmatrix} 1 & 0 \\ 0 & 1 \end{bmatrix} \begin{bmatrix} x \\ y \end{bmatrix} = \begin{bmatrix} 2 \\ 4.5 \end{bmatrix}.$$

Notice that the so-called *elementary* matrices which we use to multiply both sides are merely the coefficients of the operations given on the left.

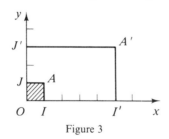

Figure 3

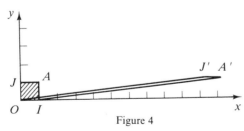

Figure 4

The elementary matrices always represent shears or stretches. For example, $\begin{bmatrix} 5 & 0 \\ 0 & 3 \end{bmatrix}$ transforms the base vectors $\begin{bmatrix} 1 \\ 0 \end{bmatrix}, \begin{bmatrix} 0 \\ 1 \end{bmatrix}$ to $\begin{bmatrix} 5 \\ 0 \end{bmatrix}, \begin{bmatrix} 0 \\ 3 \end{bmatrix}$ and is a two-way stretch with scale factors 5 parallel to the x-axis and 3 parallel to the y-axis. (See Figure 3.) $\begin{bmatrix} 1 & 10 \\ 0 & 1 \end{bmatrix}$ transforms the base vectors $\begin{bmatrix} 1 \\ 0 \end{bmatrix}, \begin{bmatrix} 0 \\ 1 \end{bmatrix}$ to $\begin{bmatrix} 1 \\ 0 \end{bmatrix}, \begin{bmatrix} 10 \\ 1 \end{bmatrix}$ giving a shear with the x-axis invariant and shearing constant 10. (See Figure 4.)

Combining all the steps gives

$$\left(\begin{bmatrix} \frac{1}{15} & 0 \\ 0 & -\frac{1}{2} \end{bmatrix}\begin{bmatrix} 1 & 10 \\ 0 & 1 \end{bmatrix}\begin{bmatrix} 1 & 0 \\ -1 & 1 \end{bmatrix}\begin{bmatrix} 5 & 0 \\ 0 & 3 \end{bmatrix}\right)\begin{bmatrix} 3 & 4 \\ 5 & 6 \end{bmatrix} = \begin{bmatrix} 1 & 0 \\ 0 & 1 \end{bmatrix}$$

so the product of the four elementary matrices is the inverse of the matrix $\begin{bmatrix} 3 & 4 \\ 5 & 6 \end{bmatrix}$.

$$\begin{bmatrix} 3 & 4 \\ 5 & 6 \end{bmatrix}^{-1} = \begin{bmatrix} \frac{1}{15} & 0 \\ 0 & -\frac{1}{2} \end{bmatrix}\begin{bmatrix} 1 & 10 \\ 0 & 1 \end{bmatrix}\begin{bmatrix} 1 & 0 \\ -1 & 1 \end{bmatrix}\begin{bmatrix} 5 & 0 \\ 0 & 3 \end{bmatrix}$$

$$= \begin{bmatrix} \frac{1}{15} & \frac{2}{3} \\ 0 & -\frac{1}{2} \end{bmatrix}\begin{bmatrix} 5 & 0 \\ -5 & 3 \end{bmatrix} = \begin{bmatrix} -3 & 2 \\ \frac{5}{2} & -\frac{3}{2} \end{bmatrix}.$$

This gives another interpretation of the algebra in terms of transformations rather than lines.

$$\begin{bmatrix} 3 & 4 \\ 5 & 6 \end{bmatrix}\begin{bmatrix} x \\ y \end{bmatrix} = \begin{bmatrix} 24 \\ 37 \end{bmatrix}$$

can be interpreted as 'the transformation $\begin{bmatrix} 3 & 4 \\ 5 & 6 \end{bmatrix}$ maps $\begin{bmatrix} x \\ y \end{bmatrix}$ to $\begin{bmatrix} 24 \\ 37 \end{bmatrix}$'.

The inverse transformation, made up of shears and stretches, maps $\begin{bmatrix} 24 \\ 37 \end{bmatrix}$ back to $\begin{bmatrix} 2 \\ 4.5 \end{bmatrix}$.

Example 4

Use elementary matrices to solve $\begin{bmatrix} 2 & 5 \\ 3 & -4 \end{bmatrix}\begin{bmatrix} x \\ y \end{bmatrix} = \begin{bmatrix} 1 \\ 13 \end{bmatrix}$ and hence give the inverse of $\begin{bmatrix} 2 & 5 \\ 3 & -4 \end{bmatrix}$.

$$\begin{bmatrix} 2 & 5 \\ 3 & -4 \end{bmatrix}\begin{bmatrix} x \\ y \end{bmatrix} = \begin{bmatrix} 1 \\ 13 \end{bmatrix}$$

Current product of
elementary matrices

Multiplying:

by $\begin{bmatrix} 3 & 0 \\ 0 & 2 \end{bmatrix}$ gives $\begin{bmatrix} 6 & 15 \\ 6 & -8 \end{bmatrix}\begin{bmatrix} x \\ y \end{bmatrix} = \begin{bmatrix} 3 \\ 26 \end{bmatrix}$; $\begin{bmatrix} 3 & 0 \\ 0 & 2 \end{bmatrix}$

by $\begin{bmatrix} 1 & 0 \\ -1 & 1 \end{bmatrix}$ gives $\begin{bmatrix} 6 & 15 \\ 0 & -23 \end{bmatrix}\begin{bmatrix} x \\ y \end{bmatrix} = \begin{bmatrix} 3 \\ 23 \end{bmatrix}$; $\begin{bmatrix} 3 & 0 \\ -3 & 2 \end{bmatrix}$

by $\begin{bmatrix} 1 & 0 \\ 0 & -\frac{1}{23} \end{bmatrix}$ gives $\begin{bmatrix} 6 & 15 \\ 0 & 1 \end{bmatrix}\begin{bmatrix} x \\ y \end{bmatrix} = \begin{bmatrix} 3 \\ -1 \end{bmatrix}$; $\begin{bmatrix} 3 & 0 \\ \frac{3}{23} & -\frac{2}{23} \end{bmatrix}$

by $\begin{bmatrix} 1 & -15 \\ 0 & 1 \end{bmatrix}$ gives $\begin{bmatrix} 6 & 0 \\ 0 & 1 \end{bmatrix}\begin{bmatrix} x \\ y \end{bmatrix} = \begin{bmatrix} 18 \\ -1 \end{bmatrix}$; $\begin{bmatrix} \frac{24}{23} & \frac{30}{23} \\ \frac{3}{23} & -\frac{2}{23} \end{bmatrix}$

by $\begin{bmatrix} \frac{1}{6} & 0 \\ 0 & 1 \end{bmatrix}$ gives $\begin{bmatrix} 1 & 0 \\ 0 & 1 \end{bmatrix}\begin{bmatrix} x \\ y \end{bmatrix} = \begin{bmatrix} 3 \\ -1 \end{bmatrix}$. $\begin{bmatrix} \frac{4}{23} & \frac{5}{23} \\ \frac{3}{23} & -\frac{2}{23} \end{bmatrix}$

Hence $x = 3$, $y = -1$.

The inverse of $\begin{bmatrix} 2 & 5 \\ 3 & -4 \end{bmatrix}$ is

$$\begin{bmatrix} \frac{1}{6} & 0 \\ 0 & 1 \end{bmatrix} \begin{bmatrix} 1 & ^-15 \\ 0 & 1 \end{bmatrix} \begin{bmatrix} 1 & 0 \\ 0 & -\frac{1}{23} \end{bmatrix} \begin{bmatrix} 1 & 0 \\ ^-1 & 1 \end{bmatrix} \begin{bmatrix} 3 & 0 \\ 0 & 2 \end{bmatrix} = \begin{bmatrix} \frac{4}{23} & \frac{5}{23} \\ \frac{3}{23} & -\frac{2}{23} \end{bmatrix}.$$

Doubtless in this case you would have been able to write down the inverse immediately and then use it to solve the equation. However, you are unlikely to be able to do the same thing in the three-dimensional case, whereas the method of this example can still be applied.

Exercise C

1 Describe the transformations given by the following matrices:

(a) $\begin{bmatrix} 1 & 0 \\ 2 & 1 \end{bmatrix}$; (b) $\begin{bmatrix} 3 & 0 \\ 0 & 1 \end{bmatrix}$; (c) $\begin{bmatrix} 1 & ^-4 \\ 0 & 1 \end{bmatrix}$; (d) $\begin{bmatrix} 5 & 0 \\ 0 & 5 \end{bmatrix}$; (e) $\begin{bmatrix} 1 & 0 \\ 0 & ^-1 \end{bmatrix}$.

2 Use elementary matrices to solve the following and to obtain the inverse transformations.

(a) $\begin{bmatrix} 2 & 3 \\ 3 & 5 \end{bmatrix} \begin{bmatrix} x \\ y \end{bmatrix} = \begin{bmatrix} 4 \\ 7 \end{bmatrix}$; (b) $\begin{bmatrix} 3 & ^-8 \\ 2 & ^-7 \end{bmatrix} \begin{bmatrix} x \\ y \end{bmatrix} = \begin{bmatrix} 5 \\ 4 \end{bmatrix}$; (c) $\begin{bmatrix} 9 & ^-2 \\ 7 & 4 \end{bmatrix} \begin{bmatrix} x \\ y \end{bmatrix} = \begin{bmatrix} 3 \\ ^-5 \end{bmatrix}$.

3 Describe the transformations given by the following matrices:

(a) $\begin{bmatrix} 1 & 3 \\ 0 & 1 \end{bmatrix}$; (b) $\begin{bmatrix} 1 & 0 \\ 0 & 3 \end{bmatrix}$; (c) $\begin{bmatrix} 0 & 1 \\ 1 & 0 \end{bmatrix}$;

(d) $\begin{bmatrix} ^-1 & 0 \\ 0 & 1 \end{bmatrix}$; (e) $\begin{bmatrix} ^-2 & 0 \\ 0 & ^-2 \end{bmatrix}$.

4 Use elementary matrices to solve the following and to obtain the inverse transformations.

(a) $\begin{bmatrix} 2 & 5 \\ 3 & 8 \end{bmatrix} \begin{bmatrix} x \\ y \end{bmatrix} = \begin{bmatrix} 4 \\ 1 \end{bmatrix}$; (b) $\begin{bmatrix} 1 & 3 \\ 3 & ^-1 \end{bmatrix} \begin{bmatrix} x \\ y \end{bmatrix} = \begin{bmatrix} 2 \\ 4 \end{bmatrix}$.

4. THREE DIMENSIONS

In this extension to three dimensions it may prove helpful to refer back to the corresponding two-dimensional sections.

Planes and normals

In three-dimensional space, in which a general point P is represented by the coordinates (x, y, z), a linear equation $ax + by + cz = d$ restricts P to lie on a plane.

For example, $2x + 3y + 4z = 12$ is a plane through $(6, 0, 0)$, $(0, 4, 0)$, $(0, 0, 3)$, and $(5, 2, ^-1)$ as these are some of the possible triples of values for x, y, z which satisfy the restriction $2x + 3y + 4z = 12$. (See Figure 5.)

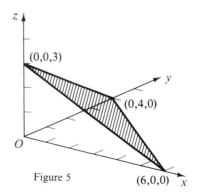

Figure 5

$2x+3y+4z = 0$ is a parallel plane through the origin.

$\begin{bmatrix} 2 \\ 3 \\ 4 \end{bmatrix} \cdot \begin{bmatrix} x \\ y \\ z \end{bmatrix} = 0$ is the same equation, but written using scalar product. This

shows that $\underset{\sim}{OP} = \begin{bmatrix} x \\ y \\ z \end{bmatrix}$ is perpendicular to $\begin{bmatrix} 2 \\ 3 \\ 4 \end{bmatrix}$. So the set of points for which

$2x+3y+4z = 0$ is a plane perpendicular to $\begin{bmatrix} 2 \\ 3 \\ 4 \end{bmatrix}$. $2x+3y+4z = 12$ is thus also

perpendicular to $\begin{bmatrix} 2 \\ 3 \\ 4 \end{bmatrix}$. Thus $2x+3y+4z = 12$ is a plane with *normal* $\begin{bmatrix} 2 \\ 3 \\ 4 \end{bmatrix}$. (See

Figure 6.)

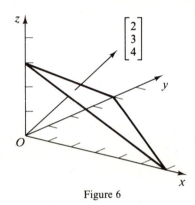

Figure 6

Note that it is not helpful to talk of the gradient of a plane, while the idea of a normal is of use for both lines and planes.

Example 5

Find the equation of the plane through $(5, {}^-4, 3)$ normal to $\begin{bmatrix} 1 \\ 3 \\ -2 \end{bmatrix}$.

Since the plane has normal $\begin{bmatrix} 1 \\ 3 \\ -2 \end{bmatrix}$, its equation is $x+3y-2z = k$. Since it passes

through $(5, {}^-4, 3)$, $5+3({}^-4)-2(3) = k$, and so $k = {}^-13$.
The equation of the plane is $x+3y-2z = {}^-13$.

The angle between two planes

The angle between two planes can conveniently be found by using normals. For

example, the plane $2x+3y+4z = 12$ has normal $\begin{bmatrix} 2 \\ 3 \\ 4 \end{bmatrix}$, while the plane

$4x+5y+6z = 22$ has normal $\begin{bmatrix} 4 \\ 5 \\ 6 \end{bmatrix}$. Since the normals are perpendicular to their planes, the angle between the planes is the same as the angle between their normals.

The angle between $\mathbf{p} = \begin{bmatrix} 2 \\ 3 \\ 4 \end{bmatrix}$ and $\mathbf{q} = \begin{bmatrix} 4 \\ 5 \\ 6 \end{bmatrix}$ is found by using scalar product:

$$\mathbf{p} \cdot \mathbf{q} = pq \cos \theta$$
$$8 + 15 + 24 = \sqrt{29} \times \sqrt{77} \cos \theta$$
$$\cos \theta = \frac{47}{\sqrt{29} \times \sqrt{77}}$$
$$\theta \approx 6°$$

The angle between the planes is 6° to the nearest degree.

Example 6
Find the angle between the plane $15x + 30y - 10z = 7$ and
(a) $6x - 2y + 3z = 8$;   (b) $^-2x + 3y + 6z = 8$;   (c) $3x + 6y - 2z = 8$.

Consider the angle between the normals using $\mathbf{p} \cdot \mathbf{q} = pq \cos \theta$ in each case:

(a) $90 - 60 - 30 = \sqrt{(15^2 + 30^2 + 10^2)} \times \sqrt{(6^2 + (^-2)^2 + 3^2)} \cos \theta$
    $\Rightarrow \cos \theta = 0 \Rightarrow \theta = 90°$;

(b) $^-30 + 90 - 60 = \sqrt{(1225)} \times \sqrt{((^-2)^2 + 3^2 + 6^2)} \cos \theta$
    $\Rightarrow \cos \theta = 0 \Rightarrow \theta = 90°$;

(c) $45 + 180 + 20 = \sqrt{1225} \times \sqrt{49} \cos \theta \Rightarrow \cos \theta = \frac{245}{35 \times 7} = 1 \Rightarrow \theta = 0°$.

In case (c) we should have noticed that the normals were in the same direction, so that the planes are parallel.

Exercise D

1   Find the equation of the plane

    (a) through $(1, 2, 3)$ with normal $\begin{bmatrix} 4 \\ 5 \\ 6 \end{bmatrix}$;

    (b) through $(7, ^-5, 2)$ with normal $\begin{bmatrix} 3 \\ 8 \\ 1 \end{bmatrix}$;

    (c) through $(4, ^-3, ^-9)$ with normal $\begin{bmatrix} 2 \\ ^-5 \\ ^-6 \end{bmatrix}$.

2   Find the angle between the planes $2x + 3y + 4z = 5$ and
    (a) $3x - 6y + 2z = 7$;   (b) $6x + 2y - 3z = 7$;   (c) $4x + 6y + 8z = 7$.

3   Find the equation of the plane

    (a) through $(3, 2, 1)$ with normal $\begin{bmatrix} 6 \\ 5 \\ 4 \end{bmatrix}$;

(b) through $(2, {}^-3, 5)$ with normal $\begin{bmatrix} 7 \\ 2 \\ 6 \end{bmatrix}$;

(c) through $(4, {}^-5, {}^-6)$ with normal $\begin{bmatrix} 9 \\ {}^-8 \\ {}^-3 \end{bmatrix}$.

4 Find the angle between the plane $x - 2y + 3z = 4$ and
   (a) $4x + 3y - 2z = 1$; (b) $z = 0$; (c) $5x + 4y + z = 3$.

5. ELIMINATION

$2x + 3y + 4z = 12$ restricts $P(x, y, z)$ to lie on a plane with normal $\begin{bmatrix} 2 \\ 3 \\ 4 \end{bmatrix}$. Similarly,

$4x + 5y + 6z = 22$ and $6x + 7y + 9z = 35$ restrict P to lie on planes with normals $\begin{bmatrix} 4 \\ 5 \\ 6 \end{bmatrix}$ and $\begin{bmatrix} 6 \\ 7 \\ 9 \end{bmatrix}$ respectively. The three equations taken simultaneously will restrict P to be at the intersection of these three planes. The point of intersection may be found by elimination.

$$\begin{aligned} E_1: &\quad 2x + 3y + 4z = 12, \\ E_2: &\quad 4x + 5y + 6z = 22, \\ E_3: &\quad 6x + 7y + 9z = 35. \end{aligned}$$

Using equation one to eliminate x from equations two and three:

$$\begin{aligned} E_1' &= E_1 & : &\quad 2x + 3y + 4z = 12, \\ E_2' &= {}^-2E_1 + E_2 & : &\quad {}^-y - 2z = {}^-2, \\ E_3' &= {}^-3E_1 \quad + E_3 & : &\quad {}^-2y - 3z = {}^-1. \end{aligned}$$

Using equation two to eliminate y from equations one and three:

$$\begin{aligned} E_1' &= E_1 + 3E_2 & : &\quad 2x \quad - 2z = 6, \\ E_2' &= E_2 & : &\quad {}^-y - 2z = {}^-2, \\ E_3' &= \quad {}^-2E_2 + E_3 & : &\quad z = 3. \end{aligned}$$

Using equation three to eliminate z from equations one and two:

$$\begin{aligned} E_1' &= E_1 \quad + 2E_3 & : &\quad 2x \quad = 12, \\ E_2' &\quad E_2 + 2E_3 & : &\quad {}^-y \quad = 4, \\ E_3' &= \quad E_3 & : &\quad z = 3. \end{aligned}$$

$x = 6$, $y = {}^-4$, $z = 3$; the planes meet at $(6, {}^-4, 3)$.

    Sometimes it is more convenient to eliminate y or z first, rather than x. Often a mixture of elimination and substitution is simplest.

Example 7
Find the equation of the plane through $(1, 2, 3)$, $(4, 5, 7)$ and $(9, 8, 6)$.

Take the equation of the plane to be $ax + by + cz = 1$.

$$\text{Through } (1, 2, 3) \ \Rightarrow \ 1a + 2b + 3c = 1;$$
$$\text{through } (4, 5, 7) \ \Rightarrow \ 4a + 5b + 7c = 1;$$
$$\text{through } (9, 8, 6) \ \Rightarrow \ 9a + 8b + 6c = 1.$$

Using equation one to eliminate a from equations two and three:

$$\left.\begin{array}{llrl}
E_1 = & E_1 & : \ a + 2b + \ 3c = & 1, \\
E_2 = & {}^-4E_1 + E_2 & : \ \ \ \ {}^-3b - \ 5c = & {}^-3, \\
E_3 = & {}^-9E_1 \ \ \ \ + E_3 : & {}^-10b - 21c = & {}^-8.
\end{array}\right\}$$

Using equations two and three to eliminate b from equation three only:

$$\left.\begin{array}{llrl}
E_1 = E_1 & & : \ a + 2b + \ 3c = & 1, \\
E_2 = & E_2 & : \ \ \ \ \ {}^-3b - \ 5c = & {}^-3, \\
E_3 = & -10E_2 + 3E_3 : & {}^-13c = & 6.
\end{array}\right\}$$

$$\left.\begin{array}{lrl}
\text{So from equation three} & c = & {}^-\tfrac{6}{13}, \\
\text{from equation two} & b = & \tfrac{23}{13}, \\
\text{from equation one} & a = & {}^-\tfrac{15}{13}.
\end{array}\right\}$$

So the equation of the plane is

$$^-\tfrac{15}{13}x + \tfrac{23}{13}y - \tfrac{6}{13}z = 1$$

or

$$^-15x + 23y - 6z = 13.$$

Note that three equations may not always have a unique solution. Figure 7 illustrates various possible configurations of three planes represented by three equations. Only in case (e) is there a unique solution.

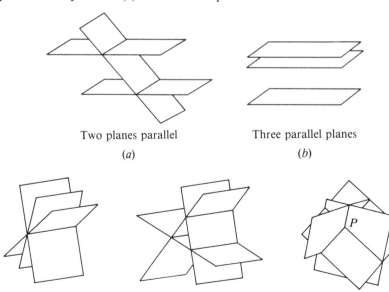

Two planes parallel

(a)

Three parallel planes

(b)

Three planes
intersecting in a line

(c)

A prism of planes

(d)

Three planes intersecting
in a unique point

(e)

Figure 7

Exercise E

1   Solve the following simultaneous equations:

$$
\begin{array}{ll}
(a) & x+2y+3z = 1, \\
& 2x+3y+4z = 3, \\
& 3x+4y+6z = 4;
\end{array}
\qquad
\begin{array}{ll}
(b) & x-2y+5z = 4, \\
& 5x-4y-2z = 5, \\
& 3x+4y-8z = 9;
\end{array}
$$

$$
\begin{array}{ll}
(c) & 2x+5y+3z = 14, \\
& 3x+6y+2z = 6, \\
& 5x+7y- z = 0;
\end{array}
\qquad
\begin{array}{ll}
(d) & 3x-4y+2z = 9, \\
& 2x+5y-3z = 8, \\
& 7x-6y+4z = 17.
\end{array}
$$

2   Find the equation of the planes through the following points:
   (a) $(1,3,3)$, $(4,3,5)$, $(7,4,6)$;
   (b) $(2,{}^-3,4)$, $(3,4,{}^-5)$, $(4,{}^-5,{}^-6)$;
   (c) $(3,4,6)$, $(2,{}^-3,4)$, $(6,7,5)$.

3   Solve the following simultaneous equations:

$$
\begin{array}{ll}
(a) & x+3y+2z = 3, \\
& 2x+4y+3z = 5, \\
& 3x+5y+6z = 9;
\end{array}
\qquad
\begin{array}{ll}
(b) & x-3y+4z = 4, \\
& 3x-4y-2z = 3, \\
& 5x+6y-3z = 18;
\end{array}
$$

$$
\begin{array}{ll}
(c) & 4x-2y+z = 3, \\
& 5x-3y+z = 2, \\
& 6x-4y+3z = 7;
\end{array}
\qquad
\begin{array}{ll}
(d) & 2x+3y-4z = 5, \\
& 3x-4y+5z = 6, \\
& 4x-5y+6z = 9.
\end{array}
$$

4   Find the equation of the plane through the following points:
   (a) $(1,2,1)$, $(2,3,4)$, $(3,4,7)$;
   (b) $(2,1,0)$, $(3,2,3)$, $(4,5,8)$;
   (c) $(3,{}^-2,1)$, $(4,{}^-3,2)$, $(5,4,{}^-3)$.

6. TRANSFORMATIONS

The elimination process of the previous section may be expressed in matrix notation.

$$
\begin{array}{l}
2x+3y+4z = 12, \\
4x+5y+6z = 22, \\
6x+7y+9z = 35.
\end{array}
\qquad
\begin{bmatrix} 2 & 3 & 4 \\ 4 & 5 & 6 \\ 6 & 7 & 9 \end{bmatrix}
\begin{bmatrix} x \\ y \\ z \end{bmatrix}
=
\begin{bmatrix} 12 \\ 22 \\ 35 \end{bmatrix}.
$$

Using equation one to eliminate x: Multiplying both sides by

$$
\begin{array}{lll}
E_1' = E_1 & : & 2x+3y+4z = 12, \\
E_2' = {}^-2E_1+E_2 & : & {}^-y-2z = {}^-2, \\
E_3' = {}^-3E_1 + E_3 : & & {}^-2y-3z = {}^-1.
\end{array}
\quad
\begin{bmatrix} 1 & 0 & 0 \\ {}^-2 & 1 & 0 \\ {}^-3 & 0 & 1 \end{bmatrix}
\text{ gives }
\begin{bmatrix} 2 & 3 & 4 \\ 0 & {}^-1 & {}^-2 \\ 0 & {}^-2 & {}^-3 \end{bmatrix}
\begin{bmatrix} x \\ y \\ z \end{bmatrix}
=
\begin{bmatrix} 12 \\ {}^-2 \\ {}^-1 \end{bmatrix}.
$$

Using equation two to eliminate y: Multiplying both sides by

$$
\begin{array}{lll}
E_1' = E_1+3E_2 & : & 2x -2z = 6, \\
E_2' = E_2 & : & {}^-y-2z = {}^-2, \\
E_3' = -2E_2+E_3 : & & z = 3.
\end{array}
\quad
\begin{bmatrix} 1 & 3 & 0 \\ 0 & 1 & 0 \\ 0 & {}^-2 & 1 \end{bmatrix}
\text{ gives }
\begin{bmatrix} 2 & 0 & {}^-2 \\ 0 & {}^-1 & {}^-2 \\ 0 & 0 & 1 \end{bmatrix}
\begin{bmatrix} x \\ y \\ z \end{bmatrix}
=
\begin{bmatrix} 6 \\ {}^-2 \\ 3 \end{bmatrix}.
$$

Using equation three to eliminate z: Multiplying both sides by

$$
\begin{array}{lll}
E_1' = E_1 +2E_3 & : & 2x = 12, \\
E_2' = E_2+2E_3 & : & {}^-y = 4, \\
E_3' = E_3 & : & z = 3.
\end{array}
\quad
\begin{bmatrix} 1 & 0 & 2 \\ 0 & 1 & 2 \\ 0 & 0 & 1 \end{bmatrix}
\text{ gives }
\begin{bmatrix} 2 & 0 & 0 \\ 0 & {}^-1 & 0 \\ 0 & 0 & 1 \end{bmatrix}
\begin{bmatrix} x \\ y \\ z \end{bmatrix}
=
\begin{bmatrix} 12 \\ 4 \\ 3 \end{bmatrix}.
$$

Solving: Multiplying both sides by

$$\begin{aligned} E_1' &= \tfrac{1}{2}E_1 \\ E_2' &= \phantom{\tfrac{1}{2}}{}^-E_2 \\ E_3' &= \phantom{\tfrac{1}{2}}E_3 \end{aligned} \quad \begin{aligned} :\;\; x &= 6, \\ :\;\; y &= {}^-4, \\ :\;\; z &= 3. \end{aligned}\right\}$$

$$\begin{bmatrix} \tfrac{1}{2} & 0 & 0 \\ 0 & {}^-1 & 0 \\ 0 & 0 & 1 \end{bmatrix} \text{ gives } \begin{bmatrix} 1 & 0 & 0 \\ 0 & 1 & 0 \\ 0 & 0 & 1 \end{bmatrix}\begin{bmatrix} x \\ y \\ z \end{bmatrix} = \begin{bmatrix} 6 \\ {}^-4 \\ 3 \end{bmatrix}.$$

Once again the elementary matrices used to multiply both sides are the coefficients for the operations on the left. The elementary matrices always represent shears or stretches. For example,

$$\begin{bmatrix} 1 & 0 & 0 \\ {}^-2 & 1 & 0 \\ {}^-3 & 0 & 1 \end{bmatrix} \text{ transforms the base vectors } \begin{bmatrix} 1 \\ 0 \\ 0 \end{bmatrix}, \begin{bmatrix} 0 \\ 1 \\ 0 \end{bmatrix}, \begin{bmatrix} 0 \\ 0 \\ 1 \end{bmatrix} \text{ to } \begin{bmatrix} 1 \\ {}^-2 \\ {}^-3 \end{bmatrix}, \begin{bmatrix} 0 \\ 1 \\ 0 \end{bmatrix}, \begin{bmatrix} 0 \\ 0 \\ 1 \end{bmatrix},$$

giving a shear with the plane $x = 0$ invariant.

As in the two-dimensional case, the product of the elementary matrices is the inverse of the original matrix, since they reduce it to the identity matrix:

$$\begin{bmatrix} 2 & 3 & 4 \\ 4 & 5 & 6 \\ 6 & 7 & 9 \end{bmatrix}^{-1} = \begin{bmatrix} \tfrac{1}{2} & 0 & 0 \\ 0 & {}^-1 & 0 \\ 0 & 0 & 1 \end{bmatrix}\begin{bmatrix} 1 & 0 & 2 \\ 0 & 1 & 2 \\ 0 & 0 & 1 \end{bmatrix}\begin{bmatrix} 1 & 3 & 0 \\ 0 & 1 & 0 \\ 0 & {}^-2 & 1 \end{bmatrix}\begin{bmatrix} 1 & 0 & 0 \\ {}^-2 & 1 & 0 \\ {}^-3 & 0 & 1 \end{bmatrix}$$

$$= \begin{bmatrix} {}^-1\tfrac{1}{2} & {}^-\tfrac{1}{2} & 1 \\ 0 & 3 & {}^-2 \\ 1 & {}^-2 & 1 \end{bmatrix}.$$

This gives another interpretation of the algebra in terms of transformations rather than planes. The transformation $\begin{bmatrix} 2 & 3 & 4 \\ 4 & 5 & 6 \\ 6 & 7 & 9 \end{bmatrix}$ sends $\begin{bmatrix} x \\ y \\ z \end{bmatrix}$ to $\begin{bmatrix} 12 \\ 22 \\ 35 \end{bmatrix}$. The inverse, made up of shears and stretches, sends $\begin{bmatrix} 12 \\ 22 \\ 35 \end{bmatrix}$ back to $\begin{bmatrix} 6 \\ {}^-4 \\ 3 \end{bmatrix}$.

Example 8
Use elementary matrices to solve

$$\begin{bmatrix} 1 & 2 & 3 \\ 4 & 5 & 7 \\ 9 & 8 & 6 \end{bmatrix}\begin{bmatrix} x \\ y \\ z \end{bmatrix} = \begin{bmatrix} 1 \\ 1 \\ 1 \end{bmatrix}$$

and hence find the inverse of $\begin{bmatrix} 1 & 2 & 3 \\ 4 & 5 & 7 \\ 9 & 8 & 6 \end{bmatrix}$.

$$\begin{bmatrix} 1 & 2 & 3 \\ 4 & 5 & 7 \\ 9 & 8 & 6 \end{bmatrix}\begin{bmatrix} x \\ y \\ z \end{bmatrix} = \begin{bmatrix} 1 \\ 1 \\ 1 \end{bmatrix}.$$

Multiplying by: Current product of elementary matrices

$$\begin{bmatrix} 1 & 0 & 0 \\ {}^-4 & 1 & 0 \\ {}^-9 & 0 & 1 \end{bmatrix} \text{ gives } \begin{bmatrix} 1 & 2 & 3 \\ 0 & {}^-3 & {}^-5 \\ 0 & {}^-10 & {}^-21 \end{bmatrix}\begin{bmatrix} x \\ y \\ z \end{bmatrix} = \begin{bmatrix} 1 \\ {}^-3 \\ {}^-8 \end{bmatrix}. \qquad \begin{bmatrix} 1 & 0 & 0 \\ {}^-4 & 1 & 0 \\ {}^-9 & 0 & 1 \end{bmatrix}$$

$$\begin{bmatrix} 1 & 0 & 0 \\ 0 & 1 & 0 \\ 0 & -10 & 3 \end{bmatrix} \text{ gives } \begin{bmatrix} 1 & 2 & 3 \\ 0 & -3 & -5 \\ 0 & 0 & -13 \end{bmatrix}\begin{bmatrix} x \\ y \\ z \end{bmatrix} = \begin{bmatrix} 1 \\ -3 \\ 6 \end{bmatrix}. \quad \begin{bmatrix} 1 & 0 & 0 \\ -4 & 1 & 0 \\ 13 & -10 & 3 \end{bmatrix}$$

$$\begin{bmatrix} 13 & 0 & 3 \\ 0 & 13 & -5 \\ 0 & 0 & 1 \end{bmatrix} \text{ gives } \begin{bmatrix} 13 & 26 & 0 \\ 0 & -39 & 0 \\ 0 & 0 & -13 \end{bmatrix}\begin{bmatrix} x \\ y \\ z \end{bmatrix} = \begin{bmatrix} 31 \\ -69 \\ 6 \end{bmatrix}. \quad \begin{bmatrix} 52 & -30 & 9 \\ -117 & 63 & -15 \\ 13 & -10 & 3 \end{bmatrix}$$

$$\begin{bmatrix} 3 & 2 & 0 \\ 0 & 1 & 0 \\ 0 & 0 & 1 \end{bmatrix} \text{ gives } \begin{bmatrix} 39 & 0 & 0 \\ 0 & -39 & 0 \\ 0 & 0 & -13 \end{bmatrix}\begin{bmatrix} x \\ y \\ z \end{bmatrix} = \begin{bmatrix} -45 \\ -69 \\ 6 \end{bmatrix}. \quad \begin{bmatrix} -75 & 36 & -3 \\ -117 & 63 & -15 \\ 13 & -10 & 3 \end{bmatrix}$$

$$\begin{bmatrix} \frac{1}{39} & 0 & 0 \\ 0 & -\frac{1}{39} & 0 \\ 0 & 0 & -\frac{1}{13} \end{bmatrix} \text{ gives } \begin{bmatrix} 1 & 0 & 0 \\ 0 & 1 & 0 \\ 0 & 0 & 1 \end{bmatrix}\begin{bmatrix} x \\ y \\ z \end{bmatrix} = \begin{bmatrix} -\frac{15}{13} \\ \frac{23}{13} \\ -\frac{6}{13} \end{bmatrix}. \quad \frac{1}{13}\begin{bmatrix} -26 & 12 & -1 \\ -39 & -21 & -5 \\ -13 & 10 & -3 \end{bmatrix}$$

Hence $x = -\frac{15}{13}$, $y = \frac{23}{13}$, $z = -\frac{6}{13}$.

The inverse of $\begin{bmatrix} 1 & 2 & 3 \\ 4 & 5 & 7 \\ 9 & 8 & 6 \end{bmatrix}$ is

$$\begin{bmatrix} \frac{1}{39} & 0 & 0 \\ 0 & -\frac{1}{39} & 0 \\ 0 & 0 & -\frac{1}{13} \end{bmatrix}\begin{bmatrix} 3 & 2 & 0 \\ 0 & 1 & 0 \\ 0 & 0 & 1 \end{bmatrix}\begin{bmatrix} 13 & 0 & 3 \\ 0 & 13 & -5 \\ 0 & 0 & 1 \end{bmatrix}\begin{bmatrix} 1 & 0 & 0 \\ 0 & 1 & 0 \\ 0 & -10 & 3 \end{bmatrix}\begin{bmatrix} 1 & 0 & 0 \\ -4 & 1 & 0 \\ -9 & 0 & 1 \end{bmatrix}$$

$$= \frac{1}{13}\begin{bmatrix} -26 & 12 & -1 \\ -39 & -21 & -5 \\ -13 & 10 & -3 \end{bmatrix}.$$

Exercise F

1 Describe the transformations given by the matrices:

(a) $\begin{bmatrix} 1 & 0 & 0 \\ 0 & 1 & 0 \\ 0 & 0 & -1 \end{bmatrix}$; (b) $\begin{bmatrix} 1 & 0 & 0 \\ 0 & 3 & 0 \\ 0 & 0 & 1 \end{bmatrix}$; (c) $\begin{bmatrix} 1 & 0 & 3 \\ 0 & 1 & 2 \\ 0 & 0 & 1 \end{bmatrix}$; (d) $\begin{bmatrix} 2 & 0 & 0 \\ 0 & 2 & 0 \\ 0 & 0 & 2 \end{bmatrix}$.

2 Use elementary matrices to solve the following and to find the inverse matrices.

(a) $\begin{bmatrix} 1 & 2 & 3 \\ 2 & 3 & 4 \\ 3 & 4 & 6 \end{bmatrix}\begin{bmatrix} x \\ y \\ z \end{bmatrix} = \begin{bmatrix} 0 \\ 0 \\ 1 \end{bmatrix}$; (b) $\begin{bmatrix} 1 & 3 & 5 \\ 3 & -5 & 7 \\ 5 & 7 & -9 \end{bmatrix}\begin{bmatrix} x \\ y \\ z \end{bmatrix} = \begin{bmatrix} 14 \\ 6 \\ 20 \end{bmatrix}$;

(c) $\begin{bmatrix} 2 & 4 & 5 \\ 3 & 5 & 7 \\ 4 & 6 & 10 \end{bmatrix}\begin{bmatrix} x \\ y \\ z \end{bmatrix} = \begin{bmatrix} 3 \\ 4 \\ 6 \end{bmatrix}$; (d) $\begin{bmatrix} 3 & -5 & 2 \\ 2 & 3 & -4 \\ 4 & -6 & 2 \end{bmatrix}\begin{bmatrix} x \\ y \\ z \end{bmatrix} = \begin{bmatrix} 5 \\ -7 \\ 4 \end{bmatrix}$.

3 Given that the inverse of $\begin{bmatrix} 2 & 3 & 4 \\ 4 & 5 & 6 \\ 6 & 7 & 9 \end{bmatrix}$ is $\begin{bmatrix} -1.5 & -0.5 & 1 \\ 0 & 3 & -2 \\ 1 & -2 & 1 \end{bmatrix}$, write down the inverses of the following and check that the product of each matrix with its inverse does give the identity matrix.

(a) $\begin{bmatrix} 4 & 5 & 6 \\ 2 & 3 & 4 \\ 6 & 7 & 9 \end{bmatrix}$; (b) $\begin{bmatrix} 4 & 2 & 3 \\ 6 & 4 & 5 \\ 9 & 6 & 7 \end{bmatrix}$; (c) $\begin{bmatrix} 5 & 4 & 6 \\ 3 & 2 & 4 \\ 7 & 6 & 9 \end{bmatrix}$; (d) $\begin{bmatrix} 3 & 1 & -2 \\ 0 & -6 & 4 \\ -2 & 4 & -2 \end{bmatrix}$.

4 Describe the transformations given by the matrices:

(a) $\begin{bmatrix} -1 & 0 & 0 \\ 0 & 1 & 0 \\ 0 & 0 & 1 \end{bmatrix}$; (b) $\begin{bmatrix} 1 & 0 & 0 \\ 0 & 2 & 0 \\ 0 & 0 & 3 \end{bmatrix}$; (c) $\begin{bmatrix} 1 & 0 & 0 \\ 2 & 1 & 0 \\ 3 & 0 & 1 \end{bmatrix}$; (d) $\begin{bmatrix} 3 & 0 & 0 \\ 0 & 3 & 0 \\ 0 & 0 & 3 \end{bmatrix}$.

5 Use elementary matrices to solve the following and to find the inverse matrices.

(a) $\begin{bmatrix} 1 & 3 & 2 \\ 2 & 5 & 4 \\ 3 & 4 & 7 \end{bmatrix}\begin{bmatrix} x \\ y \\ z \end{bmatrix} = \begin{bmatrix} 0 \\ 1 \\ 5 \end{bmatrix}$; (b) $\begin{bmatrix} 1 & -3 & 5 \\ 2 & 4 & -6 \\ 3 & -2 & -1 \end{bmatrix}\begin{bmatrix} x \\ y \\ z \end{bmatrix} = \begin{bmatrix} 4 \\ 3 \\ 3 \end{bmatrix}$;

(c) $\begin{bmatrix} 3 & 2 & 1 \\ 4 & 3 & 2 \\ 6 & 5 & 3 \end{bmatrix}\begin{bmatrix} x \\ y \\ z \end{bmatrix} = \begin{bmatrix} 0 \\ 0 \\ -1 \end{bmatrix}$; (d) $\begin{bmatrix} 0 & 3 & 5 \\ 2 & 0 & 7 \\ 9 & 8 & 0 \end{bmatrix}\begin{bmatrix} x \\ y \\ z \end{bmatrix} = \begin{bmatrix} 2 \\ 9 \\ 1 \end{bmatrix}$.

6 Given that the inverse of $\begin{bmatrix} 2 & 5 & 6 \\ 3 & 7 & 9 \\ 1 & 3 & 4 \end{bmatrix}$ is $\begin{bmatrix} -1 & 2 & -3 \\ 3 & -2 & 0 \\ -2 & 1 & -1 \end{bmatrix}$, write down the inverses of the following and check that the product of each matrix with its inverse does give the identity matrix.

(a) $\begin{bmatrix} 3 & 7 & 9 \\ 2 & 5 & 6 \\ 1 & 3 & 4 \end{bmatrix}$; (b) $\begin{bmatrix} 6 & 2 & 5 \\ 9 & 3 & 7 \\ 4 & 1 & 3 \end{bmatrix}$; (c) $\begin{bmatrix} 1 & 4 & 3 \\ 2 & 6 & 5 \\ 3 & 9 & 7 \end{bmatrix}$; (d) $\begin{bmatrix} 1 & 2 & 3 \\ -3 & -2 & 0 \\ 2 & 1 & -1 \end{bmatrix}$.

7. SUMMARY

$ax+by = c$ is a line perpendicular to the normal vector $\begin{bmatrix} a \\ b \end{bmatrix}$.

$ax+by+cz = d$ is a plane perpendicular to the normal vector $\begin{bmatrix} a \\ b \\ c \end{bmatrix}$.

The angle between lines or planes is equal to the angle between their normals and can be found by using the scalar product.

Simultaneous equations may be solved by systematic elimination.

The elimination process may be written in matrix notation and carried out using elementary matrices, which are shears or stretches.

The solution of a set of simultaneous equations may be interpreted geometrically either as the point of intersection of lines or planes or as a point whose image is given under a matrix transformation.

The nature of a matrix transformation may be found by considering the images of the base vectors.

The inverse matrix for the transformation may be found as the product of the elementary matrices used in the elimination process.

Miscellaneous exercise

1 (*a*) Solve the equations

$$3x+y \ -2z = 0,$$
$$x+5y+4z = 7,$$
$$x+y \ + \ z = 4.$$

(*b*) If these equations are written in the form of a matrix equation $\mathbf{A} \begin{bmatrix} x \\ y \\ z \end{bmatrix} = \begin{bmatrix} 0 \\ 7 \\ 4 \end{bmatrix}$ write down the matrix **A**.

(*c*) Without further working write down the value of $\mathbf{A}^{-1} \begin{bmatrix} 0 \\ 7 \\ 4 \end{bmatrix}$.   (SMP)

2 (i)
$$\begin{bmatrix} 3 & 2 & 6 \\ 5 & 6 & 2 \\ 7 & 8 & 4 \end{bmatrix}$$

The columns of this 3×3 matrix are said to be dependent because the last column depends on taking multiples of the other columns. Hence the third column can be obtained by multiplying the first column by 4, the second by $^-3$ and adding; i.e. $3x+2y = 6$, $5x+6y = 2$, $7x+8y = 4$ are all satisfied by $x = 4$ and $y = ^-3$.

Writing down three equations and solving them, or otherwise, show that the rows are also dependent by stating what multiples of the first two rows will add to give the third row.

(ii)
$$\begin{bmatrix} 1 & 6 & 4 & 1 \\ 2 & 9 & 6 & 2 \\ 3 & 8 & 5 & 2 \\ 2 & 8 & 5 & k \end{bmatrix}$$

If the columns of this 4×4 matrix are dependent, write down three equations which must be satisfied in addition to $2x+8y+5z = k$.

(*a*) Solve your equations to get the values of the multipliers x, y and z.

(*b*) Hence verify that the value of k must be 1.

(*c*) Show that the rows will then also be dependent, with the same multipliers but in a different order.   (SMP)

3 Multiply together the two matrices

$$\begin{bmatrix} 1 & 3 & 2 \\ 2 & ^-1 & 1 \\ 3 & ^-2 & ^-1 \end{bmatrix} \begin{bmatrix} 3 & ^-1 & 5 \\ 5 & ^-7 & 3 \\ ^-1 & 11 & ^-7 \end{bmatrix}.$$

Hence, or otherwise, solve the simultaneous equations

$$\left. \begin{array}{r} 3x-y \ +5z = 32, \\ 5x-7y \ +3z = ^-8, \\ x-11y+7z = 20. \end{array} \right\}$$ (SMP)

4
$$\left. \begin{array}{r} x+y+z \ = 1, \\ x-y+z \ = 2, \\ x+y+kz = 3. \end{array} \right\}$$

(*a*) Solve these three simultaneous equations when $k = 3$.

(*b*) State the value of k for which there is no solution.

Interpret your results in terms of intersections of three planes.   (SMP)

5 Verify that the plane $3x+2y+z = 6$ meets the *x*-axis at A $(2,0,0)$, and state the coordinates of the points B and C in which it meets the *y*- and *z*-axes.

Show that the plane $px+4y-3z = 12$ passes through B. Find the value of p if it also passes through A and verify that it then passes through D $(1,3,2)$.

Show that the plane $qx+ry+2z = 12$ passes through C and find the values of q and r if it also passes through A and D. (SMP)

6 Solve the simultaneous equations

$$\left.\begin{array}{rcl} 2x+5y+3z &=& 7, \\ x-2y-z &=& 3, \\ 4x+7y+2z &=& 9. \end{array}\right\}$$ (SMP)

7 Given the matrix

$$\mathbf{B} = \begin{bmatrix} 2 & ^-1 & 3 \\ 1 & 3 & ^-1 \\ 0 & ^-4 & 2 \end{bmatrix},$$

find x, y, z such that

$$\mathbf{B}\begin{bmatrix} x \\ y \\ z \end{bmatrix} = \begin{bmatrix} 6 \\ 4 \\ ^-5 \end{bmatrix}.$$ (SMP)

8 The point (x, y) is mapped onto the point (x', y') by the matrix $\mathbf{M}$, where

$$\mathbf{M} = \begin{bmatrix} 0 & ^-1 & 7 \\ 1 & 0 & 1 \\ 0 & 0 & 1 \end{bmatrix},$$

by means of the transformation

$$\begin{bmatrix} x' \\ y' \\ 1 \end{bmatrix} = \mathbf{M}\begin{bmatrix} x \\ y \\ 1 \end{bmatrix}.$$

Show that the image of the point $(2, 1)$ under the transformation represented by $\mathbf{M}$ is the point $(6, 3)$.

Given the matrices

$$\mathbf{A} = \begin{bmatrix} 1 & 0 & 3 \\ 0 & 1 & 4 \\ 0 & 0 & 1 \end{bmatrix}, \quad \mathbf{B} = \begin{bmatrix} 0 & ^-1 & 0 \\ 1 & 0 & 0 \\ 0 & 0 & 1 \end{bmatrix}, \quad \mathbf{C} = \begin{bmatrix} 1 & 0 & ^-3 \\ 0 & 1 & ^-4 \\ 0 & 0 & 1 \end{bmatrix},$$

find in the same way the image point of P under each of the transformations represented by $\mathbf{A}$, $\mathbf{B}$ and $\mathbf{C}$.

By considering a rectangle with OP as diagonal, or otherwise, describe geometrically the transformations effected by $\mathbf{A}$, $\mathbf{B}$, $\mathbf{C}$ and $\mathbf{M}$. (SMP)

Revision exercise 5

1 The image of the point P $(3, {}^-2)$ after a quarter-turn about the origin followed by the translation $\begin{bmatrix} 3 \\ 1 \end{bmatrix}$ may be found by calculating $\mathbf{A} \begin{bmatrix} 3 \\ -2 \end{bmatrix} + \begin{bmatrix} 3 \\ 1 \end{bmatrix}$.

Write down the matrix $\mathbf{A}$ and find the coordinates of the image of P.

Show that the same combination of transformations maps the point (a, b) to $(3-b, 1+a)$. Find the values of a and b which make (a, b) an invariant point, and hence describe fully the equivalent single transformation. (SMP)

2 A transformation $\mathbf{S}$ is composed of reflection in the line $y = x$ followed by a half-turn about the point $(2, 2)$. Find the coordinates of the image under $\mathbf{S}$ of the points (a) $(5, 3)$, (b) $(3, 1)$, (c) (p, q).

Show that (p, q) is an invariant point if and only if $p + q = 4$.

State fully the transformation $\mathbf{S}$ in geometrical terms. (SMP)

3 (a) A and B are the points $(1, 0)$ and $(0, 1)$. By drawing and folding, or otherwise, write down correct to one place of decimals the coordinates of the reflections of A and B in the line $y = 2x$.

(b) Hence write down the matrix $\mathbf{M}$ for this reflection.

(c) Write down similarly, correct to one place of decimals, the matrix $\mathbf{N}$ for reflection in the line $x = {}^-2y$.

(d) Explain why these two reflection lines are perpendicular.

(e) Hence, using the fact that the combination of reflections in two lines is equivalent to a rotation through twice the angle between the two lines, write down the matrix $\mathbf{R}$ for the single transformation equivalent to the combination of these two reflections in the given lines.

(f) Calculate the product of $\mathbf{M}$ and $\mathbf{N}$ and verify that it gives $\mathbf{R}$ correct to one place of decimals. (SMP)

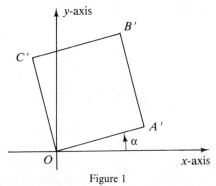

Figure 1

4 Figure 1 shows the image of the unit square after a rotation of angle α about the origin.

(a) Write down the coordinates of A' and C', and hence the matrix $\mathbf{R}_\alpha$ for this rotation.

(b) Write down the matrix $\mathbf{R}_\beta$ for a rotation of angle β about the origin.

(c) State the single transformation represented by the matrix product $\mathbf{R}_\beta \mathbf{R}_\alpha$ and complete the matrix $\mathbf{T}$ for this:

$$\mathbf{T} = \begin{bmatrix} \cos(\alpha+\beta) & \\ & \end{bmatrix}.$$

(d) Carry out the matrix multiplication $\mathbf{R}_\beta \mathbf{R}_\alpha$ and show how this gives the formula for $\cos(\alpha+\beta)$. (SMP)

5 Figure 2 shows part of a plane pattern, which extends in all directions indefinitely. Make a sketch of this and mark on it three centres of symmetry, points A, B and C, each with a different order of rotation. State the order attached to each point. State three other distinct transformations under which the pattern is invariant. (SMP)

Figure 2

6 A region R of the plane is shown in Figure 3 with A, B, C, D as the midpoints of its sides. It is to form a pattern by repeated half-turns $\mathbf{H}_A$, $\mathbf{H}_B$, $\mathbf{H}_C$, $\mathbf{H}_D$ about A, B, C, and D (considered as fixed points of the plane). Sketch part of the pattern, shading R, and labelling $\mathbf{H}_A(R)$, $\mathbf{H}_B(R)$, $\mathbf{H}_D \mathbf{H}_A(R)$.
  If $\mathbf{H}_D \mathbf{H}_A = \mathbf{H}_X \mathbf{H}_B$, where is X? (SMP)

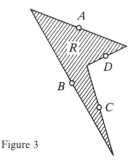

Figure 3

7 $\mathbf{X}$ is the glide consisting of translation $\begin{bmatrix} 2 \\ 0 \end{bmatrix}$ followed by reflection in the x-axis. $\mathbf{Y}$ is the glide consisting of translation $\begin{bmatrix} 0 \\ 2 \end{bmatrix}$ followed by reflection in the y-axis.

  On graph paper, allowing $^-3$ to 5 on both axes, show and label clearly the unit square $OIAJ$ and its images after the following transformations:

$$\mathbf{X}; \quad \mathbf{Y}; \quad \mathbf{XY}; \quad \mathbf{X}^2; \quad \mathbf{YXY}; \quad \mathbf{Y}^2\mathbf{XY}.$$

(a) One of these is a translation. State which one.

(b) Two of these are half-turns. Identify both, stating the coordinates of the centres of rotation.

(c) The remaining three are glides, two of which form an inverse pair. Identify the pair.

Write down a combination of the transformations **X** and **Y** which is equivalent to the translation $\begin{bmatrix} 4 \\ 8 \end{bmatrix}$.   (SMP)

8 AB, $A'B'$ are non-parallel line-segments of equal length lying in a plane but not in the same line. Show that a point P can be found such that pure rotation about P will carry A into A' and B into B'.

$x'Ox$, $y'Oy$ are rectangular axes. **T, R, M** denote transformations in the plane of the axes; **T** being a translation through 2 units in direction Ox, **R** an anticlockwise rotation through a right-angle about O, and **M** a reflection in the y-axis. The transformation written as **RT** denotes **T** followed by **R**. Find, by considering their effect on convenient points, the geometrical nature of the following: (i) **RT**; (ii) **TR**; (iii) **MT**; (iv) **TM**; (v) **TMT**; (vi) **TRT**$^{-1}$.   (SMP)

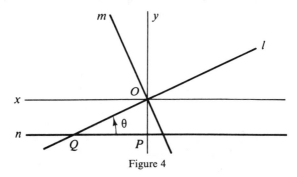

Figure 4

9 In Figure 4 the lines n and x are parallel, y is perpendicular to x, and m is perpendicular to l, which makes an angle θ with n. A reflection in any line z is denoted by **Z**, and **AB** denotes 'B then A'. Identify the plane isometries **LN, YX, ML, XN** and **YXN**. Hence identify the single isometry equivalent to a rotation of 2θ about Q followed by **M**.   (SMP)

10 The number of elements in three sets A, B, C and their intersections are given as follows:

$$n(A) = 16,$$
$$n(C) = 20,$$
$$n(A \cap B' \cap C') = 6,$$
$$n(A' \cap B \cap C') = 8,$$
$$n(A' \cap B' \cap C) = 9,$$
$$n(A \cap B \cap C) = p,$$
$$n(A' \cap B \cap C) = 2p,$$
$$n(A \cap B' \cap C) = q,$$
$$n(A \cap B \cap C') = 3.$$

Show these data on a Venn diagram.
Show that $3p+q = 11$ and $p+q = 7$.
Solve this pair of equations and hence find the value of $n(A \cup B \cup C)$.   (SMP)

11 Solve the simultaneous equations

$$\left. \begin{array}{r} 2x+6y-3z = 4, \\ x-y + z = 1, \\ 3x+y +2z = 7. \end{array} \right\}$$

Hence write down a vector $\mathbf{r}$ satisfying

$$\begin{bmatrix} 2 & 6 & ^-3 \\ 1 & ^-1 & 1 \\ 3 & 1 & 2 \end{bmatrix} \mathbf{r} = \begin{bmatrix} 2 \\ 0.5 \\ 3.5 \end{bmatrix}.$$

(SMP)

12 (a) Form the product matrix $\mathbf{AB}$, where

$$\mathbf{A} = \begin{bmatrix} ^-3 & 2 & 1 \\ 9 & ^-5 & ^-2 \\ ^-5 & 3 & 1 \end{bmatrix} \quad \text{and} \quad \mathbf{B} = \begin{bmatrix} 1 & 1 & 1 \\ 1 & 2 & 3 \\ 2 & ^-1 & ^-3 \end{bmatrix}.$$

(b) Solve the simultaneous equations

$$\begin{aligned} x+y+z &= 2, \\ x+2y+3z &= 1, \\ 2x-y-3z &= 6. \end{aligned}$$

(c) Solve the simultaneous equations

$$\begin{aligned} ^-3x+2y+z &= 8, \\ 9x-5y-2z &= ^-21, \\ ^-5x+3y+z &= 17. \end{aligned}$$

(SMP)

13 Solve the simultaneous equations

$$\begin{aligned} 7x+4y+4z &= 18, \\ 4x+y-8z &= ^-27, \\ 4x-8y+z &= 36. \end{aligned}$$

Hence write down an invariant point A for the transformation given by the matrix $\mathbf{M}$, where

$$\mathbf{M} = \frac{1}{9} \begin{bmatrix} 7 & 4 & 4 \\ 4 & 1 & ^-8 \\ 4 & ^-8 & 1 \end{bmatrix}.$$

Verify that B $(2, 1, 0)$ and O $(0, 0, 0)$ are also invariant points.
P is the point $(1, ^-2, ^-2)$. Show that OP is perpendicular to both OA and OB, and so is perpendicular to the plane OAB.
Find the image of P under $\mathbf{M}$.
Hence state the nature of the transformation $\mathbf{M}$. (SMP)

14 Let

$$\mathbf{A} = \begin{bmatrix} 1 & 0 & 1 \\ 2 & 1 & 2 \\ 0 & ^-1 & 1 \end{bmatrix}.$$

Find elementary matrices $\mathbf{E}_1$, $\mathbf{E}_2$, $\mathbf{E}_3$ such that
(a) $\mathbf{E}_1 \mathbf{A}$ has only one non-zero entry in the first column;
(b) $\mathbf{E}_2 \mathbf{E}_1 \mathbf{A}$ has only one non-zero entry in each of the first two columns;
(c) $\mathbf{E}_3 \mathbf{E}_2 \mathbf{E}_1 \mathbf{A}$ is the identity matrix.
Use these results to find $\mathbf{A}^{-1}$. (SMP)

19

Logic

1. STATEMENTS AND THEIR COMBINATIONS

If you try and do not fail then you can decode this.

$$(t \wedge \sim f) \Rightarrow d$$

We shall be concerned with statements such as 'you can decode this', here denoted by the letter d, which are either true or false. We shall not allow any possibility of any ambiguity or intermediate situation between truth and falsehood.

The *negation* of any statement s is denoted by $\sim s$. Thus $\sim d$ denotes 'you cannot decode this' or 'it is not true that you can decode this'.

If s is true then $\sim s$ will be false.

If s is false the $\sim s$ will be true.

With two statements, as with tossing two coins, there are four possible situations:

true, true; true, false; false, true; false, false.

It is convenient to set this out in a table.

a denotes 'Adam is anti-social'.

b denotes 'Belinda plays billiards'.

	a	b
1	T	T
2	T	F
3	F	T
4	F	F

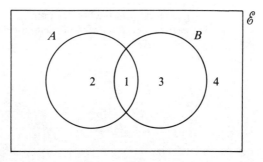

Figure 1

Notice that these correspond to the four regions of the Venn diagram shown in Figure 1. For example, region 2 consists of points in *A* but not in *B*, corresponding to statement *a* true and statement *b* false.

And

There are many ways of combining, or connecting, two statements to form a compound statement. We shall consider first the connective 'and'.

The symbol ∧ is used, conveniently looking rather like a capital letter A. Thus, with *a* and *b* defined as before, *a* ∧ *b* denotes 'Adam is anti-social and Belinda plays billiards'. Again we shall assume that a compound statement is either true or false. Which do you think *a* ∧ *b* will be? Of course, it depends on the original statements *a* and *b*. If both are true then the compound statement will also be true, but if either is false then the compound statement will also be false. This can be shown conveniently in a *truth table*:

	a	*b*	*a* ∧ *b*		∧
1	*T*	*T*	*T*		*T*
2	*T*	*F*	*F*	or merely	*F*
3	*F*	*T*	*F*		*F*
4	*F*	*F*	*F*		*F*

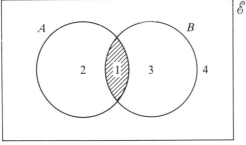

Figure 2

In the Venn diagram (Figure 2) we shade the regions corresponding to the conditions in which *a* ∧ *b* is true.

Or

The symbol ∨ is used, being a reminder of the Latin word (*vel*) for 'or'. *a* ∨ *b* denotes 'Either Adam is anti-social or Belinda plays billiards'. Can you write down the truth table for *a* ∨ *b*? There is some difficulty here because the English language is ambiguous. Compare the use of 'or' in these two questions:

> Would you like tea or coffee for breakfast?

> Do you take milk or sugar?

'Both' would be a very surprising answer to the first question, but not an unusual reply to the second. The symbol ∨ is used for the second use of 'or' – that is, $a \lor b$ is true if a is true, or b is true, or both. So the truth table is:

	a	b	$a \lor b$		$\lor$
1	T	T	T		T
2	T	F	T	or merely	T
3	F	T	T		T
4	F	F	F		F

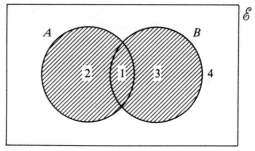

Figure 3

The corresponding Venn diagram is shown in Figure 3.

Example 1

m denotes 'this is male'; s denotes 'this is small'.
   (a) Write in English: (i) $\sim m \land s$; (ii) $\sim (m \lor \sim s)$.
   (b) Express in symbols: 'This is neither male nor large'.
   (c) Draw Venn diagrams for (a) and (b).

   (a) (i) $\sim m \land s$ denotes 'this is a small female'.
        (ii) $\sim (m \lor \sim s)$ denotes 'this is not either male or large'.
   (b) 'This is neither male nor large' is denoted by $\sim m \land s$.
   (c) The Venn diagram shown in Figure 4 illustrates all three compound statements. They are, in fact, different ways of saying the same thing.

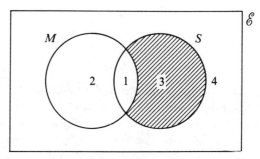

Figure 4

'Applicants must be female and sixteen or tall'.
Could you apply? There are two possible meanings (Figure 5).

(a) $(f \wedge s) \vee t$ (b) $f \wedge (s \vee t)$

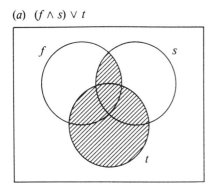 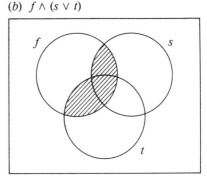

Figure 5

A tall man would qualify in the first but not the second case. The brackets are essential as $f \wedge s \vee t$ is ambiguous.

It is more difficult to be precise in English. It would be possible to use commas, but this is not usual. Indeed they are never used in legal documents as a single misplaced comma could alter the entire meaning. We shall generally write

$(f \wedge s) \vee t$ as 'both female and sixteen or tall'

$f \wedge (s \vee t)$ as 'female and either sixteen or tall'.

The convention is that 'both...and' brackets whatever is joined by 'and'. Similarly 'either...or' brackets whatever is joined by 'or'. So 'either sixteen or both tall and female' means $(s \vee (t \wedge f))$.

Sometimes there is an implicit assumption that there are simple opposites to statements. So, in Example 1, 'not small' is replaced by 'large'. This dichotomy should be assumed in the exercises that follow.

Exercise A

In questions 1–4,
a denotes 'it is alive'; b denotes 'it is bad';
c denotes 'it is cold'; d denotes 'it is dark'.

1   Write in English:
       (a) $\sim a$;   (b) $c \vee d$;   (c) $b \wedge a$;
       (d) $d \wedge \sim c$;   (e) $\sim b \vee c$;   (f) $\sim(a \vee b)$;
       (g) $\sim a \wedge \sim b$;   (h) $\sim(\sim c)$;   (i) $\sim(c \wedge \sim a)$.

2   Express in symbols:
       (a) It is not light.   (b) It is dead and bad.
       (c) It is neither good nor hot.   (d) It is hot but not alive.
       (e) It is cold and either bad or dead.

3   Draw Venn diagrams for question 1.

4   Draw Venn diagrams for question 2.

In questions 5–8

e denotes 'that is edible'; *f* denotes 'that is fat';

g denotes 'that is good'; *h* denotes 'that is hard'.

5 Write in English:

(*a*) ~ *g*;   (*b*) *f* ∧ *e*;   (*c*) *g* ∨ *h*;   (*d*) *h* ∧ ~ *e*;   (*e*) ~*f* ∧ *g*;

(*f*) ~(*h* ∧ *e*);   (*g*) ~ *h* ∨ ~ *e*;   (*h*) ~(~ *f*);   (*i*) ~(*f* ∨ ~ *e*).

6 Express in symbols:

(*a*) That is not soft.   (*b*) That is bad and inedible.

(*c*) That is neither lean nor good.   (*d*) That is hard but not edible.

(*e*) That is either fat and inedible or good.

7 Draw Venn diagrams for question 5.

8 Draw Venn diagrams for question 6.

2. TRUTH TABLES

'It is not true both that it is Monday and that there is no bread.' If *m* denotes 'it is Monday' and *b* denotes 'there is bread', then this compound statement is given by ~(*m* ∧ ~ *b*). Can you tell when this statement is true? If *m* and *b* are both true you can probably decide that the compound statement is also true. But what if *m* and *b* are both false – for example, if you are without bread on a Tuesday?

$$\sim b \qquad \text{('there is no bread') is true;}$$
$$m \wedge \sim b \qquad \text{('it is a Monday without bread') is false;}$$
$$\sim (m \wedge \sim b) \qquad \text{('it isn't a Monday without bread') is true.}$$

So the compound statement is true, not only when *m* and *b* are both true, but also when they are both false. Are these the only circumstances in which the compound statement is true? We need to consider all the four possible situations, and this is most conveniently done using a truth table.

We start by writing down the four cases for *m* and *b* and then work through as the compound statement builds up:

Stage 1		Stage 2		Stage 3	Stage 4
~(*m* ∧ ~ *b*)		~(*m* ∧ ~ *b*)		~(*m* ∧ ~ *b*)	~(*m* ∧ ~ *b*)
T	*T*	*T*	*F*	*F*	*T*
T	*F*	*T*	*T*	*T*	*F*
F	*T*	*F*	*F*	*F*	*T*
F	*F*	*F*	*T*	*F*	*T*

So the compound statement is true unless *m* is false and *b* is true. All the stages can be set out in a single table, with the columns numbered to show each stage.

4	1	3	2	1
~	(*m*	∧	~	*b*)
T	*T*	*F*	*F*	*T*
F	*T*	*T*	*T*	*F*
T	*F*	*F*	*F*	*T*
T	*F*	*F*	*T*	*F*

In fact you do not need to know what *m* and *b* denote in order to complete the truth table. So, with practice, this method can become a quick routine.

Example 2

Make out the truth table for $(a \wedge \sim b) \vee (\sim a \vee b)$.

		1	3	2	1	4	2	1	3	1
a	*b*	(*a*	∧	~	*b*)	∨	(~	*a*	∨	*b*)
T	*T*	*T*	*F*	*F*	*T*	*T*	*F*	*T*	*T*	*T*
T	*F*	*T*	*T*	*T*	*F*	*T*	*F*	*T*	*F*	*F*
F	*T*	*F*	*F*	*F*	*T*	*T*	*T*	*F*	*T*	*T*
F	*F*	*F*	*F*	*T*	*F*	*T*	*T*	*F*	*T*	*F*

In Example 2 the compound statement is true whatever the circumstances. Such a statement is said to be logically true and is called a *tautology*.

The simplest tautology is $s \vee \sim s$ (for example, 'I am stupid or I am not stupid'). Clearly this is always true.

A statement can also be *logically false*, a *contradiction*, such as, for example, 'I am short and I am not short' $(s \wedge \sim s)$. This is untrue whatever size I may be.

Exercise B

1 (*a*) Complete the following truth tables:

a	*b*	~*a*	~*b*	~*a* ∧ ~*b*	~*a* ∨ ~*b*	~*a* ∨ *b*	~(*a* ∧ ~*b*)
T	*T*						
T	*F*						
F	*T*						
F	*F*						

(*b*) For each of the above, draw a Venn diagram, shading those regions which represent the true value of the compound statement.

2 (i) Construct the truth tables for the following.
(ii) Write them in English with *r* as 'I am right', *s* as 'I am stupid'.
(iii) State which are tautologies and which contradictions.

(*a*) $r \wedge (s \wedge \sim r)$;   (*b*) $\sim (r \vee s) \wedge r$;
(*c*) $(r \wedge s) \vee (r \wedge \sim s)$;   (*d*) $s \vee \sim (s \wedge \sim r)$.

3 Find under what circumstances the following compound statements are true.

(*a*) Either today is not Monday or both I am silly and it is Monday.
(*b*) He is right or both I am right and he is wrong.
(*c*) Both this is a fiddle and that is twiddle or this is not a fiddle.
(*d*) You are never either right or both stupid and wrong.

4 Find an expression containing $\wedge$, $\vee$ and $\sim$ which is a tautology. Illustrate it in English. Alter it into a contradiction and express it in symbols.

5 (*a*) Complete the following truth tables:

c	*d*	~*c*	~*d*	~*c* ∧ *d*	*c* ∨ ~*d*	~(*c* ∨ ~*d*)
T	*T*					
T	*F*					
F	*T*					
F	*F*					

(*b*) For each of the above, draw a Venn diagram, shading those regions which represent the true value of the compound statement.

6   (i) Construct the truth tables for the following.
    (ii) Write them in English with m as 'you are mad', and n as 'I am nutty'.
    (iii) State which are tautologies and which contradictions.
        (*a*) $m \lor (n \land \sim m)$;   (*b*) $\sim(m \land n) \lor m$;
        (*c*) $(m \lor n) \land (m \lor \sim n)$;   (*d*) $m \land \sim(m \lor \sim n)$.

7   Find under what circumstances the following compound statements are true.
    (*a*) Either today is Tuesday or both you are right and it is not Tuesday.
    (*b*) She is tall or both you are tall and she is short.
    (*c*) That is a bind and both this is a chore or that is not a bind.
    (*d*) You are either right or never both stupid and wrong.

8   Find an expression containing $\land$, $\lor$ and $\sim$, which is a contradiction. Illustrate it in English. Alter it into a tautology and express it in symbols.

3. VENN DIAGRAMS

A Venn diagram can be a convenient way of checking a tautology. In Figure 6, the region corresponding to $(a \land \sim b)$ is shown shaded horizontally, and that

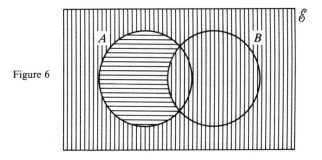

Figure 6

for $(\sim a \lor b)$ is shown shaded vertically. The union of the two sets is $\mathscr{E}$, so $(a \land \sim b) \lor (\sim a \lor b)$ is a tautology.

More generally, a Venn diagram can be as useful as a truth table in showing when a compound statement is true.

Example 3
When is $\sim a \land (b \lor a)$ true?

In Figure 7 the region for $\sim a$ is shaded horizontally, and $(b \lor a)$ is shaded

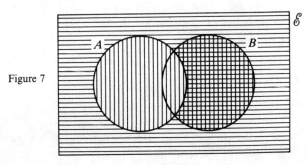

Figure 7

vertically. We see that the shading is both horizontal and vertical inside B but outside A, that is, when b is true but a is false.

Alternatively, using a truth table:

	a	b	2 ~	1 a	3 ∧	1 (b	2 ∨	1 a)
1	T	T	F	T	F	T	T	T
2	T	F	F	T	F	F	T	T
3	F	T	T	F	T	T	T	F
4	F	F	T	F	F	F	F	F

we find that only the third line is true overall, which is when a is false and b is true. For example 'I am not angry and I am either brave or angry' is only true when I am brave but not angry.

You will have noticed that $\wedge$, $\vee$ and $\sim$ correspond to intersection, union and complement for sets. This makes it quite simple to find a compound statement to match a given truth table, by using a Venn diagram.

Example 4
Find a compound statement whose truth table is *TFTT*.

This means

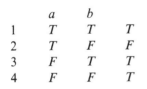

	a	b	
1	T	T	T
2	T	F	F
3	F	T	T
4	F	F	T

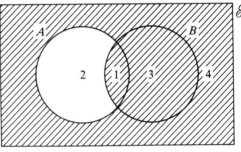

Figure 8

The corresponding regions are shown shaded in Figure 8. There are several ways of describing these regions. For example:

Region 1 together with 3 and 4:	$(a \wedge b) \vee \sim a.$
Not region 2:	$\sim(a \wedge \sim b).$
1 and 3 together with 3 and 4:	$b \vee \sim a.$

This last is probably the simplest.

In a similar way we can find compound statements to match every possible truth table.

Exercise C

1 Use (i) Venn diagrams, (ii) truth tables to find the truth sets of the following. (iii) Write then in English with r as 'he is rich', s as 'he is short'.

(a) $(r \wedge s) \wedge \sim r$; (b) $(r \vee s) \wedge \sim r$;

(c) $(r \vee s) \wedge (r \vee \sim s)$; (d) $\sim(r \wedge [s \vee \sim r])$.

2 Express in symbols and use (i) Venn diagrams, (ii) truth tables, to find under what circumstances the following are true.
  (*a*) Either it is a head or both I am observant and it is a tail.
  (*b*) Either he is right and I am wrong or I am right.
  (*c*) Both Ann and Bill are short or neither Bill nor Ann is short.
  (*d*) Either it is both in summer and in London or it is not in London.

3 (i) Find a compound statement whose truth table is:
        (*a*) *T F F T*; (*b*) *F T F T*; (*c*) *T T F T*; (*d*) *F F F T*.
  (ii) How many possible truth tables are there for a compound of two statements?

4 Use (i) Venn diagrams, (ii) truth tables, to find the truth sets of the following. (iii) Write them in English with *f* as 'she is fat', *g* as 'she is good'.
        (*a*) $(f \lor g) \lor \sim f$; (*b*) $(f \land g) \lor \sim f$;
        (*c*) $(f \land g) \lor (f \land \sim g)$; (*d*) $\sim(f \lor [g \land \sim f])$.

5 Express in symbols and use (i) Venn diagrams, (ii) truth tables, to find under what circumstances the following are true.
  (*a*) Both it is a winner and I am unobservant or it is loser.
  (*b*) She is slow or both you are slow and she is quick.
  (*c*) Either Charles or David is lazy and neither David nor Charles is lazy.
  (*d*) It is both hot and either humid or cold and not humid.

6 Find a compound statement whose truth table is:
        (*a*) *F T T F*; (*b*) *F F T T*; (*c*) *F F T F*; (*d*) *T F F F*.

4. IMPLICATION AND EQUIVALENCE

We have seen that it is possible to construct a compound statement to match any given truth table. Example 4, for instance, showed that the table *TFTT* was given by the compound statement $b \lor \sim a$.

So the connectives $\land$, $\lor$, $\sim$, (and, or, not), are sufficient for all purposes. However, some other symbols are often used.

In particular, for the truth table *TFTT*, we write $a \Rightarrow b$ (if *a* then *b*, or *a* implies *b*) instead of $b \lor \sim a$. This generally agrees with our normal usage of English. Instead of 'it is a badger or it is not an animal', we could say 'if it is an animal then it is a badger'. In English there might be some slight difference in nuance. But in logic $a \Rightarrow b$ is identical to $b \lor \sim a$ having truth table *TFTT* and the same Venn diagram. Note that it is only false when *a* is true and *b* is false.

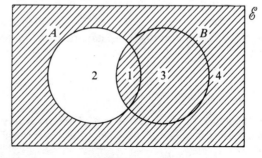

Figure 9

Example 5
Give the truth table and hence the Venn diagram for $b \Rightarrow (a \wedge \sim b)$.

	a	b		1	4	1	3	2	1
	a	b		b	$\Rightarrow$	(a	$\wedge$	$\sim$	b)
1	T	T		T	F	T	F	F	T
2	T	F		F	T	T	T	T	F
3	F	T		T	F	F	F	F	T
4	F	F		F	T	F	F	T	F

So this in fact the same as $\sim b$, giving a Venn diagram shaded in regions 2 and 4 (Figure 10).

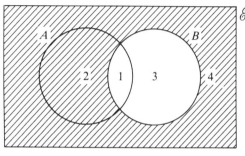

Figure 10

Logical implication

Example 6
Give the truth table for $[(a \vee b) \wedge \sim b] \Rightarrow a$.

a	b	1	2	1	3	2	1	4	1
a	b	[(a	$\vee$	b)	$\wedge$	$\sim$	b]	$\Rightarrow$	a
T	T	T	T	T	F	F	T	T	T
T	F	T	T	F	T	T	F	T	T
F	T	F	T	T	F	F	T	T	F
F	F	F	F	F	F	T	F	T	F

We find that this compound statement is always true, a tautology. In such a case we say that $(a \vee b) \wedge \sim b$ *logically implies* a. 'It was Andy or Bob, and it wasn't Bob' logically implies 'it was Andy'.

In general conversation when we use the word 'implies' or something similar, we really mean 'logically implies'. This is also the case when the implication symbol $\Rightarrow$ is used in any other area of mathematics to mean 'therefore'. That is to say, we normally tend to assume that implication is true. However, in this chapter on logic, implication will frequently be false. For example, $f \Rightarrow m$ could be 'today is Friday implies tomorrow is Monday'. So bear in mind that $\Rightarrow$ may not always be used as we intuitively expect. Just remember that $\Rightarrow$ has truth table *TFTT*, so that $f \Rightarrow m$ is always exactly the same as $m \vee \sim f$.

Equivalence

The truth table *TFFT* also has a special symbol. We write $a \Leftrightarrow b$ and say a is equivalent to b.

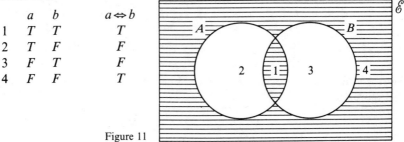

	a	b	$a \Leftrightarrow b$
1	T	T	T
2	T	F	F
3	F	T	F
4	F	F	T

Figure 11

The corresponding Venn diagram is shown in Figure 11.

We see that this is the same as $(a \wedge b) \vee ({\sim}a \wedge {\sim}b)$. That is to say, it is true when a and b are both true or when they are both false.

Example 7

Give the truth table for $({\sim}a \wedge {\sim}b) \Leftrightarrow {\sim}(a \vee b)$.

		2	1	3	2	1	4	3	1	2	1
a	b	(~	a	∧	~	b)	⇔	~	(a	∨	b)
T	T	F	T	F	F	T	T	F	T	T	T
T	F	F	T	F	T	F	T	F	T	T	F
F	T	T	F	F	F	T	T	F	F	T	T
F	F	T	F	T	T	F	T	T	F	F	F

Once again this is an example of a compound statement which is a tautology. We can say $({\sim}a \wedge {\sim}b)$ is logically equivalent to ${\sim}(a \vee b)$. This means that they have the same truth table. So in effect they are two ways of saying exactly the same thing.

'It is not Andy and it is not Bill' is exactly the same as, logically equivalent to, 'It isn't Andy or Bill'.

We use the equals sign $=$ to show that statements are logically equivalent:

$$({\sim}a \wedge {\sim}b) = {\sim}(a \vee b).$$

Otherwise in logic equivalence can be false: 'Being a girl is equivalent to being male', $g \Leftrightarrow m$. This is a valid statement but is clearly false. In the rest of mathematics $\Leftrightarrow$ is unfortunately used to mean logically equivalent. So bear in mind in this chapter that your intuitive feelings may mislead you and remember that $\Leftrightarrow$ has truth table *TFFT*.

Exercise D

1 (i) Give truth tables for the following compound statements.
   (ii) Express the statements in English taking a as 'she is an angel', b as 'she is a blonde'.

(a) $(a \wedge b) \Rightarrow b$;   (b) $a \Rightarrow (b \vee \sim a)$;
(c) $(a \vee b) \Leftrightarrow (b \wedge a)$;   (d) $(a \Rightarrow b) \Leftrightarrow (b \vee \sim a)$;
(e) $(a \wedge b) \Rightarrow \sim (a \Leftrightarrow b)$;   (f) $(a \Leftarrow b) \Leftrightarrow (\sim a \Rightarrow \sim b)$.

2   (i) Use truth tables to show the following logical equivalences.
    (ii) Express them in English stating your choice of a, b.
       (a) $(a \wedge b) = \sim (a \Rightarrow \sim b)$;
       (b) $(a \vee b) = (\sim a \Rightarrow b)$;
       (c) $(a \Leftrightarrow b) = \sim [(a \Rightarrow b) \Rightarrow \sim (b \Rightarrow a)]$.

3   (i) Show that the following are logical implications.
    (ii) Express them in English stating your choice of a, b.
       (a) $[(a \vee b) \wedge \sim b] \Rightarrow a$;
       (b) $[(a \Rightarrow b) \wedge \sim b] \Rightarrow \sim a$;
       (c) $[(a \Rightarrow b) \wedge a] \Rightarrow b$.

4   Each part in question 3 is of the form x logically implies y. For each part draw a
    Venn diagram and shade the truth set of x vertically and of y horizontally. Hence
    make a general statement about the truth sets of x and y when x logically implies y.

5   (i) Give truth tables for the following compound statements.
    (ii) Express the statements in English taking c as 'he likes cars', d as 'he can drive'.
       (a) $(c \vee d) \Rightarrow c$;   (b) $d \Rightarrow (c \wedge \sim d)$;
       (c) $(c \wedge d) \Leftrightarrow (\sim c \vee d)$;   (d) $(c \Leftrightarrow d) \Rightarrow (d \vee \sim c)$;
       (e) $(c \vee \sim d) \Leftrightarrow \sim (c \Rightarrow d)$;   (f) $(c \Leftarrow d) \Rightarrow (\sim d \Rightarrow \sim c)$.

6   (i) Use truth tables to show the following logical equivalences.
    (ii) Express them in English stating your choice of c, d.
       (a) $(c \Rightarrow d) = \sim (c \wedge \sim d)$;
       (b) $(\sim c \vee d) = (c \Rightarrow d)$;
       (c) $(c \Leftrightarrow \sim d) = (c \Rightarrow \sim d) \wedge (d \vee c)$.

7   (i) Show that the following are logical implications.
    (ii) Express them in English stating your choice of c, d.
       (a) $[(c \vee \sim d) \wedge d] \Rightarrow c$;
       (b) $[(c \Rightarrow \sim d) \wedge d] \Rightarrow \sim c$;
       (c) $[(c \Rightarrow \sim d) \wedge c] \Rightarrow \sim d$.

8   Each part in question 7 is of the form x logically implies y. For each part draw
    a Venn diagram and shade the truth set of x horizontally and of y vertically.
    Hence make a general statement about the truth sets of x and y when x logically
    implies y.

5. CONVERSE AND INVERSE

'You do maths so you are sensible', $m \Rightarrow s$.
'You are sensible so you do maths', $s \Rightarrow m$.

   Whether or not either of these compound statements is true, each is said to
be the converse of the other.

$$a \Rightarrow b \quad \text{has } converse \quad b \Rightarrow a.$$

Note that $b \Rightarrow a$ is sometimes written $a \Leftarrow b$.

Example 8
Give the truth table for $(m \Rightarrow s) \wedge (s \Rightarrow m)$.

		1	2	1	3	1	2	1
m	*s*	(*m*	$\Rightarrow$	*s*)	$\wedge$	(*s*	$\Rightarrow$	*m*)
T	*T*	*T*	*T*	*T*	*T*	*T*	*T*	*T*
T	*F*	*T*	*F*	*F*	*F*	*F*	*T*	*T*
F	*T*	*F*	*T*	*T*	*F*	*T*	*F*	*F*
F	*F*	*F*	*T*	*F*	*T*	*F*	*T*	*F*

Notice, stage 2, that $(m \Rightarrow s)$ is not the same as $(s \Rightarrow m)$, but that overall, stage 3, this truth table is the same as that for $\Leftrightarrow$.

$$(m \Rightarrow s) \wedge (s \Rightarrow m) = (m \Leftrightarrow s)$$

This is a result which you will have often used:

$$\Rightarrow \text{ and } \Leftarrow \text{ make } \Leftrightarrow.$$

But notice that this is true whether or not the individual implications are true.

Compare these two statements:
'Nitwits have wings', $n \Rightarrow w$.
'It is wingless so it is not a nitwit', $\sim w \Rightarrow \sim n$.
Each compound statement is the *inverse* of the other.
Again we may not know whether either compound statement is true.

Example 9
Give the truth table for $(n \Rightarrow w) \Leftrightarrow (\sim w \Rightarrow \sim n)$.

		1	2	1	4	2	1	3	2	1
n	*w*	(*n*	$\Rightarrow$	*w*)	$\Leftrightarrow$	($\sim$	*w*	$\Rightarrow$	$\sim$	*n*)
T	*T*	*T*	*T*	*T*	*T*	*F*	*T*	*T*	*F*	*T*
T	*F*	*T*	*F*	*F*	*T*	*T*	*F*	*F*	*F*	*T*
F	*T*	*F*	*T*	*T*	*T*	*F*	*T*	*T*	*T*	*F*
F	*F*	*F*	*T*	*F*	*T*	*T*	*F*	*T*	*T*	*F*

This is a tautology, so a proposition and its inverse are logically equivalent.

$$(n \Rightarrow w) = (\sim w \Rightarrow \sim n).$$

Sometimes care is needed when converting a compound statement into appropriate symbols.

Example 10
Find the truth table for 'Cows are male so if this animal is female it is no cow'.
c denotes 'this animal is a cow', *m* denotes 'this animal is male'. 'Cows are male' can then be written $(c \Rightarrow m)$ and the whole sentence becomes $(c \Rightarrow m) \Rightarrow (\sim m \Rightarrow \sim c)$.

		1	2	1	4	2	1	3	2	1
c	*m*	(*c*	$\Rightarrow$	*m*)	$\Rightarrow$	($\sim$	*m*	$\Rightarrow$	$\sim$	*c*)
T	*T*	*T*	*T*	*T*	*T*	*F*	*T*	*T*	*F*	*T*
T	*F*	*T*	*F*	*F*	*T*	*T*	*F*	*F*	*F*	*T*
F	*T*	*F*	*T*	*T*	*T*	*F*	*T*	*T*	*T*	*F*
F	*F*	*F*	*T*	*F*	*T*	*T*	*F*	*T*	*T*	*F*

We find, perhaps surprisingly, that this is always true.

$(c \Rightarrow m)$ logically implies $(\sim m \Rightarrow \sim c)$. In fact there is an even stronger relation between them; one is the inverse of the other, so that they are actually logically equivalent and $(c \Rightarrow m) = (\sim m \Rightarrow \sim c)$.

These subsidiary parts are merely two different ways of saying the same thing – if cows are male then cows are male – making the overall sentence true. The fact that in this case we know these parts are both false should not let our intuition prevent us from drawing the correct logical conclusion.

Exercise E

1   Express the following in symbols, stating your choices clearly. Give the truth tables and hence state the circumstances under which each is true.
   (*a*)   Cats have ears so if it is not a cat it has no ears.
   (*b*)   Pigs do not have wings so it is not a pig.
   (*c*)   If this is an elephant then it is pink or not an elephant.
   (*d*)   Males are tall so short females are not male.

2   (*a*) $(b \vee \sim a) \Rightarrow b$;   (*b*) $(b \wedge \sim a) \Rightarrow \sim a$;
    (*c*) $\sim(b \vee \sim a) \Rightarrow (a \vee \sim b)$;   (*d*) $\sim(b \wedge \sim a) \Rightarrow (\sim b \vee a)$.
   Each part is of the form $x \Rightarrow y$. In each case give the truth set
       (i) of $x \Rightarrow y$,   (ii) of the inverse $\sim y \Rightarrow \sim x$,
       (iii) of the converse $y \Rightarrow x$,   (iv) of the contrapositive $\sim x \Rightarrow \sim y$.

3   Are the following true? In each case check your conclusion by expressing the statements symbolically and using truth tables to see if they are logically equivalent.
   (*a*)   'When $\theta = 30°$ then $\sin \theta = \frac{1}{2}$' is the same as 'If $\sin \theta = \frac{1}{2}$ then $\theta = 30°$'.
   (*b*)   To prove that 'If n^2 is even, then n is also even' where n is an integer, it is enough to prove that 'If n is odd, then n^2 will also be odd'.

4   Express the following in symbols, stating your choices clearly. Give the truth tables and hence state the circumstances under which each is false.
   (*a*)   Dogs have tails so if it does not have a tail it is not a dog.
   (*b*)   Cows do not jump over moons so it is not a cow.
   (*c*)   If this is a mouse then it is white or not a mouse.
   (*d*)   Winners are happy so unhappy winners are not losers.

5   (*a*) $(a \wedge \sim b) \Rightarrow a$;   (*b*) $(a \vee \sim b) \Rightarrow \sim b$;
    (*c*) $\sim(a \wedge \sim b) \Rightarrow (b \vee \sim a)$;   (*d*) $(a \vee b) \Rightarrow \sim(b \wedge a)$.
   Each part is of the form $x \Rightarrow y$. In each case give the truth set
       (i) of $x \Rightarrow y$,   (ii) of the inverse $\sim y \Rightarrow \sim x$,
       (iii) of the converse $y \Rightarrow x$,   (iv) of the contrapositive $\sim x \Rightarrow \sim y$.

6   Are the following true? In each case check your conclusion by expressing the statements symbolically and using truth tables to see if they are logically equivalent.
   (*a*)   'When $A = 45°$ then $\tan A = 1$' is the same as
            'If $\tan A = 1$ then $A = 45°$'.
   (*b*)   To prove that 'If x is prime then $f(x)$ is composite' it is enough to prove that
            'If $f(x)$ is prime then x will be composite'. (Composite means not prime.)

6. COMBINATIONS OF STATEMENTS

So far we have restricted our attention to two statements. In this section we review this work and extend it to three statements.

We have seen that two statements a, b give rise to four possible situations, TT, TF, FT, FF. When a, b are combined into one compound statement each of these four situations may be either true or false, giving the $2^4 = 16$ outcomes below.

Situation		Outcomes															
	a b	$\vee$	$\Leftarrow$		a	$\Rightarrow$	b	$\Leftrightarrow$	$\wedge$								
1	T T	T	T	T	T	T	T	T	T	F	F	F	F	F	F	F	F
2	T F	T	T	T	T	F	F	F	F	T	T	T	T	F	F	F	F
3	F T	T	T	F	F	T	T	F	F	T	T	F	F	T	T	F	F
4	F F	T	F	T	F	T	F	T	F	T	F	T	F	T	F	T	F
		1	2	3	4	5	6	7	8	9	10	11	12	13	14	15	16

We have seen that outcome 1, logically true, is particularly important since it is independent of the original statements a, b.

These outcomes correspond to the sixteen different ways in which a Venn diagram with two sets can be shaded. These can all be described using union, intersection and complements, so, although $\Rightarrow$, $\Leftrightarrow$, $\Leftarrow$ are used as well, all these outcomes can be represented by means of the connectives $\wedge$, $\vee$, $\sim$.

Example 11
Express outcome 12, $FTFF$, in terms of $\wedge$, $\vee$, $\sim$.

See Figure 12. The outcome is only true for the second region, where a is true but b is false. So it is given by $a \wedge \sim b$, for example.

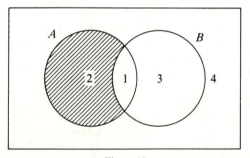

Figure 12

Binary notation

Sometimes 1 and 0 are used in place of T and F. The four possible situations then become 11, 10, 10, 00, the binary numbers from 3 to 0, while the sixteen possible outcomes are represented by the binary numbers from fifteen to zero, 1111 to 0000. This provides a convenient order for listing various possibilities and for checking that none has been missed.

Using the notation, we can give the truth table for $a \wedge {\sim} b$, for example.

		1	3	2	1
a	b	a	$\wedge$	$\sim$	b
1	1	1	0	0	1
1	0	1	1	1	0
0	1	0	0	0	1
0	0	0	0	1	0

The outcome is thus 0100 or *FTFF*. (Compare Example 11.)

Three statements

'A crud is not a big animal'; $c \Rightarrow {\sim} (b \wedge a)$,
where c denotes 'it is a crud', b 'it is big' and a 'it is an animal'.

When a third statement is combined, our previous methods still work because each connective links only two of the statements. However, there are now eight possible situations for the statements a, b, c corresponding to the binary notation for the numbers from seven to zero.

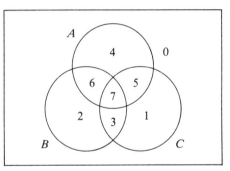

	a	b	c		a	b	c
7	T	T	T		1	1	1
6	T	T	F		1	1	0
5	T	F	T		1	0	1
4	T	F	F		1	0	0
3	F	T	T		0	1	1
2	F	T	F		0	1	0
1	F	F	T		0	0	1
0	F	F	F		0	0	0

Figure 13

We label the regions of the Venn diagram to correspond (Figure 13).

Example 12
Give the truth table and Venn diagram for $c \Rightarrow {\sim} (b \wedge a)$.

				1	4	3	1	2	1
	a	b	c	c	$\Rightarrow$	$\sim$	$(b$	$\wedge$	$a)$
7	T	T	T	T	F	F	T	T	T
6	T	T	F	F	T	F	T	T	T
5	T	F	T	T	T	T	F	F	T
4	T	F	F	F	T	T	F	F	T
3	F	T	T	T	T	T	T	F	F
2	F	T	F	F	T	T	T	F	F
1	F	F	T	T	T	T	F	F	F
0	F	F	F	F	T	T	F	F	F

Figure 14 shows the Venn diagram.

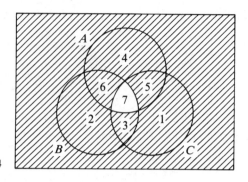

Figure 14

Thus this combined statement is true under all circumstances unless all three statements a, b, c happen to be true.

Exercise F

1 The sixteen possible outcomes at the beginning of Section 6 can be paired so that an outcome is paired with its negation. Thus a, outcome 4, is paired with $\sim a$, outcome 13. Show how the remaining outcomes pair up.

2 Express each of the last eight outcomes shown at the beginning of Section 6 in terms of a, b and the connectives $\wedge$, $\vee$, $\sim$.

3 (a) Draw up a truth table for the following.
   (b) Express each one in English taking a = 'it is angry', b = 'it is a bull' c = 'I am a coward'.
   (c) Find under what circumstances all four combined statements are true.
   (i) $(a \wedge b) \vee \sim c$;   (ii) $(a \vee c) \Rightarrow (\sim b \wedge c)$;
   (iii) $\sim (a \wedge \sim b) \Leftrightarrow (b \vee c)$;   (iv) $(c \Rightarrow a) \Leftrightarrow (\sim b \vee [c \Rightarrow b])$.

4 Express in symbols the following triple statements stating in each case the circumstances under which the statement is true.
   (a) Animals are beastly or cowardly.
   (b) Pigs do not have queer relations.
   (c) Life is short and brutish.
   (d) If there is no tea or coffee I shall perish.

5 Give, in an appropriate order, the possible situations when four statements are combined.

6 Choose statements for a, b, c and express in English:
$$[(a \Rightarrow b) \wedge (b \Rightarrow c)] \Leftrightarrow (a \Rightarrow c).$$
Find the circumstances under which this compound statement is not true.

7 (a) Draw up truth tables for the following.
   (b) Express each one in English taking d = 'I am in danger', e = 'it is enjoyable', f = 'it is free-fall'.
   (c) Find under what circumstances all four combined statements are true.
   (i) $(d \wedge e) \Rightarrow f$;   (ii) $(d \wedge \sim e) \Leftrightarrow \sim f$;
   (iii) $[\sim (d \vee e) \wedge f] \Rightarrow \sim e$;   (iv) $[(\sim d \Rightarrow \sim f) \wedge (e \Rightarrow f)] \Leftrightarrow (e \Rightarrow d)$.

8 Express in symbols the following triple statements, stating in each case the circumstances under which the statement is true.

(a) If you lose on the swings then you gain on the roundabouts.
(b) If it is not good to be alive it is bad to be married or dead.
(c) Sages agree and youths don't disagree, so youths are sages.
(d) Teachers are wise and so are owls, so teachers are owls.

7. DEDUCTION

Some combinations of statements occur frequently in argument or proof. 'The milkman comes on Tuesday. It is not a Tuesday. So the milkman will not come.'

1	2	1	3	2	4	2	1		
$[(t$	$\Rightarrow$	$m)$	$\wedge$	$\sim$	$t]$	$\Rightarrow$	$\sim$	m	$m = $ 'the milkman comes'
T	T	T	F	F	T	T	F	T	$t = $ 'it is Tuesday'
T	F	F	F	F	T	T	T	F	
F	T	T	T	T	F	F	F	T	
F	T	F	T	T	F	T	T	F	

But we find this is not a tautology; it breaks down when m is true but t is false. That is to say, the milkman might come on Wednesday. Although this sort of reasoning is sometimes heard in arguments, it leads to wrong conclusions.

'In April the buds open. The buds are not open. So it is not April.'

$$[(a \Rightarrow b) \wedge \sim b] \Rightarrow \sim a.$$

Do you think this is a correct conclusion? This, though similar to the previous example, you will find is in fact always true.

Another useful tautology is

$$[(a \Rightarrow b) \wedge (b \Rightarrow c)] \Rightarrow (a \Rightarrow c).$$

'All alligators go bathing. Bathing needs concentration. So alligators need concentration.'

We write $\qquad (a \Rightarrow b) \wedge (b \Rightarrow c)$
$$\Rightarrow \qquad (a \Rightarrow c),$$

where writing $\Rightarrow$ at the margin indicates that we are using a tautology.

This can be extended to a chain of statements:

$$(a \Rightarrow b) \wedge (b \Rightarrow c) \wedge (\sim c)$$
$$\Rightarrow \qquad (a \Rightarrow c) \qquad \wedge (\sim c)$$
$$\Rightarrow \qquad \sim a.$$

In effect we are stating that $[(a \Rightarrow b) \wedge (b \Rightarrow c) \wedge (\sim c)] \Rightarrow \sim a$ is itself a tautology without needing to go through the labour of completing an eight-row truth table. We can look on the three initial statements

$$(a \Rightarrow b), \quad (b \Rightarrow c), \quad (\sim c)$$

as premises or assumptions from which $\sim a$ can be deduced as a conclusion.

'Apes are beastly. Beastliness requires cunning. It is not cunning. Hence it is not an ape.'

To be valid a conclusion must come from the premises by a chain of logical implications. 'It is not an ape' is a valid conclusion from the first three statements (the premises of the argument). However, whether or not the conclusion is true depends on whether or not the premises are true.

Valid arguments

There are four tautologies which are particularly useful for making deductions:

$$\text{T1} \quad [(a \lor b) \land (\sim b)] \Rightarrow a;$$
$$\text{T2} \quad [(a \Rightarrow b) \land (\sim b)] \Rightarrow \sim a;$$
$$\text{T3} \quad [(a \Rightarrow b) \land (a)] \Rightarrow b;$$
$$\text{T4} \quad [(a \Rightarrow b) \land (b \Rightarrow c)] \Rightarrow (a \Rightarrow c).$$

In addition it is helpful to know

$$\text{T5} \quad (a \Rightarrow b) \quad = (\sim b \Rightarrow \sim a) \quad \text{(inverse)};$$
$$\text{T6} \quad \sim (a \lor b) = \sim a \land \sim b;$$
$$\text{T7} \quad \sim (a \land b) = \sim a \lor \sim b.$$

These may give rise to several valid conclusions. False conclusions can best be exposed by a truth table.

Example 13
'If it rained or no one came, the party would have been a failure. The party was a success. So it did not rain.' Is this correct?

$$[(r \lor n) \Rightarrow f] \land (\sim f)$$
$$\Rightarrow \quad \sim (r \lor n) \quad \text{(using T2)}$$
$$= \quad \sim r \land \sim n \quad \text{(using T6)}.$$

A valid conclusion is that it did not rain, and also that someone came.

Example 14
'If that was an insect it should not have eight legs. If that was a spider it should have eight legs.' What can be deduced?

$$(i \Rightarrow \sim e) \land (s \Rightarrow e)$$
$$= \quad (i \Rightarrow \sim e) \land (\sim e \Rightarrow \sim s) \quad \text{(using T5)}$$
$$\Rightarrow \quad i \Rightarrow \sim s \quad \text{(using T4)}.$$

A valid conclusion is that 'if it is an insect it is not a spider'.

Example 15
'When a person is young he is not very sensible. Now I am not young. So I must be very sensible.' Is this correct?

$$(y \Rightarrow \sim s) \land (\sim y)$$
$$= \quad (s \Rightarrow \sim y) \land (\sim y) \quad \text{(using T5)}.$$

This does not seem to lead to the stated conclusion.

1	3	2	1	4	2	1	5	1
[(y	$\Rightarrow$	$\sim$	s)	$\wedge$	$\sim$	y]	$\Rightarrow$	s
T	F	F	T	F	F	T	T	T
T	T	T	F	F	F	T	T	F
F	T	F	T	T	T	F	T	T
F	T	T	F	T	T	F	F	F

We see that this is not a tautology, so the conclusion is not valid.

All logical questions can be settled by truth tables. However when there are more than three statements a table requires sixteen or more lines and becomes impractical. Tautologies may be used either instead of a table or at least to simplify an expression first.

Exercise G

Check the validity of the following arguments.
Where it is of interest, comment upon the actual truth or falsity of the premises or conclusion.

1   Either that is a hex or it is a rhomb. It is not a rhomb, so it must be a hex.

2   The lines have remained parallel, so the transformation was a translation, half-turn or enlargement. It was an enlargement, I'm sure. It can't be either of the others.

3   If the probability of one head in one throw is $\frac{1}{2}$, then the probability of one head only in two throws will also be $\frac{1}{2}$. The probability of one head in one throw is $\frac{1}{2}$. So the probability of one head in two throws is $\frac{1}{2}$.

4   If $x > 4$ then $x^2 > 16$, so $x > 4$.

5   It is not possible for a number to be both prime and a multiple of three. This number is prime. It cannot be a multiple of three.

6   If I think, then I exist.
    If I exist, then I am aware of my imperfection.
    If I am aware of my imperfection, then I am aware of perfection.
    If I am aware of perfection, then perfection exists.
    So, if I think, then perfection exists.
    I do think, so perfection does exist.

7   Express each of the following arguments in symbols and determine whether any are valid.
    (*a*)   I spend my vacations in Scotland. Today I am in Wales. Therefore the vacation has not yet begun.
    (*b*)   If I am working or I have no money, I am unable to go to London. I am in London. Therefore I cannot be working.
    (*c*)   If the parts do not come or no one works then the job will be delayed. The job has been delayed. Therefore the parts did not come.

8   All people who say they have seen UFOs are liars.
    No people who are free from delusions eat haggis.
    The headmaster is honest.
    No people who don't say they have seen UFOs are given to delusions.
    What conclusion can be drawn by taking account of all these premises?

8. CIRCUITS

The flow in a circuit is controlled by a switch S which either allows the flow through, denoted by T, or fails to do so, denoted by F. A switch will be illustrated as shown in Figure 15.

Figure 15

Consider the flow in the two situations in Figure 16, in which there are two switches s, t.

(a)

$\longrightarrow s \longrightarrow t \longrightarrow$

Series switches

(b)

Parallel switches

Figure 16

Series switches				Parallel switches		
s	t	Circuit		s	t	Circuit
T	T	T		T	T	T
T	F	F		T	F	T
F	T	F		F	T	T
F	F	F		F	F	F

The tables show all possible combinations for switches s, t.

The series switches have the same table as $\wedge$. There is flow through the circuit only when there is flow through s and t.

Similarly the parallel switches have the same table as $\vee$. There is flow through the circuit when there is flow through s or t or both.

If the flow is electricity, then a bulb can be made to light up when the table is 'true'.

By combining these two circuits we can illustrate any logical expression with $\wedge, \vee$ only. For example, Figure 17 illustrates $a \wedge (b \vee c)$.

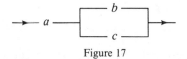

Figure 17

For $\sim$ we need a switch $\sim s$ which allows the flow through when s fails to and vice versa. For convenience of the diagram $\sim s$ will be written s'.

(a)

(b)

Figure 18

The circuit in Figure 18(a) represents $(a \wedge b) \vee a'$ for there is flow through the circuit either if a and b allow flow through or if a' allows flow through. Venn diagrams or a truth table will show that

$$(a \wedge b) \vee a' = a' \vee b.$$

So there is an alternative circuit as shown in Figure 18(b), which is simpler than the original circuit, yet has exactly the same properties.

Sometimes the circuit can aid the algebra.

Example 16

A school has options biology, chemistry, French, German with possible combinations:

$$(b \wedge f) \vee (c \wedge g) \vee (c \wedge f) \vee (b \wedge g).$$

Write these in a simpler form.

The original statement gives the circuit of Figure 19(a). There is no need to have two switches labelled c, so we can draw the equivalent circuit shown in Figure 19(b). By altering the order of the wires we can also avoid the duplication of the switch b. (See Figures 19(c) and (d).)

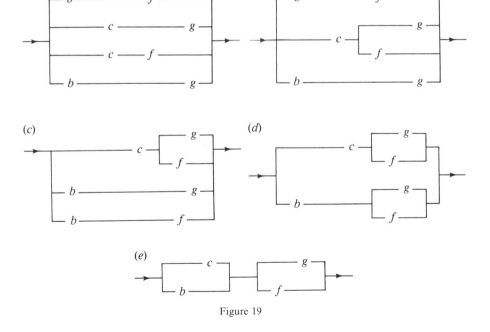

Figure 19

Removing the duplication of f and g gives Figure 19(e), which represents $(b \vee c) \wedge (f \vee g)$. So the options are either biology or chemistry with either French or German.

Exercise H

1   Draw circuits representing the following:
   (a) $(a \wedge b) \vee c$;   (b) $(a \vee b) \wedge (c \vee d)$;   (c) $a \wedge (b \vee c)$;
   (d) $(a \vee \sim b) \vee (b \wedge c)$;   (e) $a \Rightarrow b$;   (f) $a \Leftrightarrow b$.

2   What do the circuits in Figure 20 represent?

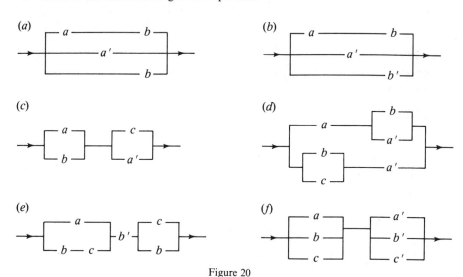

Figure 20

3   Give an alternative circuit for each part of question 2.

4   Draw a circuit for binary multiplication, ignoring any carry digit.

5   The light on a staircase goes on when the switch at the bottom or top, but not both, is used. In symbols this is $(b \wedge t') \vee (b' \wedge t)$. Draw a circuit for this and explain the connection between a staircase light and a computer. Give, in symbols and with a circuit, the arrangement for a light operated at the bottom, middle or top of a flight of stairs.

6   Four men, A, B, C, D, are preparing for isolation experiments in separate rooms. They each have a switch to show when they are ready. Write down the alternatives in symbols and draw a circuit for:
   (a) the light is on only when all four are ready.
   (b) the light is on if one or more is unready, no light is on when they are all ready.

7   In a party of explorers A and B are doctors, A and C are surveyors and B and D can cook. Two of them are to make a journey and there must be a doctor, someone who can survey and someone who can cook. Draw up a table giving possible combinations and use it to design a circuit with switches A, B, C, D which will cause a light to shine when two switches denoting a possible combination are put on.

8   Give a circuit for (a) $a \Rightarrow b$, (b) $a \Leftrightarrow b$.

9   Two players toss coins. A wins if one is a head and the other a tail. B wins if they are the same. Express this in symbols and devise a circuit to light a lamp when A wins.

10   A study bedroom has a light X at the bed and a light Y at the desk. They can never both be on at once. They are controlled by switches B, D, E by the bed, the desk and at the entrance to the room. If X is on B will turn it off. If a light is on D will turn on the other one instead. If a light is on E will turn it off, otherwise it will turn one light on. Draw up a possible eight-row truth table to show the state of B, D, E and the corresponding state of X, Y. Hence express x in terms of b, d, e and similarly express y in terms of b, d, e and give a circuit for x and also for y.

9. BOOLEAN ALGEBRA

We have already met several statements which have the same truth table and so are logically equivalent or equal. For example,

$$a \wedge b = \sim (a \Rightarrow \sim b),$$
$$a \vee b = \sim a \Rightarrow b,$$
$$a \Leftrightarrow b = \sim (a \Rightarrow b) \Rightarrow \sim (b \Rightarrow a).$$

Notice that these three also show that $\wedge$, $\vee$, $\Leftrightarrow$ can be replaced by a combination of $\sim$ and $\Rightarrow$ only. We can also express everything in terms of $\wedge$, $\vee$, $\sim$.

$$a \Rightarrow b = \sim a \vee b,$$
$$a \Leftrightarrow b = (a \wedge b) \vee (\sim a \wedge \sim b).$$

The first person to develop a system like this was George Boole who, in 1847, set down the 'laws of thought' where T stands for a logically true and F for a logically false proposition.

Law 1:	$a \wedge b = b \wedge a$	Commutative
	$a \vee b = b \vee a$	
Law 2:	$a \wedge (b \wedge c) = (a \wedge b) \wedge c$	Associative
	$a \vee (b \vee c) = (a \vee b) \vee c$	
Law 3:	$a \wedge (b \vee c) = (a \wedge b) \vee (a \wedge c)$	Distributive
	$a \vee (b \wedge c) = (a \vee b) \wedge (a \vee c)$	
Law 4:	$a \wedge a = a$	
	$a \vee a = a$	
Law 5:	$a \wedge (a \vee b) = a$	
	$a \vee (a \wedge b) = a$	
Law 6:	$\sim (\sim a) = a$	
Law 7:	$\sim (a \wedge b) = \sim a \vee \sim b$	
	$\sim (a \vee b) = \sim a \wedge \sim b$	
Law 8:	$a \wedge \sim a = F$	
	$a \vee \sim a = T$	
Law 9:	$T \wedge a = a$	
	$F \vee a = a$	
Law 10:	$F \wedge a = F$	
	$T \vee a = T$	

These laws, which are easily proved using truth tables, can be used to simplify complicated statements. They have the additional advantage of being the same as for sets when $\cap$, $\cup$, $'$, $\mathscr{E}$, ϕ are replaced by $\vee$, $\wedge$, $'$, T, F respectively. Knowledge and experience of one system will help with the other. Venn diagrams can be applied to both.

Notice that the associative law 2 means that the position of the brackets make no difference when the connectives are the same. So we can leave the brackets out and write $a \wedge b \wedge c$. Similarly for any number of statements with the same connective we may leave out the brackets and write $a \wedge b \wedge c \wedge d$, for example. The commutative law 1 also enables us to alter their order in any way and write $b \wedge a \wedge c \wedge d$ instead of $a \wedge b \wedge c \wedge d$, for example.

Example 17

Simplify $a \vee [(b \wedge \sim c) \wedge (d \wedge \sim e) \wedge (f \wedge \sim b)]$.

$$
\begin{aligned}
&a \vee [(b \wedge \sim c) \wedge (d \wedge \sim e) \wedge (f \wedge \sim b)] \\
&= a \vee [b \wedge \sim c \wedge d \wedge \sim e \wedge f \wedge \sim b] && \text{Law 2   Associative} \\
&= a \vee [b \wedge \sim b \wedge d \wedge \sim e \wedge f \wedge \sim c] && \text{Law 1   Commutative} \\
&= a \vee [\quad F \quad \wedge (d \wedge \sim e \wedge f \wedge \sim c)] && \text{Law 8} \\
&= a \vee \quad F && \text{Law 10} \\
&= a. && \text{Law 9}
\end{aligned}
$$

Although the strength of Boolean algebra lies in its generality, we shall confine its application to simpler examples.

Example 18

Simplify $(a \vee b) \wedge \sim a$.

$$
\begin{aligned}
(a \vee b) \wedge \sim a &= \sim a \wedge (a \vee b) && \text{Law 1} \\
&= (\sim a \wedge a) \vee (\sim a \wedge b) && \text{Law 3} \\
&= \quad F \quad \vee (\sim a \wedge b) && \text{Law 8} \\
&= \qquad\qquad \sim a \wedge b. && \text{Law 9}
\end{aligned}
$$

Example 19

Simplify $b \wedge \sim (a \vee \sim b)$.

$$
\begin{aligned}
b \wedge \sim (a \vee \sim b) &= \quad b \wedge (\sim a \wedge \sim \sim b) && \text{Law 7} \\
&= \quad b \wedge (\sim a \wedge b) && \text{Law 6} \\
&= \quad b \wedge \sim a \wedge b && \text{Law 2} \\
&= \sim a \wedge b \wedge b && \text{Law 1} \\
&= \sim a \wedge b && \text{Law 4}
\end{aligned}
$$

One difficulty is that there are so many different ways of representing the same thing. Each line of Example 19 gives a different representation. There is no unique simplest result and only great experience with the laws will suggest how to obtain a required result. However a change of notation can be very helpful.

If instead of $\qquad\qquad\wedge\quad\vee\quad\sim\quad T\quad F$

we write $\qquad\qquad\qquad.\quad+\quad'\quad 1\quad 0$

law 1 becomes $\qquad\qquad\qquad a.b = b.a$

$$a+b = b+a$$

Similarly law 2 and the first part of laws 3, 9, 10 are what we should get by ordinary algebra. The only things that need to be remembered are fairly simple:

Law 4: $a.a = a$ *It's Adam* and *It's Adam* is the same as *It's Adam*
Law 8: $a.a' = 0$ *It's Adam* and *It isn't Adam* is false
Law 7: $(ab)' = a'+b'$ not (*Adam* and *Ben*) = *not Adam* or *not Ben*
         $(a+b)' = a'b'$ not (*Adam* or *Ben*) = *not Adam* and *not Ben*

Furthermore we can confine ourselves to the standard procedure, which gives all solutions, of 'multiplying out, negating if necessary'. As in ordinary algebra, we often write ab for $a.b$.

Example 20
Simplify $a \vee [(b \wedge \sim c) \wedge (d \wedge \sim e) \wedge (f \wedge \sim b)]$.   (Compare Example 17.)
$$a+[(bc')(de')(fb')] = a+bb'c'de'f$$
$$= a+0.c'de'f$$
$$= a.$$

Example 21
Simplify $(a \vee b) \wedge \sim a$.   (Compare Example 18.)
$$(a+b)a' = aa'+a'b$$
$$= 0+a'b$$
$$= a'b.$$

Example 22
Simplify $b \wedge \sim (a \vee \sim b)$.   (Compare Example 19.)
$$b(a+b')' = b(a'b)$$
$$= a'b.$$

Logical puzzles

We shall end with an idealised situation where the honest are always truthful and the wicked always lie.

Example 23
Adam says to Eve: 'One of us is a liar.' What can we deduce?

Either Adam is honest and Adam or Eve is a liar: $a(a'+e')$;
or Adam lies and his statement is false: $a'(a'+e')'$.
($a = $ 'Adam is honest'; $e = $ 'Eve is honest'.)
So the situation can be represented by

$$a(a'+e')+a'(a'+e')' = aa'+ae'+a'(ae)$$
$$= 0+ae'+a'ae$$
$$= 0+ae'+0$$
$$= ae'.$$

We deduce that Adam is honest and Eve is a liar.

Example 24
Adam says that Eve is a liar; Eve admits that he is right. What can be deduced?

Both Adam is honest and Eve is a liar or Adam lies and Eve doesn't: $ae'+a'e$;
and Eve is honest and Adam is honest or Eve lies and Adam lies: $ea+e'a'$.

$$(ae'+a'e)(ea+e'a') = ae'ea+ae'e'a'+a'eea+a'ee'a'$$
$$= aee'+aa'e'+aa'e+a'ee'$$
$$= a0+0e'+0e+a'0$$
$$= 0$$

So this is a contradiction.

Example 25
Brenda says 'Ann is a liar.' Chris says 'So is Brenda.' What can be deduced?

Writing $a = $ 'Ann is honest', and so on, the two pieces of information are equivalent to $(ba'+b'a)$ and $c(a'b')+c'(a'b')'$ respectively.

$$(ba'+b'a)(c(a'b')+c'(a'b')') = (ba'+b'a)(ca'b'+c'(a+b))$$
$$= (a'b+ab')(a'b'c+ac'+bc')$$
$$= a'ba'b'c+a'bac'+a'bbc'$$
$$\quad +ab'a'b'c+ab'ac'+ab'bc'$$
$$= a'bb'c+aa'bc'+a'bbc'$$
$$\quad +aa'b'c+ab'c'+abb'c'$$
$$= 0+0+a'bc'+0+ab'c'+0$$
$$= (a'b+ab')c'$$

So Chris is a liar, and just one of Ann and Brenda is honest.

Example 26
Mr Roach, Mr Salmon and Mr Trout went fishing. They got a roach, a salmon and a trout, each having caught one of them. When questioned they replied as follows:

> Mr Roach: Mr Salmon got the trout; I caught the salmon.
> Mr Salmon: Mr Trout got the roach; I caught the trout.
> Mr Trout: Mr Salmon got the trout; I caught the salmon.

What actually happened, if each one either always tells the truth or always lies?

The three replies can be written as:

$$(R_s S_t + R'_s S'_t)(S_t T_r + S'_t T'_r)(S_t T_s + S'_t T'_s)$$

where R_s denotes Mr Roach got the salmon, and so on. In this case $R_s T_s = 0$, for example, since Mr Roach and Mr Trout cannot both have caught the salmon.

$$
\begin{aligned}
&(R_s S_t + R'_s S'_t)(S_t T_r + S'_t T'_r)(S_t T_s + S'_t T'_s) \\
&= (R_s S_t S_t T_r + R_s S_t S'_t T'_r + R'_s S'_t S_t T_r + R'_s S'_t S'_t T'_r)(S_t T_s + S'_t T'_s) \\
&= (R_s S_t T_r + 0 + 0 + R'_s S'_t T'_r)(S_t T_s + S'_t T'_s) \\
&= R_s S_t T_r S_t T_s + R_s S_t T_r S'_t T'_s + R'_s S'_t T'_r S_t T_s + R'_s S'_t T'_r S'_t T'_s \\
&= R_s T_r S_t T_r + R_s S_t S'_t T_r T'_s + R'_s S_t S'_t T'_r T_s + R'_s S'_t T'_r T'_s \\
&= 0 + 0 + 0 + R'_s S'_t T'_r T'_s.
\end{aligned}
$$

From this it is straightforward to deduce that Mr Trout caught the trout, Mr Roach caught the roach, and Mr Salmon caught the salmon. Incidentally, all three were liars!

Exercise I

1   Use Boolean algebra to prove the following:
    (a) $a \wedge (\sim a \vee b) = a \wedge b$;
    (b) $(a \Rightarrow b) \wedge \sim b = \sim a \wedge \sim b$;
    (c) $\sim (a \vee b) \wedge a = F$;
    (d) $a \wedge (a \Rightarrow b) \wedge (b \Rightarrow c) = a \wedge b \wedge c$.

2   Find the conclusions from the following:
    (a) I am a liar or you are honest.
    (b) I am honest or $2 + 2 = 4$.
    (c) I am honest and you are a liar.
    (d) Ann: We are both liars. Ben: One of us is honest.
    (e) Ann: We are all three liars. Chris: Only one of us is honest.
    (f) Ann: We are all three honest. Chris: At least one of us is not.

3   Fred, George and Harriet went shooting and between them got a bull, an inner and a magpie, each getting one of them.
    Fred said: I got the bull; George didn't.
    George said: I got the bull; Harriet got the magpie.
    Harriet said: I got the bull; Fred didn't.
    What was the true situation?

4   Messrs. Long, Short and Tall arrived by air, bus or car.
    Long said: I came by air; Short came by bus.
    Short said: I came by car; Tall came by air.
    Tall said: I came by bus; Long did not come by air.
    Who was untruthful?

5   Jack, Kate, Larry and Mary play in a quartet.
    Jack says: I play the clarinet; Kate plays the flute.
    Kate says: Larry plays the horn; Mary plays the trumpet.
    Larry says: I play the flute; Jack plays the trumpet.
    Mary says: Kate plays the horn; I play the clarinet.
    In fact each of them gives one true and one false piece of information. Find who plays what.

6   A says that B denies that C is honest.
    B says that C affirms that A is a liar.
    C says that A and B are both honest.
    Find the true situation.

10. SUMMARY

Logical symbol	Truth table	Venn diagram	Circuit	Boolean algebra

Negation $\sim a$

a	$\sim a$
T	F
F	T

Circuit: $\longrightarrow a' \longrightarrow$ Boolean: a'

And $a \wedge b$

a	b	$a \wedge b$
T	T	T
T	F	F
F	T	F
F	F	F

Circuit: $\longrightarrow a - b \longrightarrow$ Boolean: ab

Or $a \vee b$

a	b	$a \vee b$
T	T	T
T	F	T
F	T	T
F	F	F

Circuit: a, b in parallel   Boolean: $a+b$

Implication $a \Rightarrow b$ $= \sim a \vee b$

a	b	$a \Rightarrow b$
T	T	T
T	F	F
F	T	T
F	F	T

Circuit: a', b in parallel   Boolean: $a'+b$

Equivalence $a \Leftrightarrow b$

a	b	$a \Leftrightarrow b$
T	T	T
T	F	F
F	T	F
F	F	T

Circuit: $a - b$ / $a' - b'$ Boolean: $ab + a'b'$

A compound statement which is logically true in all circumstances is called a tautology.

A compound statement which is logically false in all circumstances is called a contradiction.

An implication which is a tautology is called a logical implication.

An equivalence which is a tautology is called a logical equivalence.

$a \Rightarrow b$ has converse $b \Rightarrow a$.

$a \Rightarrow b$ has inverse $\sim b \Rightarrow \sim a$.

A statement is logically equivalent to its inverse.

Complicated statements can be simplified by using the 'laws of thought', which, in the alternative notation, are:

1.	$ab = ba$;	$a+b = b+a$.
2.	$a(bc) = (ab)c$;	$a+(b+c) = (a+b)+c$.
3.	$a(b+c) = ab+ac$;	$a+bc = (a+b)(a+c)$.
4.	$aa = a$;	$a+a = a$.
5.	$a(a+b) = a$;	$a+ab = a$.
6.	$(a')' = a$.	
7.	$(ab)' = a'+b'$;	$(a+b)' = a'b'$.
8.	$aa' = 0$;	$a+a' = 1$.
9.	$1a = a$;	$0+a = a$.
10.	$0a = 0$;	$1+a = 1$.

Miscellaneous exercise

1 (*a*) An electric circuit with battery and lamp is shown in Figure 21. Does the combination correspond to $\wedge$ or $\vee$ or $\Rightarrow$?

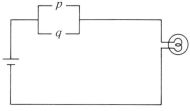

Figure 21

(*b*) Give the Boolean expression represented by Figure 22.

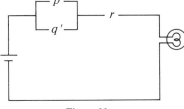

Figure 22

(*c*) Draw a circuit representing

$$(p \wedge q) \vee (p' \wedge q').$$

(*d*) Show that $[(p \wedge q) \vee (p' \wedge q)] \vee (r \wedge q')$ is equivalent to $q \vee r$, indicating clearly each stage in your reasoning.

(In the circuit diagram a switch, e.g. p, is regarded as on or off according as the corresponding statement is true or false. The truth of the combination of propositions is shown by the light being lit. p' denotes 'not p', $\wedge$ denotes 'and', $\vee$ denotes 'or'.)   (SMP)

2   By a truth table, or otherwise, show that

$$(p \vee q) \wedge \sim p \Rightarrow q,$$

whatever the propositions p and q may be.

By identifying p and q, draw the correct conclusion from the statement: 'You are not drunk or I am not sober, and I am sober'.   (SMP)

3   Construct a truth table to show that $\sim (p \vee q)$ is equivalent to $\sim p \wedge \sim q$.

Give the corresponding result for sets A and B and illustrate the set concerned in a Venn diagram.   (SMP)

4   Let p stand for the statement 'I like it', q stand for the statement 'you like it', a stand for 'I like it if you like it' and b stand for 'I do not like it if you do not like it'.

Copy and complete the following truth table:

p	q	a	b	$a \Rightarrow b$
T	T			
T	F			
F	T			
F	F			

Explain why it is not logically true that a implies b.   (SMP)

5   A student decides to study either Russian or both French and German. This proposition is denoted by $r \vee (f \wedge g)$. (a) Draw a circuit diagram to represent this proposition, and (b) draw up its truth table.

A second proposition is denoted by $\sim f \Rightarrow r$. (c) State this proposition in words, and (d) draw up its truth table.

Hence state, *with reasons*, whether or not the proposition

$$[r \vee (f \wedge g)] \Leftrightarrow [\sim f \Rightarrow r]$$

is a tautology (i.e. is always true).   (SMP)

6   The connective 'nor' is denoted by *. Thus $a * b$ is true if, and only if, neither a nor b is true.

Give the truth table for $a * b$ and for $a * a$.

Hence show that $a * a$ is logically equivalent to $\sim a$.

Thus verify that $(a * a) * (b * b)$ has the same truth table as $a \wedge b$. Show, by giving the truth table, that $\sim (a * b) \Leftrightarrow a \vee b$ is also a tautology.

Hence express $a \vee b$ using only the connective *.   (SMP)

20

Complex numbers

1. SQUARE ROOTS OF NEGATIVE NUMBERS

During the sixteenth century the Italian mathematician Cardano was trying to solve a problem which involved the following:

the sum of two numbers is 2,

and their product is also 2.

That is, find x and y such that $x+y = 2$ and $xy = 2$.

From the first equation, $y = 2-x$

and hence, using the second equation, $x(2-x) = 2$.

$$
\begin{aligned}
x(2-x) = 2 \quad &\Rightarrow \quad 2x - x^2 = 2 \\
&\Rightarrow \quad x^2 - 2x + 2 = 0 \\
&\Rightarrow \quad (x-1)^2 + 1 = 0 \\
&\Rightarrow \quad (x-1)^2 = -1 \\
&\Rightarrow \quad x - 1 = \pm \sqrt{-1}.
\end{aligned}
$$

So the two roots of the equation are $1 + \sqrt{-1}$ and $1 - \sqrt{-1}$.

But what is $\sqrt{-1}$? That question had puzzled mathematicians for some time. Cardano's problem is only one of many where this unknown number appeared.

One interesting property of $\sqrt{-1}$ is that if we assume that it has the same properties as a real number, then we can use it in calculations and arrive at reasonable conclusions.

For example, we know that

$$\sqrt{20} = \sqrt{(4 \times 5)} = 2\sqrt{5};$$

is it true that $\qquad \sqrt{-4} = \sqrt{(4 \times {}^-1)} = 2\sqrt{-1}?$

$$[2\sqrt{-1}]^2 = 2\sqrt{-1} \times 2\sqrt{-1} = 2 \times 2 \times \sqrt{-1} \times \sqrt{-1} = 4 \times {}^-1 = {}^-4,$$

so this seems to be all right.

The sum of the 'solutions' of the quadratic equation above is

$$(1 + \sqrt{-1}) + (1 - \sqrt{-1}) = 2$$

and their product is

$$(1 + \sqrt{-1})(1 - \sqrt{-1}) = 1^2 - (\sqrt{-1})^2 = 1 + 1 = 2,$$

so it appears that these two 'numbers' are indeed a solution to Cardano's problem.

Cardano used these methods successfully; so did another Italian, Bombelli.

Bombelli called such numbers 'imaginary numbers' as did the French mathematician Descartes in his book *La Géometrie*, published in 1637. Euler, writing in 1770, described them as 'impossible or imaginary numbers'. Eventually a square root of $^-1$ was seen to be useful and important and it was given its own letter – first i and, more recently, j.

Notice that $j^2 = ^-1 \Rightarrow (^-j)^2 = ^-j \times ^-j = j^2 = ^-1$,

So $j^2 = ^-1$ and $(^-j)^2 = ^-1$: we have two square roots of $^-1$, just as we have two square roots of positive numbers.

But once we have j, we also have two square roots of *any* negative number. For example,

the square roots of $^-4$ are $2j$ and ^-2j;

the square roots of $^-3$ are $j\sqrt{3}$ and $^-j\sqrt{3}$.

Notice that in this latter case we write the j first, so that it does not appear to be under the square root sign.

The solutions to Cardano's problem would be written $1+j$ and $1-j$. These are examples of *complex numbers*, which are of the general form $a+bj$, with a and b real numbers. a is called the *real part* of the number, and b is the *imaginary part*.

We can calculate with complex numbers as if they were ordinary (real) numbers, but with the extra rule that $j^2 = ^-1$.

Example 1

Simplify: (*a*) $(4+3j)+(5-7j)$; (*b*) $(4+3j) \times (5-7j)$.

(*a*) $\qquad (4+3j)+(5-7j) = 4+5+3j+^-7j = 9-4j.$

(*b*) $\qquad (4+3j) \times (5-7j) = 4(5-7j)+3j(5-7j)$
$$= 20-28j+15j-21j^2$$
$$= 20-13j+21$$
$$= 41-13j.$$

Exercise A

1 Write down the square roots of:
    (*a*) $^-9$;   (*b*) $^-100$;   (*c*) $^-5$;   (*d*) $^-\frac{9}{4}$.

2 Calculate the following:
    (*a*) $(3+4j)+(2-4j)$;   (*b*) $(2+7j)+(3-6j)$;   (*c*) $(5-2j)+(^-3-4j)$;
    (*d*) $(4-3j)+(^-4+3j)$;   (*e*) $(3+4j)-(2-4j)$;   (*f*) $(2+7j)-(3-6j)$.

3 Calculate the following:
    (*a*) $(3+4j)(2-3j)$;   (*b*) $(4+7j)(5-2j)$;   (*c*) $(^-3-5j)(2-4j)$;
    (*d*) $(5-3j)(5+3j)$;   (*e*) $(a+bj)(c+dj)$;   (*f*) $(a+bj)(a-bj)$.

4 Simplify the following:
    (*a*) j^3;   (*b*) $(7+2j)^2$;   (*c*) $(5j)^2+(5j)^4$;   (*d*) $1+j+j^2+j^3$.

5 Write down the square roots of:
    (*a*) $^-81$;   (*b*) $^-8$;   (*c*) $^-6\frac{1}{4}$;   (*d*) $^-0.01$.

6 Calculate the following:
   (a) $(7+2j)+(7-2j)$;   (b) $(7+2j)+(^-2+7j)$;   (c) $(^-4+j)-(3-2j)$;
   (d) $(6+3j)-(7+4j)$;   (e) $(^-8-\frac{1}{2}j)-(^-7+\frac{1}{2}j)$;   (f) $(5-7j)-(^-3-8j)$.

7 Calculate the following:
   (a) $j(8-5j)$;   (b) $(2+j)(8-5j)$;   (c) $(3-7j)(5-8j)$;
   (d) $^-2j(4-3j)$;   (e) $(^-3-4j)(^-3+4j)$;   (f) $(a+bj)^2$.

8 Simplify the following:
   (a) $(1+j)^2$;   (b) $1-j-j^2-j^3$;   (c) $\dfrac{1}{j}$;   (d) $(1-j)^4$.

2. THE ARGAND DIAGRAM

There is clearly a correspondence between the addition and subtraction of complex numbers and the addition and subtraction of position vectors. For example, compare

$$(2+3j)+(5-8j) = 7-5j$$

with

$$\begin{bmatrix} 2 \\ 3 \end{bmatrix} + \begin{bmatrix} 5 \\ -8 \end{bmatrix} = \begin{bmatrix} 7 \\ -5 \end{bmatrix}$$

and

$$(2+3j)-(5-8j) = {}^-3+11j$$

with

$$\begin{bmatrix} 2 \\ 3 \end{bmatrix} - \begin{bmatrix} 5 \\ -8 \end{bmatrix} = \begin{bmatrix} -3 \\ 11 \end{bmatrix}.$$

It is useful, therefore, to represent complex numbers on a plane, using Cartesian coordinates. The complex number $a+bj$ is represented by the point with position vector $\begin{bmatrix} a \\ b \end{bmatrix}$. Such a diagram is called the *Argand diagram* (after a French mathematician who wrote about it in 1806, although he was not the first to use it) or the *complex plane*. (See Figure 1.)

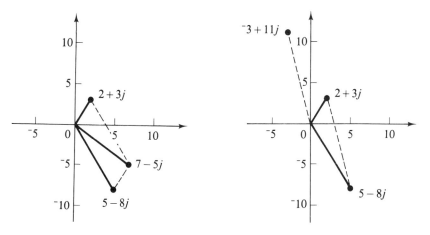

Figure 1

The x-axis (the points of which represent real numbers) is called the *real axis*, and the y-axis (the points of which represent imaginary numbers bj) is called the *imaginary axis*.

We can multiply a position vector by a real number, and this is consistent with the corresponding operation on a complex number. Compare, for example,

$$2\begin{bmatrix} 5 \\ -8 \end{bmatrix} = \begin{bmatrix} 10 \\ -16 \end{bmatrix}$$

with $2(5-8j) = 10-16j.$

In the Argand diagram, multiplication by 2 corresponds to an enlargement, centre the origin, scale factor 2. (See Figure 2.)

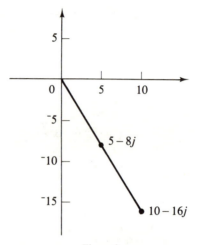

Figure 2

But we have no rule for multiplying two position vectors together to obtain another position vector, so it is not immediately obvious what would be the geometrical effect of multiplying by $2+3j$, for example.

Exercise B

1 The points A and B in the Argand diagram correspond to the complex numbers $a = 2+j$ and $b = 4-j$.
   (a) Show A and B on an Argand diagram.
   (b) Show, on the same diagram, the points P, Q and R corresponding to the complex numbers $2a$, $a+b$ and $2a-b$.

2 Write down the complex numbers corresponding to the points A, B, C, D and E in the Argand diagram in Figure 3.

   In questions 3–6 the complex numbers a, b, c, d, e, f are those represented by the points A, B, C, D, E, F respectively in the Argand diagram in Figure 3.

3 A transformation **P** is applied to the 'boat' in Figure 3 by adding $3+5j$ to each of the complex numbers a to f. Show the image of the boat on a diagram and hence describe the transformation.

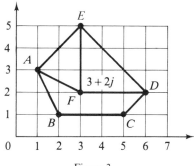

Figure 3

*4   Repeat question 3 but this time multiply each of the six complex numbers by j.

5   Repeat question 3 but this time multiply each of the six complex numbers by ^-2j.

*6   Repeat question 3 but this time multiply each of the six complex numbers by $3+4j$.

*7   Multiply $0, 1, 1+j$ and j (the vertices of the 'unit square') by $3+4j$. Show the square and its image on a diagram. Can you write down a matrix which corresponds to the same transformation?

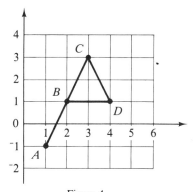

Figure 4

In questions 8–11 the complex numbers a, b, c, d are those represented by the points A, B, C, D respectively in the Argand diagram in Figure 4.

8   A transformation is applied to the 'flag' in Figure 4 by multiplying each of the complex numbers a to d by $^-2$. Show the image of the flag on a diagram and hence describe the transformation.

9   Repeat question 8 but this time multiply a to d by ^-j.

10   Repeat question 8 but this time multiply a to d by $3j$.

11   Repeat question 8 but this time multiply a to d by $2+3j$.

12   Multiply $0, 1, 1+j$ and j by $2+3j$. Show the unit square and its image on a diagram. Can you write down a matrix which corresponds to the same transformation?

3. MULTIPLICATION BY COMPLEX NUMBERS

Multiplication by j

The result of Exercise B, question 4 suggests that multiplication by j corresponds to a quarter-turn about the origin in the Argand diagram. In general, we have that

$$j(p+qj) = pj+qj^2 = {}^-q+pj,$$

so the point with position vector $\begin{bmatrix} p \\ q \end{bmatrix}$ is mapped to the point with position vector $\begin{bmatrix} {}^-q \\ p \end{bmatrix}$. (See Figure 5.) This transformation is indeed a quarter-turn about the origin, and is represented by the matrix $\mathbf{J} = \begin{bmatrix} 0 & {}^-1 \\ 1 & 0 \end{bmatrix}$, since $\begin{bmatrix} {}^-q \\ p \end{bmatrix} = \begin{bmatrix} 0 & {}^-1 \\ 1 & 0 \end{bmatrix}\begin{bmatrix} p \\ q \end{bmatrix}$.

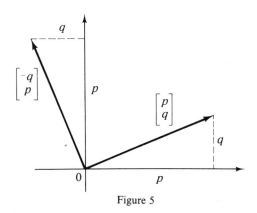

Figure 5

Multiplication by $3+4j$

Exercise B, question 7 leads us to suspect that multiplication by $3+4j$ corresponds to the transformation given by the matrix $\begin{bmatrix} 3 & {}^-4 \\ 4 & 3 \end{bmatrix}$, which appears (from question 6) to be a combination of an enlargement and a rotation. For a general complex number $p+qj$, we have that

$$(3+4j)(p+qj) = 3p+3qj+4pj+4qj^2$$
$$= (3p-4q)+(4p+3q)j,$$

so the point with position vector $\begin{bmatrix} p \\ q \end{bmatrix}$ is mapped to the point with position vector $\begin{bmatrix} 3p-4q \\ 4p+3q \end{bmatrix} = \begin{bmatrix} 3 & {}^-4 \\ 4 & 3 \end{bmatrix}\begin{bmatrix} p \\ q \end{bmatrix}$, confirming our guess. We observe that $\begin{bmatrix} 3 & {}^-4 \\ 4 & 3 \end{bmatrix} = \begin{bmatrix} 3 & 0 \\ 0 & 3 \end{bmatrix}+\begin{bmatrix} 0 & {}^-4 \\ 4 & 0 \end{bmatrix} = 3\mathbf{I}+4\mathbf{J}$, where $\mathbf{I}$ is the identity matrix and $\mathbf{J}$ the matrix for a quarter-turn, $\begin{bmatrix} 0 & {}^-1 \\ 1 & 0 \end{bmatrix}$.

Similarly, multiplication by $2+3j$ corresponds to the transformation given by the matrix $2\mathbf{I}+3\mathbf{J} = \begin{bmatrix} 2 & -3 \\ 3 & 2 \end{bmatrix}$, and, in general, multiplication by $x+yj$ corresponds to the transformation given by the matrix $x\mathbf{I}+y\mathbf{J} = \begin{bmatrix} x & -y \\ y & x \end{bmatrix}$.

Polar form

The image of the unit square under the transformation given by the matrix $\begin{bmatrix} 3 & -4 \\ 4 & 3 \end{bmatrix}$ is shown in Figure 6.

$$\begin{bmatrix} 1 \\ 0 \end{bmatrix} \rightarrow \begin{bmatrix} 3 \\ 4 \end{bmatrix} \text{ and } \begin{bmatrix} 0 \\ 1 \end{bmatrix} \rightarrow \begin{bmatrix} -4 \\ 3 \end{bmatrix}.$$

$\begin{bmatrix} 3 \\ 4 \end{bmatrix}$ and $\begin{bmatrix} -4 \\ 3 \end{bmatrix}$ have length 5 and they are perpendicular, since

$$\begin{bmatrix} 3 \\ 4 \end{bmatrix} \cdot \begin{bmatrix} -4 \\ 3 \end{bmatrix} = -12+12 = 0.$$

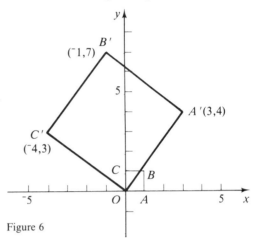

Figure 6

So the transformation maps the unit square to a square of side 5. It is a *spiral similarity*, rotating and enlarging about the origin. The scale factor of the enlargement is 5, the length, r, of the vector $\begin{bmatrix} 3 \\ 4 \end{bmatrix}$. The angle of rotation is the angle

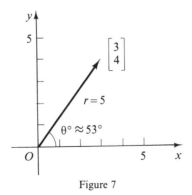

Figure 7

between this vector and the x-axis – i.e. the angle $\theta°$ shown in Figure 7.

But $(r, \theta°)$ are the polar coordinates of the point with Cartesian coordinates $(3, 4)$.

In general, if z is the complex number $x + yj$ and z is represented by the point Z in the Argand diagram, then multiplication by z corresponds to an enlargement, centre the origin, scale factor r, and a rotation about the origin through an angle θ°, where (r, θ°) are the polar coordinates of Z.

The connection between (r, θ°) and (x, y) is illustrated by Figure 8. r is called the *modulus* of z, written $|z|$; θ° is called the *argument* of z, written arg z.

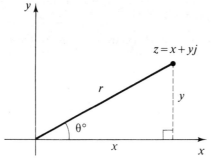

$$x = r \cos \theta^\circ \qquad y = r \sin \theta^\circ$$

$$r^2 = x^2 + y^2 \qquad \tan \theta^\circ = \frac{y}{x}$$

Figure 8

Since $x = r \cos \theta^\circ$ and $y = r \sin \theta^\circ$, we can write

$$z = x + yj = r(\cos \theta^\circ + j \sin \theta^\circ).$$

z is then said to be written in *polar form*, or *modulus–argument form*.

The effect of an enlargement, centre the origin with scale factor r, on a point with polar coordinates (s, ϕ°) is to map it to the point with polar coordinates (rs, ϕ°). The effect of a rotation about the origin through an angle θ° is to map (s, ϕ°) to $(s, \theta^\circ + \phi^\circ)$. The combined effect is to map (s, ϕ°) to $(rs, \theta^\circ + \phi^\circ)$. (See Figure 9.)

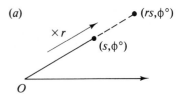

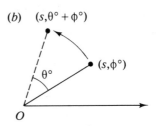

$$z \leftrightarrow (r, \theta^\circ): |z| = r, \text{ arg } z = \theta^\circ;$$

$$w \leftrightarrow (s, \phi^\circ): |w| = s, \text{ arg } w = \phi^\circ$$

$$\Rightarrow \quad zw \leftrightarrow (rs, \theta^\circ + \phi^\circ),$$

so

$$|zw| = rs = |z| \cdot |w|,$$

$$\text{arg}(zw) = \theta^\circ + \phi^\circ = \text{arg } z + \text{arg } w.$$

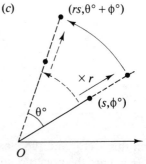

Figure 9

Example 2
Write in polar form $z = 2 - 3j$ and $w = {}^{-}2$.

$$|z| = \sqrt{(2^2 + 3^2)} = \sqrt{13} \approx 3.6.$$

From Figure 10(*a*) we see that $\tan\theta° = \frac{3}{2} \Leftarrow \theta \approx 56$.
So $\arg z \approx {}^{-}56°$,
$z \approx 3.6(\cos {}^{-}56° + j\sin {}^{-}56°)$.

From Figure 10(*b*) we see that $|w| = 2$ and $\arg w = 180°$.
$w = 2(\cos 180° + j\sin 180°)$.

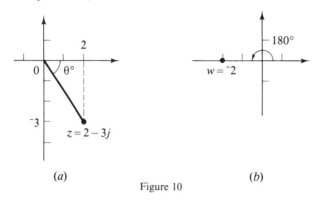

(*a*)   (*b*)

Figure 10

Example 3
The complex numbers z_1 and z_2 are represented on the Argand diagram by the points with polar coordinates $(2, 40°)$ and $(5, 80°)$. Find the complex numbers $z_1 z_2$ and z_2^2 in the form $x + yj$.

$z_1 z_2$ is represented by the point with polar coordinates $(5 \times 2, \quad 40° + 80°) = (10, 120°)$.

$$z_1 z_2 = 10 \,(\cos 120° + j\sin 120°)$$
$$\approx {}^{-}5 + 8.66j.$$

$z_2^2 = z_2 z_2$ is represented by the point with polar coordinates $(5 \times 5, \quad 80°) = (25, 160°)$.

$$z_2^2 = 25(\cos 160° + j\sin 160°) \approx {}^{-}23.5 + 8.55j.$$

Exercise C

1 Write the following complex numbers in polar form:
$$4 + 3j, \quad 5j, \quad {}^{-}5 + 12j, \quad 6 + 6j.$$

2 Find the complex numbers represented by the points with the following polar coordinates: $(10, 53.1°), \quad (7, 90°), \quad (4, {}^{-}45°), \quad (20, {}^{-}100°).$

3 If z and w are complex numbers represented on the Argand diagram by the points with polar coordinates $(2, 45°)$ and $(1\frac{1}{2}, 30°)$ draw a diagram to show z, w, zw, zw^2.

4 With z and w as defined in question 3, draw a diagram to show z, w, z^2w, z^3w and z^4w.

5   With z and w as defined in question 3:
   (a) write down the modulus and argument of zw;
   (b) find z and w in the form $x+yj$;
   (c) use your answer to (b) to find zw in the form $x+yj$;
   (d) use your answer to (a) to find zw in the form $x+yj$.

6   Write the following complex numbers in polar form:
$$8+15j, \quad {}^-4j, \quad 3-3j, \quad 7-24j.$$

7   Find the complex numbers represented by the points with the following polar coordinates:
$$(6, 60°), \quad (8, {}^-50°), \quad (9, 101°), \quad (3, 180°).$$

8   If z and w are the complex numbers represented on the Argand diagram by the points with polar coordinates $(1, 100°)$ and $(1, {}^-40°)$, draw a diagram to show z, w, zw, z^2, w^2, zw^2 and zw^3.

9   With z and w defined as in question 8, find the following numbers in the form $x+yj$: z, w, zw and z^2.

4. DIVISION AND CONJUGATES

Multiplication by $3+4j = 5(\cos 53° + j \sin 53°)$ corresponds to an enlargement with scale factor 5 and rotation through 53°, which is the transformation given by the matrix $\begin{bmatrix} 3 & {}^-4 \\ 4 & 3 \end{bmatrix}$. We would therefore expect division by $3+4j$ to correspond to the inverse transformation and the inverse matrix. The inverse of the matrix $\begin{bmatrix} 3 & {}^-4 \\ 4 & 3 \end{bmatrix}$ is $\begin{bmatrix} \frac{3}{25} & \frac{4}{25} \\ -\frac{4}{25} & \frac{3}{25} \end{bmatrix}$, which corresponds to the complex number $\frac{3}{25} - \frac{4}{25}j$. As a check, we calculate $(\frac{3}{25} - \frac{4}{25}j)(3+4j)$.

$$(\tfrac{3}{25} - \tfrac{4}{25}j)(3+4j) = \tfrac{9}{25} + \tfrac{12}{25}j - \tfrac{12}{25}j - \tfrac{16}{25}j^2$$
$$= \tfrac{9}{25} + \tfrac{16}{25} = 1.$$

So
$$\frac{1}{3+4j} = \tfrac{3}{25} - \tfrac{4}{25}j.$$

In general, since $p+qj$ corresponds to the matrix $\begin{bmatrix} p & {}^-q \\ q & p \end{bmatrix}$, $\dfrac{1}{p+qj}$ corresponds to the inverse matrix $\begin{bmatrix} \frac{p}{r^2} & \frac{q}{r^2} \\ \frac{{}^-q}{r^2} & \frac{p}{r^2} \end{bmatrix}$, where $r^2 = p^2 + q^2$,

and so
$$\frac{1}{p+qj} = \frac{p}{r^2} - \frac{q}{r^2}j = \frac{1}{p^2+q^2}(p-qj).$$

Notice that this can be obtained by multiplying by $\dfrac{p-qj}{p-qj}$, as follows:
$$\frac{1}{p+qj} = \frac{p-qj}{(p+qj)(p-qj)} = \frac{p-qj}{p^2-q^2j^2} = \frac{p-qj}{p^2+q^2}.$$

Example 4

Calculate $\dfrac{3+11j}{2+3j}$.

Using the above result, $\dfrac{1}{2+3j} = \dfrac{1}{2^2+3^2}(2-3j) = \tfrac{1}{13}(2-3j)$.

Hence

$$\dfrac{3+11j}{2+3j} = \tfrac{1}{13}(3+11j)(2-3j) = \tfrac{1}{13}(6-9j+22j-33j^2) = \tfrac{1}{13}(39+13j) = 3+j.$$

Alternatively, we could multiply by $\dfrac{2-3j}{2-3j}$ to obtain a real number as the denominator of the fraction:

$$\dfrac{3+11j}{2+3j} = \dfrac{(3+11j)(2-3j)}{(2+3j)(2-3j)} = \dfrac{6-9j+22j-33j^2}{4-9j^2} = \dfrac{39+13j}{13} = 3+j.$$

This is not the first time that a pair of numbers of the forms $x+yj$ and $x-yj$ have occurred together. Solving Cardano's problem at the beginning of the chapter gave us $1+j$ and $1-j$. In fact, solving any quadratic equation (with real coefficients) which has non-real solutions will give a pair of complex numbers related in this way.

Example 5

Solve: (*a*) $x^2-4x+13 = 0$;   (*b*) $2x^2+5x+7 = 0$.

(*a*) Here it is fairly simple to complete the square.

$$\begin{aligned}
x^2-4x+4 &= {}^-9 \\
\Rightarrow \quad (x-2)^2 &= 9j^2 \\
\Rightarrow \quad x-2 &= \pm 3j \\
\Rightarrow \quad x &= 2+3j \quad \text{or} \quad 2-3j
\end{aligned}$$

(*b*) In this case, using the formula is probably easiest.

$$a = 2, \quad b = 5, \quad c = 7$$

$$\Rightarrow \quad x = \dfrac{{}^-b\pm\sqrt{(b^2-4ac)}}{2a} = \dfrac{{}^-5\pm\sqrt{(25-56)}}{4} = \dfrac{{}^-5\pm\sqrt{(31j^2)}}{4}$$

$$\Rightarrow \quad x \approx {}^-1.25+1.39j \ \text{or} \ {}^-1.25-1.39j.$$

Because pairs of numbers $x+yj$ and $x-yj$ occur so frequently, such numbers are given a special name and are called *complex conjugates*. The conjugate of a complex number z is denoted by $z*$ (or $\bar{z}$). On the Argand diagram, $z*$ is the reflection of z in the real axis.

Notice the following properties of z and $z*$:

(1) $zz* = |z|^2$, since $(x+yj)(x-yj) = x^2-y^2j^2 = x^2+y^2$;

(2) $\dfrac{1}{z} = \dfrac{z*}{|z|^2}$;

(3) $|z*| = |z|$;

(4) $\arg z* = {}^-\arg z.$

Note also that:

(5) $\left|\dfrac{1}{z}\right| = \dfrac{1}{|z|}$;

(6) $\arg\left(\dfrac{1}{z}\right) = -\arg z$.

From (5) and (6) and the previous properties of the modulus and argument, we can deduce that

$$\left|\frac{z}{w}\right| = \frac{|z|}{|w|} \quad \text{and} \quad \arg\left(\frac{z}{w}\right) = \arg z - \arg w.$$

Example 6

If z and w are represented on the Argand diagram by the points with polar coordinates $(3, 75°)$ and $(2, 40°)$ show z, w, z^*, $\dfrac{1}{z}$ and $\dfrac{z}{w}$ on a diagram.

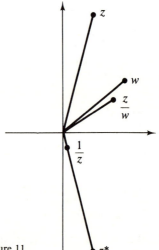

The polar coordinates of the points representing z^*, $\dfrac{1}{z}$ and $\dfrac{z}{w}$ are $(3, ^-75°)$, $(\frac{1}{3}, ^-75°)$ and $(3 \div 2, 75° - 40°) = (1\frac{1}{2}, 35°)$ respectively. The five points are shown in Figure 11.

Figure 11

Exercise D

1 Calculate the following:

$$\frac{1}{4+3j}, \quad \frac{3-j}{4+3j}, \quad \frac{j}{4+3j}, \quad \frac{6+7j}{4-3j}, \quad \frac{^-3+4j}{1+2j}.$$

2 If $|z| = 2$ and $\arg z = 30°$:

    (*a*) write down the modulus and argument of $\dfrac{1}{z}$;

    (*b*) show z and $\dfrac{1}{z}$ on an Argand diagram;

    (*c*) find z and $\dfrac{1}{z}$ in the form $x + yj$, and check that their product is 1.

3 If $z = 4 + 3j$:

    (*a*) write down z^* and find $\dfrac{1}{z}$;

(b) find the modulus and argument of (i) z, (ii) $\dfrac{1}{z}$;

(c) show z and $\dfrac{1}{z}$ on an Argand diagram.

4 If $|z| = 4$ and $\arg z = 60°$, and $w = 1+j$;

    (a) write z in the form $x+yj$;

    (b) find the modulus and argument of w and of $\dfrac{z}{w}$;

    (c) show z, w and $\dfrac{z}{w}$ on an Argand diagram;

    (d) use your answer to (a) to calculate $\dfrac{z}{w}$ in the form $x+yj$;

    (e) find $\dfrac{z}{w}$ in the form $x+yj$ using your values for $\left|\dfrac{z}{w}\right|$ and $\arg\left(\dfrac{z}{w}\right)$ obtained in (b).

5 If $z = 5+6j$, calculate:

    (a) $z+z^*$;   (b) $z-z^*$;   (c) zz^*;   (d) z^2+z^{*2}.

6 Calculate the following:

$$\frac{1}{3+5j}, \quad \frac{11+7j}{3+5j}, \quad \frac{^-3+6j}{5-4j}, \quad \frac{^-3-6j}{5+4j}.$$

7 If $|z| = 3$ and $\arg z = 40°$:

    (a) write down the modulus and argument of $\dfrac{1}{z}$;

    (b) show z and $\dfrac{1}{z}$ on an Argand diagram;

    (c) find z and $\dfrac{1}{z}$ in the form $x+yj$. Is their product equal to 1?

8 If $z = 5-3j$:

    (a) write down z^* and find $\dfrac{1}{z}$;

    (b) show z, z^* and $\dfrac{1}{z}$ on an Argand diagram.

9 If z and w are represented on the Argand diagram by the points with polar coordinates $(4, 120°)$ and $(5, 30°)$:

    (a) write z and w in the form $x+yj$;

    (b) write down the modulus and argument of $\dfrac{z}{w}$;

    (c) show z, w and $\dfrac{z}{w}$ on an Argand diagram;

    (d) find $\dfrac{z}{w}$ in the form $x+yj$.

10 If $z = x+yj$, find, in terms of x and y:

    (a) $z+z^*$;   (b) $z-z^*$;   (c) zz^*;   (d) z^2-z^{*2}.

11 Solve the equations:

    (a) $z^2+6z+10 = 0$;   (b) $w^2+5w+7 = 0$;

    (c) $3x^2-7x+6 = 0$;   (d) $z^2-2jz+8 = 0$.

12 Find two numbers whose sum is 4 and whose product is 5.

13 Solve the simultaneous equations:
$$2x + 3yj = 5,$$
$$jx + y = 5.$$

---

14 Solve the equations:
(a) $z^2 + 12z + 61 = 0$;   (b) $w^2 + 7w + 11 = 0$;
(c) $z^2 - 6z + 13 = 0$;   (d) $w^2 + 10w + 41 = 0$.

15 Solve the simultaneous equations:
$$jx + y = 0,$$
$$4x + 3y = 25.$$

5. GEOMETRY AND COMPLEX NUMBERS

Complex numbers can be used to describe loci and hence to give equations of lines, circles and other curves drawn in the Argand diagram.

Circles

(i) $|z| = 3$ means that the point Z representing z is at a distance of 3 units from the origin (Figure 12). It is therefore on a circle, centre the origin, and radius 3. Notice that the condition $|z| = 3$ can also be written as $zz^* = 9$, since $|z|^2 = zz^*$.

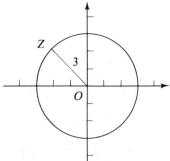

Figure 12

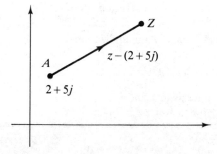

Figure 13

(ii) $z - (2 + 5j)$ is represented by $\underset{\sim}{AZ}$ where A is the point representing $2 + 5j$. (See Figure 13.) $|z - (2 + 5j)| = 3$ means that $\underset{\sim}{AZ}$ is of length 3. Z is therefore on a circle, centre $(2 + 5j)$, radius 3. (See Figure 14.)

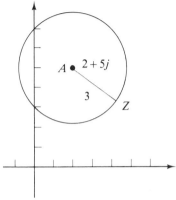

Figure 14

Straight lines

(i) $\arg z = 20°$ means that $\underset{\sim}{OZ}$ makes an angle of $20°$ with the real axis. Z is therefore on the half-line through O at $20°$ to the real axis. (See Figure 15.) The other 'half' of the line is given by $\arg z = {}^-160°$.

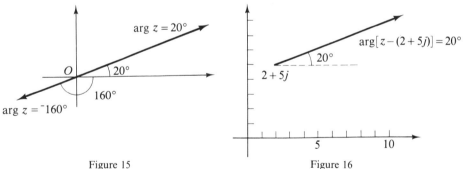

Figure 15   Figure 16

(ii) $\arg[z - (2 + 5j)] = 20°$ is the parallel half-line through A, the point representing $2 + 5j$. (See Figure 16.)

(iii) Straight lines can also be given as perpendicular bisectors. For example,

$$|z - (2 + 5j)| = |z - (4 + 3j)|$$

says that Z is equidistant from A (representing $2 + 5j$) and B (representing $4 + 3j$). Z is therefore on the mediator of A and B.

Example 7
Find the equation of the circle with centre at the point corresponding to j and with radius 3.

The equation is $|z-j| = 3$.

Example 8
Find the equation of the mediator of the points corresponding to ^-2j and 1.

The equation is $|z+2j| = |z-1|$.

Example 9
Find the centre of the circle through ^-2j, 1 and $2-j$.

The centre of the circle is a point equidistant from ^-2j, 1 and $2-j$. It is therefore the point z for which

$$|z+2j| = |z-1| = |z-(2-j)|.$$

So we must solve the simultaneous equations

$$|z+2j| = |z-1|,$$
$$|z-1| = |z-(2-j)|.$$

One way of proceeding is to use $|w|^2 = ww^*$, so that from the first equation we deduce that

$$(z+2j)(z+2j)^* = (z-1)(z-1)^*$$
$$(z+2j)(z^*-2j) = (z-1)(z^*-1)$$
$$zz^* + 2jz^* - 2jz + 4 = zz^* - z^* - z + 1$$
$$z+z^* - 2j(z-z^*) + 3 = 0.$$

Similarly from the second equation it can be deduced that

$$z+z^* + j(z-z^*) - 4 = 0.$$

Subtracting these two new equations gives

$$-3j(z-z^*) + 7 = 0$$

so

$$z-z^* = ^-\tfrac{7}{3}j.$$

Hence

$$z+z^* - 2j(^-\tfrac{7}{3}j) + 3 = 0$$

so

$$z+z^* = \tfrac{5}{3}.$$

Adding $(z-z^*)$ and $(z+z^*)$

$$2z = \tfrac{5}{3} - \tfrac{7}{3}j$$
$$z = \tfrac{5}{6} - \tfrac{7}{6}j.$$

The centre of the circle is $\tfrac{5}{6} - \tfrac{7}{6}j$.

Exercise E

1 Write down the equations of the following circles in the Argand diagram:
   (*a*) centre origin, radius 3; (*b*) centre $4j$, radius 2; (*c*) centre ^-1+j, radius 4.

2   Write down the equation of the perpendicular bisector of AB if, in the Argand diagram:
   (a) A is $2+j$, B is 3;   (b) A is ^-1-2j, B is 3;   (c) A is ^-1-2j, B is $2+j$.

3   Find the centre of the circle in the Argand diagram passing through $2+j$, 3 and ^-1-2j.

4   Show the following lines on an Argand diagram:
   (a) $z=z^*$;   (b) $z+z^* = 0$.

5   Show the following circles on an Argand diagram:
   (a) $|z-3| = 2$;   (b) $|z+1+j| = 1$;   (c) $|z+j| = 1$;   (d) $|z-3j| = 4$.

6. SUMMARY

The square roots of $^-1$ are j and ^-j. The set of complex numbers is the set $\{p+qj : p, q \text{ real numbers}\}$. Calculating with complex numbers looks like ordinary arithmetic and algebra, together with the rule that $j^2 = ^-1$.

Complex numbers can be represented on the Argand diagram, in which $p+qj$ is represented by the point with position vector $\begin{bmatrix} p \\ q \end{bmatrix}$. Addition of complex numbers corresponds to the addition of vectors. Multiplication by $p+qj$ corresponds to enlargement, scale factor r, and rotation through $\theta°$, centre the origin, where $(r, \theta°)$ are the polar coordinates of the point (p, q). The matrix representing this transformation is $\begin{bmatrix} p & ^-q \\ q & p \end{bmatrix}$. In particular, multiplication by j corresponds to a quarter-turn about the origin.

When a complex number z is written in the form $r(\cos \theta° + j \sin \theta°)$, it is said to be written in polar form. r is called the modulus of z, written $|z|$, and $\theta°$ the argument of z, $\arg z$.

zw has modulus equal to $|z| \cdot |w|$ and argument equal to $\arg z + \arg w$.

$\dfrac{z}{w}$ has modulus equal to $\dfrac{|z|}{|w|}$ and argument equal to $\arg z - \arg w$.

$z = x+yj$ and $z^* = x-yj$ are called complex conjugates.

$|z-a| = k$ is the equation of a circle centre A and radius k, where A is the point representing the complex number a.

$|z-a| = |z-b|$ is the equation of the perpendicular bisector of AB, where A and B are the points representing the complex numbers a and b.

Miscellaneous exercise

1   If $|z| = 2$ and $\arg z = 60°$:
   (a) show z, z^2 and z^3 on an Argand diagram;
   (b) find z in the form $x+yj$;
   (c) use your answer to (b) to find z^3 in the form $x+yj$.
   Can you find three cube roots of $^-8$?

2 If $w = 1+j$:

    (*a*) find w^2 in the form $x+yj$;

    (*b*) find $|w|$ and $\arg w$;

    (*c*) show w, w^2, w^3, and w^4 on an Argand diagram;

    (*d*) write down the values of $|w^4|$ and $\arg(w^4)$.

    What other values of $\arg w$ would give $\arg w^4 = 180°$ or an equivalent angle? Use your answer to find four fourth roots of $^-4$ in the form $x+yj$.

3 In the Argand diagram A, B represent the complex numbers $a = 5+2j$ and $b = 8+6j$.

    (*a*) Explain why $j(b-a)$ is represented by a vector perpendicular to $\underset{\sim}{AB}$ and of length equal to AB.

    (*b*) Hence find complex numbers c and d, represented by C and D, such that $ABCD$ is a square.

4 In Figure 17, $PQRS$ is a square. Find the complex number representing the point C, the centre of the square.

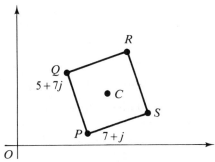

Figure 17

5 Squares are drawn outwards on the sides AB, BC, CD, DA of the quadrilateral $ABCD$ shown in Figure 18.

    Find p, q, r, s, the complex numbers represented by the centres of these squares.

    Calculate $\dfrac{r-p}{q-s}$.

    What does this tell you about the lines PR and QS?

    Can you prove this to be true for all quadrilaterals $ABCD$?

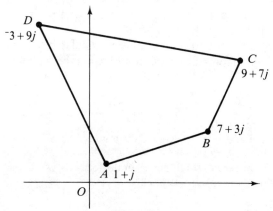

Figure 18

6   If $ABCD$ is a square, show that $a-b = d-c$. If $a-c = j(b-d)$ explain how this tells you whether the vertices are labelled in a clockwise or anticlockwise sense.

7   A, B, C, D and E are points in the Argand diagram representing $a = 5+5j$, $b = 1-7j$, $c = {}^-7-j$, $d = {}^-1-3j$ and $e = \frac{1}{2}d$.
   (a) Show that $OA = OB = OC$ where O is the origin.
   (b) Show that AD is perpendicular to BC, BD is perpendicular to CA and CD is perpendicular to AB.
   (c) Write down the complex numbers represented by A_1, B_1, C_1, the midpoints of BC, CA, AB respectively, and A_2, B_2, C_2, the midpoints of AD, BD, CD respectively.
   (d) Deduce from (a) that O is the centre of the circle ABC and use a similar method to show that E is the centre of a circle passing through A_1, B_1, C_1, A_2, B_2, C_2. What is the ratio of the radii of the two circles?

8   A, B, C are points on a circle centre O. $p = b+c$, $q = c+a$, $r = a+b$ and $d = a+b+c$. Explain why:
   (a) $OARB$ is a rhombus;
   (b) OR is perpendicular to AB;
   (c) CD is perpendicular to AB.
   Similarly show that AD and BD are perpendicular to BC and CA respectively.

9   The complex number ω has modulus 1 and argument 120°.
   (a) Show ω and ω^2 on an Argand diagram.
   (b) Find ω and ω^2 in the form $x+yj$ and show that $\omega^2+\omega+1 = 0$.
   (c) Explain why the points representing a, b and $^-a\omega-b\omega^2$ are at the vertices of an equilateral triangle in the Argand diagram. Where is the point representing $^-a\omega^2-b\omega$?

10   A triangle ABC is drawn in the Argand diagram and points P, Q, R taken (outside triangle ABC) such that the triangles BCP, CAQ and ABR are equilateral.
   (a) Use the results of question 9 to write down the complex numbers p, q, r (represented by P, Q, R) in terms of a, b, c and ω.
   (b) Show that the complex numbers $a-p$, $b-q$, $c-r$ are in the ratio $1:\omega:\omega^2$. What does this tell you about:
   (i) the lengths of AP, BQ, CR; (ii) the angles between AP, BQ and CR?

21

Groups

1. CALENDAR PROBLEM

1st January 1901 was a Monday. What day of the week was 9th August 1911?

To solve a problem like this we need to remember that every 7 days we count forward from a given date will bring us back to the same day of the week. 'Counting on 7' is an *identity operation* as far as days of the week are concerned – it has the same effect as 'counting on 0'. So do 'counting on 14', 'counting on 21', etc. In fact, we have an *equivalence class* of 'counting on' numbers which all bring us back to the original day of the week:

$$\{0, 7, 14, 21, ..., 364, ...\}.$$

Similarly, 'counting on' 8 days (or 15, or 22, ...) is the same as 'counting on' 1 day, and will take us from a Thursday to a Friday. This gives us another equivalence class:

$$\{1, 8, 15, 22, ..., 365, ...\}.$$

Notice that 'counting on' one (normal) year of 365 days is the same as 'counting on' 1 day, since 365 and 1 belong to the same equivalence class. In the same way, 'counting on' a month of 31 days is equivalent to 'counting on' 3 days, and so on.

From 1st January 1901 to 9th August 1911 is

10 years (including two leap years, 1904 and 1908),
7 months (4 of 31 days, 2 of 30 days and 1 of 28 days),
8 days.

Since 365 days is equivalent to 1 day in this context, we write

$$365 = 1$$

and similarly

$$366 = 2$$
$$31 = 3$$
$$30 = 2$$
$$28 = 0$$
$$8 = 1$$

The total number of days between the two dates is

$$(8 \times 365) + (2 \times 366) + (4 \times 31) + (2 \times 30) + (1 \times 28) + 8$$

which is equivalent to

$$(8 \times 1) + (2 \times 2) + (4 \times 3) + (2 \times 2) + (1 \times 0) + 1 = 29.$$

Since $29 = 1$, 'counting on' between the given dates is equivalent to 'counting on' 1 day; so August 9th 1911 was a Tuesday.

Exercise A

1 21st February is a Sunday in a non-leap year. In the same year, what days are:
(*a*) 21st March, (*b*) 21st June, (*c*) 21st October?

2 If 11th March is a Friday, what day is 2nd September of the same year?

3 4th March 1980 was a Tuesday. What day was 1st August 1978?

4 Christmas Day 1981 was a Friday. What day was Christmas Day 1920?

2. ISOMORPHISM

It is clear that in Section 1 we were working in arithmetic modulo 7. We are able to do this because the system of 'counting on' applied to days of the week has the same structure as arithmetic modulo 7. We describe this fact by saying that the two systems are *isomorphic* (from a Greek word meaning 'having the same shape').

If two systems are isomorphic, we can solve a problem in one system by translating it into an equivalent problem in the other system; for example, we translated the operation 'counting on 1 leap year' into the operation 'adding 2'. Having done our calculations in arithmetic modulo 7 – which is fairly easy and familiar – we translated our answer *back* into the system of 'counting on' in the days of the week.

For another example of isomorphism, consider the set {**R, S, I**}, where

R is rotation about the origin through 120° anticlockwise,
S is rotation about the origin through 120° clockwise,
I is the identity transformation.

If we use the usual operation 'followed by', it is easy to check that

$$\mathbf{RR} = \mathbf{S}, \quad \mathbf{SS} = \mathbf{R}, \quad \mathbf{RS} = \mathbf{I}, \text{ and so on.}$$

Now compare this with the set {0, 1, 2} under addition modulo 3: we have

$$1+1 = 2, \quad 2+2 = 1, \quad 1+2 = 0, \quad \text{etc.}$$

We notice that if we take the set {**R, S, I**} and replace **R** by 1, **S** by 2 and **I** by 0, at the same time changing the operation 'followed by' to 'add modulo 3', we get all the correct relations in the second system. In other words, the two systems are isomorphic: **R** corresponds to 1, **S** corresponds to 2, and **I** corresponds to 0. We may write:

$$\mathbf{R} \leftrightarrow 1$$
$$\mathbf{S} \leftrightarrow 2$$
$$\mathbf{I} \leftrightarrow 0.$$

(Check that, in this case, there is still isomorphism between the two sets when $\mathbf{S} \leftrightarrow 1$, $\mathbf{R} \leftrightarrow 2$ and $\mathbf{I} \leftrightarrow 0$.)

Exercise B

Show that the following pairs of systems are isomorphic:

1 {Reflection in the *x*-axis, the identity transformation} under the operation 'followed by', and
{1, ⁻1} under multiplication.

Which transformation corresponds to 1 and which to ⁻1?

2 {Half-turn, positive quarter-turn, negative quarter-turn, identity} (all about the origin), under the operation 'followed by', and
{0, 1, 2, 3} under addition modulo 4.

Which transformation corresponds to which number?

3 {... ¼, ½, 1, 2, 4, 8, 16, ...} under multiplication, and
{... ⁻3, ⁻2, ⁻1, 0, 1, 2, ...} under addition.

Explain how to find which number in the first set corresponds to a given number in the second set.

4 Nine counters, numbered from 1 to 9, are placed on a table with the numbers in view. In a game for two players, each in turn takes a counter and keeps it. The object of the game is to collect exactly three counters which add up to 15 (though in the course of a game a player may collect more than three counters).

Show that this game is isomorphic to 'noughts and crosses'. (Hint: construct a 3×3 magic square.)

3. PROPERTIES OF A SYSTEM

The device of solving a problem in system *A* by translating it into an equivalent problem in system *B* (which is isomorphic to system *A*) is very common and useful in mathematics, especially if system *B* is one with which we are already familiar. For this reason we need to know ways in which we can discover whether two systems are isomorphic. One way to do this is to identify a list of properties which a system might or might not possess; if two systems do not have the same (relevant) properties, they cannot be isomorphic.

A *system* is a set of elements together with one or more operations which combine two of those elements. For the moment we shall confine ourselves to systems with one operation only, such as the real numbers under multiplication, or geometrical transformations under the operation 'followed by'. You will already have met some properties which such a system might possess, but we review them here for convenience.

(i) Closure

A system is *closed* if the operation, when applied to any two members of the set, produces a member of the same set.

Example: The odd numbers are closed under multiplication, since the multiplication of two odd numbers always produces an odd number.

Counter-example: The odd numbers are not closed under addition, since if two odd numbers are added they produce an *even* number.

(ii) **An identity element**

An *identity element* is one which, when combined with any other element in the set, leaves it unaltered.

Example: The integers under multiplication have the identity element 1, since $1 \times x = x$ and $x \times 1 = x$ whichever integer x represents.

Counter-example: The even numbers under multiplication have no identity element, since there is no even number which leaves every other even number unchanged by multiplication.

(iii) **Inverses**

Two elements form an inverse pair if they combine (in both orders) to produce the identity element; each is said to be the *inverse* of the other. If an element combines with itself to produce the identity, it is said to be *self-inverse*, or, alternatively, it is said to be its own inverse.

Example: In the set of rational numbers under multiplication, the identity is the number 1; also, the numbers $\frac{1}{5}$ and 5 form an inverse pair, since $\frac{1}{5} \times 5 = 1$. Furthermore, the number 1 is self-inverse, since $1 \times 1 = 1$.

Counter-example: The set of rational numbers under multiplication contains one number which has *no* inverse – the number 0; we cannot find a number x such that $0 \times x = 1$.

(iv) **Associativity**

When *three* elements (in a given order) have to be combined two at a time, it may not matter which pair is combined first. If, for an operation $*$, $(a * b) * c = a * (b * c)$ for all sets of three elements a, b, c, then $*$ is said to be associative.

Example: The operation of addition on the positive integers is associative. We can calculate $2 + 7 + 5$ either as

$$(2+7)+5 = 9+5 = 14$$

or as $2+(7+5) = 2+12 = 14.$

Counter-example: The operation of division on the rational numbers is *not* associative.

For example, $(10 \div 5) \div 2 = 2 \div 2 = 1$

but $10 \div (5 \div 2) = 10 \div \frac{5}{2} = 4.$

(v) Commutativity

If the result of combining two elements of the set is the same for either order of the elements, the operation is said to be commutative.

Example: The operation of addition on the integers is commutative.

$$2+3 = 3+2 = 5.$$

Counter-example: The operation of subtraction on the integers is not commutative.

$$2-3 \ne 3-2.$$

Exercise C

1   In this question the notation $a * b$ will be used to mean the average of two numbers a and b, that is, $a * b = \dfrac{a+b}{2}$.

    (i) Solve the following equations:

        (a) $x * 5 = 7$;   (b) $(2 * y) * 7 = 5$;   (c) $z * (3 * 8) = -2$.

    (ii) Is it possible to find an identity element for this operation? If so, state its value, or if not, explain why an identity does not exist.

    (iii) If $(a * b) * c = a * (b * c)$, what can you deduce about a, b and c?   (MEI)

2   We define $a * b$ to mean the positive (or zero) difference between a and b. We consider this operation on the set $\{0, 1, 2, 3, 4, 5, \ldots\}$. For example, $2 * 7 = 5$, $10 * 1 = 9$.

    (i) Find numbers a, b, c such that

$$a * (b * c) \ne (a * b) * c,$$

    showing clearly that the numbers that you have given have this property.

    (ii) Show that there is an identity element.

    (iii) What can you say about the inverses of numbers in this set?

    (iv) Find the solution sets of each of the following equations:

        (a) $x * x = 4$;   (b) $6 * x = 3$;   (c) $6 * (y * 5) = 3$.   (MEI)

3   The operation $*$ is defined on the set $S = \{0, 1, 2\}$ by the formula

$$x * y = x+y-xy.$$

For example, $1 * 2 = 1+2-(1 \times 2) = 1$.

    (i) Copy and complete the following table:

$*$	0	1	2
0			
1			1
2		1	

    (ii) State whether or not S is closed under this operation.

    (iii) Calculate   (a) $1 * (1 * 2)$,   (b) $(1 * 1) * 2$,

        (c) $2 * (2 * 2)$,   (d) $(2 * 2) * 2$.

    State, with reason, whether or not your results suggest that the operation $*$ is associative on this set.

    (iv) State the identity element.

    (v) State, with reason, which element has no inverse.   (MEI)

4   The operation $*$ is defined on the positive integers by the formula

$$a * b = a^b.$$

For example, $3 * 4 = 3^4 = 81$.
  (i) Calculate (a) $2 * 4$, (b) $4 * 2$, (c) $5 * 2$, (d) $2 * 5$. State, with reason, whether or not the operation $*$ is commutative.
  (ii) Calculate (a) $(2 * 2) * 3$, (b) $2 * (2 * 3)$. State, with reason, whether or not the operation $*$ is associative.
  (iii) Find a number x such that $a * x = a$, whatever the value of a.
  (iv) For the number x found in (c), show by giving an example that $x * a$ is *not* always equal to a. (MEI)

5 (i) $A = \begin{bmatrix} a & b \\ -b & a \end{bmatrix}$ and $B = \begin{bmatrix} p & q \\ -q & p \end{bmatrix}$. Calculate **AB** and **BA**.
  (ii) The set R is the set of all 2×2 matrices of the form
$$\begin{bmatrix} m & n \\ -n & m \end{bmatrix},$$
  where m and n are integers which are not both zero.
    (a) State with reasons whether or not the set R is closed under matrix multiplication.
    (b) Is matrix multiplication commutative for matrices belonging to R? Give a reason for your answer.
    (c) Explain why no member of R can have a determinant whose value is zero.
    (d) Give an example of a matrix belonging to R, other than the identity matrix, whose inverse is also a member of R. Give the inverse of this matrix. (MEI)

6 The operation $*$ on two numbers a and b is defined by
$$a * b = a + b + ab.$$
  (i) Evaluate:
      (a) $2 * 3$;   (b) $(2 * 3) * 1$;   (c) $2 * (3 * 1)$.
  (ii) Find the identity element of the system, that is, the number e such that
$$a * e = e * a = a$$
  for every number a.
  (iii) For this value of e, find the solution sets of each of the equations:
      (a) $2 * x = e$;   (b) $x * \frac{1}{2} = e$;   (c) $(^-1) * x = e$. (MEI)

7 State whether or not the operations in questions 1–3 above are commutative.

4. GROUPS

A set with an operation which has the following four properties is called a group:
  (i) The set is closed under the operation.
  (ii) There is an identity element in the set.
  (iii) Each element in the set has an inverse in the set.
  (iv) The operation is associative.
If, in addition to the above, the operation is commutative, the group is called a *commutative group*, or *abelian group*.

  When we know that a system is a group, we can use the 'usual' processes of algebra confidently, knowing that there will always be an 'answer' within the set. For example, consider the system $\{3, 6, 9, 12\}$ under multiplication modulo 15, which has the following combination table:

× mod 15	3	6	9	12
3	9	3	12	6
6	3	6	9	12
9	12	9	6	3
12	6	12	3	9

(i)   The set is closed – no 'new' numbers appear in the table.

(ii) The identity is 6, since $6 \times x = x$ for every x in the table.

(iii) The numbers 6 and 9 are self-inverse, and the numbers 3 and 12 form an inverse pair, so every element has an inverse.

(iv) The operation is in fact associative, though this is not easy to check, since it requires verifying for every possible (ordered) set of elements a, b, c. Since multiplication and addition of positive integers are both associative, it is reasonable to assume that the same operations are associative when using any modulus, though we shall not prove this.

It follows that the system is a group – in fact, a commutative group.

Suppose now that we wish to solve the equation

$$12x = 3$$

in this system. Of course, with the table laid out before us we can 'spot' the answer by looking at the table, but a more general method is to proceed as follows:

(*a*) Multiply both sides of the equation by the inverse of 12:

$$3 \times 12x = 3 \times 3.$$

(We know this is possible because, since the system is a group, 12 *has* an inverse, and the set is closed; that is, multiplying by any number will always give an answer within the set.)

(*b*) Rewrite the left-hand side as

$$(3 \times 12)\, x.$$

(This is possible since the operation is associative.) This gives us

$$6x = 9.$$

(*c*) Since 6 is the identity, we have

$$x = 9$$

and the equation is solved.

It is worth looking at these steps again in a more general case: consider the equation

$$R * X = M$$

where R, X, M are elements of *any* group whose operation is denoted by $*$, and we wish to find X. When we are looking at a very general case like this where the operation $*$ is not specified, we often leave out the operation symbol and write simply

$$RX = M.$$

Similarly, for convenience, we often refer to the unspecified operation as if it were multiplication, and use phrases such as 'multiply both sides by Y' rather than the more clumsy 'operate on both sides by Y'. Using this language, we carry out the following steps:

(*a*) Multiply both sides on the left by the inverse of R:

$$R^{-1}RX = R^{-1}M$$

(possible because the system is closed and every element has an inverse).

(*b*) Rewrite as

$$(R^{-1}R)X = R^{-1}M$$

(possible because the system is associative).

(*c*) This gives

$$IX = R^{-1}M$$

where I is the identity.

(*d*) Since I is the identity, we have

$$X = R^{-1}M.$$

Notice that all *four* properties of a group are used in the course of the solution. Notice also that since we do not know whether the group is commutative, we must multiply both sides of the equation on the *left* at step (*a*); we cannot guarantee that $R^{-1}M = MR^{-1}$.

We have shown that in *any* group the equation $RX = M$ has the solution

$$X = R^{-1}M.$$

This can be interpreted according to the group in question:

(*a*) R and M might be a rotation and a reflection. X will then be a transformation.

(*b*) R and M might be rational numbers. X will then be a number.

(*c*) R and M might be matrices. X will then be a matrix.

But whatever the system, the above formula gives the solution to the equation, provided that the system is a group.

Exercise D

In questions 1–4, you may assume that the given system is a group.

1   In the rational numbers under multiplication, what is the inverse of $\frac{3}{4}$? Solve the equation $\frac{3}{4}x = \frac{7}{5}$.

2   In the set $\{1, 3, 7, 9\}$ under multiplication modulo 10, what is the identity? What is the inverse of 9? Solve the equation $9x = 3$.

3   In the set of 2×2 matrices with non-zero determinant, what is the inverse of $\begin{bmatrix} 2 & 3 \\ 1 & 2 \end{bmatrix}$?

Solve the equation $\begin{bmatrix} 2 & 3 \\ 1 & 2 \end{bmatrix} \mathbf{X} = \begin{bmatrix} 1 & -1 \\ 3 & 1 \end{bmatrix}$.

4 $\mathbf{M}$ is reflection in the line $x = y$, and $\mathbf{Q}$ is a positive quarter-turn about the origin. What is the inverse of $\mathbf{M}$? Find the transformation $\mathbf{X}$ such that $\mathbf{MX} = \mathbf{Q}$.

In questions 5–8, explain why we cannot deduce that $X = A^{-1}B$ from the given equation $AX = B$.

5   We are working in the set of rational numbers, and $A = 0$, $B = \frac{3}{2}$.

6 We are working in the set of 2×2 matrices, and

$$\mathbf{A} = \begin{bmatrix} 2 & 6 \\ 1 & 3 \end{bmatrix}, \quad \mathbf{B} = \begin{bmatrix} 4 & -1 \\ 2 & 2 \end{bmatrix}.$$

7 We are working in the set of 2×3 matrices, and

$$\mathbf{A} = \begin{bmatrix} 1 & 1 & 1 \\ 2 & -2 & 3 \end{bmatrix}, \quad \mathbf{B} = \begin{bmatrix} 0 & 4 & 2 \\ 1 & 0 & 3 \end{bmatrix}.$$

8 $\mathbf{A}$ is the vector $\begin{bmatrix} 2 \\ 1 \end{bmatrix}$, $\mathbf{X}$ is a vector, the operation is 'scalar product', and $B = 2$.

5. FINITE GROUPS

Many of the groups we have looked at so far contain an infinite number of elements, but we can investigate many important ideas by examining groups with small numbers of elements. We start with four different systems, each containing only four elements, each of which is a group.

A: the set $\{1, 3, 5, 7\}$ under multiplication modulo 8.

B: the set $\{2, 4, 6, 8\}$ under multiplication modulo 10.

C: the set $\{\mathbf{Q}, \mathbf{Q}^{-1}, \mathbf{H}, \mathbf{I}\}$ under 'followed by', where $\mathbf{Q}$ and $\mathbf{Q}^{-1}$ are positive and negative quarter-turns about the origin, $\mathbf{H}$ is a half-turn about the origin, and $\mathbf{I}$ is the identity.

D: the set $\{\mathbf{H}, \mathbf{I}, \mathbf{X}, \mathbf{Y}\}$ under 'followed by', where $\mathbf{H}$ and $\mathbf{I}$ are as before, and $\mathbf{X}, \mathbf{Y}$ are reflections in the x-and y-axes.

Here are the four operation tables:

A:	× mod 8	1	3	5	7
	1	1	3	5	7
	3	3	1	7	5
	5	5	7	1	3
	7	7	5	3	1

B:	× mod 10	2	4	6	8
	2	4	8	2	6
	4	8	6	4	2
	6	2	4	6	8
	8	6	2	8	4

C:		$\mathbf{Q}$	$\mathbf{Q}^{-1}$	$\mathbf{H}$	$\mathbf{I}$
	$\mathbf{Q}$	$\mathbf{H}$	$\mathbf{I}$	$\mathbf{Q}^{-1}$	$\mathbf{Q}$
	$\mathbf{Q}^{-1}$	$\mathbf{I}$	$\mathbf{H}$	$\mathbf{Q}$	$\mathbf{Q}^{-1}$
	$\mathbf{H}$	$\mathbf{Q}^{-1}$	$\mathbf{Q}$	$\mathbf{I}$	$\mathbf{H}$
	$\mathbf{I}$	$\mathbf{Q}$	$\mathbf{Q}^{-1}$	$\mathbf{H}$	$\mathbf{I}$

D:		$\mathbf{H}$	$\mathbf{I}$	$\mathbf{X}$	$\mathbf{Y}$
	$\mathbf{H}$	$\mathbf{I}$	$\mathbf{H}$	$\mathbf{Y}$	$\mathbf{X}$
	$\mathbf{I}$	$\mathbf{H}$	$\mathbf{I}$	$\mathbf{X}$	$\mathbf{Y}$
	$\mathbf{X}$	$\mathbf{Y}$	$\mathbf{X}$	$\mathbf{I}$	$\mathbf{H}$
	$\mathbf{Y}$	$\mathbf{X}$	$\mathbf{Y}$	$\mathbf{H}$	$\mathbf{I}$

It is fairly clear that all these are *group tables*. They are all closed, and we know that multiplication to a modulus, and combining transformations are both associative operations. The identity elements in A, C and D are obvious, but we need to look a little harder to see that 6 is the identity in B. Checking shows that in each table every element has an inverse, and that completes the verification that each table is a group table. We say that each system is a *group of order 4*, because it contains four elements.

If each of these systems is a group of order 4, are there any differences between them? It might have been helpful to write each table with the identity *first* in the list of border elements, but (as in *B*) we may not always know when we start which element is the identity. However, even with the tables as they are we can see one difference between *A* and *B*: in *A*, *every* element is self-inverse, whereas in table *B* only 4 and 6 are self-inverse – the other two elements form an inverse pair. The same difference can be seen between tables *C* and *D*: table *D* is like *A*, and *C* is like *B*. To show these similarities and differences more clearly, we shall rewrite all four tables with the identity element first, and in the case of *B* and *C* we shall put the other self-inverse element last:

A:	1	3	5	7
1	1	3	5	7
3	3	1	7	5
5	5	7	1	3
7	7	5	3	1

B:	6	2	8	4
6	6	2	8	4
2	2	4	6	8
8	8	6	4	2
4	4	8	2	6

C:	**I**	**Q**	**Q**$^{-1}$	**H**
I	**I**	**Q**	**Q**$^{-1}$	**H**
Q	**Q**	**H**	**I**	**Q**$^{-1}$
Q^{-1}	**Q**$^{-1}$	**I**	**H**	**Q**
H	**H**	**Q**$^{-1}$	**Q**	**I**

D:	**I**	**X**	**Y**	**H**
I	**I**	**X**	**Y**	**H**
X	**X**	**I**	**H**	**Y**
Y	**Y**	**H**	**I**	**X**
H	**H**	**Y**	**X**	**I**

Now if we examine tables *A* and *D* we see that they have *exactly* the same pattern: the diagonal of 1's in *A* corresponds to the diagonal of **I**'s in table *D*; the diagonal of 7's in *A* corresponds to the diagonal of **H**'s in *D*; the 3's in *A* are in exactly the same positions as the **X**'s in *D*; and so on. The patterns of the tables are identical, and we can say that the two systems are isomorphic. Furthermore, we can show a correspondence between the elements of the two tables:

$$1 \leftrightarrow \mathbf{I}$$
$$3 \leftrightarrow \mathbf{X}$$
$$5 \leftrightarrow \mathbf{Y}$$
$$7 \leftrightarrow \mathbf{H}.$$

In the same way, we can see that tables *B* and *C*, though different from tables *A* and *D*, are isomorphic to each other; for example:

$$6 \leftrightarrow \mathbf{I}$$
$$2 \leftrightarrow \mathbf{Q}$$
$$8 \leftrightarrow \mathbf{Q}^{-1}$$
$$4 \leftrightarrow \mathbf{H}.$$

Exercise E

1   Draw up tables for the following groups of order 2, and show that they are all
isomorphic by listing the correspondences between the elements.
   A: $\{^-1, 1\}$ under multiplication.
   B: $\{0, 1\}$ under addition modulo 2.
   C: $\{\mathbf{X}, \mathbf{I}\}$ under 'followed by', where $\mathbf{X}$ is a reflection in the x-axis and $\mathbf{I}$ is the identity
   transformation.

2   Repeat question 1 for the following groups of order 3.
   A: $\{0, 1, 2\}$ under addition modulo 3.
   B: $\{\mathbf{R}, \mathbf{S}, \mathbf{I}\}$ under 'followed by', where $\mathbf{R}$ is a 120° positive turn about the origin,
   $\mathbf{S}$ is a 240° positive turn about the origin, and $\mathbf{I}$ is the identity transformation.
   C: $\{\mathbf{L}, \mathbf{M}, \mathbf{I}\}$ under matrix multiplication, where

$$\mathbf{L} = \begin{bmatrix} 0 & ^-1 \\ 1 & ^-1 \end{bmatrix}, \quad \mathbf{M} = \begin{bmatrix} ^-1 & 1 \\ ^-1 & 0 \end{bmatrix}, \quad \mathbf{I} = \begin{bmatrix} 1 & 0 \\ 0 & 1 \end{bmatrix}.$$

3   Draw up tables for the following groups of order 4.
   P: $\{0, 1, 2, 3\}$ under addition modulo 4.
   Q: $\{1, 2, 3, 4\}$ under multiplication modulo 5.
   R: $\{1, ^-1, j, ^-j\}$ under multiplication, where $j^2 = {}^-1$.
   S: $\{1, 5, 7, 11\}$ under multiplication modulo 12.

   Three of these systems are isomorphic. Show correspondences between their
   elements. To which tables on page 435 are they also isomorphic? To which tables
   on page 435 is the fourth system isomorphic?

6. LATIN SQUARES

You will have noticed that in a group table every element appears just once in
each row and each column. It is easy to show that this must be so: suppose one
row of a group table looked like this:

	.	.	P	Q	.
	.	.	.	.	.
C	.	.	E	E	.
	.	.	.	.	.

that is, $CP = E$ and also $CQ = E$. Because it is a group, we may solve these
equations to get $P = C^{-1}E$ and $Q = C^{-1}E$; P and Q are the same element, and
we should not be showing this element twice in the border of our table!
   A table in which every element appears just once in each row and in each
column is called a *Latin square*. Every group table is a Latin square, but the
converse is not true; it is possible to construct a Latin square which is not a
group table. Here is an example:

	A	B	C	D
A	A	B	C	D
B	B	A	D	C
C	D	C	B	A
D	C	D	A	B

This is certainly a Latin square, but it is not a group table. At first glance it seems that A is the identity element, but look at the first *column*; $CA = D$, so that A is *not* the identity. There is *no* identity, so the table cannot be a group.

The idea of the Latin square helps us to build up possible group tables. We have already seen two different groups of order 4; are there any others of that order? To answer this problem, we might start to construct a group table by filling in the 'identity' row and column:

	A	B	C	D
A	A	B	C	D
B	B			
C	C			
D	D			

We know that in a group the elements are either self-inverse or occur in inverse pairs, so we deduce that, of the remaining three elements, either *all* are self-inverse, or *one* is self-inverse and the other two form an inverse pair. In the second case, we may choose to put the single remaining self-inverse element last, and then we have the following possible tables:

	A	B	C	D
A	A	B	C	D
B	B	A		
C	C		A	
D	D			A

or

	A	B	C	D
A	A	B	C	D
B	B		A	
C	C	A		
D	D			A

Each of these tables can be completed in only one way to form a Latin square, so there are only two possible groups of order 4.

Exercise F

1 Copy and complete the two tables above for groups of order 4. Which group is isomorphic to the group of symmetries of a rhombus (identity, half-turn, reflections in the diagonals)?

2 Show that there cannot be a group of order 3 in which every element is self-inverse. Deduce that there is only one possible group table pattern for a group of order 3.

3 (a) The table shows the beginning of an attempt to construct a group table of order 5 in which every element is self-inverse.

	A	B	C	D	E
A	A	B	C	D	E
B	B	A	D		
C	C		A		
D	D			A	
E	E				A

If $BC = D$, show that there is only one way to complete the table as a Latin square. By considering the triple CDE, show that the 'operation' represented by the table is not associative, and deduce that there is *no* group of order 5 in which every element is self-inverse.

(*b*) Try to construct a group table of order 5 in which *A* is the identity, *B* and *C* are self-inverse, and *D*, *E* form an inverse pair. Write *BC* as *D*. Show that it is now impossible to construct the rest of the table as a Latin square.

(*c*) What do your answers to (*a*) and (*b*) tell you about self-inverse elements in a group of order 5?

7. NON-COMMUTATIVE GROUPS

There are eight possible symmetry operations on a square: the identity (**I**), positive rotations about the centre of 90°, 180°, 270° (**Q, H, R**), and reflection in any one of four axes of symmetry (**X, Y, A, B**) (Figure 1).

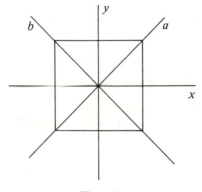

Figure 1

When combining transformations, we must remember that the square may be thought of as 'moveable', but the axes are 'fixed' on the paper. So if we apply a quarter-turn **Q** to the square, the square moves but the axes *do not move with it* – for example, the *y*-axis is still 'up the page'.

The group table is:

		Second letter (first transformation)							
		I	Q	R	H	X	Y	A	B
	I	I	Q	R	H	X	Y	A	B
	Q	Q	H	I	R	A	B	Y	X
First letter	R	R	I	H	Q	B	A	X	Y
(second transformation)	H	H	R	Q	I	Y	X	B	A
	X	X	B	A	Y	I	H	R	Q
	Y	Y	A	B	X	H	I	Q	R
	A	A	X	Y	B	Q	R	I	H
	B	B	Y	X	A	R	Q	H	I

The first thing to notice is that not all combinations are commutative, so that we have to specify which transformation is performed first. **X** followed by **Q** gives **A**, and this is shown as **QX** = **A** in the table. Check this carefully to see how to read the table. On the other hand, **Q** followed by **X** gives **B**, and this is shown as **XQ** = **B**. Because of the convention that successive transformations are written from right to left, the 'first transformation' is the 'second letter' when we write down our results in algebraic form.

The table has been divided up by broken lines in order to emphasise another property: the 64 entries in the body of the table fall naturally into four 'blocks' of 16. The top left-hand block we have already met – see page 435, table *C*. Here we have a 'group within a group', which is called a *subgroup*. However, we must not jump to the conclusion that all the blocks are subgroups; in the top right-hand one, for example, only **X**, **Y**, **A**, **B** appear in the 'body' of the block, and the same letters appear along the top border, but the letters along the left-hand border are **I**, **Q**, **R**, **H**, and we cannot have a group table with different elements along the two borders.

Exercise G

Questions 1–4 refer to the table of symmetry operations of a square shown opposite.

1 The bottom right-hand 'block' has the elements **X**, **Y**, **A**, **B** along both borders. Why is it not a group table?

2 Rewrite the whole table, placing the border elements in the order **I**, **X**, **Y**, **H**, **Q**, **R**, **A**, **B**. Hence find another subgroup of the given group.

3 Rewrite the table with the border letters in the order **I**, **A**, **B**, **H**, **Q**, **R**, **X**, **Y**. Hence find another subgroup.

4 How many subgroups can you find of order 2? Are any of these subgroups of the order 4 subgroups you have already found in questions 2 and 3?

5 An equilateral triangle has six symmetry operations: rotations about the centre of 0°, 120°, 240° (**I**, **R**, **S**) and reflections in three axes of symmetry (**X**, **Y**, **Z**) (Figure 2).

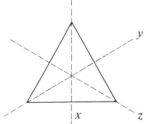

Figure 2

(*a*) Draw up a group table for these transformations, placing the border elements in the order **I**, **R**, **S**, **X**, **Y**, **Z**.
(*b*) Find a subgroup of order 3.
(*c*) How many subgroups of order 2 can you find?
(*d*) Is the group commutative?

6 Write out the table for {0, 1, 2, 3, 4, 5} under addition modulo 6. Find a subgroup of order 2 and a subgroup of order 3.

8. CYCLIC GROUPS

Consider the set $\{1, 2, 3, 4\}$ under multiplication modulo 5. By straightforward calculation we find that

$$2^2 = 4$$
$$2^3 = 3$$
$$2^4 = 1 \; (= 2^0)$$

and, of course, $\qquad 2^1 = 2.$

We see that *every* element of the group can be expressed as a power of 2. On the other hand,

$$4^1 = 4$$
$$4^2 = 1$$
$$4^3 = 4$$
$$4^4 = 1$$
$$4^5 = 4$$

$$\cdots$$

so that only the elements 1 and 4 can be expressed as powers of 4. We say that the element 2 *generates* the whole group but the element 4 does not. Check that 3 also generates the group.

If we can find an element a which generates a whole group like this, we often set out the table with the border elements in the order

$$a^0, a^1, a^2, a^3, \ldots;$$

in the above example we would write the elements in the order

$$2^0, 2^1, 2^2, 2^3,$$

that is, in the order 1, 2, 4, 3:

$\times \bmod 5$	1	2	4	3
1	1	2	4	3
2	2	4	3	1
4	4	3	1	2
3	3	1	2	4

Such a group is called a *cyclic* group. If you look carefully at each row of the table in turn, you will see that the elements in the body of the table are 'shifted' one place to the left at each stage, with the left-hand entry 'moving across' to the right. Any cyclic group can be written to show this pattern if the border elements are written as successive powers of some element which generates the whole group.

Notice that the group is of order 4, and $2^4 = 1$. In any cyclic group which is generated by an element a, the order of the group will be the smallest positive integer n such that a^n equals the identity. (n is called the *period* of a.)

We can use the relation $2^4 = 1$ to simplify expressions such as 2^9; we have

$$2^9 = 2^4 \times 2^4 \times 2^1$$
$$= 1 \times 1 \times 2$$
$$= 2.$$

Similarly, we could if we wished calculate 3×3 in the following way:

$$3 \times 3 = 2^3 \times 2^3$$
$$= 2^6$$
$$= 2^4 \times 2^2$$
$$= 1 \times 4$$
$$= 4.$$

This is rather laborious for an example taken from a simple modular arithmetic, but can be useful in slightly more complicated cases.

Exercise H

1   Write out the table for $\{0, 1, 2, 3\}$ under addition modulo 4. Examine the pattern to see if this is a cyclic group.

2   Write out the tables for (a) $\{\mathbf{I}, \mathbf{X}, \mathbf{Y}, \mathbf{H}\}$, (b) $\{\mathbf{I}, \mathbf{Q}, \mathbf{H}, \mathbf{R}\}$, where $\mathbf{X}$ and $\mathbf{Y}$ are reflections in the *x*- and *y*-axes, $\mathbf{Q}, \mathbf{H}, \mathbf{R}$ are quarter-, half- and three-quarter-turns about the origin and $\mathbf{I}$ is the identity. Is either group cyclic?

3   Write out the table for $\{1, 3, 4, 5, 9\}$ under multiplication modulo 11. Calculate 3^1, 3^2, 3^3, 3^4, 3^5 and hence show that the element 3 generates the whole group. Rewrite the original table using the elements 1, 3, 4, 5, 9 in the order suggested by your calculations of successive powers of 3 and check that the table now has the typical 'cyclic group pattern'.

4   Write out the table for $\{0, 1, 2, 3, 4, 5\}$ under addition modulo 6. Write out also the table for $\{1, 2, 3, 4, 5, 6\}$ under multiplication modulo 7. Show that this second group is *not* generated by the element 2, but *is* generated by the element 3. Rewrite the second table with the border elements in the order 1, 3, 3^2, 3^3, ... and hence show the cyclic pattern of this group.
The two groups in this question are isomorphic. Show a list of possible correspondences between the elements of the two groups.

5   The set $\{1, 3, 5, 7, 9, 11, 13, 15\}$ under multiplication modulo 16 forms a group. Show that none of the elements generates the whole group, and that this cannot therefore be a cyclic group. Nevertheless it has *subgroups* which are cyclic groups of order 4 (two of them) and subgroups of order 2 (three of them). Find these subgroups.

6   Show that the set $\{1, 2, 3, ..., 12\}$ under multiplication modulo 13 is cyclic. Write 10 and 11 as powers of the generator you have found and hence calculate 10×11 (mod 13).

9. SUBGROUPS

Consider the group formed by $\{1, 2, 4, 5, 8, 10, 11, 13, 16, 17, 19, 20\}$ under multiplication modulo 21. The group table is shown below. What subgroups can we find?

There is of course the subgroup $\{1\}$ – and indeed the whole group itself. These are usually called the *trivial* subgroups.

Can we find any subgroups of order two? Such a subgroup would have to contain the identity and the other member would have to be self-inverse. The self-inverse elements (apart from 1) are 8, 13, and 20. So there are subgroups

$$\{1, 8\}, \quad \{1, 13\}, \quad \text{and} \quad \{1, 20\}.$$

	1	2	4	5	8	10	11	13	16	17	19	20
1	1	2	4	5	8	10	11	13	16	17	19	20
2	2	4	8	10	16	20	1	5	11	13	17	19
4	4	8	16	20	11	19	2	10	1	5	13	17
5	5	10	20	4	19	8	13	2	17	1	11	16
8	8	16	11	19	1	17	4	20	2	10	5	13
10	10	20	19	8	17	16	5	4	13	2	1	11
11	11	1	2	13	4	5	16	17	8	19	20	10
13	13	5	10	2	20	4	17	1	19	11	16	8
16	16	11	1	17	2	13	8	19	4	20	10	5
17	17	13	5	1	10	2	19	11	20	16	8	4
19	19	17	13	11	5	1	20	16	10	8	4	2
20	20	19	17	16	13	11	10	8	5	4	2	1

We could then search for other cyclic subgroups. Trying each element as a generator in turn gives the following cyclic subgroups:

$$\{1, 4, 16\}$$
$$\{1, 2, 4, 8, 16, 11\}$$
$$\{1, 5, 4, 20, 16, 17\}$$
$$\{1, 10, 16, 13, 4, 19\}$$

None of the elements generates the whole group, so the group itself is not cyclic.

There is also a non-cyclic subgroup of order 4: $\{1, 8, 13, 20\}$. Each member of this group is self-inverse, and it is isomorphic to the group of symmetries of a rectangle (group D on p. 435).

Notice that we have found subgroups of orders 1, 2, 3, 4, 6 and 12, and that these numbers are all factors of 12, the order of the group. This is an example of a general result, called Lagrange's theorem, which says that the order of any subgroup must be a factor of the order of the group. We shall not prove this result here, but indicate how a proof could be constructed.

Consider the subgroup $\{1, 8, 13, 20\}$. If we multiply each member of this subgroup by a non-member (say 2), we obtain a set of four different elements of the group:

$$1 \times 2 = 2$$
$$8 \times 2 = 16$$
$$13 \times 2 = 5$$
$$20 \times 2 = 19.$$

If we now multiply each member of the subgroup by a number which isn't a member of the subgroup or of the new set which we have just obtained (say 10), we obtain a further set of four 'new' elements of the group:

$$1 \times 10 = 10$$
$$8 \times 10 = 17$$
$$13 \times 10 = 4$$
$$20 \times 10 = 11.$$

It can be shown that, in general, this procedure always produces, at each step, a set of different 'new' elements of the group, and, since we are considering finite groups, eventually we must obtain the complete group. This is the stage we have just reached in this case:

Subgroup:	$\{1, 8, 13, 20\}$
First new set:	$\{2, 16, 5, 19\}$
Second new set:	$\{10, 17, 4, 11\}$

Each member of the group appears in one of the sets above.

Each set has as many members as the subgroup has, so the order of the group must be a multiple of the order of the subgroup – that is, the order of the subgroup must be a factor of the order of the group.

Exercise I

1 For the group G of $\{1, 2, 3, 4, 5, 6\}$ under multiplication modulo 7:
   (a) Write out the table for $\{1, 2, 4\}$. Is this set a subgroup of G?
   (b) Find the smallest subgroup of G containing 2 and 3.
   (c) Find a subgroup of G, of order 2.

2 (a) Write out the table for $\{1, 3, 7, 9\}$ under multiplication modulo 20.
   (b) Show that the smallest group, with this operation, containing 1, 3, 7, 9 and 11, is of order 8.
   (c) Find a subgroup of order 4, other than $\{1, 3, 7, 9\}$, of this group of order 8.

3 Given a group of order 9, what must be the order of any non-trivial subgroup?

4 Explain why a group of order 7 cannot have any non-trivial subgroups. What other groups have this property?

5 (1 row, 3 column) matrices containing the numbers 0 and 1 only, such as $[0 \quad 1 \quad 1]$, can be added modulo 2, so that, for example,

$$[0 \quad 1 \quad 1] + [0 \quad 1 \quad 0] = [0 \quad 0 \quad 1]$$

   (a) Show that the set $\{\mathbf{I}, \mathbf{A}\}$ under this operation forms a group of order 2, where
   $$\mathbf{I} = [0 \quad 0 \quad 0] \quad \text{and} \quad \mathbf{A} = [0 \quad 0 \quad 1].$$
   (b) Show that the smallest group containing $\mathbf{I}$, $\mathbf{A}$ and $\mathbf{B} = [0 \quad 1 \quad 0]$ is of order 4. State the new element $\mathbf{C}$ and write out the table for this group.
   (c) Show that the smallest group containing $\mathbf{I}$, $\mathbf{A}$, $\mathbf{B}$, $\mathbf{C}$ and $\mathbf{D} = [1 \quad 0 \quad 0]$ is of order 8. State the new elements $\mathbf{E}$, $\mathbf{F}$ and $\mathbf{G}$, and construct the group table. Are there any cyclic subgroups of order 4?

6 Write out the table for the group $\{1, 2, 4, 5, 7, 8\}$ under multiplication modulo 9. List all the subgroups of this group. Is the group cyclic?

10. GENERATING RELATIONS

We know that the group of rotations of a square is generated by a quarter-turn $\mathbf{Q}$, and that the elements of the group are

$$\{\mathbf{I}, \mathbf{Q}, \mathbf{Q}^2, \mathbf{Q}^3\}$$

since $\mathbf{Q}^4 = \mathbf{I}$. Suppose we now include a 'new' element in the group – the reflection in the x-axis, $\mathbf{X}$. Since $\mathbf{X}^2 = \mathbf{I}$, the only new element introduced by considering powers of $\mathbf{X}$ is the element $\mathbf{X}$ itself. When we consider the products of **X** with the elements of the original group, we get **XQ**, **XQ²** and **XQ³**, so that altogether we have introduced four 'new' elements, so obtaining the complete group of symmetries.

However, we know that the operation 'followed by' on transformations is not necessarily commutative, so we ought to have considered three more elements – **QX**, **Q²X** and **Q³X**. These three elements cannot be new since they are symmetries of a square, and there are only eight such transformations. In fact they are the same as **XQ**, **XQ²** and **XQ³**, but in a different guise and in a different order. If we go back to the geometry and actually examine the transformations concerned (or use the group table on p. 438), we find that

$$XQ = Q^3X,$$
$$XQ^2 = Q^2X$$
and
$$XQ^3 = QX.$$

So far, we have written down five relations between the elements of the group: the three above, $Q^4 = I$, and $X^2 = I$. It turns out that not all of these are necessary – some can be deduced from the others. Three such relations are sufficient, provided we choose the right three. These relations are called *generating relations* for the group.

(1) $Q^4 = I$
(2) $X^2 = I$
(3) $XQ = Q^3X$

Example 1
From the generating relations listed, deduce that:
   (a) $XQ^2 = Q^2X$;   (b) $(XQ)^2 = I$.

(a)
$$XQ = Q^3X \qquad (3)$$
$$XQ^2 = Q^3XQ \qquad \text{(multiplying on the right by } Q\text{)}$$
$$= Q^3(XQ)$$
$$= Q^3(Q^3X) \qquad \text{(using (3) again)}$$
$$= Q^6X$$
$$= Q^4(Q^2X)$$
$$= Q^2X \qquad \text{using (1)).}$$

(b)
$$(XQ)^2 = (XQ)(XQ) \qquad \text{(by definition of squaring)}$$
$$= (XQ)(Q^3X) \qquad \text{(using (3))}$$
$$= XQ^4X$$
$$= X^2 \qquad \text{(using (1))}$$
$$= I \qquad \text{(using (2)).}$$

Exercise J

1   Using the relations in the last section, prove that
   (a) $QXQ = X$,   (b) $QX = XQ^3$,   (c) $(QX)(XQ^3) = I$,
   (d) $XQX = Q^3$,   (e) $XQ^2X = Q^2$,   (f) $XQ^3X = Q$.

2 **A** is reflection in $x = 0$, **B** is reflection in $x = 2$. Verify geometrically that **BA** = **T**, where **T** is the translation $\begin{bmatrix} 4 \\ 0 \end{bmatrix}$.

Using this relation, together with the relations $\mathbf{A}^2 = \mathbf{I}$ and $\mathbf{B}^2 = \mathbf{I}$, prove the following relations by algebra and verify them geometrically:
(a) **A** = **BT**, (b) **B** = **TA**, (c) (**AB**)(**BA**) = **I**, and hence that **AB** = **T**$^{-1}$.

3 Given that $R^3 = I$, $A^2 = I$, $AR = R^2A$, prove that
(a) $RAR = A$,   (b) $RA = AR^2$,   (c) $(RA)(AR^2) = I$,
(d) $ARA = R^2$,   (e) $(RA)^2 = I$.
Construct the group of order 6 whose elements are

$$I, R, S(= R^2), A, B(= AR), C(= AR^2).$$

Show that this group is isomorphic to the group of symmetries of the equilateral triangle.

4 Find a pair of generators and a set of generating relations for $\{1, 2, 4, 7, 8, 11, 13, 14\}$ under multiplication modulo 15.

5 Find a pair of generators and a set of generating relations for $\{1, 2, 4, 5, 8, 10, 11, 13, 16, 17, 19, 20\}$ under multiplication modulo 21.

11. SUMMARY

A group is a set of elements which can be combined by an operation ∗ such that:
  (i) the set is closed under the operation;
  (ii) the set contains an identity element;
  (iii) each element has an inverse;
  (iv) the operation is associative.

If, in addition, the operation is commutative, then the group is called a commutative, or abelian, group.

The order of a group is the number of elements in the group.

Two groups are isomorphic if the patterns of their tables can be exactly matched, so that each element in one table appears in exactly the same place as the corresponding element in the other table.

In any group, the equation

$$AX = B$$

has the solution $X = A^{-1}B$.

A cyclic group is one in which every element can be expressed as some power of a given single element.

If all the elements of a group can be expressed in terms of just a few given elements, these given elements are called the generators of the group. In order to build up the group, we need to know the relations between these generators.

A subgroup is a subset of a group which itself forms a group.

Miscellaneous exercise

1   The operations o and * are defined by the tables

Second element				
o	a	b	c	d
a	a	c	d	a
b	b	a	c	d
c	d	b	a	c
d	c	d	b	a

First element: rows a, b, c, d

Second element				
*	a	b	c	d
a	b	c	d	a
b	c	d	a	b
c	d	a	b	c
d	a	b	c	d

First element: rows a, b, c, d

(For example, $a \circ d = a$.)
   Give the complete solution set for x in each of the following:
   (i) $b \circ x = c$;
   (ii) $(a \circ x) * c = a$;
   (iii) $(a \circ x) * b = c$;
   (iv) $[a * (x \circ b)] \circ c = b$.   (SMP)

2   The operations o and * are defined on the sets $\{a, b, c, d)\}$ and $\{P, Q, R, S\}$ by the
   following tables:

o	a	b	c	d
a	b	a	d	c
b	a	b	c	d
c	d	c	a	b
d	c	d	b	a

*	P	Q	R	S
P	S	R	Q	P
Q	R	P	S	Q
R	Q	S	P	R
S	P	Q	R	S

   (i) Show that these two systems are isomorphic. (Listing a suitable one–one
       correspondence between the elements of the two sets will be sufficient.)
   (ii) Show that they are both groups by stating a geometrical group to which they
       are each isomorphic.
   (iii) In $\{P, Q, R, S\}$ state (*a*) the identity, (*b*) the inverse of each element.
                                                                            (SMP)

3   In this question integers are combined by multiplication modulo 14.
       (i) Construct a multiplication table for the numbers 1, 9, 11 combined in this
           way.
       (ii) Is the set $\{1, 9, 11\}$ closed under multiplication modulo 14? Explain your
           answer.
       (iii) The set $\{1, 3, x, 9, y, z\}$ is closed under multiplication modulo 14. Give the
           values of x, y and z.
       (iv) Write as one number:
               (*a*) 3×3;   (*b*) $3 \times 3 \times 3$;   (*c*) $3 \times 3 \times 3 \times 3$;
               (*d*) $3 \times 3 \times 3 \times 3 \times 3$;   (*e*) $3 \times 3 \times 3 \times 3 \times 3 \times 3$;
           where in every case the multiplication is modulo 14.   (MEI)

4   The matrix $\mathbf{A} = \begin{bmatrix} 0 & -1 \\ 1 & -1 \end{bmatrix}$.
       (*a*) Calculate $\mathbf{A}^2$ and $\mathbf{A}^3$.
       (*b*) Copy and complete the following matrix multiplication table where
       $\mathbf{I} = \begin{bmatrix} 1 & 0 \\ 0 & 1 \end{bmatrix}$.

	A	A²	I
A			
A²			
I	A		

(c) Write down the inverse of **A** as a power of **A**.

(d) The matrix $\mathbf{B} = \begin{bmatrix} 1 & -1 \\ 1 & 0 \end{bmatrix}$. Calculate $\mathbf{B}^2$.

(e) Express $\mathbf{B}^4$ in terms of **A**.

(f) What can you say about the possible values of the positive integer n if $\mathbf{B}^n = \mathbf{I}$? Give reasons for your answer.

(g) Express the inverse of **B** as a power of **B**.   (MEI)

5   (i) The group G consists of the set $\{4, 8, 12, 16\}$ under multiplication mod 20. Copy and complete the table.

	4	8	12	16
4	16	12		
8	12	4		
12				
16				

(a) State the identity element.

(b) Write down the inverse of each element.

(c) List each of the subgroups of G.

(ii) The group H consists of the elements $\{a, b, c, d\}$ under the operation $*$. Part of the combination table is given as follows:

*	a	b	c	d
a				
b		b	c	
c		c	a	
d				

State the identity element and hence copy and complete the table.

(iii) The groups G and H are isomorphic. Show a correspondence between their elements.   (SMP)

6   The 'digital sum' of an integer N is obtained by adding together the digits of N; for example, the digital sum of 358 is 16. If the digital sum is more than 9 its digits are again added, and so on until a single-figure number is obtained. This number is called the 'reduced digital sum' of the original integer N, and is denoted by $d(N)$. For example, $d(358) = 7$, $d(103) = 4$ and $d(8) = 8$.

(i) Find $d(77)$ and $d(968\,752\,847)$.

M is the set $\{1, 2, 3, 4, 5, 6, 7, 8, 9\}$, and x, y are the members of M. We define $x * y$ to be $d(xy)$, where xy means x multiplied by y.

(ii) Find $5 * 5$ and $3 * (4 * 5)$.

(iii) Show by writing out the combination table that the subset $A = \{1, 4, 7\}$ is a group under $*$.

(iv) Find p such that the subset $B = \{1, p\}$ is also a group under $*$.

(v) Show that if $x \in M$, the equation $6 * x = 3$ has more than one solution; explain briefly why this shows that M is not a group under $*$.   (MEI)

7   The combination table for the set

$$G = \{1, 3, 5, 9, 11, 13\}$$

under multiplication modulo 14 is shown.

×	1	3	5	9	11	13
1	1	3	5	9	11	13
3	3	9	1	13	5	11
5	5	1	11	3	13	9
9	9	13	3	11	1	5
11	11	5	13	1	9	3
13	13	11	9	5	3	1

(a) State the identity element.

(b) State the inverse of each element.

(c) State the two other properties this table has which make it a group table.

(d) $S = \{1, 13\}$ is a subgroup of order 2. (It is a subset with two elements, and is a group.) State any other subgroups of order 2.

(e) If we multiply each element of S by 3, we get the set $S_3 = \{3, 11\}$ which is called a *coset* of S. Write down S_1, S_5, S_9, S_{11}, S_{13} and verify that there are three different cosets altogether.

(f) Write down (i) the union, (ii) the intersection, of these three cosets.

(g) Write down a subgroup of G other than those of order 1, 2 or 6.   (SMP)

8   $e:x \to x$, $f:x \to 1-x$, $g:x \to 1/x$.

(a) Verify that $f^2 = g^2$. (b) Write down the function fg and show that $(fg)^2 = gf$. Hence (c) prove that $fgfgf = g$, and (d) express $(fg)^3$ in its simplest form.

Thus copy and complete the group table generated by f and g. Its elements are e, f, g, h, r, r^2; where $h = fgf$, $r = fg$, $r^2 = fgfg = (fg)^2$.

	e	f	g	h	r	r²
e	e	f	g	h	r	r²
f	f	e	r	r²	g	h
g	g			h		f
h	h				f	g
r	r	h		r²		e
r²	r²	g	h	f	e	r

List the elements of a subgroup of order 2 and those of a subgroup of order 3. Explain why there can be no subgroup of order 4.   (SMP)

Revision exercise 6

1 (a) Copy and complete the truth table below. ($\sim p$ means 'not-p'):

p	$\sim p$	q	(q or $\sim p$)	($p \Rightarrow q$)
T	T			
T	F			
F	T			T
F	F			T

(Note: (p or q) is understood to mean 'either p or q or both'.)
Comment on the table which you have obtained.
If ($q \Rightarrow \sim r$) is true, what can you say about ($r \Rightarrow \sim q$)?
If (q or $\sim p$) is true and also ($r \Rightarrow \sim q$) is true, show that ($p \Rightarrow \sim r$) is also true.

(b) The following argument was overheard:
'Water is wet, or I'm not an Englishman.
If our House came bottom, then water isn't wet.
So, since I'm an Englishman, our House wasn't bottom.'
By putting p = 'I'm an Englishman', q = 'water is wet', and r = 'our House came bottom', and using the results of (a), or otherwise, discuss the logic of this argument and say whether or not you would be prepared to accept it.

(SMP)

2 Construct the truth table for the tautology

$$(p \Rightarrow q) \Leftrightarrow (q \lor \sim p).$$

Hence draw a Venn diagram for $p \Rightarrow q$.
Copy and complete the circuit shown in Figure 1 for $p \Rightarrow q$, where the convention is that a two-way switch, represented by a triangle, connects the top vertex to the bottom left or right vertex according to the position of the switch.

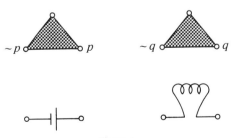

Figure 1

3 (a) Write down the condition represented by the circuit shown in Figure 2. Simplify the condition and draw the equivalent circuit.

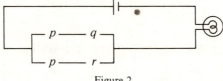

Figure 2

(b) Copy the Venn diagram in Figure 3 and shade it to illustrate the implication $p \Rightarrow q$.

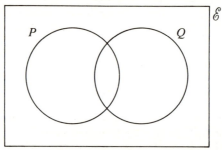

Figure 3

(c) Write down, in two forms, the formula involving statements p and q illustrated by the Venn diagram in Figure 4.

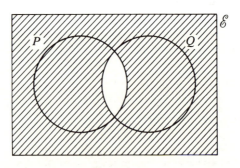

Figure 4

4 The electric circuit shown in Figure 5, with battery, lamp and switches p, q, r represents the statement

$$(p \wedge q) \vee (p \wedge r).$$

Show by a truth table that this is equivalent to $p \wedge (q \vee r)$.
   Draw the circuit which corresponds to the simplified form.

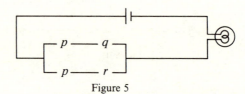

Figure 5

5 Draw circuit diagrams to illustrate (a) $p \vee (q \wedge r)$,   (b) $(p \vee q) \vee r$.
   Complete the truth table for $[p \vee (q \wedge r)] \Rightarrow [(p \vee q) \vee r]$.
   Hence show that (a) logically implies (b) and state briefly the significance of this
   in relation to their circuits.

6 Construct a combination table for $\{p, q, r, t\}$ under the operation $\Leftrightarrow$, where $r = p \Leftrightarrow q$
   and t is a tautology (that is, a statement that is always true). Does the set form a
   group under the operation?

7 In this question, * denotes the 'exclusive or': that is, $a*b$ is true if a or b is true,
   but not both.
   (a) Construct a truth table for $(a*b)*b$. To what simple statement is this
       compound statement equivalent?
   (b) If c is a contradiction (a statement that is always false) and $p*q = r$, construct
       a combination table for $\{c, p, q, r\}$ under the operation *. Does the set form
       a group?

8 When two binary digits are added (see Table 1), a 'put down' digit and a 'carry'
   digit are produced (see Tables 2 and 3). For example $1+1 = 10$ produces a 'put
   down' digit 0 and a 'carry' digit 1.

Table 1				Table 2				Table 3		
+	0	1			0	1			0	1
0	0	1		0	0	1		0	0	0
1	1	10		1	1	0		1	0	1
					'put down' digit				'carry' digit	

   (a) Interpreting 0 as 'false' and 1 as 'true', and denoting the two digits by p and
       q, write down compound statements equivalent to the 'put down' and 'carry'
       digits. Draw equivalent circuits.
   (b) Find the compound statements for the 'put down' and 'carry' digits when
       a third digit (denoted by r) is added.

9 If $(x+yj)^2 = {}^-5+12j$, find a pair of simultaneous equations for x and y. Eliminate
   y to obtain an equation for x. Hence determine the square roots of $^-5+12j$.

10 If $z = 4+3j$:
   (a) find $|z|$ and $\arg z$;
   (b) calculate z^2 and find its modulus and argument;
   (c) show z and z^2 on an Argand diagram.
   Explain the connection between your answers to (a) and (b). If $\tan \theta = \frac{3}{4}$, what is
   $\tan 2\theta$?
   Use a similar method with the complex number $a+bj$ to show that
$$\tan \theta = \frac{b}{a} \Rightarrow \tan 2\theta = \frac{2ab}{a^2 - b^2}.$$

11 $z = 1+j$ and $w = \sqrt{3}+j$. Calculate z^2 and show the representations of z, w, z^2 on
   an Argand diagram. Verify that $w^3 = 4z^2$ and write down a cube root of j.
   Calculate the modulus, m (distance from the origin), and argument, α (angle the
   position vector makes with the positive real axis), for z. Calculate z^3 and verify that
   modulus of $z^3 = m^3$ and argument of $z^3 = 3\alpha$. Make corresponding statements about
   the modulus and argument of z^2. Write down the modulus and argument of z^6 and
   hence state a square root of ^-j.   (SMP)

12 (a) Write down a cube root of 8.

(b) If $\alpha = \dfrac{-1+j\sqrt{3}}{2}$, calculate β where $\beta = \alpha^2$.

(c) Verify that $\alpha^3 = 1$, and hence write down a second cube root of 8.

(d) Write down the value of α^6.

(e) Write down the value of β^3, and hence write down a third cube root of 8.

(SMP)

13 Given $z = 3+5j$ and $w = 1+2j$, calculate (a) zw, (b) $\dfrac{z}{w}$, in the form $x+yj$, where x and y are real.

On an Argand diagram, mark and label points corresponding to $\dfrac{z}{w}$, z and zw.

(SMP)

14 (i) Find the smallest positive integers (if any) which satisfy the following equations:

(a) $2x+3 = 1 \pmod 5$;

(b) $3x+2 = 1 \pmod 6$.

(ii) Solve $x^2+x+3 = 0 \pmod 5$.   (SMP)

15 **I**, **Q**, **H** and **T** denote respectively the identity, quarter-turn, half-turn and three-quarter-turn about the origin. Write down the combination table for {**I**, **Q**, **H**, **T**} under the operation 'followed by' and the table for {0, 1, 2, 3} under addition modulo 4. Indicate clearly *two* isomorphisms between them.   (SMP)

16 The set $\{1, 2, 3, 4, 5, 6\}$ forms a group under multiplication modulo 7. Write out the combination table for this group. Describe briefly in geometrical terms a symmetry group which is isomorphic to this group; identify two subgroups, one of order 2 and one of order 3.   (SMP)

17 The numbers in the set $\{1, 2, 4, 5, 7, 8\}$ are combined by multiplication modulo 9, that is, the numbers are multiplied, the result divided by 9, and the remainder written down. For example, $4*8 = 5$, since 32 when divided by 9 has a remainder of 5.

(a) Copy and complete a 'multiplication' table for these numbers:

*	1	2	4	5	7	8
1	1	2	4	5	7	8
2	2	4	8	1		
4	4					
5						
7						
8						

(b) State (i) the inverse of 2, and (ii) the inverse of 8.

(c) Express each of the six numbers as a power of 2 (where, for example, 2^3 means $2*2*2$).

(d) Find two subsets of this set which are closed under *, such that one set contains two numbers, and the other contains three numbers.   (MEI)

18 A set P is $\{2, 4, 6, 8, 10, 12\}$. The operation * is defined as follows: if a and b are any elements of P, then $a*b$ denotes the remainder when the result of multiplying a by b is divided by 14; for example,

$$6*12 = 2.$$

(a) Find the identity element of P, that is, the element e such that

$$e * a = a$$

for every element a of the set.

(b) Find the inverse of 10, that is, the element y such that

$$y * 10 = e$$

where e is the identity element.

(c) Solve the equation

$$10 * x = 4$$

by multiplying both sides of the equation by the value of y found above.

(d) Use a similar technique to solve the equation

$$6 * x = 2.$$

(e) Find two solutions of the equation

$$x * x = 4.$$ (MEI)

19 The set M is $\{1, 3, 7\}$. An operation $*$ is defined as follows: if x and y are any elements of M, $x * y$ means the remainder when the result of multiplying x by y is divided by 10. For example, $3 * 7 = 1$.

The set M is not closed under the operation $*$. Explain what this means, and state the new element p which it is necessary to add to obtain a closed set, G.

You are given that the set G with the operation $*$ is a group. State the identity element and the inverse of each of the other elements.

The set V consists of the four matrices $\mathbf{A}$, $\mathbf{B}$, $\mathbf{C}$, $\mathbf{D}$ where

$$\mathbf{A} = \begin{bmatrix} 1 & 0 \\ 0 & 1 \end{bmatrix}, \quad \mathbf{B} = \begin{bmatrix} -1 & 0 \\ 0 & -1 \end{bmatrix},$$

$$\mathbf{C} = \begin{bmatrix} 0 & 1 \\ 1 & 0 \end{bmatrix}, \quad \mathbf{D} = \begin{bmatrix} 0 & -1 \\ -1 & 0 \end{bmatrix}.$$

A mapping f is defined from G to V as follows: $f(1) = \mathbf{A}$, $f(3) = \mathbf{B}$, $f(7) = \mathbf{C}$, $f(p) = \mathbf{D}$. Show that $f(3 * p) = f(3) f(p)$, where the right-hand side is the matrix product $\mathbf{BD}$, and find two members x and y of G such that $f(x * y) \neq f(x) f(y)$. (SMP)

20 The combination table for the set of integers $E = \{1, 2, 3, 4, 5\}$ under the operation $*$ is shown below, e.g. $2 * 3 = 1$.

$*$	1	2	3	4	5
1	4	3	5	1	2
2	5	4	1	2	3
3	2	5	4	3	1
4	1	2	3	4	5
5	3	1	2	5	4

(a) Is the set closed under $*$?

(b) State the identity element.

(c) Is $*$ commutative?

(d) State the inverse for each element.

(e) Give an example to show that $*$ is not associative.

(f) State the solution set for $5 * x = 2$.

(g) Does $p * x = q$ have a unique solution for x for all $p, q \in E$?

(h) Is the table shown above a group table?

Give a brief reason for each answer. (SMP)

Groups of order 8

21 Construct the combination table for the matrices **A, B, C, D, E, F, G, I**, under multiplication, where

$$\mathbf{A} = \begin{bmatrix} j & 0 \\ 0 & j \end{bmatrix}, \quad \mathbf{B} = \begin{bmatrix} -j & 0 \\ 0 & -j \end{bmatrix}, \quad \mathbf{C} = \begin{bmatrix} 0 & j \\ j & 0 \end{bmatrix}, \quad \mathbf{D} = \begin{bmatrix} 0 & -j \\ -j & 0 \end{bmatrix},$$

$$\mathbf{E} = \begin{bmatrix} 0 & 1 \\ 1 & 0 \end{bmatrix}, \quad \mathbf{F} = \begin{bmatrix} 0 & -1 \\ -1 & 0 \end{bmatrix}, \quad \mathbf{G} = \begin{bmatrix} -1 & 0 \\ 0 & -1 \end{bmatrix}, \quad \mathbf{I} = \begin{bmatrix} 1 & 0 \\ 0 & 1 \end{bmatrix}.$$

Find two subgroups of order 4. Which elements are of period 2?

Show that this group is isomorphic to the group of $\{1, 2, 4, 7, 8, 11, 13, 14\}$ under multiplication modulo 15.

22 Show that the following three groups are isomorphic:

G_1: $\{1, 2, 4, 7, 8, 11, 13, 14\}$ under multiplication modulo 15;

G_2: $\{1, 3, 7, 9, 11, 13, 17, 19\}$ under multiplication modulo 20;

G_3: $\left\{ \begin{bmatrix} 0 \\ 0 \end{bmatrix}, \begin{bmatrix} 1 \\ 0 \end{bmatrix}, \begin{bmatrix} 2 \\ 0 \end{bmatrix}, \begin{bmatrix} 3 \\ 0 \end{bmatrix}, \begin{bmatrix} 0 \\ 2 \end{bmatrix}, \begin{bmatrix} 1 \\ 2 \end{bmatrix}, \begin{bmatrix} 2 \\ 2 \end{bmatrix}, \begin{bmatrix} 3 \\ 2 \end{bmatrix} \right\}$ under addition modulo 4.

Why are these groups not isomorphic to the group of symmetries of the square?

23 Construct the combination table for the matrices **P, Q, R, S, T, U, V, W** under multiplication, where

$$\mathbf{P} = \begin{bmatrix} j & 0 \\ 0 & -j \end{bmatrix}, \quad \mathbf{Q} = \begin{bmatrix} -j & 0 \\ 0 & j \end{bmatrix}, \quad \mathbf{R} = \begin{bmatrix} 0 & 1 \\ -1 & 0 \end{bmatrix}, \quad \mathbf{S} = \begin{bmatrix} 0 & -1 \\ 1 & 0 \end{bmatrix},$$

$$\mathbf{T} = \begin{bmatrix} 0 & j \\ j & 0 \end{bmatrix}, \quad \mathbf{U} = \begin{bmatrix} 0 & -j \\ -j & 0 \end{bmatrix}, \quad \mathbf{V} = \begin{bmatrix} 1 & 0 \\ 0 & 1 \end{bmatrix}, \quad \mathbf{W} = \begin{bmatrix} -1 & 0 \\ 0 & -1 \end{bmatrix}.$$

How many subgroups of order 4 are there?

Why is this group not isomorphic to the group of symmetries of the square?

24 Construct the combination table for the addition, modulo 2, of the matrices

$$\mathbf{A} = \begin{bmatrix} 0 \\ 0 \\ 0 \end{bmatrix}, \quad \mathbf{B} = \begin{bmatrix} 1 \\ 0 \\ 0 \end{bmatrix}, \quad \mathbf{C} = \begin{bmatrix} 0 \\ 1 \\ 0 \end{bmatrix}, \quad \mathbf{D} = \begin{bmatrix} 0 \\ 0 \\ 1 \end{bmatrix}, \quad \mathbf{E} = \begin{bmatrix} 0 \\ 1 \\ 1 \end{bmatrix}, \quad \mathbf{F} = \begin{bmatrix} 1 \\ 0 \\ 1 \end{bmatrix},$$

$$\mathbf{G} = \begin{bmatrix} 1 \\ 1 \\ 0 \end{bmatrix}, \quad \mathbf{H} = \begin{bmatrix} 1 \\ 1 \\ 1 \end{bmatrix}.$$

How many subgroups of order 2 does the group have?

Show that the group is isomorphic to the group of symmetries of a cuboid.

22

Exponential and logarithmic functions

1. POWERS

A population of bacteria is doubling every day. If at 9 a.m. on 31st May it was 1 million, then we can set out the following table:

Date	May 29th	May 30th	May 31st	June 1st	June 2nd	June 3rd
Population at 9 a.m.	250 000	500 000	1 000 000	2 000 000	4 000 000	8 000 000

If we write x for the number of days after 9 a.m. on 31st May and $P(x)$ for the population in millions then we have

$$x\ldots \quad {}^{-}2 \quad {}^{-}1 \quad 0 \quad 1 \quad 2 \quad 3 \quad 4 \quad \ldots$$
$$P(x)\ldots \quad \tfrac{1}{4} \quad \tfrac{1}{2} \quad 1 \quad 2 \quad 4 \quad 8 \quad 16 \quad \ldots$$

so that $P(x) = 2^x$. Remember that $2^0 = 1$ and $2^{-n} = \dfrac{1}{2^n}$.

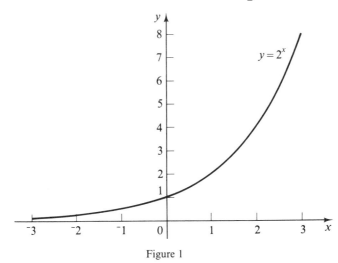

Figure 1

What is the population at 9 p.m. on 31st May? Since this corresponds to $x = \tfrac{1}{2}$, we would expect this to be $P(\tfrac{1}{2}) = 2^{\frac{1}{2}}$. From the graph in Figure 1 we can estimate this to be about 1.4, but we can calculate it as follows: $2^{\frac{1}{2}}$ is the factor by which the population increases in 12 hours. So $(2^{\frac{1}{2}})^2$ is the factor by which the

455

population increases in 24 hours. But the population doubles every 24 hours. So

$$(2^{\frac{1}{2}})^2 = 2$$

$$2^{\frac{1}{2}} = \sqrt{2} \approx 1.414.$$

Similar arguments can be used to find other powers of 2. For example, $2^{\frac{1}{3}}$ is the factor by which the population increases in 8 hours. $(2^{\frac{1}{3}})^2$ is the factor by which the population increases in 16 hours. $(2^{\frac{1}{3}})^3$ is the factor by which the population increases in 24 hours. So

$$(2^{\frac{1}{3}})^3 = 2$$

$$2^{\frac{1}{3}} = \sqrt[3]{2} \approx 1.26.$$

$2^{\frac{2}{3}}$ also corresponds to 16 hours' growth. So $2^{\frac{2}{3}} = (2^{\frac{1}{3}})^2 \approx 1.26^2 \approx 1.59$. Similarly, $2^{\frac{1}{4}} = \sqrt[4]{2}$ and $2^{\frac{3}{4}} = (\sqrt[4]{2})^3$.

In general, we define

$$N^{\frac{1}{b}} = \sqrt[b]{N}$$

for any positive integers N and b, and

$$N^{\frac{a}{b}} = (\sqrt[b]{N})^a$$

for any integer a. Note that

$$(N^{\frac{a}{b}})^b = ((\sqrt[b]{N})^a)^b = (\sqrt[b]{N})^{ab} = ((\sqrt[b]{N})^b)^a = N^a.$$

So also $$N^{\frac{a}{b}} = \sqrt[b]{(N^a)}.$$

Example 1

Write down the values of: (*a*) $81^{\frac{1}{4}}$; (*b*) $27^{\frac{2}{3}}$; (*c*) $64^{-\frac{2}{3}}$.

(*a*) $81^{\frac{1}{4}} = \sqrt[4]{81} = 3$ since $3^4 = 81$.
    Notice that since $3^4 = (3^2)^2$, $\sqrt[4]{81} = \sqrt{\sqrt{81}} = \sqrt{9} = 3$.
(*b*) $27^{\frac{2}{3}} = (\sqrt[3]{27})^2 = 3^2 = 9$.
(*c*) $64^{-\frac{2}{3}} = (\sqrt[3]{64})^{-2} = 4^{-2} = \frac{1}{16}$.

Example 2

Find: (*a*) $30^{\frac{1}{4}}$; (*b*) $20^{\frac{2}{5}}$.

(*a*) $30^{\frac{1}{4}} = \sqrt[4]{30} = \sqrt{\sqrt{30}} \approx \sqrt{5.477} \approx 2.34.$

(*b*) $20^{\frac{2}{5}}$ can be found directly from a calculator with a y^x key, as $20^{0.4}$. Alternatively it can be found by a decimal search method as follows:

$$20^{\frac{2}{5}} = \sqrt[5]{(20^2)} = \sqrt[5]{400}.$$

Now $2^5 = 32$, $3^5 = 243$ and $4^5 = 1024$, so $20^{\frac{2}{5}}$ is between 3 and 4 and apparently nearer 3 than 4.

$$3.1^5 \approx 286,$$

$$3.2^5 \approx 336,$$

$$3.3^5 \approx 391,$$

$$3.4^5 \approx 454.$$

So $20^{\frac{2}{5}}$ is between 3.3 and 3.4 and nearer 3.3 than 3.4.

$$3.31^5 \approx 397.3,$$

$$3.32^5 \approx 403.4.$$

So $20^{\frac{2}{5}}$ is about 3.315.

$$3.315^5 \approx 400.33,$$
$$3.314^5 \approx 399.73.$$

So $20^{\frac{2}{5}}$ is about 3.3145.

The procedure can be continued to any desired degree of accuracy. Using a y^x key, $20^{0.4} \approx 3.314454$.

Exercise A

1 Write down the values of:
   (a) $16^{\frac{1}{2}}$; (b) $27^{\frac{1}{3}}$; (c) $8^{\frac{1}{3}}$; (d) $36^{-\frac{1}{2}}$.

2 Write down the values of:
   (a) $4^{-\frac{1}{2}}$; (b) $8^{-\frac{1}{3}}$; (c) $16^{\frac{5}{4}}$; (d) $16^{-\frac{3}{4}}$.

3 (a) Find $\sqrt{2}$, $\sqrt{\sqrt{2}}$ and $\sqrt{\sqrt{\sqrt{2}}}$ to 3 significant figures.
   (b) Use your answers to (a) to copy and complete the following table:

x	0	$\frac{1}{8}$	$\frac{1}{4}$	$\frac{3}{8}$	$\frac{1}{2}$	$\frac{5}{8}$	$\frac{3}{4}$	$\frac{7}{8}$	1
2^x								1.83	

   (c) Plot the graph of $y = 2^x$ for $0 \leqslant x \leqslant 1$.

4 Use decimal search methods to find: (a) $10^{\frac{1}{2}}$; (b) $3^{\frac{3}{4}}$.

5 Write down the values of:
   (a) $64^{\frac{1}{2}}$; (b) $64^{\frac{1}{3}}$; (c) $64^{-\frac{2}{3}}$; (d) $64^{-\frac{1}{2}}$.

6 Write down the values of:
   (a) $100^{-\frac{1}{2}}$; (b) $32^{\frac{2}{5}}$; (c) $32^{-\frac{3}{5}}$; (d) $16^{-\frac{1}{4}}$.

7 (a) Find $\sqrt{3}$, $\sqrt{\sqrt{3}}$ and $\sqrt{\sqrt{\sqrt{3}}}$ to 3 significant figures.
   (b) Use your answer to (a) to copy and complete the following table:

x	0	$\frac{1}{8}$	$\frac{1}{4}$	$\frac{3}{8}$	$\frac{1}{2}$	$\frac{5}{8}$	$\frac{3}{4}$	$\frac{7}{8}$	1
3^x				1.51					

   (c) Plot the graph of $y = 3^x$ for $0 \leqslant x \leqslant 1$.

8 Use decimal search methods to find: (a) $5^{\frac{1}{3}}$; (b) $7^{\frac{2}{3}}$.

*9 (a) Write down the values of: (i) $64^{\frac{1}{2}}$; (ii) $64^{\frac{1}{3}}$; (iii) $64^{\frac{1}{6}}$; (iv) $64^{\frac{5}{6}}$.
   (b) Is it true or false that: (i) $64^{\frac{1}{2}} \times 64^{\frac{1}{3}} = 64^{\frac{1}{2}+\frac{1}{3}} = 64^{\frac{5}{6}}$; (ii) $64^{\frac{1}{2}} \div 64^{\frac{1}{3}} = 64^{\frac{1}{2}-\frac{1}{3}} = 64^{\frac{1}{6}}$?

*10 (a) Write down the values of: (i) $8^{\frac{1}{3}}$; (ii) 8^2.
   (b) Is it true or false that $(8^{\frac{1}{3}})^6 = 8^{\frac{1}{3} \times 6} = 8^2$?

2. INDEX LAWS

$$2^2 \times 2^3 = (2 \times 2) \times (2 \times 2 \times 2) = 2^5 = 2^{2+3}.$$

$$2^3 \div 2^5 = \frac{2 \times 2 \times 2}{2 \times 2 \times 2 \times 2 \times 2} = \frac{1}{2 \times 2} = 2^{-2} = 2^{3-5}.$$

$$(5^2)^3 = (5 \times 5) \times (5 \times 5) \times (5 \times 5) = 5^6 = 5^{2 \times 3}.$$

In general we have the following laws for any integers N, p and q.

$$(1) \quad N^p \times N^q = N^{p+q}.$$
$$(2) \quad N^p \div N^q = N^{p-q}.$$
$$(3) \quad (N^p)^q = N^{pq}.$$

These also hold if p and q are rational numbers, as we saw in some specific cases in Exercise A, questions 9 and 10. Here are some more examples:

$$27^{\frac{1}{3}} \times 27^{\frac{2}{3}} = 3 \times 9 = 27 = 27^1 = 27^{\frac{1}{3}+\frac{2}{3}}$$
$$27^{\frac{1}{2}} \div 27^{\frac{1}{6}} = 27^{\frac{1}{2}-\frac{1}{6}} = 27^{\frac{1}{3}} = 3$$
$$(81^{\frac{1}{2}})^4 = 9^4 = 6561 = 81^2 = 81^{\frac{1}{2}\times 4}$$
$$(5^{\frac{2}{3}})^6 = 5^{\frac{2}{3}\times 6} = 5^4 = 625.$$

Exercise B

Use the index laws to find:

1 $2^{\frac{1}{4}} \times 2^{\frac{3}{4}}$.   2 $9^{\frac{1}{4}} \times 9^{\frac{3}{4}}$.   3 $16^{\frac{7}{10}} \div 16^{\frac{1}{5}}$.

4 $9^{\frac{3}{4}} \div 9^{\frac{1}{4}}$.   5 $8^{\frac{1}{6}} \times 8^{\frac{1}{2}}$.   6 $8^{\frac{1}{6}} \div 8^{\frac{1}{2}}$.

7 $(3^{\frac{1}{3}})^6$.   8 $(2^{\frac{1}{5}})^{10}$.   9 $(16^{-\frac{1}{8}})^3$.

10 $(64^{-\frac{1}{3}})^{-\frac{1}{2}}$.

3. EXPONENTIAL FUNCTIONS

A function of the form $x \to N^x$ is called an *exponential function* (because x appears as an exponent, or power). We can now plot accurate graphs of exponential functions since we can find N^x for any rational number x, by using decimal search, or a y^x key on a calculator. Figure 2 shows the graph of $y = 3^x$ for $-2 \leqslant x \leqslant 3$.

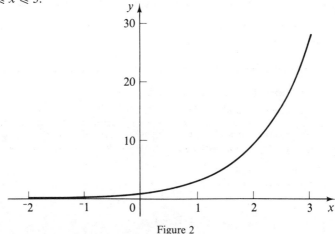

Figure 2

The derived function of $f : x \to 3^x$ at $x = 2$ can be investigated by calculating $\dfrac{3^{2+h} - 3^2}{h}$ for small values of h, using a y^x key or square roots.

If $h = \frac{1}{2}$, $3^{2+h} = 3^2 \times 3^h = 9 \times 3^{\frac{1}{2}} = 9\sqrt{3} \approx 15.59$,

$$\frac{3^{2+h} - 3^2}{h} \approx \frac{15.59 - 9}{\frac{1}{2}} = 2 \times 6.59 = 13.18.$$

If $h = \frac{1}{8}$, $3^{2+h} = 3^2 \times 3^h = 9 \times 3^{\frac{1}{8}} = 9\sqrt{\sqrt{\sqrt{3}}} \approx 10.325$,

$$\frac{3^{2+h} - 3^2}{h} \approx \frac{10.325 - 9}{\frac{1}{8}} = 8 \times 1.325 = 10.6.$$

If $h = \frac{1}{1024}$, $3^{2+h} = 3^2 \times 3^h = 9\sqrt{\sqrt{\sqrt{\ldots\sqrt{3}}}} \approx 9.009661$
 (taking the square root repeatedly 10 times),

$$\frac{3^{2+h} - 3^2}{h} \approx \frac{9.009661 - 9}{\frac{1}{1024}} = 1024 \times 0.009661 \approx 9.89.$$

Using a y^x key gives $\dfrac{3^{2+h} - 3^2}{h} \approx 9.888$ if $h = 0.0001$, and 9.887 if $h = {}^-0.0001$.

So $f'(2) = 9.89$, to 3 s.f.

From such calculations as these we obtain the following table of approximate values of $f'(x)$:

x	$^-2$	$^-1$	0	1	2
$f(x) = 3^x$	0.11...	0.33...	1	3	9
$f'(x)$	0.12	0.37	1.1	3.3	9.9

From this table it would appear that $f'(x) \approx 1.1 f(x)$. We might have expected that $f'(x)$ would be proportional to $f(x)$ since we are dealing with a growth function, for which it would seem likely that its rate of increase would be proportional to its value.

Also,

$$\frac{f(2+h) - f(2)}{h} = \frac{3^{2+h} - 3^2}{h} = \frac{3^2 \times 3^h - 3^2}{h} = \frac{3^2(3^h - 1)}{h} = \frac{3^h - 1}{h} f(2)$$

and $\dfrac{f(0+h) - f(0)}{h} = \dfrac{3^h - 1}{h}$,

so we would expect that

$$f'(2) = f'(0) f(2),$$

and more generally that

$$f'(x) = f'(0) f(x).$$

This relation holds for all exponential functions and it is therefore necessary only to find the derivative at $x = 0$ in order to find the complete derived function.

For $ f(x) = 3^x$, $ \dfrac{3^h - 3^0}{h} = \dfrac{3^h - 1}{h} \approx \begin{array}{l} 1.0987 \quad \text{if } h = 0.0001, \\ 1.0986 \quad \text{if } h = {}^-0.0001, \end{array}$

and hence $ f'(0) \approx 1.099.$

$$f'(x) \approx 1.099 f(x),$$

so that, for example, $f'(4) \approx 1.099 \times 3^4 \approx 89$.

Exercise C

1 (*a*) Copy and complete the following table. (Complete the last two columns only
   if you have a calculator with a y^x key.)

h	$\frac{1}{2}$	$\frac{1}{4}$	$\frac{1}{8}$	$\frac{1}{1024}$	0.0001	$^{-}0.0001$
2^h	1.414					
$\dfrac{2^h-1}{h}$	0.828					

(*b*) If $f(x) = 2^x$, what is your estimate for $f'(0)$?

(*c*) Use your answer to (*b*) to find $f'(4)$ to 3 significant figures.

2 Repeat question 1 for $f(x) = 5^x$.

3 Repeat question 1 for $f(x) = 8^x$.

4. THE EXPONENTIAL FUNCTION

We have seen that for an exponential function $f: x \to a^x$, $f'(x) = f'(0)f(x)$. The
larger the value of a, the more rapidly $f(x)$ will increase and the greater $f'(0)$ will
be.

From Section 3, we have that $f(x) = 3^x \Rightarrow f'(0) \approx 1.099$;
from Exercise C, question 2: $f(x) = 2^x \Rightarrow f'(0) \approx 0.693$.
So there will be a number e, between 2 and 3 for which

$$f(x) = e^x \Rightarrow f'(0) = 1.$$

For this function, therefore, $f'(x) = f(x)$. That is,

$$f(x) = e^x \Rightarrow f'(x) = e^x.$$

The value of e can be found, approximately, by considering values of
$\dfrac{f(0+h)-f(0)}{h}$ for small values of h. These should be approximately equal to $f'(0)$,
that is, 1. So

$$\frac{e^{0+h} - e^0}{h} \approx 1$$

$$e^h - 1 \approx h$$

$$e^h \approx 1+h$$

$$e \approx (1+h)^{\frac{1}{h}}.$$

We evaluate $(1+h)^{\frac{1}{h}}$ for small values of h to obtain approximations to e.

h	$(1+h)^h$
$\frac{1}{2}$	$(1+\frac{1}{2})^2 = 2.25$
$\frac{1}{4}$	$(1+\frac{1}{4})^4 \approx 2.44$ $\left.\right\}$ by repeating squaring
$\frac{1}{1024}$	$(1+\frac{1}{1024})^{1024} \approx 2.717$
0.0001	$(1.0001)^{10\,000} \approx 2.7181$ $\left.\right\}$ using a y^x key
$^{-}0.0001$	$(0.9999)^{-10\,000} \approx 2.7184$

It can be shown that, to 10 s.f., $e = 2.718\,281\,828\,5$; like π, e is an irrational number and has no recognisable pattern to the digits in the decimal places.

The function $x \to e^x$ is called *the exponential function*.

Stretches

The graph of $y = Ae^x$ is obtained from the graph of $y = e^x$ by a one-way stretch with scale factor A parallel to the y-axis, thus increasing the gradient by a factor of A.

Hence $f(x) = Ae^x \Rightarrow f'(x) = Ae^x$.

The graph of $y = e^{kx}$ is obtained from the graph of $y = e^x$ by a one-way stretch with scale factor $\dfrac{1}{k}$ parallel to the x-axis, thus increasing the gradient by a factor of k. (See Figure 3.)

Hence $f(x) = e^{kx} \Rightarrow f'(x) = ke^{kx}$.

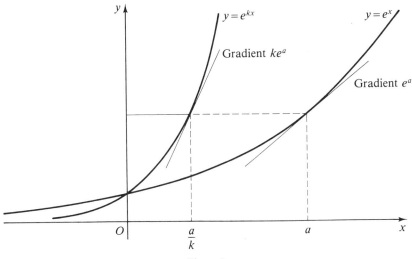

Figure 3

Combining these two results, we obtain

$$f(x) = Ae^{kx} \Rightarrow f'(x) = kAe^{kx} = kf(x).$$

Conversely,

$$f(x) = kf'(x) \Rightarrow f(x) = Ae^{kx} \quad \text{for some number } A.$$

Example 3

Write down the derived functions of:

 (a) $f : x \to 4e^x$; (b) $g : x \to e^{3x}$; (c) $h : x \to 4e^{3x}$.

 (a) $f' : x \to 4e^x$;

 (b) $g' : x \to 3e^{3x}$;

 (c) $h' : x \to 3 \times 4e^{3x} = 12e^{3x}$.

Example 4
The growth of a population is modelled by the following alternatives:
    (a) 12% a year;   (b) 6% every six months;   (c) 1% a month.
Are these alternative models equivalent?

(a) A growth rate of 12% a year means that the population will multiply by a factor of 1.12 every year.
(b) A growth rate of 6% every six months means that the population will multiply by $1.06^2 = 1.1236$ every year. This is a faster growth rate than (a).
(c) This gives a faster growth rate still: every year the population multiplies by $1.01^{12} = 1.1268$ to 5 s.f.

The procedure of Example 4 could be taken further. Suppose we divide the year into n equal time intervals and consider the effect of an increase of $\dfrac{12}{n}$% in each of those time intervals. Each year the population would increase by a factor of

$$\left(1 + \frac{12}{100n}\right)^n = (1+h)^{\frac{12}{100h}} \quad \text{where} \quad h = \frac{12}{100n}$$

$$= [(1+h)^{\frac{1}{h}}]^{0.12}.$$

As n increases, h becomes smaller, so the limit of this is $e^{0.12}$; the population would increase by a factor of $e^{0.12}$ every year. After x years the population would be given by

$$P(x) = Ae^{0.12x}$$

where A was its initial size.

$$P'(x) = 0.12Ae^{0.12x} = 0.12P(x)$$

so the population is growing *continuously* at a rate of 12% of its size at any moment.

Example 5
The population of Villanova is estimated to be growing continuously at a rate of 0.1% of its size per month. If the population on 1st January 1980 was 50000, what is it expected to be on 1st January 2000?

If $P(t)$ is the population t months after 1st January 1980, then
       $$P'(t) = 0.001P(t)$$

and hence $P(t) = Ae^{0.001t}$.

But $P(0) = 50000$ and $P(0) = Ae^0 = A$ so $A = 50000$,
       $$P(t) = 50000e^{0.001t}.$$

On 1st January 2000, $t = 240$ and the population is expected to be
       $$P(240) = 50000e^{0.24} \approx 63600.$$

Exercise D

1   Write down the derived functions of:
    (a) $f : x \to 5e^x$;   (b) $f : x \to {}^-3e^x$;   (c) $f : x \to e^{2x}$;
    (d) $f : x \to e^{-2x}$;   (e) $f : x \to 5e^{2x}$.

2   (a) Copy and complete the table below to show the effect of leaving £20 in different savings accounts.

Name of bank	Interest paid	Amount after 5 years
Yearly Bank	16% annually	£20 × 1.16⁵ = £42.01
Half and Half Bank	8% every six months	£20 × 1.08¹⁰ =
Four Seasons Bank	4% every three months	
Weakley Bank	$\frac{1}{2}$% thirty-two times a year	

   (b) The limit of the process described in (a) is continuous growth at a rate of 16% per year. This would give £20 e^{kt} after t years.
      (i) State the value of k.
      (ii) Calculate the value of this expression when $t = 5$.

*3   Explain why $\int_1^3 e^x dx = [e^x]_1^3$, and $\int_1^3 e^{2x} dx = [\frac{1}{2}e^{2x}]_1^3$.

4   Evaluate: (a) $\int_2^4 e^x dx$;   (b) $\int_0^2 3e^x dx$;   (c) $\int_1^2 e^{2x} dx$;   (d) $\int_0^1 e^{2x} dx$.

5   Write down the derived functions of:
    (a) $f : x \to 10e^x$;   (b) $f : x \to e^2 e^x$;   (c) $f : x \to e^{x+3}$;
    (d) $f : x \to 5e^{-3x}$;   (e) $f : x \to {}^-3e^{-2x}$.

6   The rate at which radioactive material decays is proportional to its mass. For polonium-84, $m'(t) = {}^-0.005 m(t)$ where $m(t)$ grams is the mass of polonium after t days.
    (a) If $m(0) = 5$, find $m(t)$.
    (b) Find the values of $m(1)$, $m(20)$, $m(100)$, $m(139)$ and $m(278)$.
    (c) Explain why 139 days is called the 'half-life'.

7   (a) A bank pays '6% interest, compounded daily' on current accounts, that is $\frac{6}{365}$% per day. What would $100 left in such an account become after 1 year?
    (b) What would $100 become in one year at a continuous growth rate of 6% per year?

8   Evaluate: (a) $\int_1^5 e^{-x} dx$;   (b) $\int_0^{20} e^{-x} dx$.
    What happens to $\int_0^q e^{-x} dx$ as q becomes larger and larger?

5. LOGARITHMIC FUNCTIONS

Figures 4 and 5 show the graphs of $y = f(x) = 2^x$ and of $y = f^{-1}(x)$.
The index laws given in Section 2 lead to some simple and useful properties of the inverse function. For example,

$$f(2) = 2^2 = 4, \quad f(3) = 2^3 = 8, \quad f(2) \times f(3) = 2^2 \times 2^3 = 2^{2+3} = f(2+3)$$

lead to

$$f^{-1}(4) = 2, \quad f^{-1}(8) = 3, \quad f^{-1}(4 \times 8) = 2+3 = f^{-1}(4) + f^{-1}(8);$$

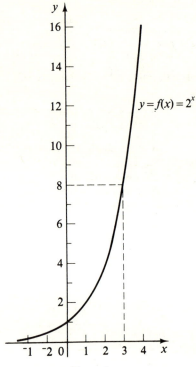

Figure 4

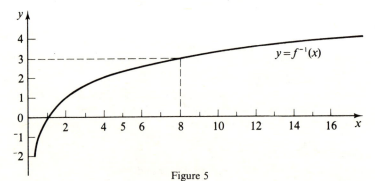

Figure 5

or, diagramatically,

$$
\begin{array}{cc}
f & f^{-1} \\
2 \rightarrow 4 & 4 \rightarrow 2 \\
3 \rightarrow 8 & 8 \rightarrow 3 \\
\end{array}
$$

$$2+3 = 5 \rightarrow 32 = 4 \times 8 = 32 \rightarrow 5$$

In general, $f^{-1}(pq) = f^{-1}(p) + f^{-1}(q)$.

Similarly, for division,

$$
\begin{array}{cc}
f & f^{-1} \\
7 \to 128 & 128 \to 7 \\
3 \to 8 & 8 \to 3 \\
7 - 3 = 4 \to 16 = 128 \div 8 = 16 \to 4
\end{array}
$$

In general, $f^{-1}(p/q) = f^{-1}(p) - f^{-1}(q)$.

For powers,

$$
\begin{array}{cc}
f & f^{-1} \\
3 \to 8 & 8 \to 3 \\
3 \times 2 = 6 \to 64 = 8^2 = 64 \to 6
\end{array}
$$

In general, $f^{-1}(p^n) = nf^{-1}(p)$.

Because the inverse function effectively replaces multiplication and division by addition and subtraction it was important historically for calculating and was given a special name, $\log_2$, an abbreviation for 'logarithm to base 2'. So for example, $\log_2 (128) = 7$ because $2^7 = 128$. A calculation such as 273×14.7 could be carried out as follows:

$$
\log_2 (273) \approx 8.093, \quad \log_2 (14.7) \approx 3.878,
$$
$$
\log_2 (273 \times 14.7) = \log_2 (273) + \log_2 (14.7) \approx 8.093 + 3.878 = 11.971,
$$
$$
273 \times 14.7 \approx 2^{11.971} \approx 4014.
$$

Of course, to use such a method requires tables of values of 2^x and $\log_2 x$.

The inverse of any exponential function is a logarithmic function. The inverse of $x \to N^x$ is written as $x \to \log_N x$. So, for example,

$$
\log_3 81 = 4 \quad \text{since } 3^4 = 81,
$$

and, in general,
$$\log_N P = a \Leftrightarrow N^a = p.$$

The most common logarithms are $\log_{10}$ (often just written log) and $\log_e$ (usually written ln, for 'natural logarithm').

The properties we met earlier in this section hold for all logarithmic functions:

(1) $\log_N(pq) = \log_N(p) + \log_N(q)$.

(2) $\log_N\left(\dfrac{p}{q}\right) = \log_N(p) - \log_N(q)$.

(3) $\log_N(p^n) = n \log_N(p)$.

Example 6

Write down the values of: (a) $\log_2 64$;   (b) $\log_4 \left(\tfrac{1}{16}\right)$.

(a) $x = \log_2 64 \Leftrightarrow 2^x = 64 \Leftrightarrow x = 6$. So $\log_2 64 = 6$.

(b) $y = \log_3 \tfrac{1}{16} \Leftrightarrow 4^y = \tfrac{1}{16} \Leftrightarrow y = {}^-2$. So $\log_4 \left(\tfrac{1}{16}\right) = {}^-2$.

Example 7
Find the values of: (*a*) $\log_6 4 + \log_6 9$; (*b*) $\log_2(\sqrt[3]{16})$.

    (*a*) $\log_6 4 + \log_6 9 = \log_6(4 \times 9) = \log_6 36 = 2$ (because $6^2 = 36$).

    (*b*) $\log_2(\sqrt[3]{16}) = \log_2(16^{\frac{1}{3}}) = \frac{1}{3}\log_2 16 = \frac{1}{3} \times 4 = \frac{4}{3}$ (because $2^4 = 16$).

Example 8
Solve the equation $5^x = 7$.
$$5^x = 7 \iff \log(5^x) = \log 7$$
$$\iff x \log 5 = \log 7$$
$$\iff x = \frac{\log 7}{\log 5} = \frac{0.845}{0.699} = 1.21.$$

(Logarithms to any base could be used; in the working above, $\log_{10}$ has been used.)

Example 9
Find $\log_5 12$ to 4 significant figures.
$$x = \log_5 12 \iff 5^x = 12$$
$$\iff x \log 5 = \log 12$$
$$\iff x = \frac{\log 12}{\log 5} = 1.544 \text{ to 4 s.f.}$$

So $\qquad\qquad\qquad\qquad \log_5 12 = 1.544$ to 4 s.f.

Exercise E

1 Write down the values of:
    (*a*) $\log_2 8$; (*b*) $\log_2 32$; (*c*) $\log_3 \frac{1}{9}$; (*d*) $\log_5 1$; (*e*) $\log_7 7$.

2 By writing $8 = 2^3$, solve the equation $8^x = 32$. Hence write down the value of $\log_8 32$. Use the same method to find:
    (*a*) $\log_8 128$; (*b*) $\log_8 4$; (*c*) $\log_4 8$; (*d*) $\log_{64} 32$.

3 Use the method of question 2 to find $\log_a b$ if $a = 2^n$ and $b = 2^m$. Explain why
$$\log_a b = \frac{\log_2 b}{\log_2 a}.$$

4 Solve the following equations to an accuracy of 3 significant figures:
    (*a*) $3^x = 17$; (*b*) $2^x = 9$; (*c*) $5^{x+3} = 100$; (*d*) $3^{2x+1} = 8$; (*e*) $x = \log_5 7$.

5 If $\log_N 12 = 1.5440$ and $\log_N 3 = 0.6826$, find:
    (*a*) $\log_N 36$; (*b*) $\log_N 4$; (*c*) $\log_N 2$; (*d*) $\log_N 0.5$; (*e*) $\log_N 432$.

6 The area of weed on a pond increases by 10% each year. After how many years will the area covered have doubled?

7 Write down the values of:
    (*a*) $\log_4 64$; (*b*) $\log_5 125$; (*c*) $\log_5 0.008$; (*d*) $\log_{10}(10^6)$; (*e*) $\log_{10}(0.001)$.

8 (*a*) What is the relationship between $\log_{10}(2^n)$ and $\log_{10} 2$?
    (*b*) Use $2^{10} \approx 1000$ to show that $\log_{10} 2 \approx 0.3$.

(c) Use $3^{12} \approx 2^{19}$ to find an approximate value for $\log_{10} 3$.

(d) Use your answers to (b) and (c) to find approximate values of:

(i) $\log_{10} 4$;   (ii) $\log_{10} 6$;   (iii) $\log_{10} 8$;   (iv) $\log_{10} 9$;   (v) $\log_{10} 5$.

9   If $\log_e N = x$, show that $x = \dfrac{\log_{10} N}{\log_{10} e}$. What is the relation between $\log_e 10$ and $\log_{10} e$?

10   Solve the following equations to an accuracy of 3 significant figures:

(a) $5^x = 1$;   (b) $3^x = 5$;   (c) $x = \log_4 3$;

(d) $x = \log_2 3$;   (e) $5^x = \frac{1}{3}$.

11   The pH of a solution is defined to be $-\log_{10}$ (molar hydrogen ion concentration).

(a) The indicator methyl orange changes colour when the hydrogen ion concentration is about 2×10^{-4}. What is the corresponding pH?

(b) Find the hydrogen ion concentration of the following:

(i) pure water at 25 °C, pH 7;

(ii) normal blood, pH 7.4.

12   The population of pinkfoot ducks on the Blue River Estuary is estimated to be decreasing by 5% each year. If the population was 100 000 in 1970, when is it expected that it will have fallen to 25 000?

6. LOGARITHMIC SCALES

The annual salary of a skilled woggle-turner was £100 in 1950 and has increased by 15% each year since then. Graphs of his salary are shown in Figures 6 and 7.

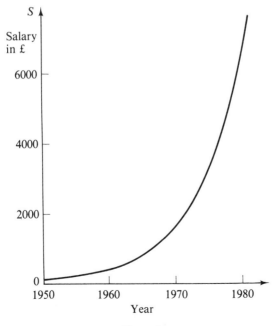

Figure 6

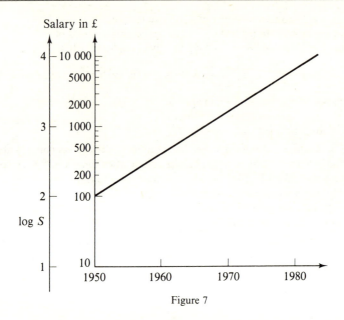

Figure 7

Notice the scale on the *S*-axis in Figure 7. It has been chosen so that every time *S* is multiplied by 10 there is the same increase on the *S*-axis. Such a scale is called *logarithmic*, since the scale is proportional to log *S*. Scales like these are used when we concentrate attention on proportionate increases. In this case *S* is increasing by the same percentage each year and in these circumstances the graph of log *S* will be a straight line (see Figure 8.)

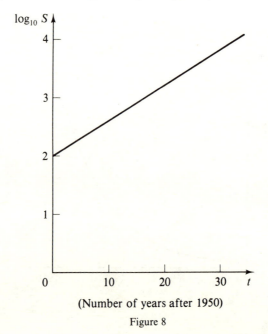

(Number of years after 1950)

Figure 8

$$S = 100 \times 1.15^t$$
$$\Rightarrow \quad \log_{10} S = \log_{10}(100) + \log_{10}(1.15^t)$$
$$= 2 + t \log_{10} 1.15$$
$$= 0.0607t + 2.$$

So the graph has gradient 0.0607 and crosses the $\log S$ axis at $(0, 2)$.

Example 9
The graph in Figure 9 has been obtained from experimental data. Find the equation connecting x and y.

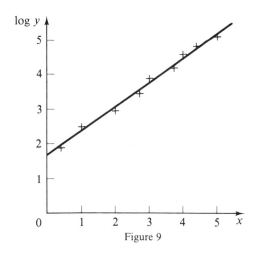

Figure 9

The straight line passes through $(0, 1.7)$ and $(5, 5.2)$. Its gradient is therefore

$$\frac{5.2 - 1.7}{5 - 0} = \frac{3.5}{5} = 0.7$$

and its equation is $\qquad \log y = 0.7x + 1.7$.

Hence $\qquad\qquad y = 10^{(0.7x + 1.7)}$
$$= (10^{0.7})^x \times 10^{1.7}$$
$$\approx 5^x \times 50$$
$$= 50 \times 5^x.$$

Example 10
Obtain the equation connecting x and y from the graph shown in Figure 10.

Notice that in this case the graph shows $\log y$ and $\log x$. The straight line passes through $(0, 0.3)$ and $(0.6, 2.1)$. Its gradient is therefore

$$\frac{2.1 - 0.3}{0.6 - 0} = \frac{1.8}{0.6} = 3$$

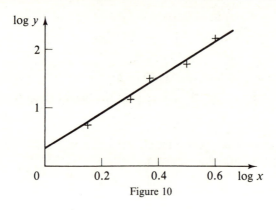

Figure 10

and its equation is
$$\log y = 3 \log x + 0.3$$
$$\approx \log x^3 + \log 2$$
$$= \log(2x^3).$$

Hence
$$y = 2x^3.$$

Exercise F

1 The table below shows the population of Sampsey Island at the censuses which are held from time to time.

Date of census	1902	1914	1919	1928	1944	1951	1964	1975
Population	5400	6100	6400	7000	8200	8800	10000	11200

It is thought that the population $P(t)$, t years after 1900, is given by the formula $P(t) = Ar^t$ for some fixed values of A and r. Draw a graph of $\log P$ against t and use the 'best-fit' straight line to find A and r.
(a) By what percentage is the population increasing each year?
(b) What was the population in 1900?
(c) What is your estimate of the population in the year 2000?

2 Repeat question 1 for the population of Small Copper butterflies on Sampsey Common given in this table:

Date	1909	1919	1931	1943	1954	1961	1971	1980
Population	3800	2800	2000	1400	1000	800	600	450

3 The table below shows the time, T seconds, for a complete swing of a pendulum of length l m.

l	0.24	0.38	0.53	0.87	0.99
T	0.98	1.24	1.46	1.87	2.00

Draw a graph of $\log T$ against $\log l$ and deduce the equation connecting T and l.

4   When water flows through a pipe, the rate at which it flows depends on the radius of the pipe, among other factors. When those factors were kept unchanged the following rates of flow were obtained for different radii of pipe:

Radius of pipe/m (r)	0.32	0.61	0.85	1.10
Rate of flow/l s^{-1} (V)	0.5	0.66	3.90	7.03

Draw a graph of $\log V$ against $\log r$ and deduce an equation connecting V and r.

7. SUMMARY

$N^{\frac{1}{b}}$ is defined to be $^b\sqrt{N}$ for any positive integer b.
$N^{\frac{a}{b}}$ is defined to be $(^b\sqrt{N})^a$ for any integer a.

$$
\left.
\begin{aligned}
N^p \times N^q &= N^{p+q} \\
N^p \div N^q &= N^{p-q} \\
(N^p)^q &= N^{pq}
\end{aligned}
\right\} \quad \text{for any rational numbers } p \text{ and } q.
$$

A function $f : x \to N^x$ is called an exponential function.
For such a function $f'(x) = f'(0)f(x)$.
The graph of any exponential function can be mapped onto the graph of another exponential function by a one-way stretch parallel to the x-axis.
There is a number e such that if $f(x) = e^x, f'(0) = 1$.

      $f(x) = e^x$ is called the exponential function.
      $f(x) = Ae^x \Rightarrow f'(x) = Ae^x.$
      $f(x) = Ae^{kx} \Rightarrow f'(x) = Ake^{kx}.$

Any exponential function can be written in the form $x \to e^{kx}$. For example, $2^x = e^{kx}$ where $e^k = 2$.
If $f'(x) = kf(x)$ then $f(x) = Ae^{kx}$.
The inverse of an exponential function is a logarithmic function.
If $N^a = p$, then $\log_N p = a$.

$$
\log_N (pq) = \log_N p + \log_N q,
$$

$$
\log_N \left(\frac{p}{q}\right) = \log_N p - \log_N q,
$$

$$
\log_N (p^n) = n \log_N p.
$$

Miscellaneous exercise

1   It was estimated in 1978 that, at the current rate, the population of Japan would have increased by 30% by the year 2000. By what percentage was the population increasing each year?

2   In the mid-1980s the population of North America was 247 million and increasing at 0.7% per year. The comparable estimate for Latin America was 360 million, with a growth rate of 2.6%. If these growth rates remain unchanged, how many years would it take for the population of Latin America to be double that of North America?

3   Show that $2^x = e^{x \ln 2}$, where $\ln t = \log_e t$. Deduce that $f(x) = 2^x \Rightarrow f'(x) = 2^x \ln 2$.

4   The graph of $y = e^x$ can be mapped onto the graph of $y = 2^x$ by a one-way stretch parallel to the x-axis. What is the scale-factor of the stretch?

Describe precisely the transformation mapping the graph of $y = \log_e x$ onto the graph of $y = \log_2 x$.

5   The loudness of a sound is often given in decibels. A decibel is one-tenth of a bel, which is the logarithm (to base ten) of the ratio of the sound intensity to the intensity of the faintest audible sound.

   (a) What is the ratio of the intensities of two sounds, a shout of 70 dB and a whisper of 30 dB?

   (b) What is the ratio of the intensities of the sound of a nearby train (80 dB) and a jet aircraft on take-off (125 dB)?

   (c) The intensity of a sound varies inversely with the square of the distance from the sound source. Bridget is standing twice as far away from a rock group as Alison. If they both measure the loudness of the sound from the group in decibels, what will be the difference in their measurements?

6   The rate at which a kettle cools is thought to be proportional to the difference between its temperature and the temperature of the air around it.

   If $f(t)$ denotes this temperature difference at time t, explain why $f'(t) = {}^-kf(t)$ for some number k. Deduce a formula for $f(t)$.

   If the surrounding air has a temperature of 15 °C and the rate of cooling is 9 °C/min when the temperature of the kettle is 98 °C, find k and determine the temperature of the kettle after 20 minutes.

23

Composite functions

1. A NEW NOTATION

We have seen that $f'(x)$ gives the gradient of $y = f(x)$. Derived functions have been found by considering what happens to

$$\frac{f(x+h)-f(x)}{h}$$

as h approaches zero.

This fraction is equal to

$$\frac{RQ}{PR} = \frac{\text{increase in } y}{\text{increase in } x} \quad \text{(see Figure 1)}$$

and this is commonly denoted by $\frac{\delta y}{\delta x}$, the δ being used as a shorthand for 'the increase in'.

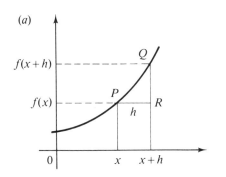
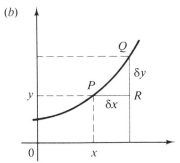

Figure 1

From this notation comes an alternative notation, $\frac{dy}{dx}$, for $f'(x)$. $\frac{dy}{dx}$ is the limit of $\frac{\delta y}{\delta x}$ as δx approaches zero. Each notation has its advantages and disadvantages. In a problem where several functions are involved, the $\frac{dy}{dx}$ notation is particularly useful, since it makes clear which pair of variables we are considering. On the other hand, $f'(3)$ is a more compact notation than '$\frac{dy}{dx}$ when $x = 3$'.

$\dfrac{d}{dx}$ can be interpreted as an instruction to differentiate, and leads to the notation $\dfrac{d^2y}{dx^2}$ for the second derivative – that is,

$$\frac{d}{dx}\left(\frac{dy}{dx}\right) = \frac{d^2y}{dx^2} = f''(x).$$

Example 1

If $y = (x+5)^3$, find $\dfrac{dy}{dx}$ and $\dfrac{d^2y}{dx^2}$.

Multiplying out　　　$y = (x+5)^3 = x^3 + 15x^2 + 75x + 125.$

So　　　　　　　　$\dfrac{dy}{dx} = 3x^2 + 30x + 75$

and　　　　　　　$\dfrac{d^2y}{dx^2} = 6x + 30.$

Exercise A

1　If $y = x^3 - 12x + 10$, write down expressions for $\dfrac{dy}{dx}$ and $\dfrac{d^2y}{dx^2}$. State the sets of values of x for which (a) $\dfrac{dy}{dx} > 0$ and (b) $\dfrac{dy}{dx} < 0$. What does this information tell you about the graph of $y = x^3 - 12x + 10$?

2　The height in metres above the ground of a ball t seconds after being thrown is given by $h = 1 + 8t - t^2$. Write down an expression for $\dfrac{dh}{dt}$. What does this represent? Find the greatest height of the ball.

3　The displacement, in millimetres, of a point on a vibrating tuning-fork is given by $x = \sin(1600t)$, where t is the time measured in seconds. How fast is the point moving when (a) $t = 0$, (b) $t = 0.2$?

4　Find expressions for $\dfrac{dy}{dx}$ in the following cases:

(a) $y = (x+3)^2$;　(b) $y = x^2 + \dfrac{1}{x}$;　(c) $y = 5x^3 - 3x^5$;

(d) $y = 3 \sin 2x$;　(e) $y = (2x-1)^3$;　(f) $y = 6 \cos \frac{1}{2}x$.

5　If $y = t^2 + \dfrac{54}{t}$, write down an expression for $\dfrac{dy}{dt}$. Find a if $\dfrac{dy}{dt} = 0$ when $t = a$. State the signs of $\dfrac{dy}{dt}$ when (a) $t < a$, (b) $t > a$. What does this tell you about the point on the graph of $t \rightarrow y$ where $t = a$?

6　The speed v m s^{-1} of a rocket t seconds after blast-off is given by $v = 7.2t - 0.1t^2$. Write down an expression for $\dfrac{dv}{dt}$. What does this represent? At what time does the rocket reach its maximum speed, and what is this speed?

7   The horizontal distance travelled by a point on a bicycle wheel is given by
$s = 500t - 33 \sin 15t$, where s is measured in centimetres and t in seconds. How fast
is the point moving when (a) $t = 6.3$, (b) $t = 9.4$?

8   Find expressions for $\dfrac{dx}{dt}$ in the following cases:

    (a) $x = 4t^5 - 2t^3$; (b) $x = 5 \cos 3t$; (c) $x = 2(t+1)^3$;
    (d) $x = 10 \sin 0.1t$; (e) $x = (3t^2 + 1)^2$; (f) $x = (3 - t)^4$.

2. TRANSLATIONS AND STRETCHES

We saw in Example 1 that if $y = (x+5)^3$ then $\dfrac{dy}{dx} = 3x^2 + 30x + 75$. This

expression for $\dfrac{dy}{dx}$ can be factorised:

$$\frac{dy}{dx} = 3x^2 + 30x + 75 = 3(x^2 + 10x + 25) = 3(x+5)^2.$$

Notice the comparison:

$$y = x^3 \quad \Rightarrow \quad \frac{dy}{dx} = 3x^2;$$

$$y = (x+5)^3 \quad \Rightarrow \quad \frac{dy}{dx} = 3(x+5)^2.$$

We should not be surprised by this result. The curve $y = (x+5)^3$ is simply the
curve $y = x^3$ translated $^-5$ units parallel to the x-axis. We would therefore expect
the gradient function to be similarly translated to give $3(x+5)^2$ instead of $3x^2$.
We saw in Chapter 9 that $\sin 2x$ oscillates twice as fast as $\sin x$, but with the
same amplitude, so that the graph of $y = \sin 2x$ is the graph of $y = \sin x$
stretched by a factor of $\frac{1}{2}$ parallel to the x-axis. We deduced that

$$y = \sin 2x \quad \Rightarrow \quad \frac{dy}{dx} = 2 \cos 2x.$$

Similarly, we might expect that $y = (2x+5)^3$ involves a stretch, as well as a
translation, of the graph of $y = x^3$ and that a factor of 2 would appear when
we differentiate. That is:

$$y = (2x+5)^3 \quad \Rightarrow \quad \frac{dy}{dx} = 3(2x+5)^2 \times 2 = 6(2x+5)^2.$$

This can be checked by multiplying out:

$$y = (2x+5)^3 = (2x)^3 + 3(2x)^2 \times 5 + 3(2x) \times 5^2 + 5^3$$
$$= 8x^3 + 60x^2 + 150x + 125$$
$$\Rightarrow \quad \frac{dy}{dx} = 24x^2 + 120x + 150$$
$$= 6(4x^2 + 20x + 25)$$
$$= 6(2x+5)^2.$$

Example 2

Find $\dfrac{dy}{dx}$ if (a) $y = \dfrac{1}{x+3}$, (b) $y = \sin(3x+1)$.

(a) $y = \dfrac{1}{x+3}$ is a translation of $y = \dfrac{1}{x}$ and so from the fact that the derivative

of $\dfrac{1}{x}$ is $\dfrac{^-1}{x^2}$, we can deduce that

$$y = \frac{1}{x+3} \quad \Rightarrow \quad \frac{dy}{dx} = \frac{^-1}{(x+3)^2}.$$

(b) In this case, a translation and stretch are involved, so we obtain

$$y = \sin(3x+1) \quad \Rightarrow \quad \frac{dy}{dx} = 3\cos(3x+1).$$

Notice that in these cases we cannot check by 'multiplying out'. The only independent check available is to consider the limit of $\dfrac{f(x+\delta x)-f(x)}{\delta x}$ in each case.

Exercise B

Write down expressions for $\dfrac{dy}{dx}$ in questions 1–5 and 6–10.

1   $y = (x-3)^2$.   2 $y = (5x-3)^2$.   3 $y = (x+5)^4$.

4   $y = (2x+5)^4$.   5 $y = (9x-4)^{10}$.

6   $y = (x+4)^3$.   7 $y = (7x+4)^3$.   8 $y = (x-3)^4$.

9   $y = (5x-3)^4$.   10 $y = (8x+13)^{15}$.

11   Check the answers to questions 1 and 2 by multiplying out and then differentiating.

12   Check the answers to questions 6 and 7 by multiplying out and then differentiating.

13   If $y = f(x) = \dfrac{1}{x+3}$, show that

$$\delta y = f(x+\delta x)-f(x) = \frac{^-\delta x}{(x+3)(x+3+\delta x)}.$$

Deduce an expression for $\dfrac{dy}{dx}$.

14   Use $a^2-b^2 = (a+b)(a-b)$ to show that if

$$y = f(x) = (3x-4)^2$$

then $$\delta y = f(x+\delta x)-f(x) = 3(6x-8+3\delta x)\,\delta x.$$

Deduce an expression for $\dfrac{dy}{dx}$.

3. THE CHAIN RULE

As Exercise B shows, the translation and stretching ideas allow us to increase considerably the range of functions we can differentiate, but there are still some relatively simple functions which we cannot differentiate, such as

$$f(x) = \frac{1}{x^2+1} \quad \text{and} \quad g(x) = (\sin x)^2.$$

We need to extend our techniques a little further to deal with these. As before, we use a polynomial to illustrate a general method.

Consider $y = (x^2+2)^3$. This can be differentiated by first multiplying out, as follows:

$$y = (x^2+2)^3 = (x^2)^3 + 3(x^2)^2 \times 2 + 3(x^2) \times 2^2 + 2^3$$
$$= x^6 + 6x^4 + 12x^2 + 8$$

$$\Rightarrow \quad \frac{dy}{dx} = 6x^5 + 24x^3 + 24x$$

$$= 6x(x^4 + 4x^2 + 4)$$
$$= 6x(x^2+2)^2.$$

If we compare this with $y = x^3 \Rightarrow \frac{dy}{dx} = 3x^2$, we see that an 'extra' $2x$ has

appeared. To see how this has arisen, we split the original function into a chain of simpler functions, and then consider the effect of increasing x by δx.

$$x \;\to\; u = x^2 + 2 \;\to\; y = u^3 = (x^2+2)^3,$$
$$x + \delta x \;\to\; u + \delta u = (x+\delta x)^2 + 2 \;\to\; y + \delta y = (u+\delta u)^3.$$

At the first stage, an increase δx in x produces an increase in u of δu, where

$$\delta u = [(x+\delta x)^2 + 2] - (x^2+2) = 2x\delta x + (\delta x)^2$$

from which we can deduce that

$$\frac{du}{dx} = 2x.$$

At the second stage, an increase δu in u produces an increase in y of δy, where

$$\delta y = (u+\delta u)^3 - u^3 = 3u^2\delta u + 3u(\delta u)^2 + (\delta u)^3$$

from which we can deduce that

$$\frac{dy}{du} = 3u^2.$$

To find $\dfrac{dy}{dx}$ it is easiest to use

$$\frac{\delta y}{\delta x} = \frac{\delta y}{\delta u} \times \frac{\delta u}{\delta x} = [3u^2 + 3u\delta u + (\delta u)^2](2x+\delta x).$$

From this it follows that

$$\frac{dy}{dx} = 3u^2 \times 2x.$$

Notice that this is equivalent to

$$\frac{dy}{dx} = \frac{dy}{du} \times \frac{du}{dx}$$

which is a general result deducible from $\dfrac{\delta y}{\delta x} = \dfrac{\delta y}{\delta u} \times \dfrac{\delta u}{\delta x}$.

Finally, we use $u = x^2 + 2$ to obtain

$$\frac{dy}{dx} = 3u^2 \times 2x = 3(x^2 + 2)^2 \times 2x = 6x(x^2 + 2)^2.$$

The general result, $\dfrac{dy}{dx} = \dfrac{dy}{du} \times \dfrac{du}{dx}$, is called the *chain rule* and can be used to differentiate composite functions (that is a function found by chaining functions together).

Example 3

Find $\dfrac{dy}{dx}$ if $y = (\sin x)^3$.

The function can be split as follows:

$$x \rightarrow \sin x \rightarrow (\sin x)^3$$

so if we write $u = \sin x$, we have $y = u^3$. Hence

$$\frac{dy}{dx} = \frac{dy}{du} \times \frac{du}{dx} = 3u^2 \cos x = 3(\sin x)^2 \cos x.$$

It is usual to write $(\sin x)^3$ as $\sin^3 x$, and so on, so the result of Example 3 can be written as

$$y = \sin^3 x \quad \Rightarrow \quad \frac{dy}{dx} = 3 \sin^2 x \cos x.$$

Exercise C

In questions 1–6 and 7–12, write down expressions for $\dfrac{dy}{dx}$.

1   $y = (x^2 - 3)^4$.   2 $y = (x^3 + 1)^2$.   3 $y = \dfrac{1}{3x+1}$.

4   $y = \dfrac{5}{x^2 - 2}$.   5 $y = \cos^4 x$.   6 $y = \cos(x^4)$.

7   $y = (5x^4 + 2)^3$.   8 $y = \left(x + \dfrac{1}{x}\right)^2$.   9 $y = \dfrac{1}{\sin x}$.

10   $y = \sin^2 x$.   11 $y = \dfrac{1}{x^2}$ (write $u = x^2$).

12   $y = \dfrac{1}{x^{10}}$ (write $u = x^{10}$).

13 If $y = u^2$, write $\dfrac{dy}{dx}$ in terms of $\dfrac{du}{dx}$.

If also $u = \sqrt{x}$, show that $y = x$ and deduce the value of $\dfrac{dy}{dx}$. Hence find $\dfrac{du}{dx}$ as a function of x.

14 If $y = e^u$, write $\dfrac{dy}{dx}$ in terms of $\dfrac{du}{dx}$.

If also $u = \log_e x$, explain why $y = x$ and deduce the value of $\dfrac{dy}{dx}$. Hence find $\dfrac{du}{dx}$ as a function of x.

4. INTEGRATION

Extending the set of functions which we can differentiate means that we have extended the set of functions which we can integrate.

Example 4
Find $\int (2x+1)^3 \, dx$.
We have to find a function whose derivative is $(2x+1)^3$. Experience with the chain rule suggests $(2x+1)^4$ as a possibility.

$$y = (2x+1)^4 \quad \Rightarrow \quad \frac{dy}{dx} = 4(2x+1)^3 \times 2 = 8(2x+1)^3.$$

This is 8 times too large, so we try $\tfrac{1}{8}(2x+1)^4$.

$$y = \tfrac{1}{8}(2x+1)^4 \quad \Rightarrow \quad \frac{dy}{dx} = \tfrac{1}{8} \times 4(2x+1)^3 \times 2 = (2x+1)^3.$$

So $\int (2x+1)^3 dx = \tfrac{1}{8}(2x+1)^4 + c.$

Exercise D

1 Expand $(2x+1)^3$ and integrate it term by term. Does your answer agree with the result of Example 4?

2 Find the following integrals:
   (a) $\int (3x-2)^8 \, dx;$ (b) $\int x(x^2+1)^4 \, dx;$
   (c) $\int \cos(2t+1) \, dt;$ (d) $\int t \sin(t^2+1) \, dt.$

3 Evaluate:
   (a) $\displaystyle\int_1^2 (x+1)^4 \, dx;$ (b) $\displaystyle\int_0^1 x(x^2+1)^4 \, dx;$
   (c) $\displaystyle\int_{0.4}^{0.6} \cos(3t-1) \, dt;$ (d) $\displaystyle\int_3^4 \frac{1}{(t+1)^2} \, dt.$

4 Find $\displaystyle\int_1^3 (3x+1)^2 \, dx$ by:
   (a) expanding and integrating term by term;
   (b) using the chain rule 'in reverse'.

5 Find the following integrals:
   (a) $\int (5x-3)^6\, dx$;   (b) $\int x^2(x^3+1)\, dx$;

   (c) $\int 3 \sin 6t\, dt$;   (d) $\int \cos t \sin^3 t\, dt$.

6 Evaluate:
   (a) $\int_1^4 (3x-2)^3\, dx$;   (b) $\int_{-1}^1 x(x^2-3)^2\, dx$;

   (c) $\int_{\pi/2}^{\pi} 4 \sin \tfrac{1}{2}t\, dt$;   (d) $\int_{0.5}^{-1} 5t \cos(t^2)\, dt$.

5. SUMMARY

δx means an increase in x.

If $y = f(x)$, $\delta y = f(x+\delta x) - f(x)$, the corresponding increase in y, and $f'(x) = \dfrac{dy}{dx}$, the limit of $\dfrac{\delta y}{\delta x}$ as δx approaches zero.

If y is a function of u, which is a function of x, then

$$\frac{dy}{dx} = \frac{dy}{du} \times \frac{du}{dx} \quad \text{(the chain rule)}.$$

Miscellaneous exercise

1 If $y = t^4 + 5t^2$ and $x = t^2$,
   (a) find $\dfrac{dy}{dt}, \dfrac{d^2y}{dt^2}, \dfrac{dx}{dt}$, and $\dfrac{d^2x}{dt^2}$;

   (b) show that $y = x^2 + 5x$ and hence find $\dfrac{dy}{dx}$ and $\dfrac{d^2y}{dx^2}$.

   Show that, in this case,

   $$\frac{dy}{dx} = \frac{dy}{dt} \div \frac{dx}{dt}$$

   and that

   $$\frac{d^2y}{dx^2} \neq \frac{d^2y}{dt^2} \div \frac{d^2x}{dt^2}.$$

2 Find the coordinates of the maximum and minimum points on the graph of $y = (x^2-1)^4$.

3 A pile of gravel is in the form of a cone in which the base diameter is equal to six times the height. If the height is x metres and the volume is V m³, show that $V = 3\pi x^3$.

   The pile grows, as a result of gravel being spilt onto the highest point, but remains the same shape. Find an expression for $\dfrac{dV}{dt}$ in terms of x and $\dfrac{dx}{dt}$.

   If gravel is spilt at a steady rate of 200 m³/h, show that $\dfrac{dx}{dt} = \dfrac{7.1}{x^2}$ m/hour approximately.

   (SMP)

4 Given that $f(x) = \dfrac{1}{x+1}$ and g is the inverse function f^{-1},
   (a) find $g(x)$ and verify that $g(\tfrac{1}{3}) = 2$;

(b) find $f'(x)$ and verify that $f'(2) = -\frac{1}{9}$;

(c) find $g'(x)$.

Calculate $g'f(2) \times f'(2)$. (SMP)

5 If $y = u^3$ write $\dfrac{dy}{dx}$ in terms of $\dfrac{du}{dx}$.

If also $u = \sqrt[3]{x}$, show that $y = x$, state the value of $\dfrac{dy}{dx}$, and deduce that $\dfrac{du}{dx} = \frac{1}{3}x^{-\frac{2}{3}}$.

6 If $y = \dfrac{1}{u}$, write $\dfrac{dy}{dx}$ in terms of $\dfrac{du}{dx}$.

If also $u = x^n$, where n is a positive integer, write down an expression for $\dfrac{du}{dx}$ and deduce the derivative of x^{-n}.

7 Differentiate e^{-x^2} and hence find $\displaystyle\int_0^1 xe^{-x^2}\,dx$.

8 Find the coordinates of the maximum and minimum points on the graph of $y = e^{\sin x}$. Sketch the graph.

24

Iteration and approximation

1. EQUATION SOLVING: DECIMAL SEARCH

Suppose we wish to solve the equation
$$f(x) = x^3 + 2x - 10 = 0.$$

We could try to factorise $f(x)$; the possible factors are $x-1$, $x+1$, $x-2$, $x+2$, $x-5$, $x+5$, $x-10$, $x+10$.

$x-1$ is a factor of $f(x)$ $\Leftrightarrow$ $f(1) = 0$. But $f(1) = {}^-7$, so $x-1$ is not a factor.

Also: $f({}^-1) = {}^-13$, so $x+1$ is not a factor;

$\qquad f(2) = 2$, so $x-2$ is not a factor;
$\qquad f({}^-2) = {}^-22$, so $x+2$ is not a factor;
$\qquad f(5) = 125$, so $x-5$ is not a factor;
$\qquad f({}^-5) = {}^-145$, so $x+5$ is not a factor;
$\qquad f(10) = 1010$, so $x-10$ is not a factor;
$\qquad f({}^-10) = {}^-1030$, so $x+10$ is not a factor.

Although we have failed to factorise $f(x)$, we can see that a solution of $f(x) = 0$ lies between 1 and 2, since $f(1) < 0$ and $f(2) > 0$.

Since $f(1) = {}^-7$ and $f(2) = 2$ we could now evaluate $f(1.5)$; but it is clear that the solution is closer to 2 than to 1, so we can try, say, 1.7:
$$f(1.7) = {}^-1.687.$$

Clearly 1.7 is close, but notice also that, since $f(1.7)$ is *negative* we can say that the solution lies between 1.7 and 2. At every stage in this process of decimal search we have an interval (or range of values) within which we can be sure that the solution lies. So far we have narrowed the interval from $[1, 2]$ to $[1.7, 2]$.

Now $f(1.8) = {}^-0.568$ giving us the interval $[1.8, 2]$. It is often useful to set out the results in a table (see Table 1).

Table 1

Current values				Trial value	
l	u	$f(l)$	$f(u)$	of x	$f(x)$
1	2	${}^-7$	2	1.7	${}^-1.687$
1.7	2	${}^-1.687$	2	1.8	${}^-0.568$
1.8	2	${}^-0.568$	2	1.85	0.031625
1.8	1.85	${}^-0.568$	0.031625	1.84	${}^-0.090496$
1.84	1.85	${}^-0.090496$	0.031625	1.847	${}^-0.00512758$
1.847	1.85	${}^-0.00512758$	0.031625	1.8475	0.000990915
1.847	1.8475	${}^-0.00512758$	0.0009909		

Notice that we tried 1.85 in row 3 because the solution is obviously closer to 1.8 than to 2. From the results entered so far we can be sure that the solution, to 4 significant figures, is 1.847 since the value of l (lower) and u (upper) in the last row agree to that accuracy. We can quickly increase the accuracy by adding more rows to the table and, at any stage, the accuracy we can be sure of is determined by the agreement of the two values at each end of the interval.

Exercise A

1 If $f(x) = x^3 - 4x^2 + 5$, show that $f(1) = 2$ and $f(2) = {}^-3$ and use decimal search to find, to 4 s.f., the solution of the equation $x^3 - 4x^2 + 5 = 0$ which lies in the interval $[1, 2]$.
   Use decimal search to find both of the other two solutions of the equation $x^3 - 4x^2 + 5 = 0$. Give your answers to 4 d.p.

2 Use decimal search to find the only real solution of the equation $x^3 + x^2 + 1 = 0$, giving your answer to 3 d.p.

3 Find the solution of $2x^3 - 4x^2 - 3 = 0$, giving your answer to 4 s.f.

4 Find the solution of $x^3 - x^2 - 5 = 0$, giving your answer to 4 s.f.

5 Find all the solutions of $3x^2 - 5x + 1 = 0$, giving your answers to 3 s.f.

6 Find the solution of $x + 1 + \dfrac{1}{x^2} = 0$, giving your answer to 4 s.f.

7 Find the solution of $x - \cos x = 0$, giving your answer to 4 s.f.

8 Find the solution of $x + \sin x - 1 = 0$, giving your answer to 4 s.f.

9 (a) Sketch graphs on the same axes of the functions
$$x \to \tan x \quad \text{and} \quad x \to 1.2x$$
   for $0 \leqslant x \leqslant 1$. (Note that, for the first, x is in radians.)
   (b) Use your graph to estimate a solution of the equation
$$5 \tan x - 6x = 0,$$
   and use decimal search to obtain this solution correct to 4 significant figures.

10 (a) For the function given by $f(x) = x^2 - 3\sqrt{x}$, show that $f(2) \approx {}^-0.2$ and $f(3) \approx 3.9$.
    (b) Use decimal search to find a solution of the equation
$$x^2 - 3\sqrt{x} - 1 = 0$$
    to 3 significant figures.
    (c) Are there any other solutions? If so, find them.

2. EQUATION SOLVING: ITERATION

What happens if you repeatedly press the cosine button on your calculator (when it is in radian mode)? You should find that the display 'settles down' to a value of about 0.7391. At this stage, an extra press of the button makes little or no difference, so if x is the number on the display, $\cos x \approx x$. You have, therefore, solved this equation to calculator accuracy.

The same idea can be used to solve other equations. For example, if
$$g(x) = \tfrac{1}{5}(x^2 + 3),$$

then
$$g(0) = 0.6$$
$$g(0.6) = 0.672$$
$$g(0.672) = 0.690\,3168$$
$$g(0.690\,3168) = 0.695\,30746$$

$$\cdots\cdots\cdots\cdots$$

$$g(0.697\,224\,36) = 0.697\,224\,36$$

so $0.697\,224\,36$ is (to 8 s.f.) a solution of the equation $x = g(x)$, that is, $x = \frac{1}{5}(x^2+3)$. This can be rewritten as $x^2-5x+3 = 0$, so $0.697\,224\,36$ is a solution of this equation also.

The equation $x^2-5x+3 = 0$ could have been solved using the quadratic formula, but we can employ the method above for other equations for which we have no formula. The first step is to rearrange the equation in the form $x = g(x)$, start with a number x_1 and then calculate $x_2 = g(x_1)$, $x_3 = g(x_2)$, $x_4 = g(x_3)$, etc. This is an *iterative process*. If the sequence 'settles down' to a number x, then for that number $x = g(x)$, and x is a solution of the original equation.

Example 1

Use an iterative process to solve the equation $x^3+7x-10 = 0$ to an accuracy of 3 decimal places.

One possible rearrangement of the equation is
$$7x = 10-x^3$$
$$x = \tfrac{1}{7}(10-x^3)$$

so we try the iteration $x_{n+1} = \frac{1}{7}(10-x_n^3)$.
$$x = 1 \;\Rightarrow\; x^3+7x-10 = {}^-2;$$
$$x = 2 \;\Rightarrow\; x^3+7x-10 = 5;$$

so there is a solution between 1 and 2, and apparently closer to 1 than 2. We will take $x_1 = 1.3$.
$$x_2 = \tfrac{1}{7}(10-1.3^3) = 1.1147,$$
$$x_3 = \tfrac{1}{7}(10-1.1147^3) = 1.2307,$$

$$\cdots\cdots\cdots\cdots$$

$$x_{15} = 1.1887,$$
$$x_{16} = 1.1886.$$

The solution is 1.189 to 3 d.p.

Notice that various other rearrangements of the equation in Example 1 are possible. For example, dividing throughout by x gives
$$x^2+7-\frac{10}{x} = 0$$
$$x^2 = \frac{10}{x}-7$$
$$x = \pm\sqrt{\left(\frac{10}{x}-7\right)}$$

Alternatively, we could write

$$x^3 = 10 - 7x$$

and hence

$$x = \sqrt[3]{(10 - 7x)},$$

or

$$x + \frac{7}{x} - \frac{10}{x^2} = 0$$

giving

$$x = \frac{10}{x^2} - \frac{7}{x}.$$

So we could have tried any of the iterative processes

$$x_{n+1} = \sqrt{\left(\frac{10}{x_n} - 7\right)}, \quad x_{n+1} = -\sqrt{\left(\frac{10}{x_n} - 7\right)},$$

$$x_{n+1} = \sqrt[3]{(10 - 7x_n)}, \quad x_{n+1} = \frac{10}{x_n^2} - \frac{7}{x_n}.$$

Many other rearrangements are possible. But several questions arise. Do all these iterations lead to a solution of the original equation? Does it matter what value of x_1 we use? If the equation has more than one solution, can the same iteration give all the solutions (by using different values of x_1)? We investigate some of these questions further in Section 3.

Exercise B

1   Consider the equation $x^3 - 9x - 5 = 0$.

   (a) Show that this can be written $x^2 = 9 + \dfrac{5}{x}$.

   (b) By taking the positive square root, show that a possible iterative formula to use to solve the equation is

$$x_{n+1} = \sqrt{\left(9 + \frac{5}{x_n}\right)}.$$

   (c) Find x_2 if x_1 is (i) 2, (ii) 3, (iii) 4.

   (d) Starting with $x_1 = 3$ write down $x_2, x_3, \ldots$, until you get two terms which agree to 5 s.f. What does this suggest as a solution of the original equation?

   (e) Substitute your value from (d) in the left side of the original equation to check that the result is close to zero.

2   (a) If $f(x) = x^3 - 9x - 5$ evaluate $f(^-3)$ and $f(^-2)$.

   (b) Explain why you expect that there will be a solution of the equation $f(x) = 0$ between $^-3$ and $^-2$.

   (c) Choosing the negative square root at stage (c) of question 1, use the iterative formula

$$x_{n+1} = -\sqrt{\left(9 + \frac{5}{x_n}\right)}$$

   to find a further solution of the equation (to 4 s.f.)

$$x^3 - 9x - 5 = 0.$$

3   Show that there is a solution of the equation $x^3 - 9x - 5 = 0$ between $^-1$ and 0.

   (a) Use the iterative formula of question 2(c) to generate a sequence of five terms, starting with $x_1 = ^-1$. Comment on your results.

(b) (i) Show that $x^3 - 9x - 5 = 0$ may be rearranged as

$$x = \frac{x^3 - 5}{9}.$$

(ii) Use the iterative formula $x_{n+1} = \frac{x_n^3 - 5}{9}$ to show that the solution of $x^3 - 9x - 5 = 0$ which lies between $^-1$ and 0 is $^-0.577$ (3 s.f.).

4 (The results of this question are used in questions 6 and 7.)

(a) Show that $x^3 - 3x + 1 = 0$ can be rearranged as

$$x = \tfrac{1}{3}(x^3 + 1).$$

(b) Use the iterative formula $x_{n+1} = \tfrac{1}{3}(x_n^3 + 1)$ to obtain the first seven terms of a sequence starting with $x_1 = {}^-1$.

(c) Repeat part (b), but with $x_1 = {}^-1.5$.

(d) Repeat part (b), but with $x_1 = 1.6$.

Comment on your results.

5 (a) Show that $x^3 - 3x + 1 = 0$ can be rearranged as $x = (3x - 1)^{\frac{1}{3}}$.

(b) Use the iterative formula $x_{n+1} = (3x_n - 1)^{\frac{1}{3}}$ to generate sequences starting with (i) $x_1 = {}^-1$, (ii) $x_1 = 2$. If the sequences converge, give their limits accurate to 4 s.f.

(c) Write down, accurate to 4 s.f., the solutions of the equation $x^3 - 3x + 1 = 0$.

6 (a) Using the same axes, draw graphs of the functions g and h, where

$$g(x) = \tfrac{1}{3}(x^3 + 1) \quad \text{and} \quad h(x) = x,$$

for values of x from $^-2$ to 2. Use equal scales for each axis. (Use as large a scale as possible.)

(b) Read off from your graphs the x-coordinates of the points where they intersect.

(c) Using the terms of the sequence obtained in question 4(c), mark on your graphs the points with the following coordinates:

$$A(x_1, x_2), \quad B(x_2, x_2), \quad C(x_2, x_3), \quad D(x_3, x_3), \quad E(x_3, x_4), \quad \text{etc.}$$

(d) Draw in the vectors $\underset{\sim}{AB}, \underset{\sim}{BC}, \underset{\sim}{CD}, \underset{\sim}{DE}$, etc.

Comment on your results.

7 (a) Using a scale of 4 cm to 1 unit for the horizontal axis and 2 cm to 1 unit for the vertical axis, draw graphs on the same axes of the functions g and h where

$$g(x) = \tfrac{1}{3}(x^3 + 1) \quad \text{and} \quad h(x) = x$$

for values of x from 0 to 4.

(b) Using the terms of the sequence obtained in question 4(d), mark on your diagram the points with coordinates

$$A(x_1, x_2), \quad B(x_2, x_2), \quad C(x_2, x_3), \quad D(x_3, x_3), \quad E(x_3, x_4), \quad \text{etc.}$$

Comment on your results.

8 (a) Show that the equation $x^3 + 15x - 3 = 0$ may be rewritten as

$$x = \frac{3 - x^3}{15}.$$

(b) Write down the iterative formula which corresponds to this rearrangement of the original equation.

(c) Show that 0 is an approximate solution of $x^3 + 15x - 3 = 0$, and, starting with $x_1 = 0$, use the iterative formula to obtain the solution of this equation, to 4 s.f.

3. CONVERGENCE AND DIVERGENCE

The iterative method of solving equations described above involves the following stages:

Stage 1:   Find an approximate solution of $f(x) = 0$.

Stage 2:   Rearrange $f(x) = 0$ into the form $x = g(x)$.

Stage 3:   Generate a sequence by taking the approximate solution found in stage 1 as the first term, x_1, and using $x_{n+1} = g(x_n)$ to generate further terms.

We can see from the results of Exercise B that the sequence we obtain in stage 3 may *converge* (the terms clearly get closer and closer to a limiting value), in which case we can determine a solution of the equation, or the sequence may *diverge* (the terms do not approach a limiting value). In the latter case, the sequence does not lead us to a solution and we have to find another rearrangement of $f(x) = 0$ into $x = h(x)$, and see whether this gives a converging sequence.

The graphs obtained in questions 6 and 7 of Exercise B help us to explain why convergence or divergence occurs. In each case we drew the graphs

$$y = x \quad \text{and} \quad y = g(x).$$

At a point of intersection of these graphs the y-coordinates are the same so we may write

$$x = g(x).$$

The x-coordinate of the point of intersection is thus the solution of the original equation.

Figure 1 shows a case where a convergent sequence is obtained. The point A has coordinates $(x_1, g(x_1))$ or, since $g(x_1) = x_2$, we can represent the coordinates

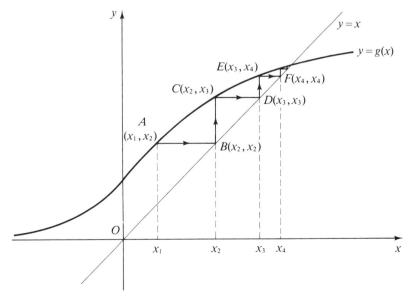

Figure 1

as (x_1, x_2). B is on the line $y = x$ and has the same y-coordinate as A. Its coordinates are thus (x_2, x_2). Similarly C has the same x-coordinate as B, but lies on $y = g(x)$. So C is the point (x_2, x_3).

As the 'staircase' is extended to D, E, F, etc., the coordinates of the points on the line $y = x$ get closer to the x-coordinate of the point of intersection of $y = x$ and $y = g(x)$. They thus converge onto the solution of $x = g(x)$.

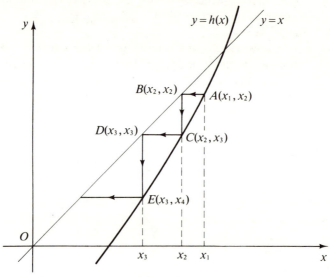

Figure 2

Figure 2 shows a divergent case. Despite the fact that x_1 is very close to the solution, x_2, x_3, etc., are increasingly further away.

Comparing Figures 1 and 2 it is easy to see that the gradient of the curve ($y = g(x)$ or $y = h(x)$) near the point of intersection with the line $y = x$ determines whether there is convergence or divergence.

In fact, near the respective points of intersection $g'(x)$ is *less than* 1, but $h'(x)$ is *greater than* 1; g gives *convergence* and h gives *divergence*. These results hold for all increasing functions. Near the point of intersection $g'(x)$ must be less than 1 to give a converging sequence. In Exercise C we shall deduce the condition for a decreasing function to give a converging sequence.

Exercise C

1   (a) Show that the equation $x^3 - 10x - 4 = 0$ has one solution between 3 and 4 and find intervals in which the other two solutions lie.
   (b) Show that the equation $x^3 - 10x - 4 = 0$ may be rearranged as

$$x = \frac{x^3 - 4}{10}.$$

   (c) Draw on the same axes the graphs of

$$y = x \quad \text{and} \quad y = \frac{x^3 - 4}{10}$$

using equal scales for each axis for values of x from $^{-}4$ to 4 and values of y from $^{-}7$ to 7.

(d) For which of the solutions would the iterative formula

$$x_{n+1} = \frac{x_n^3 - 4}{10}$$

give a converging sequence? Use it to find this solution, choosing a suitable value for x_1.

2 (a) Show that $x^3 - 10x - 4 = 0$ may be rearranged as

$$x = (10x + 4)^{\frac{1}{3}}.$$

(b) Draw on the same axes the graphs of

$$y = x \quad \text{and} \quad y = (10x + 4)^{\frac{1}{3}}$$

using equal scales for each axis for values of x from $^{-}7$ to 7 and values of y from $^{-}4$ to 4. (Hint: there is a very easy way to get the graph of the curve from question 1(c).)

(c) Use an iterative formula to obtain the other two solutions (besides the one found in question 1) of the equation $x^3 - 10x - 4 = 0$, making appropriate choices of x_1.

(d) Mark on your graphs the points (x_1, x_2), (x_2, x_2), (x_2, x_3), etc., and join them to form a 'staircase'.

3 (a) Find an integer approximation to the solution of the equation

$$x^3 - 2x^2 - 3x - 1 = 0.$$

(b) Show that the equation can be rearranged as

$$x = 2 + \frac{3}{x} + \frac{1}{x^2}.$$

(c) Draw on the same axes the graphs of

$$y = x \quad \text{and} \quad y = 2 + \frac{3}{x} + \frac{1}{x^2}$$

taking values of x in the interval $0 < x \leqslant 4$ and using equal scales for each axis.

(d) Taking $x_1 = 4$, use the iterative formula

$$x_{n+1} = 2 + \frac{3}{x_n} + \frac{1}{x_n^2}$$

to find a solution of the equation $x^3 - 2x^2 - 3x - 1 = 0$, giving your answer correct to 3 s.f. Record the values of x_2, x_3, etc.

(e) Mark on your graphs the points $A(x_1, x_2)$, $B(x_2, x_2)$, $C(x_2, x_3)$, $D(x_3, x_3)$, etc.

(f) Draw in the vectors $\underset{\sim}{AB}$, $\underset{\sim}{BC}$, $\underset{\sim}{CD}$, etc.

4 The equation $x^3 + 2x^2 - 3x - 1 = 0$ has a solution between 1 and 2 and can be rearranged in the form

$$x = {}^{-}2 + \frac{3}{x} + \frac{1}{x^2}.$$

Figure 3 shows the graphs of $y = x$ and $y = {}^{-}2 + \frac{3}{x} + \frac{1}{x^2}$.

(a) Explain why the iterative formula

$$x_{n+1} = {}^{-}2 + \frac{3}{x_n} + \frac{1}{x_n^2}$$

would not give a convergent sequence for any value of x_1 near 1.

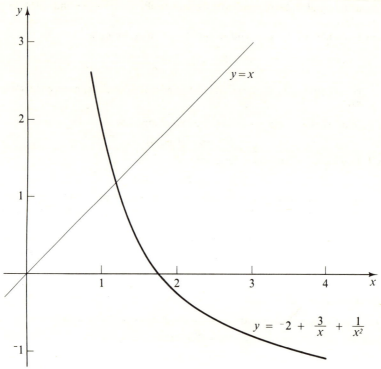

Figure 3

(b) If g is a decreasing function, when will
$$x_{n+1} = g(x_n)$$
give a converging sequence?

(c) For any function g when will
$$x_{n+1} = g(x_n)$$
give a converging sequence?

5 (a) Show that the equation $x^3 - 4x^2 + 5 = 0$ can be rewritten as
$$x = \frac{1}{4}\left(x^2 + \frac{5}{x}\right).$$

(b) By using the iterative formula, $x_{n+1} = \frac{1}{4}\left(x_n^3 + \frac{5}{x_n}\right)$ and choosing an appropriate

value for x_1, try to find the solutions of the equation $x^3 - 4x + 5 = 0$.

6 Show that the equation $x^3 - 4x^2 + 5 = 0$ can be rewritten as
$$x = 4 - \frac{5}{x^2}.$$

Hence use simple iteration to obtain a solution of the equation. Compare the rate of convergence using this method with that of question 1 of Exercise A.

7 (a) Show that a solution of the equation $x^4 - x^2 - 1 = 0$ lies between 1 and 2.

(b) Use decimal search to find this solution, accurate to 4 significant figures.

(c) Write down a further solution of this equation.

(d) By considering the equation as a quadratic in x^2, show that there are only two solutions.

(e) Investigate the iterative formula $x_{n+1} = \dfrac{1}{x_n} + \dfrac{1}{x_n^3}$ using 1 as the starting value, as a way of solving the equation.

8 (a) Show that a solution of the equation $x^3 - 4x + 2 = 0$ lies between 0 and 1.

(b) Use decimal search to find this solution, accurate to 4 significant figures.

(c) Find the other solutions of this equation, correct to 3 significant figures, by decimal search.

(d) Investigate whether any of the iterative formulae

(i) $x_{n+1} = \dfrac{x_n^3 + 2}{4}$,   (ii) $x_{n+1} = \sqrt{\left(4 + \dfrac{2}{x_n}\right)}$,   (iii) $x_{n+1} = (4x_n + 2)^{\frac{1}{3}}$,

derived from the equation will give convergent sequences and, if so, say for which of the solutions they could be used.

9 (a) Sketch graphs on the same axes of the functions

$$x \to \sin x \quad \text{and} \quad x \to 0.8x$$

for $0 \leqslant x \leqslant 1.5$.

(b) Use your graph to estimate a solution of the equation

$$5 \sin x - 4x = 0,$$

and use decimal search to obtain this solution correct to 4 significant figures.

(c) Would the iterative formula $x_{n+1} = \frac{5}{4} \sin x_n$ be suitable for finding this solution?

4. NEWTON–RAPHSON ITERATION

One of the disadvantages of the iterative procedures considered so far is that convergence, when it occurs, may be very slow. We now consider a different procedure which is often very much quicker.

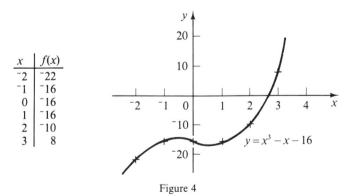

x	$f(x)$
$^-2$	$^-22$
$^-1$	$^-16$
0	$^-16$
1	$^-16$
2	$^-10$
3	8

$$y = x^3 - x - 16$$

Figure 4

Consider the equation $x^3 - x - 16 = 0$. Let $f(x) = x^3 - x - 16$. The table and the sketch graph in Figure 4 show that there is a solution between 2 and 3. The solution appears to be nearer 3 than 2. Figure 5 shows the graph in more detail between $x = 2$ and $x = 3$, together with the tangent at the point on the curve where $x = 3$.

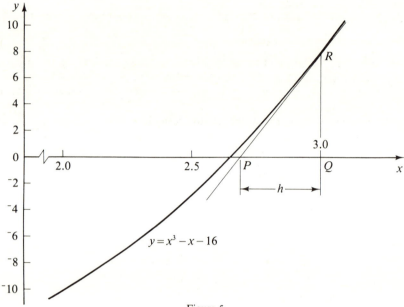

Figure 5

The point P where the tangent line intersects the x-axis will give an approximation to the solution of $f(x) = 0$. This amounts to assuming that the tangent is close to the curve when the curve crosses the x-axis. The tangent line crosses the x-axis at $3 - h$, so the problem is to calculate the value of h.

Now, since the line is tangential at $(3, 8)$, its gradient is $f'(3)$. So

$$\frac{RQ}{PQ} = f'(3).$$

But $RQ = f(3)$ and $PQ = h$,

so $\dfrac{f(3)}{h} = f'(3)$,   and $h = \dfrac{f(3)}{f'(3)}$.

$f(x) = x^3 - x - 16 \;\Rightarrow\; f'(x) = 3x^2 - 1.$ So $f(3) = 8$, $f'(3) = 26$ and $h = \frac{8}{26} \approx 0.3077$. This gives 0.3077 as the value of h, so an approximation to the solution is

$$3 - 0.3077 = 2.6923 \quad (5 \text{ s.f.}).$$

If we call the first approximation to the solution (in this case it was 3) x_1, then the second approximation, x_2, is given by

$$x_2 = x_1 - \frac{f(x_1)}{f'(x_1)}.$$

There is nothing to stop us repeating the same procedure, using the tangent at the point on the curve where $x = x_2$, and continuing in the same way until we achieve the accuracy of solution required. In general we have

$$x_{n+1} = x_n - \frac{f(x_n)}{f'(x_n)},$$

and this iterative formula bears the name of the Newton–Raphson formula.

In the example above we obtain:

$$x_1 = 3.0000000$$
$$x_2 = 2.6923076\ldots$$
$$x_3 = 2.6526394\ldots$$
$$x_4 = 2.6520105\ldots$$
$$x_5 = 2.6520103\ldots$$

At any given stage it is important to know to what accuracy the solution can be stated. We can see that x_2 and x_3 agree in their first two digits:

$$x_2 = 2.6\ldots,$$
$$x_3 = 2.6\ldots.$$

This means that we cannot give the solution to more than *one* significant figure at this stage.

If, however, we proceed as far as x_4, we can see that x_3 and x_4 agree in their first four digits:

$$x_3 = 2.652\ldots,$$
$$x_4 = 2.652\ldots,$$

so we can now give the solution to 3 s.f. as 2.65.

Looking now at x_4 and x_5 above we can see agreement in the first seven digits, and thus could state the solution accurate to 6 s.f. as 2.65201.

This shows a most important feature of the Newton–Raphson method; the rate of convergence increases as more terms are calculated. The methods previously used generally gave one extra agreeing digit per iteration. With this method the number of extra agreeing digits actually increases with each iteration.

It is interesting to represent the successive iterations graphically (Figure 6).

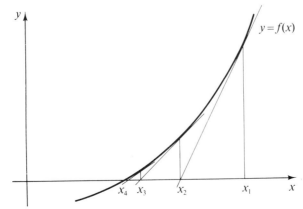

Figure 6

The rapid convergence is almost evident from the pattern formed by the tangent lines.

Example 2
Solve the equation $e^x = 2x + 1$.

An obvious solution is $x = 0$, but do others exist? We must first of all arrange the equation into the form $f(x) = 0$, giving

$$e^x - 2x - 1 = 0.$$

Now $f(1) \approx {}^-0.3$ and $f(2) \approx 2.3$, so another solution lies between 1 and 2 and is probably closer to 1 than to 2.

Differentiating $f(x) = e^x - 2x - 1$, we have

$$f'(x) = e^x - 2.$$

Starting with $x_1 = 1$, we obtain:

$x_1 = 1$ $\qquad\qquad\qquad\qquad f(x_1) = {}^-0.281\,718\,17$ $f'(x_1) = 0.718\,281\,83$

$x_2 = 1 - \dfrac{{}^-0.281\,718\,17}{0.718\,281\,83}$

$\quad = 1.392\,2112$ $\qquad\qquad\quad f(x_2) = 0.239\,315\,09$ $f'(x_2) = 2.023\,7375$

$x_3 = 1.392\,2112 - \dfrac{0.239\,315\,09}{2.023\,7375}$

$\quad = 1.273\,9572$ $\qquad\qquad\quad f(x_3) = 0.027\,057\,03$ $f'(x_3) = 1.574\,9714$

$x_4 = 1.273\,9572 - \dfrac{0.027\,057\,03}{1.574\,9714}$

$\quad = 1.256\,7778$ $\qquad\qquad\quad f(x_4) = 0.000\,524\,54$ $f'(x_4) = 1.514\,0801$

$x_5 = 1.256\,7778 - \dfrac{0.000\,524\,54}{1.514\,0801}$

$\quad = 1.256\,4313$ $\qquad\qquad\quad f(x_5) = 0.000\,000\,14$ $f'(x_5) = 1.512\,8627$

$x_6 = 1.256\,4313 - \dfrac{0.000\,000\,14}{1.512\,8627} = 1.256\,4312.$

So a solution is 1.256 43 to 6 s.f.

A sketch of the graphs of $y = e^x$ and $y = 2x + 1$ (see Figure 7) shows that there are just two points of intersection, so 0 and 1.256 43 are the only solutions.

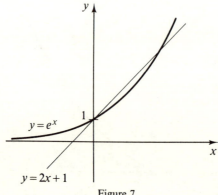

Figure 7

Exercise D

1 (*a*) Show that the positive solution of the equation
$$x^4 - 7 = 0$$
lies between 1 and 2.

(*b*) Write down the Newton–Raphson formula for the solution of this equation and use it to obtain a solution of the equation correct to 3 s.f.

2 (*a*) Draw graphs, on the same axes, of the functions
$$x \to \sin x \quad \text{and} \quad x \to \tfrac{1}{4}x.$$
From your graphs read off one solution of the equation $\sin x = \tfrac{1}{4}x$ between 0 and π.

(*b*) Write down the Newton–Raphson formula for the solution of the equation
$$4 \sin x - x = 0.$$

(*c*) Use the formula to solve the equation to 4 s.f.

3 Use the Newton–Raphson method to obtain a solution of $2x^3 + 7x + 4 = 0$ correct to 3 s.f.

4 Repeat question 3 for the equation
$$x^5 + x^3 = 9.$$

5 (*a*) Two of the solutions of the equation $2^x = x^2$ are easily found. By sketching graphs of $x \to 2^x$ and $x \to x^2$ for the interval $[-2, 1]$, estimate the third solution.

(*b*) Given that, for $f(x) = 2^x$, $f'(x) \approx 0.6931 \times 2^x$, find the third solution correct to 3 s.f.

6 (*a*) The positive solution of $x^2 - a = 0$ is $x = \sqrt{a}$. Show that the Newton–Raphson formula for the equation $x^2 - a = 0$ can be written
$$x_{n+1} = \frac{1}{2}\left\{x_n + \frac{a}{x_n}\right\}.$$

(*b*) Use this formula to obtain to 4 s.f. values for (i) $\sqrt{3}$, (ii) $\sqrt{7}$.

(*c*) Show that the Newton–Raphson formula for the equation $x^3 - a = 0$ may be written as
$$x_{n+1} = \frac{1}{3}\left\{2x_n + \frac{a}{x_n^2}\right\}.$$

(*d*) Use this formula to obtain to 4 s.f. values for (i) $\sqrt[3]{2}$, (ii) $\sqrt[3]{19}$.

(*e*) Deduce the Newton–Raphson formula for the solution of $x^m - a = 0$.

7 Two tanks simultaneously start to leak. Tank A contains $\alpha(t)$ litres and tank B contains $\beta(t)$ litres of water, where
$$\alpha(t) = 100e^{-t/10}, \quad 0 \leqslant t,$$
$$\beta(t) = 100 - 4t, \quad 0 \leqslant t \leqslant 25,$$
and t is the number of minutes for which they have been leaking.

(*a*) How many litres does each tank contain initially?

(*b*) Show that, after about $\tfrac{3}{4}$ hour, tank A will be virtually empty.

(*c*) Write down an expression which is satisfied by the values of t when the tanks contain equal volumes of water. Solve the equation by the Newton–Raphson method.

8 A dock-gate is open to a tidal river on one side. The depth of water t hours after midnight is given in metres by
$$f(t) = 20 + 10 \sin\left(\tfrac{1}{6}\pi t\right).$$

On the other side the water is supplied by a river, so that the depth t hours after midnight is given in metres by

$$h(t) = 5 + 6t,$$

up to the moment when the gates are opened.

The gates are opened when the depth of water on each side is the same. Write down an equation to express this, and use the Newton–Raphson method to find when the gates are opened.

9   Figure 8 shows three diagrams similar to Figure 6.
   (i) Copy each of the graphs (a), (b), (c). Draw in the tangent line at $(x_1, f(x_1))$ and thus show the position of x_2.
   (ii) Taking into account the signs of $f(x_1)$ and $f'(x_1)$, verify that the formula

$$x_2 = x_1 - \frac{f(x_1)}{f'(x_1)}$$

gives x_2 correctly in each case.

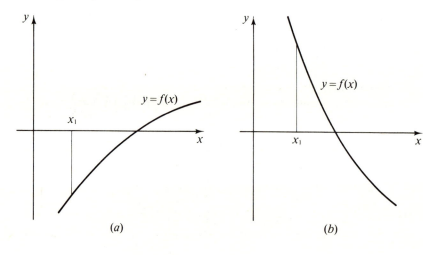

(a)   (b)

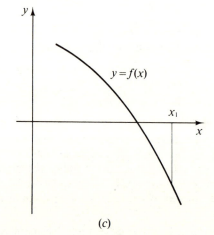

(c)

Figure 8

5. LINEAR APPROXIMATION

The Newton–Raphson method depends on using the tangent to a curve as an approximation to that curve. This idea is of more general application.

Figure 9 shows the graph of a function f together with its tangent at the point where $x = a$. If h is small, $f(a+h)$ is approximately the same as the y-coordinate of R, the point on the tangent where $x = a+h$. Since the gradient of PR is $f'(a)$ and $PQ = h$, $QR = hf'(a)$, and hence the y-coordinate of R is $f(a)+hf'(a)$. So we have the approximation

$$f(a+h) \approx f(a)+hf'(a).$$

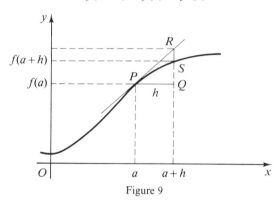

Figure 9

Example 3
Use a linear approximation, together with the fact that $\sin 45° = \cos 45° = \frac{1}{2}\sqrt{2}$, to find the approximate value of $\sin 46°$.

We use $f(a+h) \approx f(a)+hf'(a)$ with

$$f(x) = \sin x, \quad a = \tfrac{1}{4}\pi \quad \text{and} \quad h = \tfrac{1}{180}\pi \quad \text{(since } 1° = \tfrac{1}{180}\pi \text{ radians)}.$$
$$f'(x) = \cos x, \quad \text{so} \quad f'(\tfrac{1}{4}\pi) = \tfrac{1}{2}\sqrt{2}.$$
$$\sin 46° = \sin(\tfrac{1}{4}\pi + \tfrac{1}{180}\pi) \approx \sin \tfrac{1}{4}\pi + \tfrac{1}{180}\pi \cos \tfrac{1}{4}\pi$$
$$= \tfrac{1}{2}\sqrt{2} + \tfrac{1}{180}\pi \tfrac{1}{2}\sqrt{2} = 0.719 \text{ to 3 s.f.}$$

The linear approximation result can be written in the form

$$f(a+h)-f(a) \approx hf'(a).$$

The left side of this equation is the change in $f(x)$ due to a change of h in x. We could therefore write the result as

$$\delta y \approx f'(a)\,\delta x$$

or

$$\delta y \approx \frac{dy}{dx}\delta x. \quad \text{(See Figure 10.)}$$

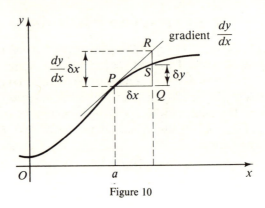

Figure 10

Example 4

The radius of a spherical balloon is increased from 11.4 cm to 11.5 cm. Calculate the approximate increase in volume.

We have
$$V = \tfrac{4}{3}\pi r^3$$

so
$$\frac{dV}{dr} = 4\pi r^2$$

and
$$\delta V \approx 4\pi r^2 \delta r$$
$$= 4\pi(11.4)^2 \, 0.1$$
$$= 163.3 \text{ to 4 s.f.}$$

So the increase in volume is about 160 cm³.

Example 5

Cast iron bath taps are to be gold plated. The casting process is followed by a grinding and smoothing process, in the course of which some metal is removed, so that the volume of the finished taps varies slightly. Calculate the percentage change in the gold required for plating due to a 1% decrease in the volume of a tap.

If we assume that the taps remain geometrically similar, then, if x is any dimension,
$$V \propto x^3 \quad \text{and} \quad A \propto x^2,$$

where V and A are the volume and surface area respectively. The amount of gold required is proportional to A. We can therefore write
$$V = k_1 x^3 \quad \text{and} \quad G = k_2 x^2$$

where k_1 and k_2 are (unknown) numbers and G is the amount of gold required.

Hence
$$\frac{dV}{dx} = 3k_1 x^2, \qquad \frac{dG}{dx} = 2k_2 x,$$

and
$$\delta V \approx 3k_1 x^2 \delta x, \qquad \delta G = 2k_2 x \delta x$$

The percentage change in $V = 100\dfrac{\delta V}{V} = \dfrac{300k_1 x^2 \delta x}{k_1 x^3} = 300\dfrac{\delta x}{x}$.

The percentage change in $G = 100\dfrac{\delta G}{G} = \dfrac{200k_2 x\delta x}{k_2 x^2} = 200\dfrac{\delta x}{x}$.

So a 1% decrease in $V \Rightarrow 300\dfrac{\delta x}{x} = {}^-1$

$\Rightarrow \dfrac{\delta x}{x} = \dfrac{{}^-1}{300}$

$\Rightarrow$ percentage change in $G = 200\dfrac{\delta x}{x} = \dfrac{{}^-200}{300} = \dfrac{{}^-2}{3}$.

About $\tfrac{2}{3}\%$ less gold is required.

Exercise E

1 Find a linear approximation to $f(x) = x^{\frac{1}{3}}$ around $x = 8$. Use your approximation to find an estimate to $8.03^{\frac{1}{3}}$. To how many figures does this estimate agree with the value given by your calculator?

$\left(\text{You may assume that } f'(x) = \dfrac{1}{3\sqrt[3]{x^2}}. \right)$

2 (a) Differentiate the function given by $f(x) = \sin x°$.
   (b) Show that $f'(30) = 0.015$, correct to 2 s.f.
   (c) Hence show that $f(30 + h) \approx 0.5 + 0.015h$.
   (d) Hence construct a table of values of $\sin x°$ at intervals of 0.1 from $x = 29$ to $x = 31$.
   (e) Compare this table with a published table and comment on the result.

3 (a) Find the equation of the tangent line to $y = \dfrac{1}{x}$ when $x = 2$.
   (b) Use the equation of the tangent line to obtain an estimate for 2.07^{-1}.

4 For the function $f : x \to x^n$ $(n \neq 0)$, write down $f'(x)$ and hence find $f'(1)$. Hence show that, for small values of x, $(1 + x)^n \approx 1 + nx$.

5 Find a linear approximation to $f(x) = x^{\frac{1}{4}}$ around $x = 81$. Use your approximation to find an estimate to $81.8^{\frac{1}{4}}$. To how many figures does this estimate agree with the value given by your calculator?
   (You may assume that $f'(x) = \tfrac{1}{4}x^{-\frac{3}{4}}$.)

6 You are given that, for the function $f : x \to \tan x°$,

$$f'(x) = \dfrac{\pi}{180(\cos x°)^2}.$$

   (a) Show that $f'(0) = 0.01745$ (4 s.f.).
   (b) Find the function $g(x)$ which is a linear approximation to $f(x)$ around $x = 0$.
   (c) Draw on the same axes the graphs of $f(x)$ and $g(x)$ for $^-30 \leqslant x \leqslant 30$.

7 Find the equation of the tangent line to the graph of the function $f : x \to x^{-2}$ at the point where $x = 1$.
   Hence calculate an estimate for $f(1.1)$.
   (You may assume that $f'(x) = {}^-2x^{-3}$.)

8   (a) Differentiate the function given by $f(x) = e^x$.

(b) Evaluate $f'(0)$.

(c) Show that, near $x = 0$, $e^x \approx 1 + x$. (Hint: $x = 0 + h$.)

(d) What is the largest value of x for which this approximation gives a value which agrees with your tables?

---

9   The radiation from a body, R, at a temperature T degrees is given by

$$R = 32.2T^4.$$

Calculate the increase in the radiation when T increases from 273 to 273.5.

10   Calculate the change in the area of a circle when its radius is increased from 7.14 cm to 7.15 cm.

11   The time of swing of a pendulum is T seconds, where $T = 2\pi \left(\dfrac{l}{g}\right)^{\frac{1}{2}}$; l cm is the length of the pendulum and $g = 9.81$. Calculate the change in the time of swing when the length is decreased from 1 metre to 99 cm.

12   The number of grams, S, of potassium chlorate that can be dissolved in 100 g of water at $T\,°C$ is given by

$$S = 3 + 0.1T + 0.0044T^2.$$

Calculate the increase in S when T increases from 10 to 11.

13   For a lens the distance v cm of the image of an object which is at a distance u cm (all distances measured from the lens) is given by

$$\frac{1}{v} + \frac{1}{u} = 0.05.$$

Find the change in v when u is decreased from 50 to 49.

14   For a given quantity of gas at constant temperature, the volume, v cm³ is related to the pressure, p pascal, by the formula $pv = 144$. Calculate the change in pressure required to reduce the volume by 0.05 cm³ when it is (a) 12 cm³, (b) 6 cm³.

15   The magnetic attraction between two magnets is given by $\dfrac{k}{r^2}$, where r is the distance between the poles. Measurements of r are subject to a 1% error. What is the corresponding variation in the force of attraction?

6. SUMMARY

Decimal search

To solve an equation $f(x) = 0$ we find the numbers l and u $(l < u)$, where $f(l)$ and $f(u)$ are of opposite sign. We then choose a number α in the interval $[l, u]$ and work out $f(\alpha)$. α then replaces either l or u according to whether $f(\alpha)$ has the same sign as $f(l)$ or $f(u)$ respectively.

(It is assumed that the graph of f is continuous for the interval $[l, u]$.)

Iterative methods

Given an equation of the form

$$f(x) = 0,$$

we rearrange it into the form
$$x = g(x),$$
and find an approximate solution, α. Then the iterative formula
$$x_{n+1} = g(x_n)$$
is used to try to obtain a sequence which converges onto the solution, starting with $x_1 = \alpha$.

Convergence and divergence

The gradient of g near the solution determines whether the iterative formula converges. It converges if $^-1 < g'(x) < 1$ for values of x near the solution.

Newton–Raphson iteration

If α is an approximate solution of the equation $f(x) = 0$, then
$$\alpha - \frac{f(\alpha)}{f'(\alpha)}$$
will, in general, be a better one. The iterative formula
$$x_{n+1} = x_n - \frac{f(x_n)}{f'(x_n)}$$
gives a sequence which converges onto the solution of $f(x) = 0$, taking $x_1 = \alpha$.

Linear approximation

The function f may be approximated near $x = a$ by the function g where
$$g(a+h) = f(a) + hf'(a).$$
The graph of g is the tangent line to the curve $y = f(x)$ at the point $(a, f(a))$. It is convenient to write $f(a+h) \approx f(a) + hf'(a)$.

The formula $\delta y \approx \dfrac{dy}{dx} \times \delta x$ is useful for finding the change in y corresponding to a small change in x.

Miscellaneous exercise

1 $f(x) = x^5$ and $g(x) = f'(x) = 5x^4$.
　　Write down the value of $f(2)$ and of $g(2)$. Hence use linear approximation to show that the value of $f(2.06)$ is about 36.8.
　　Draw a sketch of $f(x)$ and on it indicate clearly your three answers and their inter-relation.
　　A better estimate is given by $f(2.06) = f(2) + 0.06g(2.03)$. Explain, with the aid of your sketch, why this gives a better estimate.
　　Use linear approximation to show that $g(2.03) \approx 84.8$. Hence estimate the value of $(2.06)^5$, giving your answer corrected to 4 significant figures.　　　　　(SMP)

2 $f(x) = x^3 - 2x^2 - 3x - 4$.
　　(i) Calculate the values of $f(x)$ for $x = 1, 2, 3, 4$ and hence estimate a solution of the equation $f(x) = 0$ correct to the nearest integer.

(ii) Show that, by dividing both sides of the equation by $2x$ and rearranging the terms, the equation $f(x) = 0$ can be rewritten as

$$x = \frac{x^2}{2} - \frac{3}{2} - \frac{2}{x}.$$

This yields an iterative formula

$$x_{n+1} = \frac{x_n^2}{2} - \frac{3}{2} - \frac{2}{x_n}$$

Starting with $x_1 = 1$, calculate x_2 and x_3.

(iii) Similarly, by dividing both sides of the equation by x^2 and rearranging the terms, find another iterative formula of the form

$$x_{n+1} = 2 + \ldots + \ldots$$

Again starting with $x_1 = 1$, calculate the next two iterations.

(iv) Again, starting with the approximation $x_1 = 1$, use the Newton–Raphson method to calculate the next two approximations to a solution of $f(x) = 0$.

(v) Which of these three methods would have given the correct solution ultimately?   (SMP)

3 Draw the graph of $f : x \rightarrow 4^x$, plotting integer values of x for $^-1 \leqslant x \leqslant 2$.

The derivative of 4^x is $k \cdot 4^x$, where k is a constant.

Use linear approximation to explain why $f(a+2) \approx (1+2k) \cdot f(a)$.

Hence, by taking a suitable value for a, estimate the value of k.

Use your value of k and a similar linear approximation to estimate $f(1.2)$.

By considering the shape of your graph, state whether $f(a+2)$ is, in fact, greater than or less than $(1+2k) \cdot f(a)$, and hence deduce whether your estimate of k is too great or too small.   (SMP)

4 Show that $x^2 - 4x + 3 = 0 \iff x = 4 - 3/x$.

Draw, on the same axes with $^-3$ to 10 for domain and range, the graphs of $y = 4 - 3/x$ and of $y = x$.

State the quadratic equation whose roots are the points of intersection of the graphs.

Figure 11 shows a flow-chart for finding a root of this equation.

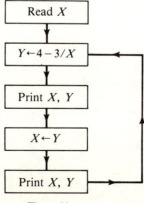

Figure 11

Take $X = 0.5$ and write down the first five pairs of values printed out. Treating each pair of values as the coordinates of a point, plot these five points on your graph, joining each to the next by a straight line. (Note that the effect is to move successively horizontally from curve to line, then vertically from line to curve.) Without further calculation show on your graph the next three points which would be given by this process, labelling them A, B, C.

Write down, correct to 2 significant figures, the hundredth pair of coordinates that would be printed. (SMP)

5 (a) Given that $f(x) = \sin x°$, express $x°$ in radians and find $f'(x)$.

(b) Show that the gradient of the graph of $f(x) = \sin x°$ at $x = 30$ is $\frac{1}{360}\pi\sqrt{3}$.

(c) A ladder of length 4 units is leaning against a vertical wall. The ladder makes an angle of $\theta°$ with the horizontal and reaches a distance h units up the wall (Figure 12). Express h in terms of $\theta°$ and evaluate h when $\theta = 30$.

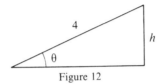

Figure 12

(d) If θ is now changed to 33, use your answers to (b) and (c) to find an approximation for the new value of h, leaving your answer in terms of π. (SMP)

6 An iterative relation to calculate the reciprocal of any given number A, without division, is shown in the flow diagram (Figure 13).

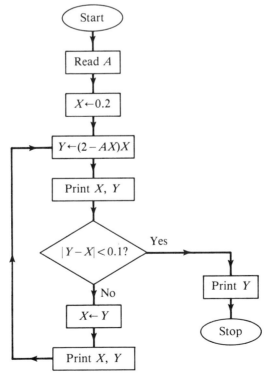

Figure 13

Throughout this question take A to be 2, and give all answers to one place of decimals.

Write down the seven pairs of values of X and Y that would be printed out.

On the same axes, plot the parabola $y = 2x - 2x^2$ and the line $y = x$, for $x = -0.5$, $0, 0.5, 1, 1.5$, with a scale of 8 cm to one unit. Show clearly on your graph each of the seven points (X, Y) that would be printed out from the flow diagram and label them $P_1, Q_1, P_2, Q_2, P_3, Q_3, P_4$. Verify that those with label P lie on the parabola and those with label Q lie on the line.

State corresponding coordinates for P_1, Q_1, P_2 if, instead of $X \leftarrow 0.2$, we start with $X \leftarrow 1.2$ and show them on your graph. Hence state the range of starting values of X for which the iteration does not give the reciprocal of A.   (SMP)

7   $f(x) = (7.4)^x - 0.089$.

  (i) A first approximation to the root of $f(x) = 0$ is -1.

    (a) Calculate $f(-1)$ to 2 significant figures.

    (b) Given that the derivative of $(7.4)^x$ may be taken to be $2 \times (7.4)^x$, show that $f'(-1) = 0.27$ approximately.

    (c) Show that the Newton–Raphson process gives -1.2 as the next approximation.

    (d) Use logarithms to calculate $f(-1.2)$, showing your working clearly.

  (ii) The flow diagram in Figure 14 is for calculating the root more accurately.

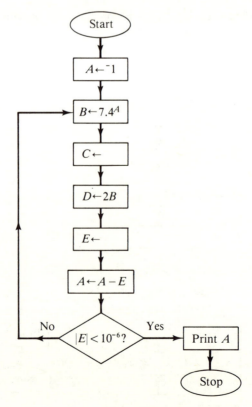

Figure 14

(a) State to how many significant figure accuracy the root would be calculated.

(b) What should be the two incomplete instructions? (SMP)

8 The flow chart in Figure 15 is designed to print out the coordinates of points on a graph of a polynomial function, $y = p(x)$.

Write down (a) the function p, and (b) the values of x at which the function is evaluated.

Modify the flow chart so as to evaluate the function at intervals of 0.5 from $x = 0$ to $x = 4$. (SMP)

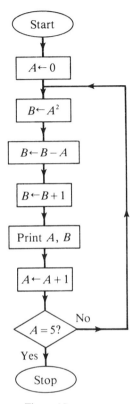

Figure 15

9 The curve $y = 4 - x^2$ cuts the axes at B $(0, 4)$ and C $(2, 0)$. Calculate the area in the first quadrant bounded by the curve and the axes. Find the equation of the tangent to the curve at C and show that the coordinates of the point A in which it meets the other axis are $(0, 8)$. Hence show that the area in the first quadrant bounded by the curve and the axes is $\frac{2}{3}$ of that bounded by this tangent and the axes.

(SMP)

10 (a) Given that $f'(x) = \dfrac{1}{1+x}$, write down the values of $f'(1)$ and $f'(2)$.

(b) If you are also given that $f(1) = 3$, show, by linear approximation, that $f(2) \approx 3.5$.

(c) Use a trapezium to estimate the area under the graph of $f'(x)$ from $x = 1$ to $x = 2$. Hence give a second estimate of $f(2)$.

11 (a) Calculate the y-coordinate of the point where $x = 16$ on the curve with equation $y = x^{\frac{1}{4}}$.

(b) Show that the gradient of the curve at this point is approximately 0.031.

(c) Hence write down an estimate for $\sqrt[4]{19}$. Illustrate your method on a sketch graph.

12 Sketch a graph of $x^4 - 6$ for $^-2 \leqslant x \leqslant 3$.

If α is a first approximation for $\sqrt[4]{6}$, use the Newton–Raphson formula to show that, in general, a second and better approximation is

$$\frac{3}{4}\left(1 + \frac{2}{\alpha^4}\right)\alpha.$$

A flow diagram for this purpose is shown in Figure 16. Explain the meanings of ↑ and *.

What first approximation for α is being taken in this flow diagram? Calculate the second approximation in this case (i.e. the first number to be printed).

Calculate the second approximations yielded by the flow diagram when the first approximations taken for α are (a) $\frac{1}{2}$, (b) 2.

From your graph and results estimate, to the nearest integer, the range of values for which the second approximation will be *worse* than the first approximation.

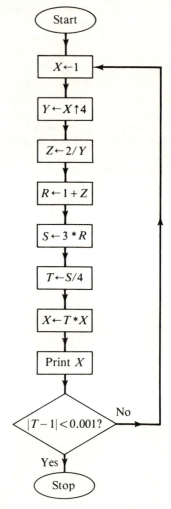

Figure 16

Revision exercise 7

1 Differentiate the following functions:

    (a) $f(x) = (2x - 5)^3$;   (b) $f(x) = \left(\dfrac{x+5}{2}\right)^6$;   (c) $f(x) = (7 - 3x)^5$;

    (d) $f(x) = (x^2 + 1)^3$;   (e) $f(x) = (5x^3 - 2)^4$;   (f) $f(x) = (x^3 + x^2 + 1)$.

2 In Groatia the rate of inflation is 9% per year. After how many years will the value of money have halved? (When the value of money is halved you need twice as much as you did before.)

3 A government department wishes to reduce its spending to one-quarter of its current level over a five-year period. What annual rate of reduction is necessary?

4 Find the gradient and equation of the tangent to the curve $y = x^5$ at the point where $x = 1$.

    Making use of this as a linear approximation, estimate the values of (a) 1.04^5 and (b) $\sqrt[5]{(1.15)}$.   (SMP)

5 For each derivative given find a possible function $f(x)$.

    (a) $f'(x) = 6(2x + 7)^2$;   (b) $f'(x) = 15(3x - 1)^4$;   (c) $f'(x) = 4(7x - 2)^3$;

    (d) $f'(x) = 2x(1 + x^2)^3$;   (e) $f'(x) = x^2(x^3 + 4)^3$;   (f) $f'(x) = e^x(e^x + 2)^5$.

6 A slice in the form of a sector of angle α radians is taken from a circular cake of uniform thickness (Figure 1). This slice is cut in two by a straight cut so that there is a segment (shown shaded in the diagram) containing the icing on the outside and a triangular piece. If these pieces have the same volume show that α must satisfy the equation $\sin \alpha - \frac{1}{2}\alpha = 0$.

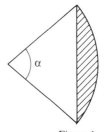

Figure 1

    Starting with $\alpha = \frac{1}{2}\pi$ as an approximate solution apply the Newton–Raphson method once to get a better approximation. Give your answer to the nearest degree.   (SMP)

7 Use the Newton–Raphson method to solve the equations:

    (a) $\cos x = x^2$;   (b) $e^x = \dfrac{6 + x}{5}$.

8 A population is increasing by 18% each year.

    (a) After n years, by what factor will the population have been multiplied?

    (b) Find how long it takes for the population to be doubled.

    (c) How long ago (assuming that the law of growth has remained the same) was the population a quarter of its present size?   (SMP)

9   Differentiate the following functions:

    (a) $f(x) = \sin(2x + \frac{1}{3}\pi)$;   (b) $f(x) = \cos(\frac{1}{6}\pi - 3x)$;   (c) $f(x) = (\sin x)^2$;

    (d) $f(x) = \cos x^2$;   (e) $f(x) = \dfrac{1}{(3x-1)^2}$;   (f) $f(x) = \dfrac{1}{\sin x}$;

    (g) $f(x) = \dfrac{1}{(2-5x)^3}$;   (h) $f(x) = \dfrac{1}{(\cos x)^2}$.

10   On squared paper, taking a scale of 1 cm to one unit on both axes, draw the graph of $y = 5\cos x$ for x in the interval $-\pi \leqslant x \leqslant 2\pi$.

    On the same diagram draw a suitable line and hence read off, as accurately as you can, the solutions to the equation

$$x + 5\cos x = 0.$$

    Use Newton–Raphson iteration to improve your answers to 3 significant figure accuracy.

11   The proportion of radioactive carbon-14 in dead wood is halved after 6000 years. If there are $M(0)$ kg of carbon-14 in a sample of wood at the end of its life, show that there will be $M(t)$ kg after t years, where

$$M(t) = M(0)\,(\tfrac{1}{2})^{t/6000}.$$

    A sample is found to have 75% of the activity of a similar living piece. How long ago did it die?

12   Find: (a) $\displaystyle\int \frac{1}{(5x-7)^2}\,dx$;   (b) $\int x(x^2+4)^4\,dx$;   (c) $\int x^2(5-x^3)^6\,dx$;

    (d) $\int x\sqrt{(x^2+1)}\,dx$;   (e) $\int x\sin x^2\,dx$;   (f) $\displaystyle\int \frac{\cos\sqrt{x}}{\sqrt{x}}\,dx$;

    (g) $\displaystyle\int \frac{3x^2+1}{(x^3+x+9)^3}\,dx$;   (h) $\int (\sin x)^4\cos x\,dx$.

13   The BBC broadcasts Radio 1 at frequencies of 1090 kHz and 1050 kHz (corresponding to wavelengths of 275 m and 285 m); Radio 2 on 690 kHz and 910 kHz (434 m and 330 m), and Radio 3 on 1210 kHz (247 m). The corresponding values of the capacitance in a radio tuning circuit are given in the table below:

f (frequency in kHz)	690	910	1050	1090	1210
C (capacitance in picofarad)	42.0	24.2	18.1	16.8	13.7

    Draw a graph of $\log f$ against $\log C$ and deduce the equation connecting f and C.

14   (a) (i) Differentiate, with respect to x, the function

$$f : x \to (x^2-5)^3.$$

    (ii) Make use of the chain rule to show that the derived function of

$$g : x \to \frac{1}{\cos^2 x}$$

    is $g' : x \to k\sin x\,(\cos x)^{-3}$, and state the value of k.

    (b) Using the results in (a), or otherwise, find

        (i) $\int x(x^2-5)^2\,dx$,

        (ii) $\displaystyle\int_0^{\pi/3} \frac{\tan x}{\cos^2 x}\,dx$.   (SMP)

15 Functions f and g are defined by the mappings

$$f : x \to x^2, \quad g : x \to \sin x.$$

(i) Give as mappings the composite functions

(a) fg and (b) gf.

(ii) If $h(x) = fg(x)$, find $h'(\frac{1}{4}\pi)$.

(SMP)

16 (a) Draw on the same axes the graphs of

$$y = x \quad \text{and} \quad y = \frac{1}{5x^2} + 3$$

for values of x from 1 to 10. Read off an integer approximation to the x-coordinate of their point of intersection.

(b) Show that the equation $5x^3 - 15x^2 - 1 = 0$ may be rewritten in the form

$$x = \frac{1}{5x^2} + 3$$

and state a suitable iterative formula for finding the positive solution of the equation. Explain whether you would expect convergence or divergence from this formula starting with a suitable first term and why.

(c) Find the solution of $5x^3 - 15x^2 - 1 = 0$ correct to 4 s.f.

17 The amount of light reaching a film when a photograph is taken is determined by the shutter speed, s, and the aperture, f.

To take a particular view a photographer may use any of the following combinations:

s	$\frac{1}{60}$	$\frac{1}{125}$	$\frac{1}{250}$	$\frac{1}{500}$	$\frac{1}{1000}$
f	22	16	11	8	5.6

Draw a graph of $\log s$ against $\log f$ and find an equation connecting s and f.

18 (a) If $f(\theta) = \sin^2 \theta$, calculate the values of θ in the interval $0 \leqslant \theta \leqslant \frac{1}{2}\pi$ for which $f(\theta) = f'(\theta)$.

(b) Sketch the graphs of $f(\theta)$ and $f'(\theta)$ against θ, using the same axes for both. Indicate clearly points on the graph that enable you to check your answers to part (a).

19 In one year I expect the value of my car to fall by 16%. If however, as a result of inflation, car prices rise by 10%, show that the cash value of my car will have fallen by about $7\frac{1}{2}\%$ during the year.

If the cash value of the car continues to fall each year by about $7\frac{1}{2}\%$ of its value at the beginning of the year, the percentage fall in three consecutive years is not $22\frac{1}{2}\%$ (i.e. 3 times $7\frac{1}{2}\%$). Give the reason for this and say whether the percentage fall is greater than or less than $22\frac{1}{2}\%$.

Calculate the constant annual percentage fall which would result in a fall of $22\frac{1}{2}\%$ over three years.

(SMP)

20 (i) Differentiate:

(a) $f(x) = e^{10x}$; (b) $f(x) = e^{(x-1)/2}$; (c) $f(x) = 2^x$.

(ii) Find:

(a) $\int e^{3x} \, dx$; (b) $\int e^{7-5x} \, dx$; (c) $\int 10^x \, dx$.

21 A supply ship, floating on a heavy swell, is moored beside a stationary oil rig. A crate is being unloaded from the oil rig by a crane 250 m above the sea-bed onto the boat vertically below. The deck of the boat is $200 + 25 \sin\left(\frac{1}{4}\pi t\right)$ metres above the sea-bed t seconds after docking. The crate is winched down at a speed of 2 m s^{-1} starting when the boat moors.

Write down a function which gives the height of the crate above the sea-bed after t seconds.

Hence show that when the crate hits the deck t satisfies the equation

$$50 - 2t = 25 \sin\left(\tfrac{1}{4}\pi t\right).$$

Use the Newton–Raphson method to obtain a solution to this equation accurate to 4 s.f.

Answers

Chapter 1   Kinematics and differentiation

Exercise B

2   0.2 m, 0.6 m, 1.0 m, 1.4 m, 1.8 m.

Exercise C

1 $h = 10^{-4}$: average speed $= 30.0005$ m s^{-1}.

2 $h = 10^{-4}$: average speed $= 29.9995$ m s^{-1}.

3 $h = 10^{-4}$: average speed $= 30.0003$ m s^{-1}.

4 $h = 10^{-4}$: average speed $= 5.9997$ m s^{-1}.

Exercise E

1 $s' : t \to 3t^2$.   **2** $s' : t \to 4t^3$.   **3** $s' : t \to 2t$.

4   (*b*) $s' : t \to nt^{n-1}$.

5   (*a*)   5.01, 80.08, 405.27, 1280.64 (to 2 d.p.).

6   (*a*)   6.02, 192.24, 1459.22, 6147.84 (to 2 d.p.).

Exercise F

1   (*a*) $t \to 12t^{11}$;   (*b*) $t \to 72t^8$;   (*c*) $t \to 80t^3$;   (*d*) $t \to 84t^5$;   (*e*) $t \to 0$.

2   (*a*) $x \to 2x + 5x^4$;   (*b*) $x \to 8x^7 + 7x^6$;   (*c*) $x \to 70x^{69} + 100x^{99}$;   (*d*) $x \to 46x^{45}$.

3   (*a*) $t \to 55t^{10} + 18t^5$;   (*b*) $t \to 36t^3 - 24t^7$;   (*c*) $t \to 120t^{14} + 6t^{11}$;   (*d*) $t \to 120t^2$.

4   (*a*) $h' : t \to 20 - 10t$;   (*b*)   20, 10, 0, 10, 20 m s^{-1}.

5   (*a*) $t \to 13t^{12}$;   (*b*) $t \to {}^{-}75t^4$;   (*c*) $t \to 72t^7$;   (*d*) $t \to 8t^9$;   (*e*) $t \to 0$.

6   (*a*) $x \to 4x^3 + 3x^2$;   (*b*) $x \to 10x^9 + 6x^5$;   (*c*) $x \to 65x^{64} + 40x^{39}$;   (*d*) $x \to 30x^{29}$.

7   (*a*) $t \to 60t^{14} + 6t^2$;   (*b*) $t \to 35t^4 - 44t^{10}$;   (*c*) $t \to 6t^{17} + 28t^6$;   (*d*) $t \to 100t^3$.

8   (*a*) $0.6t^2$;   (*b*)   0.6, 2.4, 5.4 m s^{-1}.

9   150 s.

10   20 m s^{-1};   after 3 s;   2 m s^{-1}.

11   (*a*)   450 km, 6000 m s^{-1};   (*b*)   25 s.

Chapter 2   The quadratic function

Exercise B

1   $x^2 + 2x + 1$.

2   $x^2 - 6x + 9$.

3   $y^2 + 8y + 16$.

4   $z^2 - 12z + 36$.

5   $t^2 + 3t + 2\frac{1}{4}$.

6   $k^2 - 5k + 6\frac{1}{4}$.

7   $x^2 - 2x + 1$.

8   $x^2 + 4x + 4$.

9   $y^2 - 8y + 16$.

10   $z^2 + 10z + 25$.

11   $t^2 - t + \frac{1}{4}$.

12   $x^2 - 7x + 12\frac{1}{4}$.

13   $(x + 7)^2$.

14   $(y - 2)^2$.

15   $(z + 1)^2$.

16   $(x + 2\frac{1}{2})^2$.

17   $(t - 5\frac{1}{2})^2$.

18   $(k + \frac{1}{2})^2$.

19   $(x - 4)^2$.

20   $(y + 9)^2$.

21   $(z + 6)^2$.

22   $(x + 1\frac{1}{2})^2$.

23   $(t - 2\frac{1}{2})^2$.

24   $(k + 3\frac{1}{2})^2$.

Exercise C

1   $(x + 7)^2 - 9$.

2   $(y - 4)^2 - 14$.

3   $(z + 1)^2 - 4$.

4   $(t + 3)^2 - 11$.

5   $(k - 3)^2 - 11$.

6   $(x - 2)^2 + 6$.

7   $(x + 2\frac{1}{2})^2 - \frac{1}{4}$.

8   $(y - 4\frac{1}{2})^2 - 4\frac{1}{4}$.

9   $(x + 5)^2 - 15$.

10   $(y - 3)^2 - 4$.

11   $(z - 2)^2 - 3$.

12   $(t + 1)^2 + 1$.

13   $(k - 6)^2 - 41$.

14   $(x + 4)^2 - 17$.

15   $(x - 1\frac{1}{2})^2 - \frac{1}{4}$.

16   $(y + \frac{1}{2})^2 - 1\frac{1}{4}$.

Exercise D

1   (a) $x = 1$;   (b) $(1, {}^-2)$, minimum.

2   (a) $x = {}^-2$;   (b) $({}^-2, 1)$, maximum.

3   (a) $x = {}^-3$;   (b) $({}^-3, 3)$, minimum.

4   (a) $x = 2$;   (b) $(2, {}^-3)$, maximum.

5   (a) $x = {}^-4$;   (b) $({}^-4, {}^-1)$, minimum.

6   (a) $x = 2$;   (b) $(2, 0)$, maximum.

7   (a) $x = 1$;   (b) $(1, 1)$, minimum.

8   (a) $x = {}^-3$;   (b) $({}^-3, {}^-2)$, minimum.

9   (a) $x = 2\frac{1}{2}$;   (b) $(2\frac{1}{2}, {}^-1\frac{3}{4})$, maximum.

10   (a) $x = 1\frac{1}{2}$;   (b) $(1\frac{1}{2}, {}^-1\frac{1}{4})$, minimum.

11   (a) $x = 3\frac{1}{2}$;   (b) $(3\frac{1}{2}, {}^-13\frac{3}{4})$, maximum.

12   (a) $x = 4\frac{1}{2}$;   (b) $(4\frac{1}{2}, {}^-8)$, minimum.

Exercise E

1   2, 4.

2   No solutions.

3   4.

4   ${}^-1$, 3.

5   ${}^-4$, 2.

6   ${}^-1$, 5.

7   No solutions.

8   1.

Exercise F

1 ⁻3, 3.	**2** ⁻2, 4.	**3** 0.35, 5.65.
4 3.59, 6.41.	**5** ⁻2, 4.	**6** ⁻3, ⁻1.
7 ⁻2, 10.	**8** 1, 2.	**9** ⁻0.45, 4.45.
10 ⁻4.15, 1.15.	**11** ⁻0.62, 9.62.	**12** 0.21, 4.79.
13 ⁻6, 2.	**14** ⁻3, 1.	**15** ⁻3.73, ⁻0.27.
16 ⁻4.61, 2.61.	**17** ⁻5, 1.	**18** ⁻2.
19 ⁻0.74, 6.74.	**20** ⁻1, 2.	**21** ⁻7, 2.
22 ⁻6.65, 1.65.	**23** ⁻4.41, ⁻1.59.	**24** 0.38, 2.62.

Exercise G

1 ⁻0.5, ⁻2.	**2** $-\frac{2}{3}$, ⁻1.	**3** 2, 0.25.
4 ⁻1.78, 0.28.	**5** ⁻0.92, 1.52.	**6** ⁻1.62, 0.62.
7 ⁻1.05, 0.85.	**8** 0.43, 1.64.	**9** ⁻6.38, 0.63.
10 $-\frac{1}{2}$, $\frac{2}{3}$.	**11** ⁻2, $\frac{3}{4}$.	**12** $-\frac{1}{2}$, 3.
13 $-\frac{1}{2}$, 3.	**14** ⁻2, $\frac{1}{2}$.	**15** ⁻0.80, 1.55.
16 ⁻4.16, 2.16.	**17** ⁻0.91, ⁻0.29.	**18** 0.30, 27.70.

Exercise H

1 (a) $x^2-3x-10$; (b) $x^2+3x-10$; (c) $x^2-7x+10$;
(d) $3x^2+13x+4$; (e) $3x^2-11x-4$; (f) $3x^2+11x-4$;
(g) $3x^2-13x+4$.

2 (a) $(x+2)(x+3)$; (b) $(x+6)(x-4)$; (c) $(x-8)(x+2)$;
(d) $x(x-8)$; (e) $(x-4)(x+1)$; (f) $(2x+1)(x+3)$;
(g) $(2x+3)(x+1)$; (h) $(2x-1)(x+3)$; (i) $x(3x-1)$;
(j) $(2x+1)(x-4)$; (k) $(3x-2)(x-4)$.

3 (a) $x=0$ or 5; (b) $y=$ ⁻3 or 1; (c) $z=$ ⁻5 or 2;
(d) $t=-\frac{1}{2}$ or ⁻2; (e) $x=-\frac{1}{3}$ or 3; (f) $y=$ ⁻2 or $\frac{1}{2}$;
(g) $x=$ ⁻2 or 9; (h) $x=$ ⁻2 or 3; (i) $z=-\frac{1}{2}$ or 1;
(j) $t=\frac{1}{3}$ or 3; (k) $x=$ ⁻1 or 2; (l) $y=$ ⁻4 or 2;
(m) $x=3$.

4 (a) y^2-y-6; (b) y^2+y-6; (c) y^2-5y+6;
(d) $2z^2+z-3$; (e) $2z^2-z-3$; (f) $2z^2-5z-3$;
(g) $2z^2+5z-3$.

5 (a) $(t-1)(t-6)$; (b) $(t+1)(t-6)$; (c) $(t-3)(t+2)$;
(d) $(x+3)(x-5)$; (e) $(y+8)(y-4)$; (f) $(2z+1)(z+5)$;
(g) $(2z+5)(z+1)$; (h) $(3x-2)(x+1)$; (i) $(5y+2)(y-1)$;
(j) $(3t+4)(t-1)$; (k) $(x+1)(7x-5)$.

6 (a) $x=$ ⁻1 or 4; (b) $y=3$ or 7; (c) $z=0$ or ⁻7;
(d) $t=-\frac{1}{5}$ or 3; (e) $t=1$ or $1\frac{1}{3}$; (f) $x=-\frac{2}{3}$ or 3;
(g) $y=\frac{1}{2}$ or 10; (h) $z=$ ⁻5 or 2; (i) $x=-\frac{2}{7}$ or 3;
(j) $y=$ ⁻$2\frac{1}{2}$ or ⁻2.

7   (a) $(3x+2)(2x-3)$;   (b) $(2y+3)(4y+3)$;   (c) $(4z+3)(3z-2)$.

8   (a) $x = 1\frac{1}{2}$ or 2;   (b) $x = {}^-4$ or 2;   (c) $x = {}^-\frac{1}{3}$ or 2.

Miscellaneous exercise

1   (a) $x = {}^-7$ or 3;   (b) $y = {}^-6.80$ or 2.80;   (c) $x = {}^-6.13$ or 1.63;
    (d) $y = {}^-6$ or $1\frac{1}{2}$;   (e) $z = {}^-1\frac{2}{3}$ or ${}^-3$;   (f) $z = {}^-2.54$ or 3.54;
    (g) $t = {}^-4$ or 5;   (h) $x = \frac{1}{2}$ or $1\frac{1}{2}$;   (i) $x = 0.35$ or 2.15;
    (j) $t = 6.35$ or 11.65;   (k) $y = {}^-1.4$ or 0.2.

2   (a) $(2,0)$, $(3,0)$;   (b) $(0,{}^-6)$;   0.7, 4.3.

3   ${}^-4.2$, 1.2; $\{x: {}^-4.2 < x < 1.2\}$.

4   $(2,{}^-1)$; $k = 4$.

5   $({}^-1.5, {}^-1.75)$; $x = {}^-2.82$ or ${}^-0.18$; $\{x: x \leqslant {}^-4\} \cup \{x: x \geqslant 1\}$.

6   (a) $(1,{}^-4)$;   (b) $x = 1$; $x = {}^-1.24$ or 3.24.

7   (a) $({}^-1,0)$, $(2,0)$;   (b) $(0,{}^-2)$; $x = {}^-0.73$ or 2.73.

8   $x = 4$, $y = 8$ or $x = {}^-2$, $y = 2$.

9   (a) $(0,0)$, $(2,6)$;   (b) $(1,2)$;   (c) no intersection.

10   (a) 32;   (b) 14;   (c) ${}^-7$ or 6;   (d) 11 or ${}^-11$.

11   $x = 6$.

12   $r = 8$.

13   Total surface area $= (x^2 + 33x + 70)$ cm²; $x = 10$.

14   (d) $35x - 2x^2 = 48$; $x = 1.5$.

Chapter 3   *Vectors and relative motion*

Exercise A

1   (a) $\underset{\sim}{EU}$ or $\underset{\sim}{DR}$;   (b) $\underset{\sim}{EO}$;   (c) $\underset{\sim}{BD}$ or $\underset{\sim}{RO}$ or $\underset{\sim}{AC}$.

2   (a) $\begin{bmatrix} 4 \\ 1 \end{bmatrix}$;   (b) $\begin{bmatrix} -1 \\ -3 \end{bmatrix}$;   (c) $\begin{bmatrix} -5 \\ 1 \end{bmatrix}$.

3   (a) 4.12;   (b) 3.16;   (c) 5.10.

4   $\begin{bmatrix} 4 \\ 2 \end{bmatrix}$; $\underset{\sim}{AS}$, $\underset{\sim}{DF}$, $\underset{\sim}{CE}$ or $\underset{\sim}{QU}$.

5   $\underset{\sim}{OA}$ or $\underset{\sim}{TS}$.

6   (a) $\underset{\sim}{AB}$;   (b) $\underset{\sim}{DB}$ or $\underset{\sim}{EC}$;   (c) $\underset{\sim}{OA}$ or $\underset{\sim}{DO}$.

7   (a) $\begin{bmatrix} 1 \\ 3 \end{bmatrix}$;   (b) $\begin{bmatrix} -3 \\ -1 \end{bmatrix}$;   (c) $\begin{bmatrix} -2 \\ -2 \end{bmatrix}$.

8   (a) 3.16;   (b) 3.16;   (c) 2.83.

9   $\begin{bmatrix} 3 \\ 1 \end{bmatrix}$; $\underset{\sim}{BD}$ and $\underset{\sim}{CE}$.

10   $\underset{\sim}{BO}$.

11   $\underset{\sim}{BC}$.

Exercise B

1 (*a*) **v**; (*b*) **w**; (*c*) **u**; (*d*) **r**; (*e*) **x**; (*f*) **x**.

2 (*a*) $\begin{bmatrix} -1 \\ -3 \end{bmatrix} + \begin{bmatrix} 2 \\ 0 \end{bmatrix} = \begin{bmatrix} 1 \\ -3 \end{bmatrix}$; (*b*) $\begin{bmatrix} 2 \\ 0 \end{bmatrix} + \begin{bmatrix} 4 \\ -3 \end{bmatrix} = \begin{bmatrix} 6 \\ -3 \end{bmatrix}$; (*c*) $\begin{bmatrix} -1 \\ -3 \end{bmatrix} + \begin{bmatrix} 1 \\ 2 \end{bmatrix} = \begin{bmatrix} 0 \\ -1 \end{bmatrix}$;

(*d*) $\begin{bmatrix} -1 \\ -3 \end{bmatrix} + \begin{bmatrix} 5 \\ 0 \end{bmatrix} = \begin{bmatrix} 4 \\ -3 \end{bmatrix}$; (*e*) $\begin{bmatrix} 4 \\ -3 \end{bmatrix} + \begin{bmatrix} 1 \\ 2 \end{bmatrix} = \begin{bmatrix} 5 \\ -1 \end{bmatrix}$; (*f*) $\begin{bmatrix} 5 \\ 0 \end{bmatrix} + \begin{bmatrix} 0 \\ -1 \end{bmatrix} = \begin{bmatrix} 5 \\ -1 \end{bmatrix}$.

3 (*a*) **e**; (*b*) **f**; (*c*) **c**; (*d*) **g**; (*e*) **h**; (*f*) **h**.

4 (*a*) $\begin{bmatrix} 1 \\ 3 \end{bmatrix} + \begin{bmatrix} -2 \\ -2 \end{bmatrix} = \begin{bmatrix} -1 \\ 1 \end{bmatrix}$; (*b*) $\begin{bmatrix} 1 \\ 3 \end{bmatrix} + \begin{bmatrix} 3 \\ 1 \end{bmatrix} = \begin{bmatrix} 4 \\ 4 \end{bmatrix}$; (*c*) $\begin{bmatrix} 1 \\ 3 \end{bmatrix} + \begin{bmatrix} 2 \\ -2 \end{bmatrix} = \begin{bmatrix} 3 \\ -1 \end{bmatrix}$;

(*d*) $\begin{bmatrix} -2 \\ -2 \end{bmatrix} + \begin{bmatrix} 3 \\ 1 \end{bmatrix} = \begin{bmatrix} 1 \\ -1 \end{bmatrix}$; (*e*) $\begin{bmatrix} -2 \\ -2 \end{bmatrix} + \begin{bmatrix} 4 \\ 4 \end{bmatrix} = \begin{bmatrix} 2 \\ 2 \end{bmatrix}$; (*f*) $\begin{bmatrix} 3 \\ 1 \end{bmatrix} + \begin{bmatrix} -1 \\ 1 \end{bmatrix} = \begin{bmatrix} 2 \\ 2 \end{bmatrix}$.

Exercise C

1 (*a*) $\underaccent{\sim}{AE}$ or $\underaccent{\sim}{FD}$ or $\underaccent{\sim}{HK}$; (*b*) $\underaccent{\sim}{FE}$; (*c*) $\underaccent{\sim}{EF}$.

2 (*a*) $\underaccent{\sim}{QG}$; (*b*) $\underaccent{\sim}{QL}$; (*c*) $\underaccent{\sim}{QK}$; (*d*) $\underaccent{\sim}{QL}$; (*e*) $\underaccent{\sim}{QD}$; (*f*) $\underaccent{\sim}{QL}$;
   (*g*) $\underaccent{\sim}{QM}$; (*h*) $\underaccent{\sim}{QF}$; (*i*) $\underaccent{\sim}{QN}$; (*j*) $\underaccent{\sim}{QH}$; (*k*) $\underaccent{\sim}{QE}$.

3 (*a*) **c** − **a**; (*b*) **b** − **a**; (*c*) **c** − **b**.

4 (*a*) $\underaccent{\sim}{QR}$; (*b*) $\underaccent{\sim}{QS}$; (*c*) $\underaccent{\sim}{QU}$; (*d*) $\underaccent{\sim}{QQ}$; (*e*) $\underaccent{\sim}{QV}$; (*f*) $\underaccent{\sim}{QP}$; (*g*) $\underaccent{\sim}{QT}$.

5 (*a*) **x** − **y**; (*b*) **z** − **y**; (*c*) **z** − **x**.

Exercise D

1 (*a*) $\begin{bmatrix} 50 \\ -500 \end{bmatrix}$; (*b*) $\begin{bmatrix} -50 \\ 500 \end{bmatrix}$.

2 150 km h^{-1} on 143°.

3 40 knots on 255°.

4 $\begin{bmatrix} 0 \\ 6 \end{bmatrix}$ cm s^{-1}; 10 s.

5 5 m s^{-1} on 143°.

6 77 km h^{-1} on 112°.

Exercise E

1 113°, 130 km h^{-1}.

2 288°, 19 knots.

3 81 km h^{-1}.

4 022°, 5.4 km h^{-1}, 3 minutes, 250 m.

5 078°, 48 km h^{-1}.

6 1650 km h^{-1}.

Exercise F

1 105°, 145 km h^{-1}, 42 min.

2 (*a*) 41°; (*b*) 90°.

3 211°, 243 km h^{-1}.

4 (*a*) 60 km south; (*b*) 19 min; (*c*) 252°.

5 41° to the upstream bank, 45 s.

6 019°, 3.3 h.

7 (*a*) 75 min; (*b*) 012°, 37 min; (*c*) 168°.

8 (*a*) 58 m; (*b*) 15° to the upstream bank or 55° to the downstream bank; (*c*) 53°
   to the upstream bank.

Miscellaneous exercise

1 (i) (a) $\begin{bmatrix} 4 \\ -1 \end{bmatrix}$; (b) $\begin{bmatrix} 2 \\ -\frac{1}{2} \end{bmatrix}$; (c) $\begin{bmatrix} -1 \\ -1\frac{1}{2} \end{bmatrix}$.

   (ii) (a) $\mathbf{a}+\mathbf{b}$; (b) $\frac{1}{2}\mathbf{a}+\frac{1}{2}\mathbf{b}$; (c) $\frac{1}{2}\mathbf{b}-\frac{1}{2}\mathbf{a}$.

2 (i) (a) $\mathbf{q}$; (b) $2\mathbf{q}$; (c) $2\mathbf{q}-\mathbf{p}$.

   (ii) (a) $\underset{\sim}{AC}$ or $\underset{\sim}{FD}$; (b) $\underset{\sim}{QB}$ or $\underset{\sim}{DC}$ or $\underset{\sim}{FA}$; (c) $\underset{\sim}{AE}$ or $\underset{\sim}{BD}$.

   (iii) X is the midpoint of CD.

3 42 km h^{-1} on a bearing 042°.

4 112°; 072°, 5 h 45 min.

5 (a) 28 km, 195°; (b) 200°, 3 h 20 min.

6 (a) $\begin{bmatrix} 80 \\ 80 \end{bmatrix}$ km h^{-1}, 113 km h^{-1} north-east; (b) $\begin{bmatrix} 85 \\ -10 \end{bmatrix}$ km h^{-1}; (c) 120°, $\begin{bmatrix} 69 \\ -40 \end{bmatrix}$.

7 (a) 41°; 0.9 km.

Chapter 4   Polynomials and curve sketching

Exercise A

1 $\{x : x < 1\} \cup \{x : x > 3\}$.

2 $\{x : x \leqslant ^-1\} \cup \{x : x \geqslant 1\frac{1}{2}\}$.

3 $\varnothing$.

4 $\{x : ^-3 < x < ^-2\} \cup \{x : x > 1\}$.

5 $\{x : ^-2 < x < \frac{1}{3}\} \cup \{x : x > 3\}$.

6 $\{x : x < 2\} \cup \{x : x > 5\}$.

7 $\{x : ^-2 \leqslant x \leqslant \frac{1}{2}\}$.

8 $\{x : ^-2 \leqslant x \leqslant 3\} \cup \{x : x \geqslant 4\}$.

9 $\{x : x > \frac{1}{2}, x \neq 3\}$.

10 $\{x : x < 2\}$.

Exercise B

1 (a) 2; (b) 6; (c) 1, 5.

2 (a) 3; (b) 1; (c) 0, $^-7$.

3 (a) 5; (b) 0; (c) 0, $^-6$.

4 (a) 4; (b) 5; (c) 1, $^-3$.

5 (a) 5; (b) $^-3$; (c) 4, 1.

6 (a) 2; (b) $^-2$; (c) 1, $^-7$.

7 (a) 3; (b) 0; (c) 4, $^-7$.

8 (a) 7; (b) 19; (c) 0, $^-1$.

9 (a) 3; (b) 13; (c) $^-17$, $^-19$.

10 (a) 3; (b) 3; (c) $^-2$, 9.

11 x^3+2x^2-5x-6.

12 $2+7x+2x^2-3x^3$.

13 $6x^3-17x^2+15x-4$.

14 $x^3-13x+12$.

15 $6x^3-47x^2+57x+140$.

16 $9+27x-x^2-3x^3$.

17 $3x^3+4x^2-17x-6$.

18 $2x^3+x^2-18x-9$.

19 x^3-3x^2-x+3.

20 $t^3+7t^2-25t-175$.

21 x^3+x^2-6x.

22 x^3+3x^2-x-3.

23 $2x^3+x^2-8x-4$.

24 $3x^3-2x^2-11x+10$.

25 $4x^3+4x^2-9x-9$.

26 $3x^3-11x^2+8x+4$.

27 x^4-10x^2+9.

Exercise C

1 $5x^2 - 5x - 5$; $9 - 7x - 5x^2 - 2x^3$.

2 $3x^5 + 7x^4 - 11x^3 + 5x^2 + 7x + 1$; $3x^5 - 7x^4 + 7x^3 + 11x^2 + 3x - 15$.

3 $5t^3 + 3t^2 + 8t + 6$; $5t^3 - 3t^2 - 4t - 4$.

4 $13 + x + 5x^2 - 7x^3 + 8x^4$; $1 - 5x + 5x^2 + 7x^3 - 2x^4$.

5 $4 + t - 11t^2 + 3t^3$; $t^3 + 3t^2 + t - 4$.

6 $2z^4 + 3z^3 + 3z^2 - z - 5$; $^-3z^3 + z^2 + z - 7$.

7 $P + Q = 5x^4 + 2x^3 - 3x^2 + 7x + 3$; $P - Q = 5x^4 - 2x^3 - 3x^2 - 3x - 5$;
   $(P + Q) + (P - Q) = 10x^4 - 6x^2 + 4x - 2$.

8 $6z^5 - 5z^4 + 4z^3 - 3z^2 + 2z - 1$; $6z^5 - 5z^4 + 3z^2 - 2z + 1$.

9 $18x^4 - 8x^3 + 7x^2 - 2x + 34$; $^-7x^2$.

Exercise D

1 $2x^3 - 15x^2 + 31x - 15$.

2 $x^4 - x^3 - 9x^2 + 2x + 2$.

3 $20t - 13t^2 + t^3 + t^4 - t^5$.

4 $6x^3 - 13x^2 + 13x + 6$.

5 $4x - 12x^2 - 9x^3 + 9x^4 + x^5 - x^6$.

6 $3z^6 + z^4 - 6z^3 + 3z^2 - 2z + 1$.

7 (*a*) $^-3$; (*b*) 3; (*c*) 16; (*d*) $^-7$.

8 (*a*) $x^2 + 2x + 1$; (*b*) $x^3 + 3x^2 + 3x + 1$; (*c*) $x^4 + 4x^3 + 6x^2 + 4x + 1$;
   (*d*) $x^5 + 5x^4 + 10x^3 + 10x^2 + 5x + 1$.

9 (*a*) $16x^4 - 32x^3 + 24x^2 - 8x + 1$; (*b*) $27 - 27x + 9x^2 - x^3$;
   (*c*) $27x^3 + 54x^2 + 36x + 8$.

10 (*a*) $a^2 + 2ab + b^2$; (*b*) $a^3 + 3a^2b + 3ab^2 + b^3$.

Exercise E

1 (*a*) $1 + 6x + 15x^2 + 20x^3 + 15x^4 + 6x^5 + x^6$;
   (*b*) $1 + 7x + 21x^2 + 35x^3 + 35x^4 + 21x^5 + 7x^6 + x^7$;
   (*c*) $1 + 8x + 28x^2 + 56x^3 + 70x^4 + 56x^5 + 28x^6 + 8x^7 + x^8$.

2 (*a*) $a^5 + 5a^4x + 10a^3x^2 + 10a^2x^3 + 5ax^4 + x^5$;
   (*b*) $a^6 + 6a^5x + 15a^4x^2 + 20a^3x^3 + 15a^2x^4 + 6ax^5 + x^6$.

3 (*a*) $64 + 48x + 12x^2 + x^3$; (*b*) $8 - 12x + 6x^2 - x^3$;
   (*c*) $81x^4 + 108x^3 + 54x^2 + 12x + 1$; (*d*) $1 - 5x + 10x^2 - 10x^3 + 5x^4 - x^5$;
   (*e*) $81 - 216x + 216x^2 - 96x^3 + 16x^4$.

4 $16 - 32x + 24x^2 - 8x^3 + x^4$; 13.0321.

5 $243 + 405x + 270x^2 + 90x^3 + 15x^4 + x^5$; 251.2087.

6 $64 - 192x + 240x^2 - 160x^3 + 60x^4 - 12x^5 + x^6$; 58.456, 3 s.f.

7 $1 - 15x + 90x^2 - 270x^3 + 405x^4 - 243x^5$; 0.9704.

Exercise F

1 $x = {}^-2\frac{1}{2}$ or 3.	**2** $x = \frac{1}{2}$.	**3** $x = 0$ or 7.
4 $x = {}^-3$, 1 or 4.	**5** $x = {}^-2$, $\frac{1}{2}$ or 2.	**6** $x = {}^-5$, ${}^-2$ or 2.
7 $x = 0$, ${}^-\sqrt{3}$ or $\sqrt{3}$.	**8** $x = 0$, 2 or 3.	**9** $x = {}^-2$ or $\frac{1}{3}$.
10 $x = {}^-2$.	**11** $x = {}^-1\frac{1}{2}$ or 0.	**12** $x = {}^-7$, ${}^-2$ or $1\frac{1}{2}$.
13 $x = {}^-3$, ${}^-2$ or 3.	**14** $x = {}^-3$, ${}^-2$ or $\frac{1}{2}$.	**15** $x = 0$.
16 $x = {}^-2$, 0 or $\frac{1}{2}$.		

Exercise G

1 $x^2 + 5x + 4$.	**2** $x^2 - 3x + 1$.	**3** $x = {}^-3$, ${}^-2$ or 2.
4 $x = {}^-\frac{1}{2}$ or 3.	**5** $x^2 - 3x + 5$.	**6** $x^2 + 5$.
7 $x = {}^-3$, ${}^-1$ or $\frac{1}{2}$.	**8** $x = \frac{1}{3}$.	**9** $x = {}^-4$, ${}^-1$ or 1.
10 $x = \frac{1}{2}$, ${}^-\sqrt{5}$ or $\sqrt{5}$.	**11** $x = {}^-2$ or 4.	

Exercise H

1 $x = {}^-3$, ${}^-1$ or 1.	**2** $x = 2$.	**3** $x = {}^-2$ or 3.
4 $x = {}^-2$, $\frac{1}{2}$ or 3.	**5** $x = {}^-2$, ${}^-1$ or 3.	**6** $x = {}^-2$ or 2.
7 $x = {}^-1$, $\frac{1}{2}$ or 4.	**8** $x = {}^-2$, $\frac{1}{3}$ or $\frac{1}{2}$.	

Miscellaneous exercise

1 (a) (ii) $x = 1$, 2 or 3; (iii) $x^3 - 6x^2 + 11x - 6$.
(b) (ii) $x = {}^-2$, 0 or 2; (iii) $x^3 - 4x$.
(c) (ii) $x = {}^-1$, $\frac{1}{2}$ or $\frac{2}{3}$; (iii) $6x^3 + 5x^2 - 3x - 2$.
(d) (ii) $x = {}^-1$ or 2; (iii) $x^3 - 3x^2 + 4$.
(e) (ii) $x = {}^-1$ or 1; (iii) $x^4 + x^2 - 2$.
(f) (ii) $x = {}^-1$ or 2; (iii) $2 + x - x^2$.
(g) (ii) $x = {}^-2$, ${}^-1$, 0 or 1; (iii) $x^4 + 2x^3 - x^2 - 2x$.

2 (a) $\{x : 1 < x < 2\} \cup \{x : x > 3\}$. (b) $\{x : x \leqslant {}^-2\} \cup \{x : 0 \leqslant x \leqslant 2\}$.
(c) $\{x : {}^-1 < x < \frac{1}{2}\} \cup \{x : x > \frac{2}{3}\}$. (d) $\{x : x \leqslant {}^-1$ or $x = 2\}$.
(e) $\{x : x < {}^-1\} \cup \{x : x > 1\}$. (f) $\{x : x < {}^-1\} \cup \{x : x > 2\}$.
(g) $\{x : x < {}^-2\} \cup \{x : {}^-1 < x < 0\} \cup \{x : x > 1\}$.

3 (a) $p^3 + 3p^2q + 3pq^2 + q^3$; (b) $a^5 + 5a^4x + 10a^3x^2 + 10a^2x^3 + 5ax^4 + x^5$;
(c) $1 - 4x + 6x^2 - 4x^3 + x^4$; (d) $1 + 6x + 12x^2 + 8x^3$;
(e) $81 - 432x + 864x^2 - 768x^3 + 256x^4$.

4 1.104.

5 0.9792.

6 (a) 6; (b) no other roots; (c) ${}^-3$, 3, 2; (d) ${}^-4$, ${}^-2$, 3.

7 (a) $x = {}^-2$, 2, ${}^-1$ or 1; (b) $x = {}^-3$, 0 or 2; (c) $x = {}^-\frac{4}{3}, \frac{4}{3}$, or ${}^-1$;
(d) $x = {}^-3$, ${}^-1$, 2 or 4.

Chapter 5   Differentiation and gradients

Exercise A

1 $t \to 9t^8$.   **2** $t \to 12t^2$.   **3** $t \to 35t^6 + 12t$.

4 $t \to 18t + 6$.   **5** $t \to 24t^2 - 3$.   **6** $t \to 4t^3 + 6t^2$.

7 $t \to {}^-7t^{-2}$.   **8** $t \to {}^-6t^{-3}$.   **9** $t \to \dfrac{-4}{t^3}$.

10 $t \to \dfrac{-2}{t^2} - \dfrac{6}{t^4}$.   **11** $t \to 11t^{10}$.   **12** $t \to 10t^4$.

13 $t \to 15t^4 + 15t^2$.   **14** $t \to 50t - 10$.   **15** $t \to 648t^2 - 7t^{-2}$.

16 $t \to 12t^3 + 12t^2 + 7$.   **17** $t \to 36t^8 - 2t^{-3}$.   **18** $t \to \dfrac{-8}{t^2} + \dfrac{8}{t^3}$.

19 $t \to 1 - \dfrac{1}{t^2}$.   **20** $t \to 2t + 2t^{-3}$.

21 (*a*) 5 m s^{-1} upwards;   (*b*) 5 m s^{-1} downwards; 13.25 m.

22 2.04 s; 20.4 m.

23 8 m min^{-1}; 203 m.

24 (*a*) 1 m s^{-1} upwards;   (*b*) 2 m s^{-1} downwards; 9.8 m.

Exercise B

1 $3x^2$.   **2** $\dfrac{-20}{x^3}$.   **3** $2\pi r$.   **4** $6V - \dfrac{5}{V^2}$.

5 $v : t \to 3 - 10t$; $a : t \to {}^-10$.

6 $v : t \to 3 + \frac{1}{2}t$; $a : t \to \frac{1}{2}$; 0.5 m s^{-2}.

7 $^-16$; 4 m s^{-1}, 1.75 m s^{-1}, $^-32$ m s^{-1}; at $t = 0$; $^-8$ m s^{-1}.

8 $v : t \to 4t^3 - 1$; $a : t \to 12t^2$.

9 4 m s^{-1}; $^-4$ m s^{-2}.

10 $^-1, 0, 0$; at $t = 1$; $^-2$ m s^{-1}.

11 $v : t \to 8 - t$; $a : t \to {}^-1$; 32 m; $^-1$ m s^{-2}.

12 $(1.75, 0.6125)$.

13 (*a*) 1 s;   (*b*) 2 s; no.

Exercise C

1 7.   **2** 13.   **3** $^-5$.   **4** $\frac{3}{4}$.   **5** 16.

6 11.   **7** 1.   **8** $^-8$.   **9** 4π.   **10** 24.

11 7, 5, 1.   **12** 22, $^-2$, $^-5$, 7.   **13** 51, 24; 6, $^-6$.

14 3.   **15** $\frac{1}{3}$ and $^-\frac{1}{3}$.   **16** $^-3$ and 2.

17 19, $^-1$, 3.   **18** $^-12$, 6, 0, 20.   **19** $^-4$, 4; 0, $^-8$.

20 $^-1$.   **21** 1 and $^-1$.   **22** $^-2$ and 1.

23 $y = 4x - 4$.   **24** $y = 4 - x$.   **25** $y = 12x - 15$.

26 $y = 12 - 8x$.	**27** $y = 52x - 33$.	**28** $32y = 22 - 3x$.
29 $y = 8x - 8$.	**30** $y = 6x + 2$.	**31** $y = 2x + 3$.
32 $y = {}^-3x - 5$.	**33** $y = 2176 - 720x$.	**34** $16y = 20 - 7x$.

Exercise D

1 (*a*) $(0, 0)$ minimum; (*b*) $(0, 5)$ maximum, $(4, {}^-59)$ minimum.

2 (*a*) $(1, 0)$ minimum; (*b*) $({}^-1, {}^-2)$ maximum, $(1, 2)$ minimum.

3 (*a*) $(0, 0)$ minimum; (*b*) $({}^-\frac{1}{2}, 2)$ maximum, $(\frac{1}{2}, 0)$ minimum.

4 (*a*) $({}^-1, 0)$ maximum, $(1, {}^-4)$ minimum; (*b*) $(\frac{1}{2}, 3)$ minimum.

5 (*a*) $(3, {}^-7)$ minimum; (*b*) $({}^-1, 2)$ maximum, $(1, {}^-3)$ minimum.

6 (*a*) $(0, 0)$ minimum, $(2, 4)$ maximum; (*b*) $({}^-1, 1)$ maximum, $(1, 5)$ minimum.

7 (*a*) $(0, 0)$ minimum; (*b*) $({}^-2, 16)$ maximum, $(2, {}^-16)$ minimum.

8 (*a*) $(\frac{1}{2}, \frac{3}{4})$ minimum; (*b*) $({}^-\frac{1}{2}, 3)$ minimum.

9 $(\frac{1}{2}, 12)$ minimum.

10 $({}^-3, {}^-2187)$ minimum, $(0, 0)$ maximum, $(1, {}^-11)$ minimum.

11 $({}^-2, 24)$ maximum, $(1, {}^-3)$ minimum.

12 40, 3000; 23 000, 0.

Exercise E

1 1250 m².	2 50.	3 Side of square $= \frac{1}{6}$ m.
4 0.59 m³.	5 50; £9.	6 350 cm².
7 4 cm × 12 cm × 6 cm.	8 (*a*) 0.074 m³; (*b*) 0.094 m³.	
10 144 cm².	11 0.074 m³.	

Miscellaneous exercise

1 111 m s⁻¹, 64 m s⁻².

2 (*a*) 6.9; (*b*) $\dfrac{15 + 12h + 2h^2}{2(2 + h)}$.

3 (*a*) $x \to 2x + 1$; (*b*) $x \to \dfrac{{}^-1}{x^2} + 1$; (*c*) $x \to 2x + 2$;

   (*d*) $x \to \dfrac{{}^-3}{x^4} + \dfrac{2}{x^3}$.

4 (*a*) $y = 13x - 16$; (*b*) $y = 6x + 22$; (*c*) $y = \frac{1}{2} - \frac{1}{16}x$, $y = {}^-\frac{1}{2} - \frac{1}{16}x$.

5 $y = 1 - 6x$; $y = 8x - 9$; none.

6 $\frac{1}{4}$.

7 9 cm².

9 $^-1$ minimum, 7 maximum; $y = 6x + 3$; $y = {}^-1$, $y = 7$.

10 225 m²; 450 m².

11 $y = kx - k^2$.

12 2.92 s.

13 $a = 1, k = 1$.

14 $x = \dfrac{a}{\sqrt{2}}$, area $= a^2$.

Revision exercise 1

1 0.167 m s^{-1}.

2 $(^-1, ^-3)$; (*a*) $(0, ^-2)$; (*b*) $(0.73, 0), (^-2.73, 0)$; (*c*) $(^-3.65, 4), (1.65, 4)$.

3 0.851.

4 $\begin{bmatrix} 4 \\ 2 \end{bmatrix}, \begin{bmatrix} ^-1 \\ 2 \end{bmatrix}$; $\underline{QC} = \mathbf{a} + \mathbf{b}$; $\underline{OP}$ or $\underline{PC}$, $\underline{BP}$ or $\underline{PA}$; 5, 5; diagonals of a rhombus.

5 $^-\frac{1}{3}, 1$; $y = x - 1, (^-1, ^-2)$.

6 $(^-0.65, 0), (4.65, 0)$.

7 $x = ^-1$ or $\frac{1}{3}$.

8 (*a*) $^-18$ m s^{-1}; (*b*) $t = 1.5$, 30 m s^{-2}; (*c*) $t = \frac{1}{4}$, $^-18.75$ m s^{-1}, 4.625 m.

9 50.3.

10 (*b*) $a + 6b$; (*c*) $\sqrt{5}, 6\sqrt{5}$.

11 0.0349.

12 $(^-3.5, 0), (1, 0)$; $(^-2, 27), (1, 0)$.

13 $x = ^-0.24$ or 4.24; $(2, 5)$; $k = 5$.

14 (*a*) 166°; (*b*) 48 km h^{-1}; (*c*) 166 km.

15 80 m^2.

16 (*a*) $x = ^-1.79$ or 2.79; (*d*) $y = 5x - 14$, $y = ^-\frac{1}{3}x + 2$; Q $(2.8, 0)$, R $(6, 0)$; (*e*) 1.6.

17 (*a*) 80 min; (*b*) 53.3 km; (*c*) 60 km h^{-1}.

18 082°, 1.44 h; 265°.

19 (*a*) $r + h = 9$; (*b*) $2r(9 - r)$; (*c*) 40.5 cm^2; (*d*) 339 cm^3.

20 (*a*) (i) $8x^3 - 12x^2 - 2x + 3$; (ii) $8x^3 + 12x^2 + 6x + 1$; (*b*) $^-24x^2 - 8x + 2$.

21 $(2x - 5)^2 - 16$; $x = 1\frac{1}{2}$ or $4\frac{1}{2}$; $(2x - 1)(2x - 9)$.

22 $y = 13x - 59$; $(2, ^-15)$.

23 14.4 knots on a bearing of 118°; 8.84 nautical miles.

24 17 km min^{-1}, 6.75 km min^{-2}.

25 $\pi r + x = 200$; 31.8 m, 100 m.

Chapter 6   Integration

Exercise A

1 (*a*) 26, $t^3 + k$; (*b*) 14, $t^2 + 3t + k$; (*c*) $^-92$, $4t - \frac{5}{4}t^4 + k$; (*d*) $3376\frac{4}{5}$, $\frac{2}{5}t^5 + \frac{1}{2}t^8 + k$.

2 (*a*) 16.8 m; (*b*) 5 s.

3 $v(t) = 10t$, $s(t) = 5t^2$; 55 m, 60 m s^{-1}.

4 (a) $32-0.1t^2$;   (b) 17.9 s, 382 m.

5 264 m.

6 313 m.

7 (a) $\frac{1}{6}t^6+k$;   (b) $\frac{3}{5}t^5-\frac{1}{2}t^4+k$;   (c) $\frac{1}{3}t^3+\frac{1}{2}t^2+t+k$.

8 (a) 609;   (b) 48;   (c) $^-267$;   (d) 81090.

9 $v(t)=6t$, $h(t)=3t^2$;   21 m;   48 m, 24 m s^{-1}.

10 7 s.

11 $v(t)=18-2t^2$;   3 s; 36 m.

12 4 s, 1067 m.

13 286 m.

14 (a) $\frac{1}{8}t^8+k$;   (b) $2t^2-7t+k$;   (c) $\frac{1}{2}t^6-\frac{1}{3}t^{12}+k$.

Exercise B

1 9, 5, 13.

2 $6\frac{2}{3}$, $6\frac{2}{3}$, $9\frac{1}{3}$.

3 $t=1.6$, 4.4; $1\frac{2}{3}$, 14.5.

4 $t<1$, $2<t<4$; $^-0.42$;   11.08.

5 $^-\frac{1}{6}$, $\frac{1}{6}$.

6 (a) $\frac{1}{3}b^3-\frac{1}{3}a^3$, $\frac{1}{3}a^3-\frac{1}{3}b^3$;   (b) $3b-\frac{1}{2}b^2-3a+\frac{1}{2}a^2$, $3a-\frac{1}{2}a^2-3b+\frac{1}{2}b^2$.

7 12, 16, 16.

8 $^-2.67$, 6, 15.

9 $t=0.59$, 3.4; $^-2\frac{2}{3}$, 4.88.

10 $t<1$, $3<t<4$;   18, $28\frac{2}{3}$.

11 $^-4\frac{1}{2}$, $4\frac{1}{2}$.

12 (a) $\frac{1}{4}b^4-\frac{1}{4}a^4$, $\frac{1}{4}a^4-\frac{1}{4}b^4$;   (b) $5b-\frac{1}{2}b^2-5a+\frac{1}{2}a^2$, $5a-\frac{1}{2}a^2-5b+\frac{1}{2}b^2$.

Exercise C

1 $8\frac{2}{3}$.   **2** 69.   **3** 36.

4 9, $^-4$, 13.   **5** $1<x<4$; 15.   **6** 100.

7 (a) 24, 21, 45;   (b) 4, $^-1$, 3.

8 (a) $2x^3-4x+k$;   (b) $\frac{1}{2}x^4+\frac{1}{2}x^2+k$.

9 $152\frac{1}{4}$.   **10** 62.   **11** 45.

12 16, $^-4$, 20.   **13** $2<x<3$; $9\frac{1}{2}$.   **14** 39.

15 (a) 30, 40, 70;   (b) 9, $^-4$, 5.

16 (a) $\frac{1}{2}x^2-3x^3+k$;   (b) $x+\frac{1}{2}x^2+\frac{1}{6}x^3+k$.

Miscellaneous exercise

1 (a) 3 s;   (b) $9\frac{1}{3}$ m.   **2** (a) 21;   (b) $^-0.06$.

3 $2\frac{2}{3}$.   **4** $1\frac{1}{3}$.   **5** $4\frac{2}{3}$.

6 $8\frac{2}{3}$.   **7** $6\frac{2}{3}$; $^-3.9$, 2.9.

Chapter 7   Oscillations

Exercise A

1 (*a*) 0.87;   (*b*) ⁻0.87;   (*c*) ⁻0.5;   (*d*) 0.87;   (*e*) ⁻0.87;   (*f*) ⁻0.5.

2 (*a*) 0.940;   (*b*) 0.342;   (*c*) ⁻0.342;   (*d*) 0.940;   (*e*) ⁻0.940;   (*f*) ⁻0.940.

3 (*a*) 0.64;   (*b*) ⁻0.64;   (*c*) ⁻0.77;   (*d*) 0.64;   (*e*) ⁻0.77;   (*f*) 0.64.

4 (*a*) 0.98;   (*b*) 0.17;   (*c*) 0.98;   (*d*) ⁻0.17;   (*e*) ⁻0.17;   (*f*) ⁻0.17.

5 (*a*) 1;   (*b*) 0;   (*c*) ⁻1;   (*d*) 0.

6 (*a*) ⁻0.5;   (*b*) 0;   (*c*) 0.5;   (*d*) 1.

Exercise B

2   53, 127, 413, 487.   3   ⁻314, ⁻226, 46, 134.   4   ⁻34, ⁻146.

5   42, 138.   6   229, 311.   8   43, 137.

9   27, 153, 387, 513.   10   ⁻122, ⁻58.   11   28, 152.

12   188, 352.

Exercise C

1 (*a*) 90 s;   (*b*) 8;   (*c*) 10, 35, 100, 125, 190, 215, 280, 305.

2 (*a*) 60 s;   (*b*) 12;   (*c*) 7, 23, 67, 83, 127, 143, 187, 203, 247, 263, 307, 323.

3 (*a*) 23, 67, 203, 247;   (*b*) 71, 109, 191, 229, 311, 349;   (*c*) 106, 254.

4 (*a*) 3 s;   (*b*) 1.6, 2.9, 4.6, 5.9, 7.6, 8.9.

5 (*a*) 4 s;   (*b*) 2.1, 3.9, 6.1, 7.9.

6 (*a*) 5, 31, 77, 103, 149, 175;   (*b*) 28, 40, 73, 85, 118, 130, 163, 175;   (*c*) 81.

Exercise D

1   0.02 s; 18000.

2   $\theta = 15 \sin 120t°$;   (*a*) 15°;   (*b*) 0°;   (*c*) ⁻15°;   (*d*) 13°; 0.2 s.

3   $h = 4 \sin 30t°$;   (*a*) 10 a.m.;   (*b*) 4 p.m.; 2 m; 1.28 p.m.

4   $x = 50 \sin 180t°$;   (*a*) 1 s;   (*b*) 0.5 s;   (*c*) 1.5 s;   (*d*) $\frac{1}{6}$ s;   (*e*) $\frac{5}{6}$ s;   (*f*) $1\frac{1}{6}$ s.

5   $\theta = 10 \sin 360t°$;   (*a*) 0°;   (*b*) 0°;   (*c*) 9.5°;   (*d*) 5.9°; 0.048 s, 0.452 s.

6   $d = 30 \sin(\frac{360}{13}t)°$;   5 m; 10.30 a.m.

Exercise E

2 (*a*) 72, 148;   (*b*) 40, 60, 160, 180, 280, 300;   (*c*) 294;   (*d*) 40, 80, 160, 200, 280, 320.

3 (*a*) $\theta = 20 \sin 120t°$;   (*b*) 0.75 s; $20 \sin 120(t+0.75)$;   (i) 0.5 s;   (ii) 1 s.

5   ⁻155, ⁻35, 25, 145;   (*b*) no solution;   (*c*) 74;   (*d*) ⁻141.5, ⁻78.5, ⁻51.5, 38.5, 101.5, 128.5.

6   $h = 6 \sin 28.8(t+0.25)°$;   12.05 a.m., 5.40 a.m.

Exercise F

1 $d = 8 \cos 90t°$.

2 (*c*) (i) 0.82; (ii) ⁻0.82; (iii) ⁻0.82; (iv) 0.82.

4 (*a*) 60, 300; (*b*) 142, 218; (*c*) 17, 163, 197, 343;
(*d*) 90, 270; (*e*) no solution; (*f*) 115, 145, 295, 325.

5 (*a*) ⁻77, 77; (*b*) ⁻160, ⁻80, ⁻40, 40, 80, 160; (*c*) ⁻160, ⁻10, 20, 170.

6 $h = 8 \cos 30t°$.

7 (*a*) ⁻0.94; (*b*) 0.94; (*c*) ⁻0.94; (*d*) 0.94.

9 (*a*) 32, 328; (*b*) 180; (*c*) 44.3, 75.7, 164.3, 195.7, 284.3, 315.7;
(*d*) 80; (*e*) 70, 160, 250, 340; (*f*) 49, 88.3, 169, 208.3, 289, 328.3.

10 (*a*) ⁻29, 29; (*b*) ⁻156, ⁻84, ⁻36, 36, 84, 156; (*c*) ⁻146, ⁻166.

Exercise G

2 (*a*) ⁻0.7; (*b*) 0.7; (*c*) ⁻0.7; (*d*) ⁻0.7.

3 (*a*) 167, 347; (*b*) 0, 180, 360; (*c*) 75, 255; (*d*) 21, 111, 201, 291; (*e*) 244.

5 (*a*) 2.9; (*b*) ⁻2.9; (*c*) ⁻2.9; (*d*) 2.9.

6 (*a*) ⁻153, 27; (*b*) ⁻57, 123; (*c*) ⁻161, 19; (*d*) ⁻129, ⁻69, ⁻9, 51, 111, 171;
(*e*) ⁻165, ⁻75, 15, 105.

Miscellaneous exercise

2 (*a*) ⁻165, ⁻145, ⁻45, ⁻25, 75, 95; (*b*) ⁻124, ⁻16, 56, 164; (*c*) 107.

5 (*a*) 10, 110, 130; (*b*) 40, 80, 160;
$\{x : 10 \leqslant x \leqslant 40\} \cup \{x : 80 \leqslant x \leqslant 110\} \cup \{x : 130 \leqslant x \leqslant 160\}$.

6 $h = 5 \sin 28.8(t+2)° + 7$; (*a*) 11.2 m; (*b*) before 5.36 a.m., between 8.09 a.m.
and 6.06 p.m., or after 8.39 p.m.

7 0.0023 s; $y = 0.5 \cos 158\,400\ t°$.

Chapter 8   Triangles

Exercise A

1	45 cm².	**2**	22 cm².	**3**	620 cm².
4	17 cm².	**5**	43°, 137°.	**6**	38°, 22°.
7	50 cm².	**8**	18 cm².	**9**	3.3 cm².
10	2.2 cm².	**11**	36°, 144°.	**12**	47°, 73°.

Exercise B

1	16.3 cm.	**2**	31° or 149°.	**3**	5.39 cm.
4	4.38 m, 3.58 m.	**5**	35°.	**6**	30°, 8.04.
7	(*a*) 66°; (*b*) 39° or 7°; (*c*) 31° or 15°.			**8**	11.6 mm.
9	2.86 m.	**10**	38° or 142°.	**11**	15°, 7.89 m.
12	37° or 143°.	**13**	14.3 m, 12.5 m.	**14**	3.65 cm, 5.73 cm.

Exercise C

1 118 cm. **2** 6.8 cm. **3** 29°, 47°, 104°.

4 (*a*) 29 nautical miles; (*b*) 58 n.m.; (*c*) 29*t* n.m.

5 5, $\sqrt{5}$, $\sqrt{8}$; 10°, 8°. **6** 9.2 km. **7** 62.

8 96 km. **9** 28°, 43°, 109°. **10** 8.1 km.

11 13, 17, $\sqrt{500}$; 49°, 95°. **12** 121 km.

Exercise D

1 (*a*) 5; (*b*) 36; (*c*) ⁻12. **2** (*a*) 42°; (*b*) 138°. **3** (*a*) 52°; (*b*) 71°.

4 (*a*) 3; (*b*) 17; (*c*) $\begin{bmatrix} 1 \\ 5 \\ 3 \end{bmatrix}$; (*d*) 20.

5 (*a*) $\begin{bmatrix} 1 \\ -2 \\ -2 \end{bmatrix}$, $\begin{bmatrix} -3 \\ 2 \\ -6 \end{bmatrix}$; (*b*) 3, 7; (*c*) 5; (*d*) 76°.

6 (*a*) 11; (*b*) ⁻15; (*c*) 8. **7** (*a*) 78°; (*b*) 169°. **8** (*a*) 78°; (*b*) 90°.

9 (*a*) 8; (*b*) ⁻1; (*c*) $\begin{bmatrix} -2 \\ 5 \end{bmatrix}$; (*d*) 7.

10 (*a*) $\begin{bmatrix} 4 \\ 1 \end{bmatrix}$, $\begin{bmatrix} -1 \\ 4 \end{bmatrix}$; (*b*) $\sqrt{17}$, $\sqrt{17}$; (*c*) 0; (*d*) 90°.

11 0. **12** (*a*) $\sqrt{69}$; (*b*) 69. **13** 3.

14 (*a*) 9.64; (*b*) 4.10; (*c*) ⁻4.10; (*d*) ⁻5.64; (*e*) ⁻3.42.

15 18.8, 4.10; 7.3 at 18° to **a**; 22.9.

16 6.89, ⁻4.10; 5.8 at 29° to **c**; 2.79.

17 41.5. **18** 0, 0.

19 (*a*) (5, 0, 0), (0, 5, 5), (5, 0, 5); (*b*) $\begin{bmatrix} -5 \\ 5 \\ 5 \end{bmatrix}$, $\begin{bmatrix} 0 \\ 0 \\ -5 \end{bmatrix}$; (*c*) 55°; (*d*) 71°.

Exercise E

1 ⁻6. **3** *a* = 3, *b* = ⁻1.

4 (i) (*a*) 3*x* − *y* = 3; (*b*) *x* + 3*y* = 11; (ii) (*a*) 2*x* + *y* = 3; (*b*) *x* − 2*y* = 14.

5 (*a*) 82°; (*b*) 90°. **6** ⁻4. **8** 2½.

9 (i) (*a*) 7*x* + 5*y* = 2; (*b*) 5*x* − 7*y* = 12; (ii) (*a*) 2*y* = 3*x* − 8; (*b*) 2*x* + 3*y* + 12 = 0.

Miscellaneous exercise

1 64°. **2** (*a*) 13; (*b*) $\sqrt{26}$, $\sqrt{29}$; 12.1.

3 (*a*) $\begin{bmatrix} 4 \\ 0 \\ 3 \end{bmatrix}$, $\begin{bmatrix} -1 \\ 2 \\ 2 \end{bmatrix}$; 5, 3; (*b*) 82°.

4 (*a*) 0.7; (*b*) $-\frac{4}{11}$. **5** 26, 19, 45°.

6 (a) $\begin{bmatrix} 0.6 \\ 0.8 \end{bmatrix}$; (b) (i) $^-5$; (ii) 1.8.

7 $3x - y - 5z = 0$.

8 (c) 67 m; (e) 77 m.

9 145 s.

10 2.58 km; 2.38 km.

11 25, 39, $\sqrt{1696}$, 77°.

12 (b) 6.05.

13 104°, 115 km h^{-1}.

14 13 knots, 153°.

15 133°, 21 min.

Chapter 9   *The sine and cosine functions*

Exercise A

3

x	0	30	60	90	120	150	180...
$\dfrac{k(x)}{0.01745}$	1.000	0.866	0.499	$^-0.001$	$^-0.501$	$^-0.867$	$^-1.000...$

Exercise B

1 (ii) (a) 45°; (b) 120°; (c) 135°; (d) 30°; (e) 270°.

2 (a) 85.9°; (b) 120°; (c) 390°; (d) 166°; (e) 1500°.

3 (a) $\frac{1}{8}\pi$; (b) $\frac{9}{20}\pi$; (c) $\frac{4}{5}\pi$; (d) $\frac{21}{20}\pi$; (e) $\frac{33}{10}\pi$.

4 (a) 2.217; (b) 8.412; (c) 1.592; (d) 1.993; (e) 5.133.

5 (a) 0.5; (b) 0.866; (c) $^-0.654$; (d) 1; (e) 0.841; (f) 318; (g) 0.0998; (h) 0.0100; (i) $^-0.000185$.

6 (a) 0; (b) 0; (c) $^-1$; (d) $^-1$; (e) $^-1$.

7 (ii) (a) $22\frac{1}{2}°$; (b) 60°; (c) 150°; (d) $157\frac{1}{2}°$; (e) 240°.

8 (a) 74.5; (b) 138°; (c) 203°; (d) 270°; (e) 18400°.

9 (a) $\frac{3}{8}\pi$; (b) $\frac{2}{5}\pi$; (c) $\frac{5}{8}\pi$; (d) $\frac{7}{5}\pi$; (e) $\frac{9}{4}\pi$.

10 (a) 0.805; (b) 2.007; (c) 3.264; (d) 3.702; (e) 7.994.

11 (a) 0.866; (b) 0.414; (c) $^-0.5$; (d) 0.540; (e) 318; (f) 0.909; (g) 1.00; (h) 0.500.

12 (a) 1; (b) 1; (c) $^-1$; (d) 0; (e) 0.

13 1.20, 1.94.

14 0.927, 2.214, 7.210, 8.497.

15 $^-1.721$, 1.721.

16 3.343, 6.082.

17 $\frac{1}{2}\pi$; (b) 0; (c) 0.

18 1.047, 5.236.

19 $^-6.112$, $^-3.312$, 0.171, 2.971.

20 6.484, 12.366, 19.050, 24.932.

21 2.094, 4.189.

22 (a) 0; (b) $\frac{1}{2}\pi$; (c) $\frac{1}{4}\pi$.

Exercise C

1 (a) 26.2 m; (b) 24.4 m; (c) 5.95 m; (d) 23 m.

2 (a) 1260 cm²; (b) 1350 cm².

3 (a) (i) 4190 mi; (ii) 9600 mi; (iii) 698 mi; (b) (i) 15°; (ii) 25.13.

4 1.81.

5 (a) 18.85 cm; (b) 0.127; (c) 2.4 cm.

Exercise D

1 (*b*) $(\frac{5}{18}\pi, \frac{1}{2})$;   (*c*) (i) $^-0.886$;   (ii) $^-2.598$;   (*e*) $3 \cos 3x$.

2 (*a*) $4 \cos 4x$;   (*b*) $7 \cos 7x$;   (*c*) $\frac{1}{2} \cos \frac{1}{2}x$;
   (*d*) $\frac{1}{180}\pi \cos \frac{1}{180}\pi x$;   (*e*) $6 \cos 2x$;   (*f*) $3 \cos \frac{1}{2}x$.

3 (*b*) $(\frac{1}{12}\pi, 0.866)$;   (*c*) (i) 0.5;   (ii) 2;   (*e*) $4 \cos 4x$.

4 (*a*) $5 \cos 5x$;   (*b*) $3 \cos 3x$;   (*c*) $\frac{1}{4} \cos \frac{1}{4}x$;
   (*d*) $\frac{1}{10} \cos \frac{1}{10}x$;   (*e*) $10 \cos 0.1x$;   (*f*) $\cos 10x$.

Exercise E

1 (*a*) $x \to {}^-8 \sin 8x$;   (*b*) $x \to {}^-12 \sin 4x$;   (*c*) $x \to 2 \cos 2x - 3 \sin 3x$;
   (*d*) $x \to {}^-\frac{1}{5} \sin \frac{1}{5}x$;   (*e*) $x \to {}^-\frac{1}{2} \sin \frac{1}{10}x$.

2 (*a*) $\frac{1}{3} \sin 3x + k$;   (*b*) $^-\frac{1}{10} \cos 10x + k$;   (*c*) $\dfrac{1}{\pi} \sin \pi x - \dfrac{1}{\pi} \cos \pi x + k$;
   (*d*) $^-\frac{2}{3} \cos 3x - 2 \sin 2x + k$.

3 13.

4 (*a*) (i) 1;   (ii) $^-1$;   (*b*) (i) 0;   (ii) 2.

5 (*b*) 0.866;   (*c*) 0.289.

6 (*a*) $^-\frac{5}{6}\pi \sin \frac{1}{6}\pi t$;   (*b*) (i) 2.27 m h^{-1};   (ii) 2.62 m h^{-1};   (iii) $^-1.31$ m h^{-1}.

7 (*a*) $x \to {}^-5 \sin 5x$;   (*b*) $x \to {}^-28 \sin 7x$;   (*c*) $x \to {}^-3 \sin 3x - 5 \cos 5x$;
   (*d*) $x \to {}^-\frac{1}{12} \sin \frac{1}{12}x$;   (*e*) $x \to {}^-\frac{3}{4} \sin \frac{1}{12}x$.

8 (*a*) $\frac{1}{6} \sin 6x + k$;   (*b*) $^-\frac{1}{8} \cos 8x + k$;   (*c*) $\sin 2x - \frac{1}{2} \cos 6x + k$;
   (*d*) $^-2 \cos \frac{1}{8}x - \frac{1}{6} \sin 3x + k$.

9 17.

10 (*a*) (i) 2;   (ii) $^-2$;   (*b*) (i) 1;   (ii) $^-2$.

11 (*b*) 0.134;   (*c*) 0.268.

12 (*a*) $v(t) = \frac{1}{4}\pi \cos \frac{1}{4}\pi t$;   (*b*) $t = 2, 6, 10, \ldots$;   (*c*) $a(t) = {}^-\frac{1}{16}\pi^2 \sin \frac{1}{4}\pi t$;
    (*d*) $t = 0, 4, 8, \ldots$.

Miscellaneous exercise

1 (*a*) 3;   (*b*) $^-3\pi \sin \pi t$;   (*c*) $t = 0.5$;   (*d*) $\dfrac{3}{\pi} \sin \pi t, t = 1$;   (*e*) $\dfrac{3}{\pi}$.

2 (*a*) $2r + r\theta$;   (*b*) $\frac{1}{2}r^2\theta$; $\frac{9}{32}$.

3 0.956; $p = {}^-\frac{1}{2}$ or $\pi - \frac{1}{2}$, etc.

5 $x = 0, \frac{1}{2}\pi; 0$;   (*a*) (i) $24\pi^2$;   (ii) $\frac{3}{8}\pi^2$;   (*b*) $\displaystyle\int_0^{\frac{5}{8}\pi} 12\pi^2 \sin x\, dx - \int_0^{\frac{5}{8}\pi} 9x(2x - \pi)\, dx$.

6 (*a*) $\frac{1}{2}r$;   (*b*) $\left(\dfrac{2r+1}{4}\right)^2$;   (*d*) $\frac{1}{2}$;   (*e*) 2.

7 (*a*) $\frac{1}{2}r^2 \sin \theta$;   (*b*) $\frac{1}{2}r^2\theta$;   (*c*) $\frac{1}{2}r^2(\theta - \sin \theta)$; $\theta = \frac{1}{2}\pi$.

8 (*a*) $\frac{1}{180}\pi \cos x°$;   (*c*) $h = 4 \sin \theta°$, 2;   (*d*) $2 + \dfrac{\pi^2\sqrt{3}}{5400}$.

9 (*a*) (i) 0;   (ii) 0.23;   (*b*) 0.22;   (*c*) $v(t) = 2t - \cos t$.
   (*e*) $a(t) = 2 + \sin t$, 1.

10 (*a*) $^-t^{-2} + \sin t$.

12 (a) ⁻2.356, ⁻0.262, 0.785, 2.880; (b) 0; (c) no solution.

13 (a) (i) 12.2; (ii) 24; (b) on 14th February ($t \approx 1.48$).

14 (a) (i) 2π; (ii) 2π; (iii) π; (iv) π; (v) 6π; (vi) 1.

(b) $0.8 \sin 0.8x$; (i) 0.57 m; (ii) 0.47 m; $\dfrac{2}{5\pi}$.

Chapter 10   *Kinematics*

Exercise A

1 $\begin{bmatrix} 12 \\ 5 \end{bmatrix}$; yes; 13, 067°.

2 $\begin{bmatrix} 7 \\ 24 \end{bmatrix}$; yes; 25, 016°.

3 $\begin{bmatrix} 2t+1 \\ 1 \end{bmatrix}$; no.

4 $\begin{bmatrix} 2 \\ -2 \end{bmatrix}$; yes; 2.83, 135°.

5 $\begin{bmatrix} 4 \\ 7 \end{bmatrix}$; yes; 8.06, 030°.

6 $\begin{bmatrix} -4 \\ 8 \end{bmatrix}$; yes; 8.94, 333°.

7 $\begin{bmatrix} 2t+1 \\ 2t+1 \end{bmatrix}$; no.

8 $\begin{bmatrix} 2t+3 \\ 1 \end{bmatrix}$; no.

Exercise B

1 $\begin{bmatrix} 1 \\ 2t \end{bmatrix}$; $\sqrt{(1+4t^2)}$; never at rest.

2 $\mathbf{v}(t) = \begin{bmatrix} 2t \\ 1 \end{bmatrix}$; $\mathbf{a}(t) = \begin{bmatrix} 2 \\ 0 \end{bmatrix}$.

3 (a) $t = 6$; (b) $t = 5$.

4 $\mathbf{v}(t) = \begin{bmatrix} 12t^2-2 \\ 2t-4 \end{bmatrix}$; $\mathbf{a}(t) = \begin{bmatrix} 24t \\ 2 \end{bmatrix}$.

5 $t = 1$; $\begin{bmatrix} -1 \\ 2 \end{bmatrix}$.

6 $\begin{bmatrix} 2t \\ -1 \end{bmatrix}$; $\sqrt{(4t^2+1)}$ m/s.

7 $\begin{bmatrix} -6\sin 2t \\ 6\cos 2t \end{bmatrix}$, $\begin{bmatrix} -12\cos 2t \\ -12\sin 2t \end{bmatrix}$; $t = 0, 1.57$.

8 $\begin{bmatrix} 3t^2 \\ -1 \\ t^2 \end{bmatrix}$, $\begin{bmatrix} 6t \\ 2 \\ t^3 \end{bmatrix}$.

9 2.17 m s⁻¹.

10 $\begin{bmatrix} 3\cos 3t \\ -3\sin 3t \end{bmatrix}$, $\begin{bmatrix} -9\sin 3t \\ -9\cos 3t \end{bmatrix}$; 9; $t = 1.83$.

Exercise C

1 $\mathbf{p}(t) = \begin{bmatrix} t \\ \frac{1}{2}t^2 \end{bmatrix}$, $\mathbf{a}(t) = \begin{bmatrix} 0 \\ 1 \end{bmatrix}$, $\mathbf{p}(1) = \begin{bmatrix} 1 \\ \frac{1}{2} \end{bmatrix}$.

2 $\mathbf{p}(t) = \begin{bmatrix} \frac{1}{2}t^2+2\frac{1}{2} \\ 3-t^2 \end{bmatrix}$, $\mathbf{a}(t) = \begin{bmatrix} 1 \\ -2 \end{bmatrix}$.

3 $\mathbf{p}(t) = \begin{bmatrix} \frac{1}{2}\sin 2t \\ t^2 \end{bmatrix}$, $\mathbf{a}(t) = \begin{bmatrix} -2\sin 2t \\ 2 \end{bmatrix}$, $\mathbf{p}(1) = \begin{bmatrix} 0.455 \\ 1 \end{bmatrix}$.

4 $\mathbf{p}(t) = \begin{bmatrix} t+5 \\ 2 \end{bmatrix}$, $\mathbf{a}(t) = \mathbf{0}$, $\mathbf{p}(1) = \begin{bmatrix} 6 \\ 2 \end{bmatrix}$.

5 $\mathbf{p}(t) = \begin{bmatrix} 3+t-t^2 \\ t^2+2t-9 \end{bmatrix}$, $\mathbf{a}(t) = \begin{bmatrix} -2 \\ 2 \end{bmatrix}$, $\mathbf{p}(1) = \begin{bmatrix} 3 \\ -6 \end{bmatrix}$.

6 $\mathbf{p}(t) = \begin{bmatrix} \frac{1}{3}\sin 3t \\ \frac{1}{2}-\frac{1}{2}\cos 2t \end{bmatrix}$, $\mathbf{a}(t) = \begin{bmatrix} -3\sin 3t \\ 2\cos 2t \end{bmatrix}$, $\mathbf{p}(1) = \begin{bmatrix} 0.047 \\ 0.708 \end{bmatrix}$.

7 $\mathbf{p}(t) = \begin{bmatrix} t^2+t-1 \\ \frac{1}{2}+t-\frac{1}{2}t^2 \end{bmatrix}$, $\mathbf{a}(t) = \begin{bmatrix} 2 \\ -1 \end{bmatrix}$.

8 $\mathbf{a}(t) = \begin{bmatrix} 0 \\ -0.1\sin t \end{bmatrix}$, $\mathbf{p}(t) = \begin{bmatrix} 0.25t \\ 0.1\sin t \end{bmatrix}$;   16° above horizontal.

9 $\mathbf{a}(t) = \begin{bmatrix} 12t-2 \\ 2 \end{bmatrix}$, $\mathbf{p}(t) = \begin{bmatrix} 2t^3-t^2+4 \\ t^2+3t-1 \end{bmatrix}$.

10 $\mathbf{p}(t) = \begin{bmatrix} 2t+1 \\ -5t^2 \end{bmatrix}$, $\mathbf{v}(t) = \begin{bmatrix} 2 \\ -10t \end{bmatrix}$;   never.

11 $\mathbf{p}(t) = \begin{bmatrix} \frac{1}{6}t^3 \\ \frac{1}{6}t^3+t \end{bmatrix}$, $\mathbf{v}(t) = \begin{bmatrix} \frac{1}{2}t^2 \\ \frac{1}{2}t^2+1 \end{bmatrix}$;   at $t=0$.

12 $\mathbf{v}(t) = \begin{bmatrix} 5-3\cos t \\ 1+3\sin t \end{bmatrix}$, $\mathbf{p}(t) = \begin{bmatrix} 5t-3\sin t \\ t-3\cos t+3 \end{bmatrix}$.

Exercise D

1 $\mathbf{v}(t) = \begin{bmatrix} \frac{1}{2}-\frac{1}{2}t^2 \\ 5 \\ t-1 \end{bmatrix}$.

2 $\mathbf{v}(t) = \begin{bmatrix} t+6 \\ \frac{1}{2}t^2+3 \end{bmatrix}$ or $\begin{bmatrix} t-6 \\ \frac{1}{2}t^2-3 \end{bmatrix}$.

3 $\mathbf{a}(t) = \begin{bmatrix} 1 \\ 2t \\ -1 \end{bmatrix}$, $\mathbf{p}(t) = \begin{bmatrix} \frac{1}{2}t^2-t-\frac{3}{2} \\ \frac{1}{3}t^3 \\ t-\frac{1}{2}t^2+2 \end{bmatrix}$;   2.45 m s^{-2}.

4 (a) $\begin{bmatrix} 7 \\ 1-t \\ 4-10t \end{bmatrix}$;   (b) $\begin{bmatrix} 7 \\ 0 \\ -6 \end{bmatrix}$, 9.2 m s^{-1}, $\begin{bmatrix} 7 \\ \frac{1}{2} \\ 1 \end{bmatrix}$;   (c) 49°.

5 20.1 m s^{-1}; $k = 0.2$; $\begin{bmatrix} t^2-5 \\ \frac{1}{2}t^2+2 \\ 0.1t^2-0.4 \end{bmatrix}$.

6 $t = 25$; 3.6 m s^{-1}.

7 $\mathbf{v}(t) = \begin{bmatrix} -20\sin 20t \\ 0.4t \\ 20\cos 20t \end{bmatrix}$, $\mathbf{a}(t) = \begin{bmatrix} -400\cos 20t \\ 0.4 \\ -400\sin 20t \end{bmatrix}$;   (a) 20 m s^{-1};   (b) 20.4 m s^{-1}.

Exercise E

1 $\begin{bmatrix} 9 \\ 4 \end{bmatrix}$.

2 4.

3 4.24.

4 $\begin{bmatrix} 24 \\ -44 \end{bmatrix}$.

5 2.65.

6 $\begin{bmatrix} 3\frac{1}{3} \\ -2\frac{2}{3} \end{bmatrix}$.

7 $\begin{bmatrix} -1.3 \\ 3.1 \end{bmatrix}$.

8 $\begin{bmatrix} 0 \\ -2 \\ -8 \end{bmatrix}$.

9 $\mathbf{A} = \mathbf{0}$.

10 $\begin{bmatrix} 16 \\ -28 \end{bmatrix}$.

11 $\begin{bmatrix} -3 \\ 12 \end{bmatrix}$.

12 $\begin{bmatrix} 1 \\ -1.5 \end{bmatrix}$.

Exercise F

1 5.1 m, 2.04 s.

2 24 m s^{-1}.

3 5.1 s.

4 40 m.

5 0.45 s.

6 15.8 m s^{-1}, 89°.

7 44 m.

8 9.9 m s^{-1}.

9 After 2 s; 10.4 m from the bottom.

10 0.48 m below the centre; 16.3 m s^{-1}.

11 44 m, 3 s.

12 0.26 s.

13 $\begin{bmatrix} 9.9 \\ 5.0 \end{bmatrix}$ m.

14 1.7 s, 31 m, 3.6 m.

15 13 m s^{-1} horizontally, 2.8 m s^{-1} vertically.

16 22.3 km, 28 s.

17 2.8 m, 16.

18 $\dfrac{2v}{g}\sin\theta°, \dfrac{v^2}{2g}(\sin\theta°)^2\dfrac{2v^2}{g}\sin\theta°\cos\theta°$.

Miscellaneous exercise

1 21.8 cm s^{-1} on a bearing of 323°; 7 s.

2 22.6 m, 1.63 s.

3 $2\mathbf{i}+12\mathbf{j}$; (*b*) $\mathbf{i}+3\mathbf{j}$; (*c*) $\mathbf{i}$; (*d*) 16; (*e*) $t = 8$; (*f*) 83°.

4 (*a*) $\begin{bmatrix} 1 \\ 4 \end{bmatrix}$.

5 (*a*) 10.3 m s^{-1}; (*b*) 31 m at 15° above the horizontal;
(*c*) 65° below the horizontal; (*d*) after 1.8 s.

6 $\begin{bmatrix} 5 \\ 9 \end{bmatrix}$ m s^{-1}; 10.3 m s^{-1}.

7 $\begin{bmatrix} 10 \\ -9.4 \end{bmatrix}$ m s^{-1}; $\begin{bmatrix} 30 \\ 15.9 \end{bmatrix}$ m.

8 (*a*) $\begin{bmatrix} 20t \\ 35t-5t^2 \end{bmatrix}$; (*c*) 37°.

9 (*a*) 15 m; (*b*) 2.3 s; (*d*) 49.3 m.

10 (*a*) $3\mathbf{i}+4\mathbf{j}+10t\mathbf{k}$; (*b*) 323°; (*d*) 22.4 m; (*e*) 4 s; (*f*) $10\mathbf{k}$, no.

11 (i) 10 m s^{-1}; (ii) (*a*) 5.8 s; (*b*) 8.7 m s^{-1}.

12 (*a*) 3 m s^{-1}; (*c*) 119°; (*d*) $\begin{bmatrix} 2t-1 \\ t+2 \\ 7t-2 \end{bmatrix}$; (*e*) after $\frac{9}{14}$ s.

13 (*a*) 71 m s^{-1}; (*b*) 45°; (*c*) $^{-}$10**j**; (*d*) 50t**i** + (70t − 5t²)**j**; (*e*) *t* = 7;
(*f*) 245; (*g*) no.

14 (*a*) 12 m s^{-1}; (*b*) 27 m; (*c*) falling.

Revision exercise 2

1 $x = \frac{3}{4}$; $x = ^{-}0.281$ or 1.78; $^{-}0.281 < x < 1.78$.

2 (*a*) 15, 16, 17; (*b*) 11, 12, 13; (*c*) 7, 8, 9 or $^{-}$7, $^{-}$8, $^{-}$9.

3 025°, 158 s.

4 (*a*) 2.24, 1.41, 3.16; (*c*) 79°, 18°, 63°.

5 (*b*) $12t^2 + 12tm + 4m^2$; $12t^2$.

6 $6M - \dfrac{5}{M^2}$; (*a*) 14; (*b*) 16.4; (*c*) 30.65.

7 4, $^{-}$8, 0; (*a*) 1.75; (*b*) 0.25.

8 $k = \frac{1}{2}m$, m^2.

9 (*a*) 23°, 337°; (*b*) 63°, 297°; (*e*) 123°, 237°.

10 (*a*) 8.8°; (*b*) 1.5 *s*.

11 (*a*) 15, 30, 105, 120; (*b*) $52\frac{1}{2}$, $82\frac{1}{2}$, $142\frac{1}{2}$, $172\frac{1}{2}$;
$\{x: 0 \leqslant x \leqslant 15\} \cup \{x: 30 \leqslant x \leqslant 52\frac{1}{2}\} \cup \{x: 82\frac{1}{2} \leqslant x \leqslant 105\} \cup$
$\{x: 120 \leqslant x \leqslant 142\frac{1}{2}\} \cup \{x: 172\frac{1}{2} \leqslant x \leqslant 180\}$.

12 $YZ = 12.6$, $X = 59°$, $Z = 74°$ or $YZ = 6.7$, $X = 27°$, $Z = 106°$.

13 (*a*) 74; (*b*) 10.77, 11.40; (*c*) 53°.

14 (*a*) $a = 0.7$; (*b*) $b = ^{-}1$.

15 (*a*) $\begin{bmatrix} 9.33 \\ 25.64 \end{bmatrix}$ m s^{-1}; (*b*) 34 m; (*c*) falling.

16 (*a*) 3**i** + 9**j**; (*b*) **i** + **j**; (*c*) **i**; (*d*) 9; (*e*) *t* = 6; (*f*) 81°.

17 (*a*) $^{-}\frac{7}{8}\pi$, $^{-}\frac{3}{8}\pi$, $\frac{1}{8}\pi$, $\frac{5}{8}\pi$; (*b*) $^{-}\frac{23}{24}\pi$, $^{-}\frac{7}{24}\pi$, $\frac{1}{24}\pi$, $\frac{17}{24}\pi$; (*c*) $^{-}\sqrt{2}$, $\sqrt{2}$; (*d*) $\frac{1}{2}$.

18 *t* = 1 or 5; *t* = 3. (*a*) *v*(2) = 18, *a*(2) = 12; (*b*) *v*(6) = $^{-}$30, *a*(6) = $^{-}$36.

19 (1, 0) maximum; $(\frac{7}{3}, ^{-}\frac{32}{27})$, minimum.

20 $^{-}$10.

21 (*a*) $8x^2 - 12x - 6$; (*b*) $6x^2 - 15x + 4$; (*c*) $16ac + 11bc + 5ad + 14bd$;

(*d*) $10pq + 21q^2$; (*e*) $2 + \dfrac{x}{y} + \dfrac{y}{x}$; (*f*) $ab^2 + ba^2 + \dfrac{a^3}{b} + \dfrac{b^3}{a}$;

(*g*) $\dfrac{p^4}{q} - \dfrac{q^3}{p^2}$; (*h*) $\dfrac{1}{6} + \dfrac{x}{5} - \dfrac{y}{5} - \dfrac{x}{2y} + \dfrac{y}{3x}$.

22 (*a*) $ab(c^2 + a + b)$; (*b*) $fg^2(f-1)(f^2+f-5)$; (*c*) $(x+1)(x+4)$; (*d*) $(x+6)^2$;
(*e*) $(3x+1)^2$; (*f*) $(x-1)(3x-2)$; (*g*) $(4x+1)(2x-3)$.

23 (*a*) (i) $(x+3)^2 + 2$; (iii) 2; (iv) $x = ^{-}3$; (v) none.
(*b*) (i) $(x+2)^2 - 3$; (iii) $^{-}$3; (iv) $x = ^{-}2$; (v) $^{-}$3.7, $^{-}$0.3.
(*c*) (i) $(x-4)^2 + 1$; (iii) 1; (iv) $x = 4$; (v) none.

(d) (i) $(x+2.5)^2-3.25$; (iii) $^-3.25$; (iv) $x=^-2.5$; (v) $^-4.3, ^-0.7$.
(e) (i) $(x-0.5)^2+2.75$; (iii) 2.75; (iv) $x=0.5$; (v) none.

25 16.6 m s^{-1} at 65° to the bank.

26 037°, 4 m s^{-1}.

27 045°, 33 min.

28 7.1 km.

29 $\sqrt{8}, \begin{bmatrix} 7 \\ 5 \end{bmatrix}$.

30 (a) $\mathbf{v}(t) = \begin{bmatrix} 12t^2-48t \\ 4t-1 \end{bmatrix}$, $\mathbf{a}(t) = \begin{bmatrix} 24t-48 \\ 4 \end{bmatrix}$; (b) no;
(c) $t=\frac{1}{4}$, 42.2 m s^{-2}; $t=0$, 48.2 m s^{-2}; $t=4$, 48.2 m s^{-2}.

31 $\mathbf{v}(t) = \begin{bmatrix} t^2+1 \\ 3t+1 \end{bmatrix}$, $\mathbf{p}(t) = \begin{bmatrix} \frac{1}{3}t^3+t+2\frac{2}{3} \\ \frac{3}{2}t^2+t-\frac{1}{2} \end{bmatrix}$.

32 7; $(^-\frac{36}{7}, \frac{785}{7})$.

Chapter 11 Probability

Exercise A

1 $p(A)=\frac{1}{2}$, $p(B)=\frac{3}{13}$, $p(C)=\frac{1}{4}$, $p(\sim A)=\frac{1}{2}$, $p(\sim B)=\frac{10}{13}$,
$p(\sim C)=\frac{3}{4}$, $p(A \text{ and } B)=\frac{3}{26}$, $p(B \text{ and } C)=\frac{3}{52}$, $p(A \text{ or } C)=\frac{3}{4}$, $p(B \text{ or } \sim C)=\frac{21}{26}$.

2 $p(D)=\frac{1}{6}$, $p(\sim E)=\frac{5}{9}$, $p(F)=\frac{1}{9}$, $p(\sim F)=\frac{8}{9}$, $p(D \text{ and } E)=\frac{1}{36}$,
$p(D \text{ or } E)=\frac{11}{36}$, $p(\sim E \text{ or } F)=\frac{31}{36}$.

3 $p(V)=\frac{5}{26}$, $p(\sim T)=\frac{17}{26}$, $p(V \text{ and } T)=\frac{2}{13}$, $p(V \text{ or } T)=\frac{5}{13}$, $p(\sim V \text{ and } T)=\frac{5}{26}$.

4 $p(\sim G)=\frac{13}{30}$, $p(L)=\frac{4}{15}$, $p(G \text{ or } L)=\frac{2}{3}$, $p(\sim G \text{ and } L)=\frac{1}{10}$, $p(G \text{ and } \sim L)=\frac{2}{5}$.

Exercise B

1 $p(A|B)=\frac{1}{4}$, $p(B|A)=\frac{3}{13}$, $p(B|C)=\frac{3}{13}$.

2 $p(D|F)=\frac{1}{4}$, $p(F|D)=\frac{1}{6}$, $p(F|\sim D)=\frac{1}{10}$.

3 $p(V|T)=\frac{4}{9}$, $p(T|V)=\frac{4}{5}$, $p(T|\sim V)=\frac{5}{21}$.

4 $p(G|L)=\frac{5}{8}$, $p(L|G)=\frac{5}{17}$, $p(\sim L|\sim G)=\frac{10}{13}$.

Exercise C

1 (a) (i) $\frac{81}{256}$; (ii) $\frac{9}{16}$; (b) (i) $\frac{3}{10}$; (ii) $\frac{9}{16}$; $\frac{12}{35}$.

2 $\frac{20}{39}$; $\frac{14}{33}$.

3 (a) (i) $\frac{25}{289}$; (ii) $\frac{5}{17}$; (b) (i) $\frac{5}{68}$; (ii) $\frac{5}{17}$; $\frac{1}{20}$.

4 $\frac{55}{171}$; $\frac{28}{153}$. **5** $\frac{1}{2}$. **6** $\frac{2}{7}$.

7 $\frac{1}{3}$. **8** $\frac{7}{11}$. **9** $\frac{21}{46}$.

10 $\frac{1}{3}$. **11** $\frac{1}{2}$. **12** $\frac{5}{12}$.

13 (a) $\frac{1}{6}$; (b) $\frac{7}{30}$; (c) $\frac{5}{14}$. **14** 0.125.

Exercise D

1 (*a*) 0.49; (*b*) 0.42; (*c*) 0.09.

2 (*a*) 0.027; (*b*) 0.189; (*c*) 0.441; (*d*) 0.343.

3 (*a*) 0.0081; (*b*) 0.0756; (*c*) 0.2646; (*d*) 0.4116; (*e*) 0.2401.

4 (*a*) p^2; (*b*) $2pq$; (*c*) q^2.

5 (*a*) q^3; (*b*) $3pq^2$; (*c*) $3p^2q$; (*d*) p^3.

6 (*a*) q^4; (*b*) $4pq^3$; (*c*) $6p^2q^2$; (*d*) $4p^3q$; (*e*) p^4.

Exercise E

1 0.104.

2 (*a*) 0.178; (*b*) 0.356; (*c*) 0.297; (*d*) 0.169.

3 0.302.   **4** 0.5.   **5** 0.10.

6 0.151.   **7** (*a*) 0.140; (*b*) 0.072.   **8** 0.343.

9 0.353.   **10** 1.   **11** 11.

12 $\frac{2}{7}$.

Miscellaneous exercise

1 0.384, 0.137.

2 (*b*) 0.070.

3 (i) 0.176; (ii) 0.302; 1 or 2.

4 (*a*) $9 \times 2^{13} \times 10^{-5}$; (*b*) $821 \times 2^5 \times 10^{-5}$; (*c*) $97 \times 2^{10} \times 10^{-5}$.

5 (i) (*a*) 0.03; (*b*) 0.27; (ii) (*a*) 0.48; (*b*) 0.92; (iii) $\frac{1}{17}$; (iv) $\frac{27}{34}$.

6 (i) $\frac{5}{32}$; (ii) $\frac{13}{16}$; $\frac{24}{81}$.

7 $\frac{5}{7}$.

Chapter 12   Statistics

Exercise A

1 Mean = 49; range = 26; i.q.r. = 12; m.a.d. = 5.8.

2 Mean = 1.63 m; range = 0.34 m; i.q.r. = 0.15 m; m.a.d. = 0.085 m.

3 Mean = 3; range = 12; i.q.r. = 3; m.a.d. = 2.2.

4 Mean = 17; range = 58; i.q.r. = 22.5; m.a.d. = 13.8.

Exercise B

1 9.0, 3.0.   **2** 12.5, 3.   **3** 11.6, 5.6.

4 A 23.8, 6.5; B 23.75, 30.2.   **5** 116.3, 7.0.

Exercise C

1 2.8 (mean = 5.7).   **2** 2.8 (mean = 6.7).   **3** 28 (mean = 57).

4 14 (mean = 31.6).

Exercise D

1	9.6, 4.1.	2	11.6, 4.1.	3	28.9, 12.2.
4	1.096, 0.041.	5	18, 7.8.	6	25, 7.8.
7	37, 15.7.	8	3.8, 0.78.		

Exercise E

1	4.3, 1.7.	2	2.5 km, 1.2 km.	3	117.3, 6.2.
4	7.0, 2.2.	5	47.6 s, 1.0 s.	6	38, 15.

Exercise G

1 130 g, 22 g; 94%.   2 35 m.p.h., 11 m.p.h.; 75%.
3 21 tonne, 8 tonne; 38%.

Exercise H

1	46.4.	2	9.7.	3	13.6.
4	95.5.	5	69.1.	6	8.6.

Exercise I

1	(*a*) 7; (*b*) 16.	2	20.	3	50 min.
4	8.36, 3 min.	5	(*a*) 12; (*b*) 23.	6	52 min.
7	(*a*) 62; (*b*) 39.	8	251 g.		

Miscellaneous exercise

1 (ii) 0.37;   (iii) £22, £9.
2 (*b*) 0.25;   (*c*) 43 cm;   (*d*) 9.7 cm.
3 21 months, 2 months; (*a*) 0.006;   (*b*) 34%;   (*c*) 25.6 months.
4 1.75 m; 62%.
5 Standard deviation = 2.0 s.
6 For a Normal population: 50, 49, 42, 25, 8, 1, 0.

Chapter 13   *Correlation*

Exercise B

1	0.14.	2	0.6.	3	0.67.
4	⁻0.2.	5	0.53.	6	0.27.

Exercise C

1 Not significant at 5% level.   2 Not significant at 5% level.
3 Significant at 1% level.   4 Not significant at 5% level.

5 Not significant at 5% level.   **6** Not significant at 5% level.

7 $\frac{1}{24}$.   **8** 0.005.

Miscellaneous exercise

1 Significant at 5% level.   **2** (a) 0.5; (b) 0.57; (c) 5%.

3 0.81; 0.76; Mrs B, but not Mrs A.   **4** 0; 0.3.

Revision exercise 3

1 (i) (a) $\frac{1}{6}$; (b) $\frac{5}{6}$; (c) 0.016; (ii) (a) $\frac{1}{6}$; (b) $\frac{29}{36}$; (c) 0.0007.

2 0.6.

3 0.2; (a) 0.32; (b) 0.18; (c) 0.5; 0.64.

4 (a) (i) $\frac{1}{170}$; (ii) $\frac{5}{272}$; (iii) $\frac{15}{136}$; (b) $\frac{15}{136}$.

5 (i) (a) 0.07776; (b) 0.92224; (c) 0.3456; (d) 0.01024; (ii) (a) $\frac{1}{42}$; (b) $\frac{41}{42}$; (c) $\frac{10}{21}$; (d) 0.

6 (i) (a) 0.1; (b) 0.04; (c) 0; (ii) (a) 0.86; (b) $\frac{2}{7}$.

7 (a) 0.738; (b) 0.211; 0.805.

8 (a) 0.6; (b) 0.75.

9 0.395; (a) 0.176; (b) 0.205.

10 0.784. (i) 0.054; (ii) 0.946; (iii) 0.216.

11 (i) 0.263; (ii) 0.351. 0.121.

12 (i) $\frac{1}{3}$; (ii) $\frac{1}{7}$; (iii) $\frac{1}{21}$; (iv) $\frac{3}{7}$; (v) 0.164.

13 4.8, 2.6.

14 (a) 19; (b) 3.2; (i) 1.4%; (ii) 15.

15 (i) 60, 15; (ii) 0.212; (b) 92.

16 6.7% ; 1.042 kg.

17 0.065 mm; 23%.

18 (a) 30%; (b) 65%; (c) 26; (e) $^{+}1$; (f) 1.45; (g) 0.3%.

19 5.5%; (a) 3.8 g; (b) 503.9 g.

20 $p = 0.5, q = 1.05; m = 52, s = 23$; 1.06; 76.

21 (a) 106; (b) 75 and 125; (c) 38.

22 (a) *CBADEFIHG, ADBCEGHFI*; (b) *E*; (d) 0.05; (e) 34; (f) less than 0.001, 0.

23 (a) $^{-}\frac{5}{7}$; (b) no.

24 24; 2134, 1243; 400 000.

25 (i) 0.73; (iii) 9.

Chapter 14   *Dynamics of a particle*

Exercise C

1 $\begin{bmatrix} -\frac{1}{3} \\ \frac{2}{3} \end{bmatrix}$ m s^{-2}.   **2** $\begin{bmatrix} 8 \\ 0 \\ -12 \end{bmatrix}$ N.   **3** $\begin{bmatrix} 4000 \\ 3000 \end{bmatrix}$ N.

4 $\begin{bmatrix} 0.25 \\ 0.125 \\ 0.375 \end{bmatrix}$ m s^{-2}; $\begin{bmatrix} 1 \\ 0.5 \\ 1.5 \end{bmatrix}$ m s^{-1}. **5** $\begin{bmatrix} 1.5 \\ 2.5 \end{bmatrix}$ N. **6** $\begin{bmatrix} 0.5 \\ 1.5 \\ 0.25 \end{bmatrix}$ m s^{-2}.

7 $\begin{bmatrix} 400 \\ 600 \end{bmatrix}$ N.   **8** 2 kg; $k = 8$;   9.2 N.

9 $\begin{bmatrix} 4 \\ 3 \end{bmatrix}$ m s^{-2}; $\begin{bmatrix} 32 \\ 24 \end{bmatrix}$ m s^{-1}.   **10** $\begin{bmatrix} t \\ 0.5t \\ 1.5t \end{bmatrix}$ m s^{-2}; $\begin{bmatrix} 4.5 \\ 2.25 \\ 6.75 \end{bmatrix}$ m s^{-1}.

Exercise D

1 120 N.

2 4.2×10^{-10} N; 7100 s.

3 2000 N.

4 47° to initial direction.

5 60000 N; $^{-}0.1$ m s^{-2}.

6 0.6 N.

7 1040 N.

8 90 N.

9 1800 m s^{-2}; yes.

10 2 kg.

11 1875 N at 37° to initial direction.

12 1000 N; 50 s.

13 35000 N.

14 5000 N, 2000 N.

15 102 N.

Exercise E

1 6.7×10^{-9} N.

2 1.22×10^6 kg.

3 9800 N.

4 0.102 kg.

5 137 N.

7 147000 N.

8 25 kg.

Exercise F

1 77 m s^{-1}.

2 540 N.

3 18 m s^{-1}.

4 402.5 N.

5 16 kg.

6 7000 N.

7 0.8 m.

8 592 N.

9 4 N; 3.2 s.

10 (*a*) 700 N;   (*b*) 300 N.

Exercise G

1 7 m s^{-2}.

2 0.245 m s^{-2}; 18 s.

3 30°.

4 14 m s^{-1}; 2.8 s.

5 21°.

6 7 m s^{-2}.

7 33 kg.

8 17 N.

9 40 N.

10 9.9 N.

Miscellaneous exercise

1 (*a*) 0.06 N;   (*b*) 9 s;   (*c*) 6 s.

2 $\begin{bmatrix} 2 \\ 1 \end{bmatrix}$ m s^{-2}. $\begin{bmatrix} 4 \\ 16 \end{bmatrix}$ m, $\begin{bmatrix} 5 \\ 6 \end{bmatrix}$ m s^{-1}.

3 0.433 N, 5.25 N.

4   0.2 m s^{-2}; 3.2 m s^{-2}.

5   9600 N; 39°.

6   (a) $\begin{bmatrix} 1 \\ -2 \end{bmatrix}$ m s^{-2};   (c) $\begin{bmatrix} 36 \\ 0 \end{bmatrix}$ m;   (d) after 5 s.

7   7 s, 18 m s^{-1}.

8   (a) $\begin{bmatrix} 20t \\ 35t - 5t^2 \end{bmatrix}$;   (c) 37°;   (d) vertically downwards.

9 $\mathbf{v} = (6t + 2)\mathbf{i} - 8t\mathbf{j}$; $\mathbf{a} = 6\mathbf{i} - 8\mathbf{j}$; 53° below x-axis.

Chapter 15   *Interaction*

Exercise A

1   6000 N, 3000 N.   **2**   2.4 m s^{-2}, 1700 N.   **3**   20 000 N, 2.25 m s^{-2}.
4   10 750 N, 9250 N.   **5**   4000 N.   **6**   4.4 N, 1.9 N.
7   1500 N, 500 N.   **8**   610 N.

Exercise B

1   2 m s^{-2}, 240 N, 1.4 s.   **2**   55 N.
3   0.4 N, 0.8 N, 1.2 N, 1.6 N, 2.0 N.   **4**   37 N, 0.55 s.
5   0.53 m s^{-2}, 2.5 m s^{-1}.   **6**   87.2 N, 1.09 m s^{-2}.
7   9.8 N, 92.7 N.   **8**   866 N, 250 N.
9   0.46 m s^{-2}, 16 400 N, 26 500 N; 52 500 N.   **10**   75 N, 6.0 m s^{-2}.

Exercise C

1 $\begin{bmatrix} 12 \\ 42 \end{bmatrix}$ N s, $\begin{bmatrix} 3 \\ 10.5 \end{bmatrix}$ m s^{-1}.   **2** $\begin{bmatrix} 6.5 \\ 1 \end{bmatrix}$ N s.   **3** $\begin{bmatrix} 6 \\ 1.7 \end{bmatrix}$ m s^{-1}.

4   7.5 m s^{-1} in the direction of the impulse.

5   3600 N s.   **6** $\begin{bmatrix} 2 \\ 12 \end{bmatrix}$ N s, 3.04 N.   **7** $\begin{bmatrix} 20 \\ -25 \end{bmatrix}$ N s.

8 $\begin{bmatrix} 14 \\ -11 \end{bmatrix}$ m s^{-1}.   **9** $\begin{bmatrix} 6 \\ 11 \end{bmatrix}$ N s.

10   5 m s^{-1} in the direction of the impulse.

Exercise D

1   67 s, 1 s.
2   (a) 0.7 m s^{-1};   (b) 7 m s^{-1}.
3   13 kg; 130 N s; 130 N.
4   0.47 N s at 38° to the original direction; 2.35 N at 38° to the original direction.
5 $\begin{bmatrix} 0.5 \\ 1 \end{bmatrix}$ N s, 2.8 m s^{-1}.

6   45 m s^{-1}.

7 240 N s.

8 4400 N.

9 1.7 s.

10 210 m s^{-1} at 17° to the original direction.

11 9.8 m s^{-1}.

12 35%.

Exercise E

1 2.1 m s^{-1}.

2 1 m s^{-1}.

3 42 km h^{-1}.

4 0.06 m s^{-1}.

5 0.12 m s^{-1}.

6 15 m s^{-1}.

7 $\begin{bmatrix} 1.8 \\ 1 \end{bmatrix}$ m s^{-1}, $\begin{bmatrix} 4.2 \\ -1 \end{bmatrix}$ m s^{-1}.

8 1.55 m s^{-1}.

9 12 m s^{-1}, 4800 N.

10 6.5 m s^{-1} at 27° to original direction of the lighter skater.

11 (*a*) 0.8 m s^{-1}; (*b*) 0.75 m s^{-1}.

12 3×10^4 m s^{-1}.

13 (*a*) 0.43 m s^{-1}; (*b*) 2.7 m s^{-1} at 51° to original direction of the lighter lump.

14 (*a*) 43 m s^{-1}; (*b*) 46 m s^{-1}.

15 2 m s^{-1}.

Miscellaneous exercise

1 1.8 N s, 1.2 N s.

2 $\begin{bmatrix} 20 \\ -40 \end{bmatrix}$ N s, $\begin{bmatrix} 10 \\ -20 \end{bmatrix}$ N.

3 17 m s^{-1}; 64° to the direction from which it came.

4 (*a*) 1200 N; (*b*) 200 N.

5 $\begin{bmatrix} 1.8 \\ 0.8 \end{bmatrix}$ m s^{-1}; $\pm \begin{bmatrix} 2.4 \\ 8.4 \end{bmatrix}$ N.

6 0.5 m s^{-2}; 600 N.

7 51 km h^{-1} at 61° to the car's original direction.

8 192 m; 480 N towards *A*.

9 30 N s, 090°; 15 m s^{-1}, 090°; 65 m s^{-1}.

10 $\begin{bmatrix} 0.5 \\ 2 \\ 3 \end{bmatrix}$ m s^{-1}; $\pm \begin{bmatrix} 7.5 \\ 0 \\ -15 \end{bmatrix}$ N.

Revision exercise 4

1 (ii) (a) 9120 N;   (b) 1824 N.

2 (a) $\begin{bmatrix} 1.25 \\ 2 \\ 4.25 \end{bmatrix}$ m s^{-1};   (b) $\begin{bmatrix} -2 \\ 0 \\ -10 \end{bmatrix}$ N s;   (c) 40°.

3 50 N s at 317°; 17 m s^{-1}.

4 60 N; 3 m s^{-2}, 250 N; 0.5 m s^{-2}.

5 $\begin{bmatrix} 2t-1 \\ 2-2t \end{bmatrix}$ m s^{-1}; $t = 3$.

6 2.8 m s^{-1}.

7 (a) 0.16 m s^{-2};   (c) 3200 N.

8 (i) 5 N s;   (ii) 127° to the original direction;   (iii) by 14 N s.

9 0.6 m s^{-2}, 2700 N; 1960 N.

10 (b) $\begin{bmatrix} 1.2 \\ 1.2 \end{bmatrix}$.

11 (c) 3.0 m s^{-2}.

12 (b) 730 N;   (c) 8.6 s.

13 (i) 2150 N, 2290 N;   (ii) 268 N, 9.2 m s^{-2}.

14 (b) 4 m s^{-1};   (d) 17 N s at 135°;   (e) 5.7 m s^{-1}.

16 (a) 5 m s^{-1};   (b) 25 m s^{-1};   (c) 13 m s^{-1}.

17 12 N; 1.58 s; 2 m.

18 1000**i** − 1000**j**; 30°.

19 0.33 m s^{-2}.

20 2 m s^{-1}; heavier: 3.5 m s^{-1}, lighter: 5.2 m s^{-1}.

Chapter 16   *Plane transformations*

Exercise A

1 (a) (i) Rotation of 108° about *A*;   (ii) reflection in *AO*;
   (b) (i) Rotation of 144° about *O*;   (ii) glide with axis parallel to *BC*;
   (c) (i) Rotation of 36° about the point of intersection of *AB* and *CD*;
       (ii) reflection in *EO*.

4 (a), (b) and (d) are true.

5 (b), (c) and (d) are true.

8 (a) (i) Rotation of 120° about *P*;   (ii) reflection in *PS*;
   (b) (i) rotation of 240° about *O*;   (ii) glide with axis parallel to *QR*;
   (c) (i) rotation of 60° about *X*, the point of intersection of *PQ* and *RS*;
       (ii) reflection in *XO*.

10 (b) and (c) are true.

11 (c) and (d) are true.

Exercise B

1 $(-374, 486)$.

2 $\mathbf{R}^2 = \begin{bmatrix} -0.5 & -0.866 \\ 0.866 & -0.5 \end{bmatrix}$, $\mathbf{R}^3 = \begin{bmatrix} 1 & 0 \\ 0 & 1 \end{bmatrix}$.

3 $\begin{bmatrix} -0.9 & 0.5 \\ 0.5 & 0.9 \end{bmatrix}$.

4 (a) $\begin{bmatrix} -0.8 & 0.6 \\ 0.6 & 0.8 \end{bmatrix}$; (b) $\begin{bmatrix} -0.8 & -0.6 \\ -0.6 & 0.8 \end{bmatrix}$; (c) $\begin{bmatrix} 0.8 & 0.6 \\ 0.6 & -0.8 \end{bmatrix}$; (d) $\begin{bmatrix} 0.8 & -0.6 \\ -0.6 & -0.8 \end{bmatrix}$.

6 $\begin{bmatrix} -0.2 & 0.6 \\ -2.4 & 2.2 \end{bmatrix}$.

7 (a) Rotation of $16°$; (b) reflection in $7y = x$; (c) rotation of $^-16°$; (d) reflection in $y = 7x$.

8 $(406, 605)$.

9 $\mathbf{R}^2 = \begin{bmatrix} -0.5 & 0.866 \\ -0.866 & -0.5 \end{bmatrix}$.

10 $\begin{bmatrix} -0.9 & 0.4 \\ 0.4 & 0.9 \end{bmatrix}$.

11 (a) $\begin{bmatrix} -0.6 & 0.8 \\ 0.8 & 0.6 \end{bmatrix}$; (b) $\begin{bmatrix} 0.6 & 0.8 \\ 0.8 & -0.6 \end{bmatrix}$; (c) $\begin{bmatrix} -0.6 & -0.8 \\ -0.8 & 0.6 \end{bmatrix}$; (d) $\begin{bmatrix} 0.6 & -0.8 \\ -0.8 & -0.6 \end{bmatrix}$.

13 $\begin{bmatrix} 0.4 & 0.2 \\ -1.8 & 1.6 \end{bmatrix}$.

14 (a) Reflection in $y = \frac{9}{13}x$; (b) rotation of $21°$; (c) rotation of $201°$; (d) rotation of $291°$.

Exercise C

1 $\dfrac{1}{\sqrt{17}}\begin{bmatrix} 1 & -4 \\ 4 & 1 \end{bmatrix}$.

2 $\begin{bmatrix} 1 & -2 \\ 2 & 1 \end{bmatrix}$.

3 $\dfrac{1}{\sqrt{2}}\begin{bmatrix} 1 & -1 \\ 1 & 1 \end{bmatrix}$.

4 $\dfrac{1}{\sqrt{10}}\begin{bmatrix} 3 & 1 \\ -1 & 3 \end{bmatrix}$.

5 $\dfrac{1}{\sqrt{10}}\begin{bmatrix} 1 & 3 \\ -3 & 1 \end{bmatrix}$.

6 $\dfrac{1}{\sqrt{130}}\begin{bmatrix} 11 & -3 \\ 3 & 11 \end{bmatrix}$.

7 $\dfrac{1}{\sqrt{5}}\begin{bmatrix} 1 & -2 \\ 2 & 1 \end{bmatrix}$, $\dfrac{1}{\sqrt{26}}\begin{bmatrix} 1 & -5 \\ 5 & 1 \end{bmatrix}$.

8 $\dfrac{1}{\sqrt{50}}\begin{bmatrix} 1 & -7 \\ 7 & 1 \end{bmatrix}$.

9 $\begin{bmatrix} 1 & -3 \\ 3 & 1 \end{bmatrix}$.

10 $\dfrac{1}{\sqrt{2}}\begin{bmatrix} 1 & 1 \\ -1 & 1 \end{bmatrix}$.

11 $\dfrac{1}{\sqrt{5}}\begin{bmatrix} 2 & 1 \\ -1 & 2 \end{bmatrix}$.

12 $\dfrac{1}{\sqrt{13}}\begin{bmatrix} 3 & -2 \\ 2 & 3 \end{bmatrix}$.

13 $\dfrac{1}{\sqrt{170}}\begin{bmatrix} 13 & -1 \\ 1 & 13 \end{bmatrix}$.

14 $\dfrac{1}{\sqrt{10}}\begin{bmatrix} 1 & -3 \\ 3 & 1 \end{bmatrix}$, $\dfrac{1}{\sqrt{17}}\begin{bmatrix} 1 & -4 \\ 4 & 1 \end{bmatrix}$.

Exercise D

1 $\begin{bmatrix} -0.6 & 0.8 \\ 0.8 & 0.6 \end{bmatrix}$.
   2 $\begin{bmatrix} 0.6 & -0.8 \\ -0.8 & -0.6 \end{bmatrix}$.
   3 $\begin{bmatrix} -1 & 0 \\ 0 & -1 \end{bmatrix}$.

4 $\begin{bmatrix} 0.4 & 0.2 \\ -1.8 & 1.6 \end{bmatrix}$.
   5 $\begin{bmatrix} 1.98 & -0.14 \\ -0.14 & 1.02 \end{bmatrix}$, $\begin{bmatrix} 1.02 & 0.14 \\ 0.14 & 1.98 \end{bmatrix}$, $\begin{bmatrix} 2 & 0 \\ 0 & 2 \end{bmatrix}$.

6 $\begin{bmatrix} 0.6 & 0.8 \\ 0.8 & -0.6 \end{bmatrix}$.
   7 $\begin{bmatrix} -0.6 & -0.8 \\ -0.8 & 0.6 \end{bmatrix}$.
   8 $\begin{bmatrix} -1 & 0 \\ 0 & -1 \end{bmatrix}$.

9 $\begin{bmatrix} -0.2 & 0.6 \\ -2.4 & 2.2 \end{bmatrix}$.
   10 $\begin{bmatrix} \frac{38}{13} & -\frac{5}{13} \\ -\frac{5}{13} & \frac{14}{13} \end{bmatrix}$, $\begin{bmatrix} \frac{14}{13} & \frac{5}{13} \\ \frac{5}{13} & \frac{38}{13} \end{bmatrix}$, $\begin{bmatrix} 3 & 0 \\ 0 & 3 \end{bmatrix}$.

Exercise E

1 (*a*) $(0,0)$; rotation of $37°$;   (*b*) reflection in $y = 3x$;   (*c*) reflection in $x = 3y$.

2 Area invariant; shear with $y = 2x$ invariant.

3 One-way stretch, scale factor 3, with $y = 3$ invariant.

4 (*a*) One-way stretch, scale factor 5, with $x = 7y$ invariant.
   (*b*) Shear with shearing constant 5 and $x = 7y$ invariant.
   (*c*) Reflection in $x = 7y$.

5 (*a*) $(0,0)$; rotation of $286°$;   (*b*) reflection in $3y = 4x$;   (*c*) reflection in $4y = 3x$.

6 Area invariant; shear with $y = 7x$ invariant.

7 One-way stretch, scale factor 4, with $2y = 3x$ invariant.

8 (*a*) Shear with shearing constant $^-3$ and $2x + y = 0$ invariant.
   (*b*) One-way stretch, scale factor 3, with $2x + 7y = 0$ invariant.
   (*c*) Rotation of $16°$ about $(0,0)$.

Exercise F

1 $\begin{bmatrix} x \\ y \end{bmatrix} \rightarrow \begin{bmatrix} y-1 \\ 11-x \end{bmatrix}$.
   2 $\begin{bmatrix} x \\ y \end{bmatrix} \rightarrow \begin{bmatrix} 0.5x - 0.866y + 4.598 \\ 0.866x + 0.5y - 1.964 \end{bmatrix}$.

3 $\begin{bmatrix} x \\ y \end{bmatrix} \rightarrow \begin{bmatrix} -0.8x + 0.6y - 3 \\ 0.6x + 0.8y + 1 \end{bmatrix}$.
   4 $\begin{bmatrix} x \\ y \end{bmatrix} \rightarrow \begin{bmatrix} 0.22x - 0.98y + 1.17 \\ -0.98x - 0.22y + 1.47 \end{bmatrix}$.

5 (*a*) $\begin{bmatrix} x \\ y \end{bmatrix} \rightarrow \begin{bmatrix} 4.92x - 0.56y - 2.24 \\ -0.56x + 1.08y + 0.32 \end{bmatrix}$; $P'\,(11.96, \,^-0.28)$, $Q'\,(1, 3)$.

   (*b*) $\begin{bmatrix} x \\ y \end{bmatrix} \rightarrow \begin{bmatrix} 0.58x + 0.06y + 0.24 \\ -2.94x + 1.42y + 1.68 \end{bmatrix}$; $P'\,(2.04, \,^-5.72)$, $Q'\,(1, 3)$.

6 (*a*) $(11.3, 85.5)$;   (*b*) $(175, 68)$;   (*c*) $(30.2, 82.1)$.

7 $\begin{bmatrix} x \\ y \end{bmatrix} \rightarrow \begin{bmatrix} 9x - 40 \\ 9y + 56 \end{bmatrix}$.
   8 $\begin{bmatrix} x \\ y \end{bmatrix} \rightarrow \begin{bmatrix} y-1 \\ 9-x \end{bmatrix}$.

9 $\begin{bmatrix} x \\ y \end{bmatrix} \rightarrow \begin{bmatrix} 0.5x - 0.866y + 3.232 \\ 0.866x + 0.5y - 1.598 \end{bmatrix}$.
   10 $\begin{bmatrix} x \\ y \end{bmatrix} \rightarrow \begin{bmatrix} -\frac{15}{17}x + \frac{8}{17}y + \frac{24}{17} \\ \frac{8}{17}x + \frac{15}{17}y - \frac{6}{17} \end{bmatrix}$.

11 $\begin{bmatrix} x \\ y \end{bmatrix} \rightarrow \begin{bmatrix} \frac{5}{13}x - \frac{12}{13}y + \frac{16}{13} \\ -\frac{12}{13}x - \frac{5}{13}y + \frac{24}{13} \end{bmatrix}$.
   12 $\begin{bmatrix} x \\ y \end{bmatrix} \rightarrow \begin{bmatrix} 4x - 18 \\ 4y + 15 \end{bmatrix}$.

13 (a) $\begin{bmatrix} x \\ y \end{bmatrix} \rightarrow \begin{bmatrix} -0.8x+0.6y-0.6 \\ 0.6x+0.8y+0.2 \end{bmatrix}$; $P'\,(^-5,6),\ Q'\,(2,7)$.

  (b) $\begin{bmatrix} x \\ y \end{bmatrix} \rightarrow \begin{bmatrix} 3.885x-0.577y-1.731 \\ 0.577x+1.115y+0.346 \end{bmatrix}$; $P'\,(24.3,^-1.5),\ Q'\,(2,7)$.

  (c) $\begin{bmatrix} x \\ y \end{bmatrix} \rightarrow \begin{bmatrix} 0.514x+0.081y+0.405 \\ -2.319x+1.486y+2.432 \end{bmatrix}$; $P'\,(4.2,^-15),\ Q'\,(2,7)$.

Exercise G

1   $\begin{bmatrix} x \\ y \end{bmatrix} \rightarrow \begin{bmatrix} -0.6x+0.8y+0.6 \\ 0.8x+0.6y+7.2 \end{bmatrix}$.

2   $(14.52, 27.36)$.

3   $y = 7x+5,\ \begin{bmatrix} 2 \\ 14 \end{bmatrix}$.

4   $\begin{bmatrix} x \\ y \end{bmatrix} \rightarrow \begin{bmatrix} y-1 \\ x+5 \end{bmatrix}$; $y = x+3,\ \begin{bmatrix} 2 \\ 2 \end{bmatrix}$.

5   $(^-1, 2)$.

6   $\begin{bmatrix} x \\ y \end{bmatrix} \rightarrow \begin{bmatrix} -0.8x+0.6y-1 \\ 0.6x+0.8y+7 \end{bmatrix}$.

7   $(20, 18.76)$.

8   $3y = 4x-12,\ \begin{bmatrix} 3 \\ 4 \end{bmatrix}$.

9   $\begin{bmatrix} x \\ y \end{bmatrix} \rightarrow \begin{bmatrix} -0.8x+0.6y+4.6 \\ 0.6x+0.8y+1.8 \end{bmatrix}$; $y = 3x-6,\ \begin{bmatrix} 1 \\ 3 \end{bmatrix}$.

Miscellaneous exercise

1   $\mathbf{A} = \begin{bmatrix} 1 & 0 \\ 2 & 1 \end{bmatrix}$; x-axis invariant, $(0,1),\ (^-1,1)$.

4   $A'\,(1.6, 1.2)$, $B'\,(^-1.8, 2.4)$; $6x+17y = 30$; rotation of $37°$ about O; area $= 3$; $C\,(4,^-3)$.

6   $\mathbf{X} = \begin{bmatrix} 1 & 0 \\ 0 & -1 \end{bmatrix}$; $\mathbf{Z} = \begin{bmatrix} 0 & 1 \\ 1 & 0 \end{bmatrix}$; $\mathbf{R} = \begin{bmatrix} 0.5 & -0.866 \\ 0.866 & 0.5 \end{bmatrix}$; $\mathbf{M} = \begin{bmatrix} 0.866 & -0.5 \\ 0.5 & 0.866 \end{bmatrix}$.

7   (b) Reflection in $y = x \tan \theta$;

  (c) $\mathbf{M} = \begin{bmatrix} 1 & 0 \\ 0 & -1 \end{bmatrix}$, $\mathbf{R} = \begin{bmatrix} \cos 2\theta & -\sin 2\theta \\ \sin 2\theta & \cos 2\theta \end{bmatrix}$, $\mathbf{RM} = \begin{bmatrix} \cos 2\theta & \sin 2\theta \\ \sin 2\theta & -\cos 2\theta \end{bmatrix}$.

8   (a) Rotation of 2θ about O;   (b) $\begin{bmatrix} \cos 2\theta & -\sin 2\theta \\ \sin 2\theta & \cos 2\theta \end{bmatrix}$;

  (c) (i) $\begin{bmatrix} 1 & 0 \\ 0 & -1 \end{bmatrix}$;   (ii) $\begin{bmatrix} 1 & 0 \\ 0 & -1 \end{bmatrix}$;   (d) $\begin{bmatrix} \cos 2\theta & \sin 2\theta \\ \sin 2\theta & -\cos 2\theta \end{bmatrix}$.

9   Rotation: $\begin{bmatrix} x \\ y \end{bmatrix} \rightarrow \begin{bmatrix} \frac{1}{13}(12x+5y-4) \\ \frac{1}{13}(^-5x+12y+58) \end{bmatrix}$;   glide: $\begin{bmatrix} x \\ y \end{bmatrix} \rightarrow \begin{bmatrix} -0.6x+0.8y+9.8 \\ 0.8y+0.6x-3.4 \end{bmatrix}$.

10   Area $= 1$.

Chapter 17 *Isometries*

Exercise A

2 Rotation through $^-2\theta$; rotation of $0°$.

3 (*a*) Reflection in $x = 0$; (*b*) reflection in $x = 0$; (*c*) reflection in $y = 0$.

4 (*a*) x-axis, $\begin{bmatrix} -16 \\ 0 \end{bmatrix}$; (*b*) $x + y = 8$, $\begin{bmatrix} -8 \\ 8 \end{bmatrix}$.

5 (*a*) Glide with axis $y = 4$, throw $\begin{bmatrix} 3 \\ 0 \end{bmatrix}$; (*b*) rotation through $90°$ about $(1\frac{1}{2}, 3\frac{1}{2})$.

6 (*a*) *l* and *m* are coincident or perpendicular; (*b*) *l*, *m* and *n* are concurrent.

8 $^-2\mathbf{d}$; **0**.

9 (*a*) Reflection in $x = 4$; (*b*) reflection in $x = ^-4$; (*c*) reflection in $x = 4$.

10 (*a*) y-axis, $\begin{bmatrix} 0 \\ -20 \end{bmatrix}$; (*b*) $y = x + 10$, $\begin{bmatrix} -10 \\ -10 \end{bmatrix}$.

11 (*a*) Glide with axis $x = 6$, throw $\begin{bmatrix} 0 \\ -3 \end{bmatrix}$; (*b*) rotation through $90°$ about $(^-\frac{1}{2}, 9\frac{1}{2})$.

12 (*a*) Angle of rotation is $0°$ or translation is **0**; (*b*) centre of rotation lies on mirror line.

Exercise B

1 (*a*) $120°$ about *Y* then $120°$ about *X*; (*b*) $240°$ about *X* then $240°$ about *Y*; (*c*) none.

2 (*a*) $(0, ^-2\sqrt{3})$; (*b*) $(0, 6\sqrt{3})$.

5 (*a*) Half-turn about $(5, ^-2)$; (*b*) translation $\begin{bmatrix} 10 \\ 14 \end{bmatrix}$.

6 Rotation of $37°$ about $(^-17\frac{1}{2}, 2\frac{1}{2})$.

7 (*a*) $120°$ about *X* then $120°$ about *Z*; (*b*) $240°$ about *Z* then $240°$ about *X*; (*c*) none.

8 (*a*) $(4\sqrt{3}, 0)$; (*b*) $(1.865, 0)$; (*c*) translation $\begin{bmatrix} 0 \\ 8 \end{bmatrix}$.

11 (*a*) Half-turn about $(^-4, ^-3)$; (*b*) translation $\begin{bmatrix} -16 \\ 20 \end{bmatrix}$.

12 Rotation of $127°$ about $(^-2\frac{1}{2}, 3\frac{3}{4})$.

Exercise C

1 (*a*) **mm**; (*b*) **tm**; (*c*) **t**; (*d*) **tmm**; (*e*) **tmm**; (*f*) **t2**.

2 (*a*) **m2**.

3 (*a*) **tm**; (*b*) **mm**.

4 **tmm**.

5 (*a*) **tmm**; (*b*) **tm**.

6 (*a*) 60; (*b*) 360.

7   (a) **tm**;   (b) **t**;   (c) **mm**;   (d) **tmm**;   (e) **t**;   (f) **mm**.

8   (a) **m2**.

9   (a) **mm**;   (b) **tm**.

10   **tmm**.

11   (a) **m2**;   (b) **tmm**.

12   (a) 6;   (b) 36.

Exercise D

1   **t→p1**; **t2→p2**; **g→pg**; **mm→pm**; **m2→pmg**; **tm→pm**; **tmm→pmm**.

2   (a) **p2**;   (b) **pgg**;   (c) **pmm**;   (d) **cmm**.

3   (a) **p4m**;   (b) **p6m**;   (c) **p6m**.

5   (a) **p2**;   (b) **cmm**;   (c) **pgg**;   (d) **p4g**;   (e) **cmm**;   (f) **pmm**.

9   Figure 28: (a) **cm**;   (b) **pgg**;   (c) **p1**;   (d) **pg**;   (e) **p3**;   (f) **p2**;   (g) **p4g**;
          (h) **cm**;   (i) **p4**;   (j) **p4g**.

    Figure 29: (i) (a) **p1**;   (b) **pg**;   (c) **p3m1**;   (d) **p3**;
            (ii) (a) **p1**;   (b) **pg**;   (c) **p31m**;   (d) **p6**.

Miscellaneous exercise

1   (i) Rotation of 50° about the origin;
    (ii) rotation of $^-50°$ about the origin;
    (iii) rotation of $^-100°$ about the origin. $(\mathbf{BABA})^{-1} = \mathbf{ABAB}$.

2   **K** is reflection in m^*.

3   **SR** = half-turn about $(0,0)$;   **RS** = half-turn about $(1,1)$.

4   *B* is the centre of symmetry.

5   $\begin{bmatrix} 6 \\ 4 \end{bmatrix}$.

6   $\mathbf{Q}^{-1} = \mathbf{Q}^3$; $\mathbf{T}^{-1} = \mathbf{Q}^2\mathbf{T}\mathbf{Q}^2$; $\mathbf{Q}^2\mathbf{T}\mathbf{Q}$.

8   (ii) Direct, 90°;   (iii) (a) *P*;   (b) *P*;   (c) *Q*;   (iv) rotation of 90° about *Q*.

9   (a) $\begin{bmatrix} 0 \\ 1 \end{bmatrix}$;   (b) translation $\begin{bmatrix} 0 \\ 2 \end{bmatrix}$;   (c) $x = {}^-1$;

    (d) translation $\begin{bmatrix} 0 \\ 2 \end{bmatrix}$, translation $\begin{bmatrix} -4 \\ 2 \end{bmatrix}$;   (e) $\mathbf{TMT^5M}$.

10   (a) **M, MG, GM** are self-inverse, $\mathbf{G}^{-1} = \mathbf{MGM}$;   (b) $\mathbf{MG^2M}$;

    (c) $\begin{bmatrix} 6 \\ 0 \end{bmatrix}$;   (d) (i) $\mathbf{G^2M}$;   (ii) $\mathbf{G^{20}MG}$;   (iii) $\mathbf{MG^7M}$.

12   *a* = reflection in axis 24;
    b = rotation of 180° about axis through *O*; *ba*.

Chapter 18   *Equations and matrices*

Exercise A

1 (*a*) $3x+4y = 11$; (*b*) $2x+5y+10 = 0$; (*c*) $5x-2y+29 = 0$.

2 (*a*) $\begin{bmatrix} 2 \\ 1 \end{bmatrix}$; (*b*) $\begin{bmatrix} 3 \\ -6 \end{bmatrix}$; (*c*) $\begin{bmatrix} 4 \\ 2 \end{bmatrix}$; (*d*) $\begin{bmatrix} 5 \\ -10 \end{bmatrix}$; (*e*) $\begin{bmatrix} 6 \\ 3 \end{bmatrix}$.

  (*a*), (*c*), (*e*) are parallel and perpendicular to (*b*) and (*d*) which are parallel.

3 (*a*) $77°$; (*b*) $42°$.

4 $99°$, $73°$, $8°$.

5 (*a*) $3x+6y = 51$; (*b*) $5x-2y+32 = 0$; (*c*) $4x+7y+25 = 0$.

6 (*a*) $\begin{bmatrix} 1 \\ 3 \end{bmatrix}$; (*b*) $\begin{bmatrix} 3 \\ -9 \end{bmatrix}$; (*c*) $\begin{bmatrix} 2 \\ 6 \end{bmatrix}$; (*d*) $\begin{bmatrix} 6 \\ -2 \end{bmatrix}$; (*e*) $\begin{bmatrix} 6 \\ 2 \end{bmatrix}$.

  (*a*) and (*c*) are parallel and perpendicular to (*d*). (*b*) and (*e*) are perpendicular.

7 (*a*) $81°$; (*b*) $80°$.

8 $95°$, $81°$, $3°$.

Exercise B

1 (*a*) $x = {}^-1, y = 2$; (*b*) $x = \frac{13}{16}, y = -\frac{1}{32}$; (*c*) $x = 4.2, y = 4.4$.

2 (*a*) $x+4y = 23$; (*b*) $2x+y+2 = 0$.

3 (*a*) ${}^-1$; (*c*) 1.

4 (*a*) $x = 2, y = {}^-1$; (*b*) $x = \frac{11}{16}, y = \frac{17}{32}$; (*c*) $x = 2, y = 2$.

5 (*a*) $5x-y = 13$; (*b*) $x+2y = 3$.

6 (*a*) ${}^-4$; (*c*) 4.

Exercise C

1 (*a*) Shear with the *y*-axis invariant;
  (*b*) one-way stretch parallel to the *x*-axis, scale factor 3;
  (*c*) shear with the *x*-axis invariant;
  (*d*) enlargement, scale factor 5;
  (*e*) reflection in the *x*-axis.

2 (*a*) $\begin{bmatrix} -1 \\ 2 \end{bmatrix}, \begin{bmatrix} 5 & -3 \\ -3 & 2 \end{bmatrix}$; (*b*) $\begin{bmatrix} 0.6 \\ -0.4 \end{bmatrix}, \begin{bmatrix} 1.4 & -1.6 \\ 0.4 & -0.6 \end{bmatrix}$; (*c*) $\begin{bmatrix} 0.04 \\ -1.32 \end{bmatrix}, \begin{bmatrix} 0.08 & 0.04 \\ -0.14 & 0.18 \end{bmatrix}$.

3 (*a*) Shear with the *x*-axis invariant;
  (*b*) one-way stretch, scale factor 3, with the *x*-axis invariant;
  (*c*) reflection in $y = x$;
  (*d*) reflection in the *y*-axis;
  (*e*) enlargement with scale factor ${}^-2$.

4 (*a*) $\begin{bmatrix} 27 \\ -10 \end{bmatrix}, \begin{bmatrix} 8 & -5 \\ -3 & 2 \end{bmatrix}$; (*b*) $\begin{bmatrix} 1.4 \\ 0.2 \end{bmatrix}, \begin{bmatrix} 0.1 & 0.3 \\ 0.3 & -0.1 \end{bmatrix}$.

Exercise D

1 (*a*) $4x+5y+6z = 32$; (*b*) $3x+8y+z = {}^-17$; (*c*) $2x-5y-6z = 77$.

2 (a) 84°; (b) 81°; (c) 0°.

3 (a) $6x+5y+4z = 32$; (b) $7x+2y+6z = 38$; (c) $9x-8y-3z = 94$.

4 (a) 67°; (b) 37°; (c) 90°.

Exercise E

1 (a) $x = 2, y = 1, z = {}^-1$; (b) $x = 3, y = 2, z = 1$;
  (c) $x = 62, y = {}^-40, z = 30$; (d) $x = 3, y = {}^-2, z = {}^-4$.

2 (a) $x-3y = {}^-5$; (b) $11x+y+2z = 27$; (c) $13x-7y+18z = 119$.

3 (a) $x = 1, y = 0, z = 1$; (b) $x = 3, y = 1, z = 1$;
  (c) $x = 1, y = 2, z = 3$; (d) $x = 2, y = {}^-5, z = {}^-4$.

4 (a) $x-y = {}^-1$; (b) $2x+y-z = 5$; (c) $x-3y-4z = 5$.

Exercise F

1 (a) Reflection in $z = 0$;
  (b) one-way stretch, scale factor 3, with $y = 0$ invariant;
  (c) shear with $z = 0$ invariant;
  (d) enlargement with scale factor 2.

2 (a) $\begin{bmatrix} 1 \\ -2 \\ 1 \end{bmatrix}, \begin{bmatrix} -2 & 0 & 1 \\ 0 & 3 & -2 \\ 1 & -2 & 1 \end{bmatrix}$; (b) $\begin{bmatrix} 3 \\ 2 \\ 1 \end{bmatrix}, \dfrac{1}{206}\begin{bmatrix} -2 & 31 & 23 \\ 31 & -17 & 4 \\ 23 & 4 & -7 \end{bmatrix}$;

  (c) $\begin{bmatrix} -1 \\ 0 \\ 1 \end{bmatrix}, \begin{bmatrix} -4 & 5 & -1\frac{1}{2} \\ 1 & 0 & -\frac{1}{2} \\ 1 & -2 & 1 \end{bmatrix}$; (d) $\begin{bmatrix} 10 \\ 11 \\ 15 \end{bmatrix}, \begin{bmatrix} 9 & 1 & -7 \\ 10 & 1 & -8 \\ 12 & 1 & -9\frac{1}{2} \end{bmatrix}$.

3 (a) $\begin{bmatrix} -0.5 & -1.5 & 1 \\ 3 & 0 & -2 \\ -2 & 1 & 1 \end{bmatrix}$; (b) $\begin{bmatrix} 1 & -2 & 1 \\ -1.5 & -0.5 & 1 \\ 0 & 3 & -2 \end{bmatrix}$;

  (c) $\begin{bmatrix} 3 & 0 & -2 \\ -0.5 & -1.5 & 1 \\ -2 & 1 & 1 \end{bmatrix}$; (d) $\begin{bmatrix} -1 & -1.5 & -2 \\ -2 & -2.5 & -3 \\ -3 & -3.5 & -4.5 \end{bmatrix}$.

4 (a) Reflection in $x = 0$;
  (b) two-way stretch, with scale factor 2 parallel to the y-axis and scale factor 3 parallel to the z-axis;
  (c) shear with $x = 0$ invariant;
  (d) enlargement with scale factor 3.

5 (a) $\begin{bmatrix} 3 \\ -1 \\ 0 \end{bmatrix}, \begin{bmatrix} -19 & 13 & -2 \\ 2 & -1 & 0 \\ 7 & -5 & 1 \end{bmatrix}$; (b) $\begin{bmatrix} 2 \\ 1 \\ 1 \end{bmatrix}, \begin{bmatrix} \frac{1}{3} & \frac{13}{48} & \frac{1}{24} \\ \frac{1}{3} & \frac{1}{3} & -\frac{1}{3} \\ \frac{1}{3} & \frac{7}{48} & -\frac{5}{24} \end{bmatrix}$;

  (c) $\begin{bmatrix} 1 \\ -2 \\ 1 \end{bmatrix}, \begin{bmatrix} 1 & 1 & -1 \\ 0 & -3 & 2 \\ -2 & 3 & -1 \end{bmatrix}$; (d) $\begin{bmatrix} 1 \\ -1 \\ 1 \end{bmatrix}, \dfrac{1}{269}\begin{bmatrix} -56 & 40 & 21 \\ 63 & -45 & 10 \\ 16 & 27 & -6 \end{bmatrix}$.

6 (a) $\begin{bmatrix} 2 & -1 & -3 \\ -2 & 3 & 0 \\ 1 & -2 & 1 \end{bmatrix}$; (b) $\begin{bmatrix} -2 & 1 & 1 \\ -1 & 2 & -3 \\ 3 & -2 & 0 \end{bmatrix}$;

  (c) $\begin{bmatrix} -3 & -1 & 2 \\ 1 & -2 & 1 \\ 0 & 3 & -2 \end{bmatrix}$; (d) $\begin{bmatrix} -2 & -5 & -6 \\ -3 & -7 & -9 \\ -1 & -3 & -4 \end{bmatrix}$.

Miscellaneous exercise

1 (a) $x = 2\frac{1}{2}, y = {}^{-}1\frac{1}{2}, z = 3$;   (b) $\begin{bmatrix} 3 & 1 & {}^{-}2 \\ 1 & 5 & 4 \\ 1 & 1 & 1 \end{bmatrix}$;   (c) $\begin{bmatrix} 2\frac{1}{2} \\ {}^{-}1\frac{1}{2} \\ 3 \end{bmatrix}$.

2 (i) $\frac{1}{4}, 1\frac{1}{4}$;   (ii) (a) $x = 1, y = {}^{-}2, z = 3$.

3 $\begin{bmatrix} 16 & 0 & 0 \\ 0 & 16 & 0 \\ 0 & 0 & 16 \end{bmatrix}$; $x = {}^{-}2, y = 3\frac{1}{4}, z = 8\frac{1}{4}$.

4 (a) $x = \frac{1}{2}, y = {}^{-}\frac{1}{2}, z = 1$;   (b) $k = 1$.

5 B $(0, 3, 0)$,   C $(0, 0, 6)$; $p = 6$; $q = 6, r = \frac{2}{3}$.

6 $x = 3, y = {}^{-}1, z = 2$.

7 $x = {}^{-}2, y = 3\frac{1}{2}, z = 4\frac{1}{2}$.

8 $(5, 5), ({}^{-}1, 2), ({}^{-}1, 3)$;

A: translation $\begin{bmatrix} 3 \\ 4 \end{bmatrix}$;   **B**: quarter-turn about $(0, 0)$;

C: translation $\begin{bmatrix} {}^{-}3 \\ {}^{-}4 \end{bmatrix}$;   **M**: quarter-turn about $(3, 4)$.

Revision exercise 5

1 $\mathbf{A} = \begin{bmatrix} 0 & {}^{-}1 \\ 1 & 0 \end{bmatrix}$; $(5, 4)$; $a = 1, b = 2$; quarter-turn about $(1, 2)$.

2 (a) $(1, {}^{-}1)$;   (b) $(3, 1)$;   (c) $(4 - q, 4 - p)$;   reflection in $x + y = 4$.

3 (b) $\begin{bmatrix} {}^{-}0.8 & 0.6 \\ 0.6 & 0.8 \end{bmatrix}$;   (c) $\begin{bmatrix} 0.8 & {}^{-}0.6 \\ {}^{-}0.6 & {}^{-}0.8 \end{bmatrix}$;   (e) $\begin{bmatrix} {}^{-}1 & 0 \\ 0 & {}^{-}1 \end{bmatrix}$.

4 (a) $\begin{bmatrix} \cos\alpha & {}^{-}\sin\alpha \\ \sin\alpha & \cos\alpha \end{bmatrix}$;   (b) $\begin{bmatrix} \cos\beta & {}^{-}\sin\beta \\ \sin\beta & \cos\beta \end{bmatrix}$;   (c) rotation through $(\alpha + \beta)$.

6 $X = C$.

7 (a) $\mathbf{X}^2$;   (b) $\mathbf{XY}$: $(1, {}^{-}1)$, $\mathbf{Y}^2\mathbf{XY}$: $(1, 1)$;   (c) $\mathbf{YXY}$ and $\mathbf{X}$. $\mathbf{Y}^4\mathbf{X}^2$.

8   (i) Quarter-turn about $({}^{-}1, 1)$;
   (ii) quarter-turn about $(1, 1)$;
   (iii) reflection in $x = {}^{-}1$;
   (iv) reflection in $x = 1$;
   (v) reflection in y-axis;
   (vi) quarter-turn about $(2, 0)$.

9 $\mathbf{LN} = $ rotation of 2θ about Q; $\mathbf{YX} = $ half-turn about O;
$\mathbf{ML} = $ half-turn about O; $\mathbf{XN} = $ translation of $2\underset{\sim}{PO}$;
$\mathbf{YXN} = $ glide with axis y and throw $2\underset{\sim}{PO}$.
The glide $\mathbf{YXN}$.

10 $p = 1, q = 6$; 35.

11 $x = \frac{1}{2}, y = 1\frac{1}{2}, z = 2$; $\mathbf{r} = \begin{bmatrix} 0.25 \\ 0.75 \\ 1 \end{bmatrix}$.

12 (a) $\begin{bmatrix} 1 & 0 & 0 \\ 0 & 1 & 0 \\ 0 & 0 & 1 \end{bmatrix}$;   (b) $x = 2, y = 1, z = {}^{-}1$;   (c) $x = 4, y = 17, z = {}^{-}14$.

13 $x = 2, y = ^-3, z = 4$; $A\ (2, ^-3, 4)$; $P'\ (^-1, 2, 2)$; reflection in the plane OAB.

14 $\mathbf{E_1} = \begin{bmatrix} 1 & 0 & 0 \\ ^-2 & 1 & 0 \\ 0 & 0 & 1 \end{bmatrix}$, $\mathbf{E_2} = \begin{bmatrix} 1 & 0 & 0 \\ 0 & 1 & 0 \\ 0 & 1 & 1 \end{bmatrix}$, $\mathbf{E_3} = \begin{bmatrix} 1 & 0 & ^-1 \\ 0 & 1 & 0 \\ 0 & 0 & 1 \end{bmatrix}$.

$\mathbf{A^{-1}} = \begin{bmatrix} 3 & ^-1 & ^-1 \\ ^-2 & 1 & 0 \\ ^-2 & 1 & 1 \end{bmatrix}$.

Chapter 19 *Logic*

Exercise A

1 (*a*) It is dead. (*b*) It is dark or cold. (*c*) It is bad and alive.
(*d*) It is dark and warm. (*e*) It is good or cold.
(*f*) It is neither alive nor bad. (*g*) It is dead and good.
(*h*) It is not warm. (*i*) It is not both cold and dead.

2 (*a*) $\sim(\sim d)$; (*b*) $\sim a \wedge b$; (*c*) $\sim(\sim b \vee \sim c)$; (*d*) $\sim c \wedge \sim a$;
(*e*) $c \wedge (b \vee \sim a)$.

5 (*a*) That is bad. (*b*) That is fat and edible.
(*c*) That is good or hard. (*d*) That is hard and inedible.
(*e*) That is thin and good. (*f*) That is not both hard and edible.
(*g*) That is soft or inedible. (*h*) That is not thin.
(*i*) That is neither fat nor inedible.

6 (*a*) $\sim(\sim h)$; (*b*) $\sim g \wedge \sim e$; (*c*) $\sim(\sim f \vee g)$; (*d*) $h \wedge \sim e$;
(*e*) $(f \wedge \sim e) \vee g$.

Exercise B

1 (*a*)

a	b	$\sim a$	$\sim b$	$\sim a \wedge \sim b$	$\sim a \vee \sim b$	$\sim a \vee b$	$\sim(a \wedge \sim b)$
T	T	F	F	F	F	T	T
T	F	F	T	F	T	F	F
F	T	T	F	F	T	T	T
F	F	T	T	T	T	T	T

2 (*a*) (i) $FFFF$. (ii) I am right and I am both stupid and wrong.
(iii) Contradiction.
(*b*) (i) $FFFF$. (ii) I am not either right or stupid and I am right.
(iii) Contradiction.
(*c*) (i) $TTFF$. (ii) I am right and stupid or I am right and not stupid.
(*d*) (i) $TTTT$. (ii) I am stupid or I am not both stupid and wrong.
(iii) Tautology.

3 (*a*) True unless it is Monday and I am not silly.
(*b*) True unless we are both wrong.
(*c*) True unless this is a fiddle and that is not a twiddle.
(*d*) Only true if you are wrong and not stupid.

5 (*a*)

c	*d*	~*c*	~*d*	~*c* ∧ *d*	*c* ∨ ~*d*	~(*c* ∨ ~*d*)
T	*T*	*F*	*F*	*F*	*T*	*F*
T	*F*	*F*	*T*	*F*	*T*	*F*
F	*T*	*T*	*F*	*T*	*F*	*T*
F	*F*	*T*	*T*	*F*	*T*	*F*

6 (*a*) (i) *T T T F*. (ii) You are mad or both I am nutty and you are not mad.
  (*b*) (i) *T T T T*. (ii) Either it is not the case that you are mad and I am nutty or you are mad. (iii) Tautology.
  (*c*) (i) *T T F F*. (ii) Both you are mad or I am nutty and you are mad or I am not nutty.
  (*d*) (i) *F F F F*. (ii) You are mad and neither you are mad nor am I nutty. (iii) Contradiction.

7 (*a*) True unless it is not Tuesday and you are wrong.
  (*b*) True unless you and she are both short.
  (*c*) True unless this is not a chore and that is not a bind.
  (*d*) True unless you are stupid and wrong.

Exercise C

1 (*a*) (ii) *F F F F*. (iii) He is both rich and short and he is poor.
  (*b*) (ii) *F F T F*. (iii) He is either rich or short and he is poor.
  (*c*) (ii) *T T F F*. (iii) He is both rich or short and rich or tall.
  (*d*) (ii) *F T T T*. (iii) It is not true that he is rich and either short or poor.

2 (*a*) True unless I am not observant and it is a tail.
  (*b*) True unless we are both wrong.
  (*c*) True when both are short or both are tall.
  (*d*) True unless it is in London and not in summer.

3 (*a*) (*a* ∧ *b*) ∨ (~ *a* ∧ ~ *b*); (*b*) ~ *b*; (*c*) *a* ∨ ~ *b*; (*d*) ~ *a* ∧ ~ *b*.

4 (*a*) (ii) *T T T T*. (iii) Either she is fat or good or she is thin.
  (*b*) (ii) *T F T T*. (iii) Either she is fat and good or she is thin.
  (*c*) (ii) *T T F F*. (iii) Either she is fat and good or she is fat and bad.
  (*d*) (ii) *F F F T*. (iii) It is not the case that either she is fat or she is good and thin.

5 (*a*) True unless it is a winner and I am observant.
  (*b*) True unless you and she are both quick.
  (*c*) Never true.
  (*d*) True if it is hot and humid.

6 (*a*) (*a* ∧ ~ *b*) ∨ (~ *a* ∧ *b*); (*b*) ~ *a*; (*c*) ~ *a* ∧ *b*; (*d*) *a* ∧ *b*.

Exercise D

1 (*a*) (i) *T T T T*. (ii) Blonde angels are blonde.
  (*b*) (i) *T F T T*. (ii) Angels are blonde or not angels.
  (*c*) (i) *T F F T*. (ii) She is an angel or blonde is equivalent to being a blonde angel.
  (*d*) (i) *T T T T*. (ii) Angels are blonde is equivalent to she is blonde or not an angel.

(e) (i) $T\,T\,T\,T$. (ii) If she is an angel who is not blonde then being an angel is not equivalent to being blonde.

(f) (i) $T\,T\,T\,T$. (ii) Blondes are angels is equivalent to non-angels not being blonde.

5 (a) (i) $T\,T\,F\,T$. (ii) If he likes cars or can drive then he likes cars.

   (b) (i) $F\,T\,F\,T$. (ii) If he can drive then he likes cars and cannot drive.

   (c) (i) $T\,T\,F\,F$. (ii) He likes cars and can drive is equivalent to he hates cars or can drive.

   (d) (i) $T\,T\,T\,T$. (ii) If liking cars is equivalent to being able to drive then either he can drive or he hates cars.

   (e) (i) $F\,T\,T\,F$. (ii) Liking cars or being unable to drive is equivalent to liking cars not implying being able to drive.

   (f) (i) $T\,F\,T\,T$. (ii) If being able to drive implies liking cars then not being able to drive implies hating cars.

Exercise E

1 (a) True except for an animal with ears which is not a cat.

   (b) True except for a pig without wings.

   (c) True except for an elephant which is not pink.

   (d) Always true.

2 (a) (i) $T\,T\,T\,F$; (ii) $T\,T\,T\,F$; (iii) $T\,T\,T\,T$; (iv) $T\,T\,T\,T$.

   (b) (i) $T\,T\,T\,T$; (ii) $T\,T\,T\,T$; (iii) $T\,T\,T\,F$; (iv) $T\,T\,T\,F$.

   (c) (i) $T\,T\,T\,T$; (ii) $T\,T\,T\,T$; (iii) $F\,T\,T\,F$; (iv) $F\,T\,T\,F$.

   (d) (i) $F\,F\,T\,T$; (ii) $F\,F\,T\,T$; (iii) $T\,T\,T\,T$; (iv) $T\,T\,T\,T$.

3 (a) False; (b) True.

4 (a) Always true.

   (b) True except for a cow which does not jump over a moon.

   (c) True except for a mouse which is not white.

   (d) Always true.

5 (a) (i) $T\,T\,T\,T$; (ii) $T\,T\,T\,T$; (iii) $F\,T\,T\,T$; (iv) $F\,T\,T\,T$.

   (b) (i) $F\,T\,T\,T$; (ii) $F\,T\,T\,T$; (iii) $T\,T\,T\,T$; (iv) $T\,T\,T\,T$.

   (c) (i) $F\,T\,T\,T$; (ii) $F\,T\,T\,T$; (iii) $T\,F\,T\,T$; (iv) $T\,F\,T\,T$.

   (d) (i) $F\,T\,T\,T$; (ii) $F\,T\,T\,T$; (iii) $T\,T\,T\,F$; (iv) $T\,T\,T\,F$.

6 (a) False. (b) True.

Exercise F

1 Outcome x is paired with outcome $17-x$.

2 Outcome $9 \sim (a \wedge b)$; $10\ (a \wedge \sim b) \vee (\sim a \wedge b)$; $11 \sim b$; $12\ a \wedge \sim b$; $13 \wedge \sim a$; $14 \sim a \wedge b$; $15 \sim (a \vee b)$; $16\ a \wedge \sim a$.

3 (i) (a) $T\,T\,F\,T\,F\,T\,F\,T$; (b) It is an angry bull or I am not a coward.

   (ii) (a) $F\,F\,T\,F\,F\,T\,T\,T$; (b) If it's angry or I am a coward then it's not a bull and I am a coward.

   (iii) (a) $T\,T\,F\,T\,T\,T\,T\,F$; (b) It's not an angry cow is equivalent to it's a bull or I am a coward.

   (iv) (a) $T\,T\,T\,T\,F\,T\,F\,T$; (b) If I'm a coward then it's angry is equivalent to it's a cow or if I'm a coward then it's a bull.

   (c) It is a peaceful bull and I am brave.

7 (i) (*a*) *T F T T T T T*; (*b*) If I am in danger and it is enjoyable then it is free-fall.

   (ii) (*a*) *T T F T T T T*; (*b*) If I am in danger and it is not enjoyable then it is not free-fall.

   (iii) (*a*) *T T T T T T T T*; (*b*) If neither I am in danger nor is it enjoyable and it is free-fall then it is not enjoyable.

   (iv) (*a*) *T F T T T T F T*; (*b*) Being safe implies it is not free-fall and if it is enjoyable then it is free-fall is equivalent to being enjoyable implies that I am in danger.

Exercise G

1 Valid.

2 Not valid (unless the 'or' is exclusive).

3 Valid.

4 Not valid.

5 Valid.

6 Valid.

7 (*a*) Valid. (*b*) Valid. (*c*) Not valid.

8 The headmaster does not eat haggis.

Exercise H

2 (*a*) $(a \land b) \lor \sim a \lor b$; (*b*) $(a \land b) \lor \sim a \lor \sim b$; (*c*) $(a \lor b) \land (c \lor \sim a)$; (*d*) $[a \land (b \lor \sim a)] \lor [(b \lor c) \land \sim a]$; (*e*) $[a \lor (b \land c)] \land \sim b \land (c \lor b)$; (*f*) $(a \lor b \lor c) \land (\sim a \lor \sim b \lor \sim c)$.

Exercise I

2 (*a*) You and I are both honest. (*b*) I am honest.

   (*c*) At least one of us is a liar. (*d*) Ann is a liar, Bill is honest.

   (*e*) Ann and Bill are liars. Chris is honest. (*f*) Ann is a liar. Chris is honest.

3 Fred got the bull, George the magpie and Harriet the inner.

4 Long and Short.

5 Either: Jack (trumpet), Kate (flute), Larry (horn), Mary (clarinet).
   Or: Jack (clarinet), Kate (horn), Larry (flute), Mary (trumpet).

6 All three are liars.

Miscellaneous exercise

1 (*a*) $\lor$; (*b*) $(p \lor \sim q) \land r$.

2 You are not drunk.

4

a	*b*	$a \Rightarrow b$
T	*T*	*T*
T	*F*	*F*
F	*T*	*T*
F	*F*	*T*

5

f	g	r	$r \vee (f \wedge g)$	$\sim f \Rightarrow r$
T	T	T	T	T
T	T	F	T	T
T	F	T	T	T
T	F	F	F	T
F	T	T	T	T
F	T	F	F	F
F	F	T	T	T
F	F	F	F	F

6 $a \vee b = (a*b)*(a*b)$.

Chapter 20   *Complex numbers*

Exercise A

1 (a) $3j$, ^-3j;   (b) $10j$, ^-10j;   (c) $j\sqrt{5}$, $^-j\sqrt{5}$;   (d) $\frac{3}{2}j$, $^-\frac{3}{2}j$.

2 (a) 5;   (b) $5+j$;   (c) $2-6j$;   (d) 0;   (e) $1+8j$;   (f) $^-1+13j$.

3 (a) $18-j$;   (b) $34+27j$;   (c) $^-26+2j$;   (d) 34;   (e) $ac-bd+(ad+bc)j$.
   (f) a^2+b^2.

4 (a) ^-j;   (b) $45+28j$;   (c) 600;   (d) 0.

5 (a) $9j$, ^-9j;   (b) $2\sqrt{2}j$, $^-2\sqrt{2}j$;   (c) $\frac{5}{2}j$, $^-\frac{5}{2}j$;   (d) $0.1j$, $^-0.1j$.

6 (a) 14;   (b) $5+9j$;   (c) ^-7+3j;   (d) ^-1-j;   (e) ^-1-j;   (f) $8+j$.

7 (a) $5+8j$;   (b) $21-2j$;   (c) $^-41-59j$;   (d) ^-6-8j;   (e) 25;   (f) a^2-b^2+2abj.

8 (a) $2j$;   (b) 2;   (c) ^-j;   (d) $^-4$.

Exercise B

2 $a = 1+3j$, $b = 2+j$, $c = 5+j$, $d = 6+2j$, $e = 3+5j$.

3 $a' = 4+8j$, $b' = 5+6j$, $c' = 8+6j$, $d' = 9+7j$, $e' = 6+10j$, $f' = 6+7j$;
   translation $\begin{bmatrix} 3 \\ 5 \end{bmatrix}$.

4 $a' = ^-3+j$, $b' = ^-1+2j$, $c' = ^-1+5j$, $d' = ^-2+6j$, $e' = ^-5+3j$,
   $f' = ^-2+3j$; rotation of $90°$ about O.

5 $a' = 6-2j$, $b' = 2-4j$, $c' = 2-10j$, $d' = 4-12j$, $e' = 10-6j$, $f' = 4-6j$;
   rotation of $90°$ and enlargement, scale factor 2, centre O.

6 $a' = ^-9+13j$, $b' = 2+11j$, $c' = 11+23j$, $d' = 10+30j$, $e' = ^-11+27j$,
   $f' = 1+18j$; rotation of $53°$ and enlargement, scale factor 5, centre O.

7 0, $3+4j$, ^-1+7j, ^-4+3j; $\begin{bmatrix} 3 & ^-4 \\ 4 & 3 \end{bmatrix}$.

8 $a = 1-j$, $b = 2+j$, $c = 3+3j$, $d = 4+j$; $a' = ^-2+2j$, $b' = ^-4-2j$,
   $c' = ^-6-6j$, $d' = ^-8-2j$; enlargement, scale factor $^-2$, centre O.

9 $a' = ^-1-j$, $b' = 1-2j$, $c' = 3-3j$, $d' = 1-4j$; rotation of $^-90°$, centre O.

10 $a' = 3+3j$, $b' = {}^-3+6j$, $c' = {}^-9+9j$, $d' = {}^-3+12j$; rotation of 90°, enlargement scale factor 3, centre O.

11 $a' = 5+j$, $b' = 1+8j$, $c' = {}^-3+15j$, $d' = 5+14j$; rotation of 56°, enlargement, scale factor 3.6, centre O.

12 $0, 2+3j, {}^-1+5j, {}^-3+2j$; $\begin{bmatrix} 2 & {}^-3 \\ 3 & 2 \end{bmatrix}$.

Exercise C

1 $5(\cos 37° + j \sin 37°)$, $5(\cos 90° + j \sin 90°)$, $13(\cos 113° + j \sin 113°)$, $8.5(\cos 45° + j \sin 45°)$.

2 $6+8j$, $7j$, $2.83-2.83j$, $^-3.47-19.70j$.

3 zw $(3, 75°)$; zw^2 $(4.5, 105°)$.

4 z^2w $(6, 120°)$; z^3w $(12, 165°)$; z^4w $(24, 210°)$.

5 (a) $3, 75°$;   (b) $1.41+1.41j$, $1.30+0.75j$;   (c) $0.78+2.9j$;
   (d) $3 \cos 75° + 3j \sin 75° = 0.78+2.9j$.

6 $17(\cos 62° + j \sin 62°)$, $4(\cos {}^-90° + j \sin {}^-90°)$, $4.24(\cos {}^-45° + j \sin {}^-45°)$, $25(\cos {}^-74° + j \sin {}^-74°)$.

7 $3+5.20j$, $5.14-6.13j$, $^-1.72+8.83j$, $^-3$.

8 zw $(1, 60°)$; z^2 $(1, 200°)$; w^2 $(1, {}^-80°)$; zw^2 $(1, 20°)$; zw^3 $(1, {}^-20°)$.

9 $z = {}^-0.174+0.985j$; $w = 0.766-0.643j$; $zw = 0.5+0.866j$; $z^2 = {}^-0.940-0.342j$.

Exercise D

1 $0.16-0.12j$, $0.36-0.52j$, $0.12+0.16j$, $0.12+1.02j$, $1+2j$.

2 (a) $\frac{1}{2}, {}^-30°$;   (c) $1.732+j$, $0.433-0.25j$.

3 (a) $4-3j$, $0.16-0.12j$;   (b) (i) $5, 37°$;   (ii) $0.2, {}^-37°$.

4 (a) $2+3.464j$;   (b) $1.414, 45°$; $2.828, 15°$;   (d) $2.732+0.732j$.

5 (a) 10;   (b) $12j$;   (c) 61;   (d) $^-22$.

6 $\frac{3}{34}-\frac{5}{34}j$, $2-j$, $^-\frac{39}{41}+\frac{18}{41}j$, $^-\frac{39}{41}-\frac{18}{41}j$.

7 (a) $\frac{1}{3}, {}^-40°$;   (c) $2.30+1.93j$, $0.255-0.214j$.

8 (a) $5+3j$, $\frac{5}{34}+\frac{3}{34}j$.

9 (a) $^-2+3.464j$, $4.33+2.5j$;   (b) $0.8, 90°$;   (d) $0.8j$.

10 (a) $2x$;   (b) $2yj$;   (c) x^2+y^2;   (d) $4xyj$.

11 (a) ^-3+j, ^-3-j;   (b) $^-2.5+0.866j$, $^-2.5-0.866j$;
   (c) $1.167+0.799j$, $1.167-0.799j$;   (d) $4j, {}^-2j$.

12 $2+j$, $2-j$.

13 $x = 1-3j$, $y = 2-j$.

14 (a) ^-6+5j, ^-6-5j;   (b) $^-2.382, {}^-4.618$;   (c) $3+2j$, $3-2j$;   (d) ^-5+4j, ^-5-4j.

15 $x = 4+3j$, $y = 3-4j$.

Exercise E

1 (a) $|z| = 3$; (b) $|z-4j| = 2$; (c) $|z+1-j| = 4$.
2 (a) $|z-2-j| = |z-3|$; (b) $|z+1+2j| = |z-3|$; (c) $|z+1+2j| = |z-2-j|$.
3 $1-j$.

Miscellaneous exercise

1 (b) $1+1.732j$; (c) $^-8$, $^-2$, $1+1.732j$, $1-1.732j$.
2 (a) $2j$; (b) $1.414, 45°$; (d) $4, 180°$; $135°$, $^-135°$, $^-45°$; $1+j$, $1-j$, ^-1+j, ^-1-j.
3 (b) $c = 4+9j$, $d = 1+5j$.
4 $9+5j$.
5 $p = 5-j$, $q = 10+4j$, $r = 4+14j$, $s = ^-5+3j$. j; $PR = QS$ and PR is perpendicular to QS.
7 (c) $a_1 = ^-3-4j$, $b_1 = ^-1+2j$, $c_1 = 3-j$, $a_2 = 2+j$, $b_2 = ^-5j$, $c_2 = ^-4-2j$. (d) $2:1$.
9 (b) $^-0.5+0.866j$, $^-0.5-0.866j$.
10 (a) $p = ^-c\omega - b\omega^2$, $q = ^-a\omega - c\omega^2$, $r = ^-b\omega - a\omega^2$; (b) (i) equal; (ii) $120°$.

Chapter 21   *Groups*

Exercise A

1 (a) Sunday; (b) Monday; (c) Thursday.
2 Friday.
3 Tuesday.
4 Saturday.

Exercise B

1 $1 \leftrightarrow$ identity, $^-1 \leftrightarrow$ reflection in the x-axis.
2 $0 \leftrightarrow$ identity, $2 \leftrightarrow$ half-turn, $1, 3 \leftrightarrow$ quarter-turns.
3 $2^x \leftrightarrow x$.

Exercise C

1 (i) (a) $x = 9$; (b) $y = 4$; (c) $z = ^-9\frac{1}{2}$; (ii) no; (iii) $a = c$.
2 (ii) 0; (iii) all numbers self-inverse; (iv) (a) $\varnothing$; (b) $\{3,9\}$; (c) $\{2,8,14\}$.
3 (ii) Yes; (iii) (a) 1; (b) 1; (c) 2; (d) 2; yes; (iv) 0; (v) 1.
4 (i) (a) 16; (b) 16; (c) 25; (d) 32; no;
   (ii) (a) 64; (c) 256; no; (iii) 1.
5 (i) $\begin{bmatrix} ap-bq & aq+bp \\ ^-aq-bp & ap-bq \end{bmatrix}$; (ii) (a) yes; (b) yes.
6 (i) (a) 11; (b) 23; (c) 23; (ii) 0; (iii) (a) $^-\frac{2}{3}$; (b) $^-\frac{1}{3}$; (c) no solution.
7 (i) Yes; (ii) Yes; (iii) Yes.

Exercise D

1 $\frac{4}{3}$; $x = \frac{28}{15}$.

2 1; 9; $x = 7$.

3 $\begin{bmatrix} 2 & ^-3 \\ ^-1 & 2 \end{bmatrix}$; $\mathbf{X} = \begin{bmatrix} ^-7 & ^-5 \\ 5 & 3 \end{bmatrix}$.

4 $\mathbf{M}$; $\mathbf{X} = \mathbf{MQ}$ = reflection in the *x*-axis.

Exercise E

1 $1 \leftrightarrow 0 \leftrightarrow \mathbf{I}$; $^-1 \leftrightarrow 1 \leftrightarrow \mathbf{X}$.

2 $0 \leftrightarrow \mathbf{I} \leftrightarrow \mathbf{I}$; $1, 2 \leftrightarrow \mathbf{R}, \mathbf{S} \leftrightarrow \mathbf{L}, \mathbf{M}$.

3 P, Q, R; $0 \leftrightarrow 1 \leftrightarrow 1$; $2 \leftrightarrow 4 \leftrightarrow ^-1$; $1, 3 \leftrightarrow 2, 3 \leftrightarrow j, ^-j$.
   Tables B, C; tables A, D.

Exercise F

1

	A	B	C	D
A	A	B	C	D
B	B	A	D	C
C	C	D	A	B
D	D	C	B	A

	A	B	C	D
A	A	B	C	D
B	B	D	A	C
C	C	A	D	B
D	D	C	B	A

The first table.

3 Only the identity is self-inverse.

Exercise G

2 $\{\mathbf{I}, \mathbf{X}, \mathbf{Y}, \mathbf{H}\}$.   3 $\{\mathbf{I}, \mathbf{A}, \mathbf{B}, \mathbf{H}\}$.   4 5.

5 (*a*)

	I	R	S	X	Y	Z
I	I	R	S	X	Y	Z
R	R	S	I	Z	X	Y
S	S	I	R	Y	Z	X
X	X	Y	Z	I	R	S
Y	Y	Z	X	S	I	R
Z	Z	X	Y	R	S	I

(*b*) $\{\mathbf{I}, \mathbf{R}, \mathbf{S}\}$; (*c*) 3; (*d*) No.

6 $\{0, 3\}$, $\{0, 2, 4\}$.

Exercise H

1 Cyclic.

2 (*b*) Cyclic.

5 $\{1, 3, 9, 11\}$, $\{1, 5, 9, 13\}$; $\{1, 7\}$, $\{1, 9\}$, $\{1, 15\}$.

6 6.

Exercise I

1 (a) Yes; (b) G; (c) $\{1, 6\}$.

2 (c) $\{1, 9, 11, 19\}$ or $\{1, 9, 13, 17\}$.

3 3.

4 Groups of prime order.

5 (b) $\mathbf{C} = [0\ \ 1\ \ 1]$; (c) $\mathbf{E} = [1\ \ 1\ \ 0]$, $\mathbf{F} = [1\ \ 0\ \ 1]$, $\mathbf{G} = [1\ \ 1\ \ 1]$. No cyclic subgroups of order 4.

6 Non-trivial subgroups: $\{1, 4, 7\}$, $\{1, 8\}$. The group is cyclic, generated by 2 or 5.

Exercise J

4 For example: $\{2, 14\}$, $2^4 = 14^2 = 1$, $2 \times 14 = 14 \times 2$.

5 For example: $\{2, 20\}$, $2^6 = 20^2 = 1$, $2 \times 20 = 20 \times 2$.

Miscellaneous exercise

1 (i) $\{c\}$; (ii) $\varnothing$; (iii) $\{a, d\}$; (iv) $\{a\}$.

2 (i) $b \leftrightarrow S$; $a \leftrightarrow P$; $c, d \leftrightarrow Q, R$; (ii) Rotational symmetries of a square; (a) S; (b) $P^{-1} = P$, $Q^{-1} = R$, $R^{-1} = Q$, $S^{-1} = S$.

3 (ii) Yes; (iii) 5, 11, 13; (iv) (a) 9; (b) 13; (c) 11; (d) 5; (e) 1.

4 (a) $\mathbf{A}^2 = \begin{bmatrix} -1 & 1 \\ -1 & 0 \end{bmatrix}$, $\mathbf{A}^3 = \begin{bmatrix} 1 & 0 \\ 0 & 1 \end{bmatrix}$; (c) $\mathbf{A}^2$; (d) $\mathbf{B}^2 = \begin{bmatrix} 0 & -1 \\ 1 & -1 \end{bmatrix}$;
(e) $\mathbf{A}^2$; (f) n is a multiple of 6; (g) $\mathbf{B}^5$.

5 (i) (a) 16; (b) $4^{-1} = 4$, $8^{-1} = 12$, $12^{-1} = 8$, $16^{-1} = 16$; (c) $\{16\}$, $\{4, 16\}$, G;
(ii) b; (iii) $16 \leftrightarrow b$, $4 \leftrightarrow a$; 8, $12 \leftrightarrow c$, d.

6 (i) 5, 2; (ii) 7, 6; (iv) $p = 8$; (v) $x = 2$, 5 or 8.

7 (a) 1; (b) $1^{-1} = 1$, $3^{-1} = 5$, $5^{-1} = 3$, $9^{-1} = 11$, $11^{-1} = 9$, $13^{-1} = 13$;
(d) none; (e) $S_1 = S_{13} = \{1, 13\}$, $S_5 = S_9 = \{5, 9\}$, $S_3 = S_{11} = \{3, 11\}$;
(f) (i) G; (ii) $\varnothing$; (g) $\{1, 9, 11\}$.

8 (b) $fg : x \to 1 - \dfrac{1}{x}$; (d) e.

Revision exercise 6

1 (a) (q or $\sim p$) and ($p \Rightarrow q$) are logically equivalent. ($r \Rightarrow \sim q$) is also true.

3 (a) $(p \wedge q) \vee (p \wedge r) = p \wedge (q \vee r)$; (c) $\sim (p \wedge q) = \sim p \vee \sim q$.

6 Yes.

7 (a) $T\,T\,F\,F$; a; (b) Yes.

8 (a) $(p \wedge \sim q) \vee (\sim p \wedge q)$, $p \wedge q$;
(b) $(p \wedge q \wedge r) \vee (p \wedge \sim q \wedge \sim r) \vee (\sim p \wedge q \wedge \sim r) \vee (\sim p \wedge \sim q \wedge r)$,
$(q \wedge r) \vee (r \wedge p) \vee (p \wedge q)$.

9 $x^4 + 5x^2 - 36 = 0$; $2 + 3j$, $-2 - 3j$.

10 (a) 5, $37°$; (b) $7 + 24j$, 25, $74°$; $\tan 2\theta = \frac{24}{7}$.

11 $z^2 = 2j$; $\frac{1}{2}\sqrt{3}+\frac{1}{2}j$ is a cube root of j; $m = \sqrt{2}$, $\alpha = 45°$. $z^3 = {}^-2+2j$. $|z^2| = m^2$, $\arg(z^2) = 2\alpha$. $|z^6| = 8$, $\arg(z^6) = 270°$. $\dfrac{1}{\sqrt{2}}({}^-1+j)$ is a square root of ${}^-j$.

12   (a) 2;   (b) $\dfrac{{}^-1-j\sqrt{3}}{2}$;   (c) ${}^-1+j\sqrt{3}$;   (d) 1;   (e) 1; ${}^-1-j\sqrt{3}$.

13   (a) ${}^-7+11j$;   (b) $2.6-0.2j$.

14   (i) (a) $x = 4$;   (b) no solution;   (ii) $x = 1$ or 3.

15 $\mathbf{I}\leftrightarrow0$, $\mathbf{Q}\leftrightarrow1$, $\mathbf{H}\leftrightarrow2$, $\mathbf{T}\leftrightarrow3$; $\mathbf{I}\leftrightarrow0$, $\mathbf{Q}\leftrightarrow3$, $\mathbf{H}\leftrightarrow2$, $\mathbf{T}\leftrightarrow1$.

16 $\{1, 6\}, \{1, 2, 4\}$.

17   (b) (i) 5;   (ii) 8;   (c) $1 = 2^6$, $2 = 2^1$, $4 = 2^2$, $5 = 2^5$, $7 = 2^4$, $8 = 2^3$;   (d) $\{1, 8\}, \{1, 4, 7\}$.

18   (a) 8;   (b) 12;   (c) $x = 6$;   (d) $x = 12$;   (e) $x = 2$ or 12.

19 $p = 9$. Identity $= 1$. $1^{-1} = 1$, $3^{-1} = 7$, $7^{-1} = 3$, $9^{-1} = 9$.

20   (a) Yes;   (b) 4;   (c) No;   (d) each element is self-inverse;   (f) $\{3\}$;   (g) Yes;   (h) No.

21 $\{\mathbf{I, A, B, G}\}, \{\mathbf{I, C, D, G}\}, \{\mathbf{I, E, F, G}\}$; $\mathbf{E, F}$ and $\mathbf{G}$.

23   3.

24   7.

Chapter 22   *Exponential and logarithmic functions*

Exercise A

1   (a) 4;   (b) 3;   (c) 16;   (d) $\frac{1}{6}$.

2   (a) $\frac{1}{2}$;   (b) $\frac{1}{2}$;   (c) 32;   (d) $\frac{1}{8}$.

3   (a) 1.41, 1.19, 1.09;   (b) 1, 1.09, 1.19, 1.30, 1.41, 1.54, 1.68, 1.83, 2.

4   (a) 1.5849;   (b) 2.0801.

5   (a) 8;   (b) 4;   (c) $\frac{1}{512}$;   (d) $\frac{1}{4}$.

6   (a) 0.1;   (b) 4;   (c) $\frac{1}{8}$;   (d) $\frac{1}{2}$.

7   (a) 1.73, 1.32, 1.15;   (b) 1, 1.15, 1.32, 1.51, 1.73, 1.99, 2.28, 2.62, 3.

8   (a) 1.3797;   (b) 3.6593.

9   (a) (i) 8;   (ii) 4;   (iii) 2;   (iv) 32;   (b) (i) true;   (ii) true.

10   (a) (i) 2;   (ii) 64;   (b) true.

Exercise B

1   2.   **2**   27.   **3**   4.

4   3.   **5**   4.   **6** $\frac{1}{2}$.

7   9.   **8**   4.   **9** $\frac{1}{4}$.

10   2.

Exercise C

1 (*b*) 0.693; (*c*) 11.1.   **2** (*b*) 1.61; (*c*) 1010.   **3** (*b*) 2.08; (*c*) 8520.

Exercise D

1 (*a*) $5e^x$; (*b*) $^-3e^x$;   (*c*) $2e^{2x}$; (*d*) $^-2e^{-2x}$; (*e*) $10e^{2x}$.

2 (*a*) £42.01, £43.18, £43.82, £44.42; (*b*) (i) 0.16; (ii) £44.51.

4 (*a*) 47.209; (*b*) 19.167; (*c*) 23.605; (*d*) 3.195.

5 (*a*) $10e^x$; (*b*) e^2e^x; (*c*) e^{x+3}; (*d*) $^-15e^{-3x}$; (*e*) $6e^{-2x}$.

6 (*a*) $m(t) = 5e^{-0.005t}$; (*b*) 4.98, 4.52, 3.03, 2.50, 1.25.

7 (*a*) $106.18; (*b*) $106.18.

8 (*a*) 0.361; (*b*) $1 - 2.06 \times 10^{-9}$.

Exercise E

1 (*a*) 3; (*b*) 5; (*c*) $^-2$; (*d*) 0; (*e*) 1.

2 $\frac{5}{3}$; (*a*) $\frac{7}{3}$; (*b*) $\frac{2}{3}$; (*c*) $\frac{3}{2}$; (*d*) $\frac{5}{6}$.

3 $\dfrac{m}{n}$.

4 (*a*) $x = 2.58$; (*b*) $x = 3.17$; (*c*) $x = {}^-0.139$; (*d*) $x = 0.446$; (*e*) $x = 1.21$.

5 (*a*) 2.2266; (*b*) 0.8614; (*c*) 0.4307; (*d*) $^-0.4307$; (*e*) 3.7706.

6 After 7 years.

7 (*a*) 3; (*b*) 3; (*c*) $^-3$; (*d*) 6; (*e*) $^-3$.

8 (*c*) 0.475; (*d*) (i) 0.6; (ii) 0.775; (iii) 0.9; (iv) 0.95; (v) 0.7.

9 $\log_{10} e . \log_e 10 = 1$

10 (*a*) $x = 0$; (*b*) $x = 1.46$; (*c*) 0.792; (*d*) 1.58; (*e*) $^-0.683$.

11 (*a*) 3.7; (*b*) (i) 10^{-7}; (ii) 4×10^{-8}.

12 1997.

Exercise F

1 $A = 5300, r = 1.01$; (*a*) 1%; (*b*) 5300; (*c*) 14300.

2 $A = 5000, r = 0.97$; (*a*) $^-3\%$; (*b*) 5000; (*c*) 250.

3 $T \approx 2\sqrt{l}$.

4 $V \approx 6r^4$.

Miscellaneous exercise

1 1.2%.

2 17 years.

4 1.4427; one-way stretch with scale factor 1.4427 and the *x*-axis invariant.

5 (*a*) 2.5:1; (*b*) 1:2.8; (*c*) 6 dB.

6 $f(t) = Ae^{-kt}$; $k = \frac{9}{83}$, 24 °C.

Chapter 23   Composite functions

Exercise A

1 $3x^2 - 12, 6x$; (*a*) $x > 2, x < {}^-2$; (*b*) $^-2 < x < 2$; maximum at $x = {}^-2$, minimum at $x = 2$.

2 $8 - 2t$; 17 m.

3 (*a*) 1.6 m s^{-1}; (*b*) 1.45 m s^{-1}.

4 (*a*) $2x + 6$; (*b*) $2x - \dfrac{1}{x^2}$; (*c*) $15x^2 - 15x^4$; (*d*) 6 cos 2x; (*e*) $24x^2 - 24x + 6$;

(*f*) $^-3$ sin $\tfrac{1}{2}x$.

5 $2t - \dfrac{54}{t^2}$; $a = 3$; (*a*) negative; (*b*) positive; minimum.

6 36 s, 130 m s^{-1}.

7 (*a*) 0.21 m s^{-1}; (*b*) 9.61 m s^{-1}.

8 (*a*) $20t^4 - 6t^2$; (*b*) $^-15$ sin 3t; (*c*) $6t^2 + 12t + 6$; (*d*) cos 0.1t;
(*e*) $36t^3 + 12t$; (*f*) $^-108 + 108t - 36t^2 + 4t^3$.

Exercise B

1 $2(x - 3)$.

2 $10(5x - 3)$.

3 $4(x + 5)^3$.

4 $8(2x + 5)^3$.

5 $90(9x - 4)^9$.

6 $3(x + 4)^2$.

7 $21(7x + 4)^2$.

8 $4(x - 3)^3$.

9 $20(5x - 3)^3$.

10 $120(8x + 13)^{14}$.

13 $\dfrac{^-1}{(x + 3)^2}$.

14 $6(3x - 4)$.

Exercise C

1 $8x(x^2 - 3)^3$.

2 $6x^2(x^3 + 1)$.

3 $\dfrac{^-3}{(3x + 1)^2}$.

4 $\dfrac{^-10x}{(x^2 - 2)^2}$.

5 $^-4$ cos^3 x sin x.

6 $^-4x^3$ sin x^4.

7 $60x^3(5x^4 + 2)^2$.

8 $2\left(x + \dfrac{1}{x}\right)\left(1 - \dfrac{1}{x^2}\right)$.

9 $\dfrac{^-\cos x}{\sin^2 x}$.

10 2 sin x cos x.

11 $\dfrac{^-2}{x^3}$.

12 $\dfrac{^-10}{x^{11}}$.

13 $2u\dfrac{du}{dx}$; 1; $\dfrac{1}{2\sqrt{x}}$.

14 $e^u\dfrac{du}{dx}$; 1; $\dfrac{1}{x}$.

Exercise D

1 $2x^4 + 4x^3 + 3x^2 + x + c$.

2 (*a*) $\tfrac{1}{27}(3x - 2)^9 + c$; (*b*) $\tfrac{1}{10}(x^2 + 1)^5 + c$; (*c*) $\tfrac{1}{2}$ sin $(2t + 1) + c$;
(*d*) $^-\tfrac{1}{2}$ cos $(t^2 + 1) + c$.

3 (*a*) 42.2; (*b*) 3.1; (*c*) 0.1729; (*d*) 0.05.

4 104.

5 (a) $\frac{1}{35}(5x-3)^7+c$; (b) $\frac{1}{6}(x^3+1)^2+c$; (c) $-\frac{1}{2}\cos 6t+c$; (d) $\frac{1}{4}\sin^4 t+c$.

6 (a) 833.25; (b) 0; (c) 5.657; (d) 1.485.

Miscellaneous exercise

1 (a) $4t^3+10t$, $12t^2+10$, $2t$, 2; (b) $2x+5$, 2.

2 Maximum at $(0,1)$, minima at $(1,0)$ and $(^-1,0)$.

3 $\dfrac{dV}{dt}=9\pi x^2\dfrac{dx}{dt}$.

4 (a) $\dfrac{1}{x}-1$; (b) $\dfrac{-1}{(x+1)^2}$; (c) $\dfrac{-1}{x^2}$; 1.

5 $3u^2\dfrac{du}{dx}$; 1.

6 $\dfrac{-1}{u^2}\dfrac{du}{dx}$; nx^{n-1}; $\dfrac{-n}{x^{n+1}}$.

7 $^-2xe^{-x^2}$; 0.316.

8 Maxima at $(\frac{1}{2}\pi+2n\pi,\ e)$; minima at $\left(\frac{3}{2}\pi+2n\pi,\ \dfrac{1}{e}\right)$.

Chapter 24 Iteration and approximation

Exercise A

1 1.382; $^-1$, 3.6180.   **2** $^-1.466$.   **3** 2.287.

4 2.116.   **5** 1.18, 0.205, $^-1.38$.   **6** $^-1.466$.

7 0.7391.   **8** 0.5110.   **9** (b) 0.6954.

10 (b) 2.37; (c) no.

Exercise B

1 (c) (i) 3.3912; (ii) 3.2660; (iii) 3.2016; (d) 3.2466.

2 (a) $^-5$, 5; (c) $^-2.670$.

3 (a) $^-1$, $^-2$, $^-2.5495$, $^-2.6531$, $^-2.6675$.

4 (b) $x_7=0.3472935$; (c) $x_7=0.3472755$; (d) $x_7=2125395.6$.

5 (b) $^-1.879$, 1.532; (c) $^-1.879$, 0.3473, 1.532.

6 (b) 0.35, 1.53, $^-1.88$.

8 (c) 0.1995.

Exercise C

1 (d) $^-0.4067$.

2 (c) 3.3460, $^-2.9392$.

3 (a) 3; (d) 3.08.

4   (b) $g'(x) > ^-1$;   (c) $|g'(x)| < 1$.

5   (b) One solution is 1.382.

6 $^-1$, 3.618.

7   (b) 1.272;   (c) $^-1.272$.

8   (b) 0.5392;   (c) 1.68, $^-2.21$;   (d) (i) and (ii) converge for 0.5392;
(iii) converges for all three solutions.

9   (b) 1.131;   (c) yes.

Exercise D

1   (b) 1.63.

2   (c) 2.475.

3 $^-0.529$.

4   1.43.

5   (a) 2, 4;   (b) $^-0.766$.

6   (b) (i) 1.732;   (ii) 2.646;   (d) (i) 1.260;   (ii) 2.668;
(e) $x_{n+1} = \dfrac{1}{m}\left[(m-1)x_n + \dfrac{a}{x_n^{m-1}}\right].$

7   (a) 100 litres;   (c) $t = 22.3$.

8   3.58 a.m.

Exercise E

1 $f(8+h) \approx 2 + \frac{1}{12}h$; 2.0025.

2   (a) $\frac{1}{180}\pi \cos x°$.

3   (a) $y = 1 - \frac{1}{4}x$;   (b) 0.4825.

4 nx^{n-1}; n.

5 $f(81+h) \approx 3 + \frac{1}{108}h$; 3.0074074.

6 $g(x) = 0.01745x$.

7 $y = 3 - 2x$; 0.8.

8   (a) $f'(x) = e^x$;   (b) 1.

9 1.3×10^9.

10   0.45 cm².

11 $^-0.984$ s.

12   0.19 g.

13   0.4.

14   (a) 0.05;   (b) 0.2.

15   2%.

Miscellaneous exercise

1 $f(2) = 32$, $g(2) = 80$; 37.09.

2   (i) 3;   (ii) $^-3$, $3\frac{2}{3}$;   (iii) 9, 2.3827;   (iv) $^-1$, 0;   (v) (iii).

3 7.5, 10; greater, too great.

4 (3.0, 3.0).

5 (a) $f'(x) = \frac{1}{180}\pi \cos x°$;   (c) 2;   (d) $2 + \frac{\pi^2\sqrt{3}}{5400}$.

7 (i) (a) 0.046;   (d) 0.001 56.

8 (a) $p(x) = x^2 - x + 1$;   (b) 0, 1, 2, 3, 4.

9 $5\frac{1}{3}$; $y = 8 - 4x$.

10 (a) $\frac{1}{2}, \frac{1}{3}$;   (c) 0.42; 3.42.

11 (a) 2;   (c) 2.093.

12 2.25;   (a) 12.375;   (b) 1.6875; $^{-}1 \leqslant x \leqslant 1$.

Revision exercise 7

1 (a) $6(2x-5)^2$;   (b) $3\left(\dfrac{x+5}{2}\right)^5$;   (c) $^{-}15(7-3x)^4$;   (d) $6x(x^2+1)^2$;

(e) $60x^2(5x^3-2)^3$;   (f) $3x^2+2x$.

2 After 8 years.

3 24%.

4 5, $y = 5x - 4$;   (a) 1.2;   (b) 1.03.

5 (a) $(2x+7)^3+c$;   (b) $(3x-1)^5+c$;   (c) $^{-}15(7-3x)^4$;   (d) $6x(x^2+1)^2$;
(e) $\frac{1}{12}(x^3+4)^4+c$;   (f) $\frac{1}{6}(e^x+2)^6+c$.

6 25°.

7 (a) 0.82413;   (b) 0.21801.

8 (a) 1.18^n;   (b) about 4 years;   (c) about 8 years.

9 (a) $2\cos(2x+\frac{1}{3}\pi)$;   (b) $3\sin(\frac{1}{6}\pi-3x)$;   (c) $2\sin x \cos x$;   (d) $^{-}2x\sin x^2$;

(e) $\dfrac{^{-}6}{(3x-1)^3}$;   (f) $\dfrac{^{-}\cos x}{\sin^2 x}$;   (g) $\dfrac{15}{(2-5x)^4}$;   (h) $\dfrac{2\sin x}{\cos^3 x}$.

10 1.98, $^{-}1.31$.

11 2490 years ago.

12 (a) $\dfrac{-1}{5(5x-7)}+c$;   (b) $\frac{1}{10}(x^2+4)^5+c$;   (c) $^{-}\frac{1}{21}(5-x^3)^7+c$;   (d) $\frac{1}{3}(x^2+1)^{\frac{3}{2}}+c$;

(e) $^{-}\frac{1}{2}\cos x^2+c$;   (f) $2\sin\sqrt{x}+c$;   (g) $\dfrac{-1}{2(x^3+x+9)^2}+c$;   (h) $\frac{1}{5}\sin^5 x+c$.

13 $C = \dfrac{2\times10^7}{f^2}$.

14 (a) (i) $f:x \to 6x(x^2-5)^2$;   (ii) $k = 2$;   (b) $\frac{1}{6}(x^2-5)^3+c$;   (iii) 1.5.

15 (i) (a) $fg:x \to \sin^2 x$;   (b) $gf:x \to \sin x^2$;   (ii) 1.

16 (c) 3.022.

17 $s = \dfrac{f^2}{31\,000}$.

18   (a) 0, 1.107.

19   Less; 8.15%.

20   (i) (a) $10e^{10x}$;   (b) $\frac{1}{2}e^{(x-1)/2}$;   (c) 0.693×2^x.
        (ii) (a) $\frac{1}{3}e^{3x} + c$;   (b) $-\frac{1}{5}e^{7-5x} + c$;   (c) $0.4343 \times 10^x + c$.

21   31.32.

Formulae and tables

ALGEBRA

Complex numbers

$$(a+bj) \times (c+dj) = (ac-bd)+(bc+ad)j.$$

In the Argand diagram r is called the modulus of z ($|z|$) and θ the argument of z ($\arg z$).

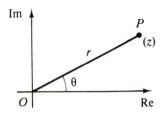

Quadratic equations

The roots of $ax^2+bx+c = 0$ are

$$x = \frac{-b \pm \sqrt{(b^2-4ac)}}{2a}$$

Group

A set of elements, $a, b, \ldots$, with an operation $*$, having the following properties:
- (i) *closure*: if a, b are elements of the set, so is $a*b$;
- (ii) *associativity*: $(a*b)*c$ is the same element as $a*(b*c)$;
- (iii) *identity element*: there is an element e such that, for all a, $a*e = e*a = a$;
- (iv) *inverse elements*: for each element a there is an element a^{-1} for which

$$a*a^{-1} = a^{-1}*a = e.$$

GEOMETRY

The matrix $\begin{bmatrix} \cos\theta & -\sin\theta \\ \sin\theta & \cos\theta \end{bmatrix}$ represents a positive rotation through an angle θ about the origin.

TRIGONOMETRY

For any angle θ, $\sin^2\theta+\cos^2\theta = 1$.
For any angles A and B:

$$\sin(A+B) = \sin A \cos B+\cos A \sin B,$$
$$\sin(A-B) = \sin A \cos B-\cos A \sin B,$$
$$\cos(A+B) = \cos A \cos B-\sin A \sin B,$$
$$\cos(A-B) = \cos A \cos B+\sin A \sin B.$$

$$\sin 2A = 2 \sin A \cos A;$$
$$\cos 2A = \cos^2 A - \sin^2 A = 2 \cos^2 A - 1 = 1 - 2 \sin^2 A.$$

In any triangle, $\dfrac{a}{\sin A} = \dfrac{b}{\sin B} = \dfrac{c}{\sin C}$

$a^2 = b^2 + c^2 - 2bc \cos A$ and two similar formulae.

Circular measure

π radians $= 180$ degrees.

VECTORS

Scalar (dot) product

$$\mathbf{a} . \mathbf{b} = ab \cos \theta,$$

where a, b are the magnitudes of the vectors $\mathbf{a}$, $\mathbf{b}$; θ is the angle between their directions.

If $\mathbf{a}$ is perpendicular to $\mathbf{b}$, $\mathbf{a} . \mathbf{b} = 0$.

In terms of components, $\mathbf{a} . \mathbf{b} = a_1 b_1 + a_2 b_2 + a_3 b_3$.

CALCULUS

Logarithms $\quad y = \log_a x \quad \Leftrightarrow \quad x = a^y$

Differentiation and integration

$f(x)$	$f'(x)$		$f(x)$	$\int f(x)\,dx$
x^n	nx^{n-1}	where n is any real constant	x^n	$\dfrac{x^{n+1}}{n+1} + C \ (n \neq -1)$
$\dfrac{1}{x+a}$	$\dfrac{^-1}{(x+a)^2}$		$\dfrac{1}{(x+a)^2}$	$\dfrac{^-1}{x+a} + C$

When x is in circular measure

$\sin x$	$\cos x$		$\cos x$	$\sin x + C$
$\cos x$	$-\sin x$		$\sin x$	$-\cos x + C$

Chain rule for composite functions

If $y = f(u)$ and $u = g(x)$, so that $y = fg(x)$, $\dfrac{dy}{dx} = f'(u) . g'(x) = \dfrac{dy}{du} . \dfrac{du}{dx}$.

Linear approximation

Near $x = a$: $\qquad f(a+h) \approx f(a) + hf'(a)$.

If $y = f(x)$, $\delta y \approx f'(x) . \delta x$.

The line through (x_1, y_1) with gradient m is $y - y_1 = m(x - x_1)$.

Newton–Raphson formula

If α is an approximate solution of $f(x) = 0$, then, subject to certain conditions, a better approximation is $\alpha - \dfrac{f(\alpha)}{f'(\alpha)}$.

MECHANICS

Motion with uniform acceleration

$$\mathbf{v} = \mathbf{u} + t\mathbf{a},$$
$$\mathbf{r} = t\mathbf{u} + \tfrac{1}{2}t^2\mathbf{a},$$
$$v^2 = u^2 + 2\mathbf{a}.\mathbf{r},$$
$$\mathbf{r} = \tfrac{1}{2}t(\mathbf{u} + \mathbf{v}).$$

Newton's law of force

$$\mathbf{F} = m\mathbf{a}.$$

Units

force $\mathbf{F}$	mass m	acceleration a
newton (N)	kg	m/s²

The weight of a mass of m kg is mg newton, where $g \approx 9.8$.

Momentum

Momentum is defined as $m\mathbf{v}$.
Change of momentum = impulse causing it;

$$m\mathbf{v} - m\mathbf{u} = t\mathbf{F}, \text{ for a constant force } \mathbf{F}.$$

STATISTICS

Mean

$$\bar{x} = \frac{\sum x}{N},$$

where the sum is taken over the whole set, and N is the total number of values, or

$$\bar{x} = \frac{\sum xf}{\sum f},$$

where the sum is taken over all classes and f is the frequency of the value x.

Standard deviation

Standard deviation, s, is given by

$$s^2 = \frac{\Sigma (x - \bar{x})^2}{N} = \frac{\Sigma x^2 f}{\Sigma f} - \left(\frac{\Sigma xf}{\Sigma f}\right)^2.$$

Pascal's triangle

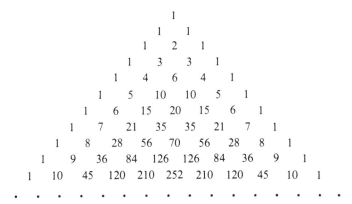

Rank correlation

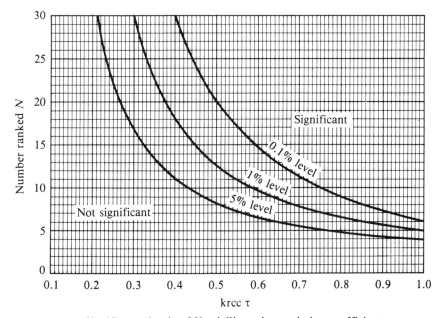

Significance levels of Kendall's rank correlation coefficient

Kendall's rank correlation coefficient (KRCC) between two different orders of N objects is given by

$$\tau = \frac{\text{no. of agreements in order} - \text{no. of disagreements}}{\frac{1}{2}N(N-1)},$$

all pairs being compared.

Normal distribution

The percentage of the Normal population lying below the value given, measured in units of s.d. above the mean.

.	0.0	0.1	0.2	0.3	0.4	0.5	0.6	0.7	0.8	0.9
0	50.0	54.0	57.9	61.8	65.5	69.1	72.6	75.8	78.8	81.6
1	84.1	86.4	88.5	90.3	91.9	93.3	94.5	95.5	96.4	97.1
2	97.7	98.2	98.6	98.9	99.2	99.4	99.5	99.7	99.7	99.8
3	99.87	99.90	99.93	99.95	99.97	99.98	99.98	99.99	99.99	100

Index